BASIC
ENGINEERING
THERMODYNAMICS

McGRAW-HILL
BOOK COMPANY
New York
St. Louis
San Francisco
Auckland
Düsseldorf
Johannesburg
Kuala Lumpur
London
Mexico
Montreal
New Delhi
Panama
Paris
São Paulo
Singapore
Sydney
Tokyo
Toronto

MARK W. ZEMANSKY, Ph.D.
Professor Emeritus of Physics
The City College of the City University of New York

MICHAEL M. ABBOTT, Ph.D.
Associate Professor of Chemical Engineering
Rensselaer Polytechnic Institute

HENDRICK C. VAN NESS, D.Eng.
Professor of Chemical Engineering
Rensselaer Polytechnic Institute

Basic Engineering Thermodynamics

SECOND EDITION

This book was set in Modern by Maryland Composition Incorporated.
The editors were B. J. Clark and J. W. Maisel;
the cover was designed by Ben Kann;
the production supervisor was Leroy A. Young.
New drawings were done by Eric G. Hieber Associates Inc.
Kingsport Press, Inc., was printer and binder.

Library of Congress Catalogiog in Publication Data

Zemansky, Mark Waldo, date
Basic engineering thermodynamics.

Includes bibliographical references.
1. Thermodynamics. I. Abbott, Michael M., joint author. II. Van Ness, Hendrick C., joint author. III. Title.

TJ265.Z54 1975 621.4′021 74-13559
ISBN 0-07-072815-1

BASIC ENGINEERING THERMODYNAMICS

6 7 8 9 10 11 12 K P K P 8 9 8 7 6 5 4 3 2 1

CONTENTS

PREFACE

It has been our purpose to prepare a textbook for an initial course in thermodynamics for engineers of any discipline. Such a course must stress the fundamental principles of thermodynamics and, at the same time, be of sufficient breadth to prepare students for a variety of subsequent courses in which the principles are applied to different kinds of systems. Thus, while paying ample attention to simple fluid systems, we have also included careful treatments of solid systems under stress, surfaces, cells, and electric and magnetic systems.

We have been guided in our presentation of the fundamental principles of thermodynamics by Zemansky's "Heat and Thermodynamics," a widely used introductory text of some forty years standing for students of science. The present work is in some measure an engineering adaptation of that work. Since this work is designed for use in engineering curricula, emphasis is placed on a variety of applications of technological significance.

The main body of the text is devoted to the principles and applications of classical thermodynamics, an area of study which in no way depends on the atomistic nature of matter. The final chapter presents an elementary discussion of statistical thermodynamics, which owes its existence to atomic and quantum theory. Its purpose is to provide an introduction to a branch of knowledge in which may be found a deeper insight into the nature of the laws of thermodynamics.

Although many individuals have contributed in one way or another, by question or comment, to the quality of this second edition of "Basic Engineering Thermodynamics," no one has approached the helpfulness of Howard E. Cyphers, Associate Professor of Mechanical Engineering at Rensselaer Polytechnic Institute, who studied the manuscript with an eagle eye and directed our attention to numerous errors and infelicities.

MARK W. ZEMANSKY
MICHAEL M. ABBOTT
HENDRICK C. VAN NESS

LIST OF SYMBOLS

Capital Italics

A	Helmholtz function; area
B, B'	Second virial coefficient
$\hat{B}$	Dimensionless second virial coefficient
B^0, B^1	Functions of T_r in the Pitzer correlation for B
C, C'	Heat capacity; third virial coefficient
D, D'	Fourth virial coefficient
E	Young's modulus; energy
E'	Unavailable energy
E_K	Kinetic energy
E_P	Gravitational potential energy
F	Force; fluid friction; degrees of freedom
G	Gibbs function
H	Enthalpy
H'	Enthalpy in ideal-gas state
$\tilde{H}$	Generalized enthalpy
I	Current

K Equilibrium constant
L Length
M General designation of thermodynamic function; molecular weight
M' General designation of thermodynamic function in ideal-gas state
Mt General designation of total thermodynamic property of a system
N Number of particles
N_0 Avogadro's number
P Pressure; probability
P_c Critical pressure
P_r Reduced pressure
Q Heat
R Universal gas constant; electric resistance
S Entropy
S' Entropy in ideal-gas state
S_{total} Total entropy of a system and its surroundings
T Absolute temperature
T_c Critical temperature
T_r Reduced temperature
U Internal energy
V Volume
V_c Critical volume
V_r Reduced volume
W Work
W_s Shaft work
X Generalized displacement coordinate
Y Generalized force coordinate
Z Compressibility factor
Z_c Critical compressibility factor

Lowercase Italics

a Acceleration
c Soundspeed
d Sign for an exact differential
e Electromotive force
f As subscript, identifies liquid water
g Acceleration of gravity; degeneracy; as subscript, identifies water vapor
h Planck's constant; height
i As subscript, identifies a chemical species or an energy level
j Moles of electrons transferred in a cell per mole of reaction advancement

k	Boltzmann's constant; proportionality constant
l	Length; as superscript, identifies a liquid phase
m	Mass; number of chemical species
n	Number of moles
n_q	Number of ensembles in quantum state q
p_i	Partial pressure
q	Electric charge; as subscript, identifies a quantum state
r	Radius; compression ratio
s	As superscript, identifies a solid phase
t	Celsius or Fahrenheit temperature; as superscript, identifies a total system property
u	Velocity
v	Potential difference; as superscript, identifies a vapor phase
x	Distance; quality
x_i	Mole fraction of species i in liquid phase
y_i	Mole fraction of species i in vapor phase
z	Elevation above a datum level

Script Capitals

$\mathcal{B}$	Magnetic induction
$\mathcal{C}$	Curie constant
$\mathcal{D}$	Electric displacement
$\mathcal{E}$	Electric field strength
$\mathfrak{F}$	Faraday's constant
$\mathcal{H}$	Magnetic field strength
$\mathcal{M}$	Total Magnetization
$\mathcal{N}$	Number of systems in an ensemble
$\mathcal{P}$	Total Polarization
$\mathcal{S}$	Ensemble entropy
$\mathcal{U}$	Ensemble internal energy
$\mathcal{Z}$	System partition function

Special Symbols

ln	Natural logarithm (base e)
log	Common logarithm (base 10)
$\mathbf{M}$	Mach number
$\Delta H'$	Residual enthalpy $\equiv H' - H$
$\Delta M'$	Residual property in general $\equiv M' - M$
$\Delta S'$	Residual entropy $\equiv S' - S$
ΔG°	Standard Gibbs function change of reaction $\equiv \Sigma \nu_i G_i^\circ$

ΔH° Standard heat of reaction $\equiv \Sigma \nu_i H_i^\circ$
$\Delta H^{\alpha\beta}$ Latent heat of phase change $\equiv H^\beta - H^\alpha$, also $\Delta H_{\alpha\beta}$
$\Delta S^{\alpha\beta}$ Entropy change of phase change $\equiv S^\beta - S^\alpha$, also $\Delta S_{\alpha\beta}$
$\Delta V^{\alpha\beta}$ Volume change of phase change $\equiv V^\beta - V^\alpha$, also $\Delta V_{\alpha\beta}$
ΔS_{total} Total entropy change of a system and its surroundings
T^* Magnetic temperature
$| \ |$ Designates an absolute value
$\{ \ \}$ Designates a $\{$set$\}$ of values
$\dot{}$ Designates a time derivative or rate
$'$ Designates a value for the ideal-gas state
$^\circ$ Designates the standard state

Greek Letters

α Linear expansivity; as superscript, identifies a phase
β Volume expansivity; as superscript, identifies a phase
γ Ratio of heat capacities, C_P/C_V; surface tension
δ Sign for an inexact differential; linear compressibility
Δ Finite-difference sign
ϵ Strain; reaction coordinate
ϵ_0 Permittivity of vacuum
η Efficiency
θ Empiric temperature
κ Isothermal compressibility
κ_S Adiabatic compressibility
μ Joule-Kelvin coefficient; chemical potential
μ_0 Permeability of vacuum
ν_i' Stoichiometric coefficient
ν_i Stoichiometric number
ν $\equiv \Sigma \nu_i$
$\prod$ Continuous-product sign
π Number of phases
ρ Density
$\sum$ Summation sign
σ Stress; deviation
τ Time
χ_e Electric susceptibility
χ_m Magnetic susceptibility
ω Acentric factor; number of ensemble arrangements; coefficient of performance

BASIC
ENGINEERING
THERMODYNAMICS

1

TEMPERATURE

1-1 Macroscopic Point of View

The application of scientific principles to the solution of any real problem must necessarily start with a separation of a restricted region of space or a finite portion of matter from its surroundings. The portion which is set aside (in the imagination) and on which attention is focused is called the *system*, and everything outside the system which has a direct bearing on its behavior is known as the *surroundings*. When a system has been chosen, the next step is to describe it in terms of quantities related to the behavior of the system or its interactions with the surroundings, or both. There are in general two points of view that may be adopted, the *macroscopic* and the *microscopic*.

Let us take as a system the contents of a cylinder of an automobile engine. A chemical analysis would show a mixture of hydrocarbons and air before explosion, and after the mixture had been ignited there would be combustion products describable in terms of certain chemical compounds. A statement of the relative amounts of these substances is a description of the *composition* of the system. At any moment, the system whose composition has just been described occupies a certain *volume*, depending on the position of the piston. The volume can be easily measured and, in the laboratory, is recorded automatically by means of an appliance coupled to the piston. Another quantity that is indispensable in the description of our system is the *pressure* of the gases in the cylinder. After explosion this pressure is large; after exhaust it is small. In the laboratory, a pressure gauge may be used to measure the changes of pressure and to make an automatic record as the engine operates. Finally, there is one more quantity without which we should have no adequate idea of the operation of the engine: the *tem-*

perature. As we shall see, in many instances, it can be measured just as simply as the other quantities.

We have described the materials in a cylinder of an automobile engine by specifying four quantities: composition, volume, pressure, and temperature. These quantities refer to the gross characteristics, or large-scale properties, of the system and provide a *macroscopic description.* They are therefore called *macroscopic coordinates.* The quantities that must be specified to provide a macroscopic description of other systems are, of course, different; but macroscopic coordinates in general have the following characteristics in common:

1 They involve no special assumptions concerning the structure of matter.
2 They are few in number.
3 They are suggested more or less directly by our sense perceptions.
4 They can in general be directly measured.

In short, a macroscopic description of a system involves the specification of a *few fundamental measurable properties* of a system. Although the macroscopic point of view is the one adopted in thermodynamics, it should be understood that the microscopic point of view is of great value and that it may lead to a deeper insight into the principles of thermodynamics. This point of view is taken in the branch of science called *statistical mechanics,* a subject considered briefly in the final chapter. We indicate here merely the distinction between the two points of view by giving a simple microscopic description of a gas in a containing vessel.

1-2 Microscopic Point of View

We assume that the gas consists of an enormous number N of particles called molecules, all having the same mass and each moving with a velocity independent of the others. The position of any molecule is specified by the three cartesian coordinates x, y, and z, and the velocity by the three components v_x, v_y, and v_z. Therefore, to describe the position and velocity of a molecule, we need six numbers. A microscopic description of the state of the gas consists of the specification of these six numbers for each of the N molecules.

We need not pursue the matter further to understand that a microscopic description involves the following characteristics:

1 Assumptions are made concerning the structure of matter; e.g., the existence of molecules is assumed.

2 Many quantities must be specified.
3 The quantities specified are not suggested by our sense perceptions.
4 These quantities cannot be measured.

1-3 Macroscopic versus Microscopic

Although it might seem that the two points of view are hopelessly different and incompatible, there is, nevertheless, a relation between them, and when both points of view are applied to the same system, they must lead to the same conclusions. The relation between the two points of view lies in the fact that the few directly measurable properties whose specification constitutes the macroscopic description are really averages over a period of time of a large number of microscopic characteristics. For example, the macroscopic quantity pressure is the average rate of change of momentum due to all the molecular collisions made on a unit of area. Pressure, however, is a property that is perceived by our senses. We feel the effects of pressure. Pressure was experienced, measured, and used long before scientists and engineers had reason to believe in the existence of molecular impacts. If molecular theory is changed or even discarded at some time in the future, the concept and meaning of pressure will likely remain. Herein lies an important distinction between the macroscopic and microscopic points of view. The few measurable macroscopic properties are as sure as our senses. They will remain unchanged as long as our senses remain the same. The microscopic point of view, however, goes much further than our senses. It postulates the existence of molecules, their motion, collisions, etc. It is constantly being changed, and we can never be sure that the assumptions are justified until we have compared some deduction made on the basis of these assumptions with a similar deduction based on observed macroscopic behavior.

1-4 Scope of Thermodynamics

It has been emphasized that a description of the gross characteristics of a system by means of a few of its measurable properties, suggested more or less directly by our sense perceptions, constitutes a macroscopic description. Such descriptions are the starting point of all investigations in all branches of science and engineering. For example, in dealing with the mechanics of a rigid body, we adopt the macroscopic point of view in that only the external aspects of the rigid body are considered. The position of its center of mass is specified with reference to coordinate axes at a particular time.

Position and time and a combination of both, such as velocity, represent some of the macroscopic quantities used in mechanics, and are called *mechanical coordinates.* The mechanical coordinates serve to determine the potential and the kinetic energy of the rigid body with reference to the coordinate axes, i.e., the kinetic and the potential energy of the body as a whole. These two types of energy constitute the *external,* or *mechanical, energy* of the rigid body. It is the purpose of mechanics to find such relations between the position coordinates and the time as are consistent with Newton's laws of motion.

In thermodynamics, however, attention is directed to the *interior* of a system. A macroscopic point of view is nevertheless adopted, and emphasis is placed on those macroscopic quantities which have a bearing on the internal state of a system. It is the function of experiment to determine the quantities that are necessary and sufficient for a description of such an internal state. *Macroscopic quantities having a bearing on the internal state of a system are called thermodynamic coordinates.* Specification of a sufficient number of such coordinates serves to determine the internal state of a system and, in particular, its *internal energy.* A system that may be described in terms of thermodynamic coordinates is called a *thermodynamic system.* It is the purpose of thermodynamics to find general relations connecting the internal energy and other internal properties of a system with the thermodynamic coordinates and to relate changes in the thermodynamic state of a system to its interactions with its surroundings. The unifying principles with which all such considerations must be consistent are known as the laws of thermodynamics.

A wide variety of thermodynamic systems is of interest. A pure vapor such as steam constitutes the working medium of a power plant. A reacting mixture of gasoline and air powers an automotive engine. A vaporizing liquid such as ammonia provides refrigeration. The expanding gases in a nozzle propel a rocket. Other examples include the stressed members of structures, the surface region of an emulsion, a fuel cell for the generation of electricity, and thermoelectric devices.

1-5 Thermal Equilibrium

We have seen that a macroscopic description of a gaseous mixture may be given by specifying such quantities as the composition, the mass, the pressure, and the volume. Experiment shows that for a given composition and for a constant mass many different values of pressure and volume are possible. If the pressure is kept constant, the volume may vary over a wide range of values, and vice versa. In other words, the pressure and the volume are independent coordinates. Similarly, experiment shows that for a wire of constant mass the tension and the length are independent coordinates, whereas in the case of a surface film, the surface tension and the area may be varied independently. Some systems that, at first sight, seem quite complicated, such as an electric cell with two different electrodes and an electrolyte, may still be described with the aid of only two independent coordinates. On the other hand, some systems composed of a number of homogeneous parts require the specification of two independent coordinates for each homogeneous part. The essential role of experiment in determining the number and nature of the independent variables is particularly to be noted. Details of various thermodynamic systems and their thermodynamic coordinates will be given in Chap. 2. For the present, to simplify our discussion, we shall deal only with systems of constant mass and composition, each requiring *only one pair* of independent coordinates for its description. This involves no essential loss of generality and results in a considerable saving of words. In referring to any nonspecified system, we shall use the symbols Y and X for the pair of independent coordinates.

A state of a system in which Y and X have definite values which remain constant so long as the external conditions are unchanged is called an *equilibrium* state. Experiment shows that the existence of an equilibrium state in one system depends on the proximity of other systems and on the nature of the wall separating them. Walls are said to be either adiabatic or diathermic. If a wall is *adiabatic* (see Fig. 1-1*a*), a state Y, X for system A and Y', X' for system B may coexist as equilibrium states for *any* attainable values of the four quantities, provided only that the wall is able to withstand the stress associated with the difference between the two sets of coordinates. Thick layers of polystyrene foam, asbestos, felt, etc., are good experimental approximations to adiabatic walls; they are in fact excellent

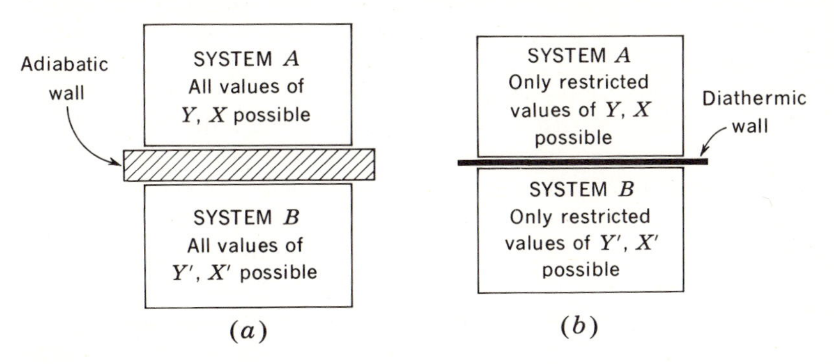

Fig. 1-1
Properties of adiabatic and diathermic walls.

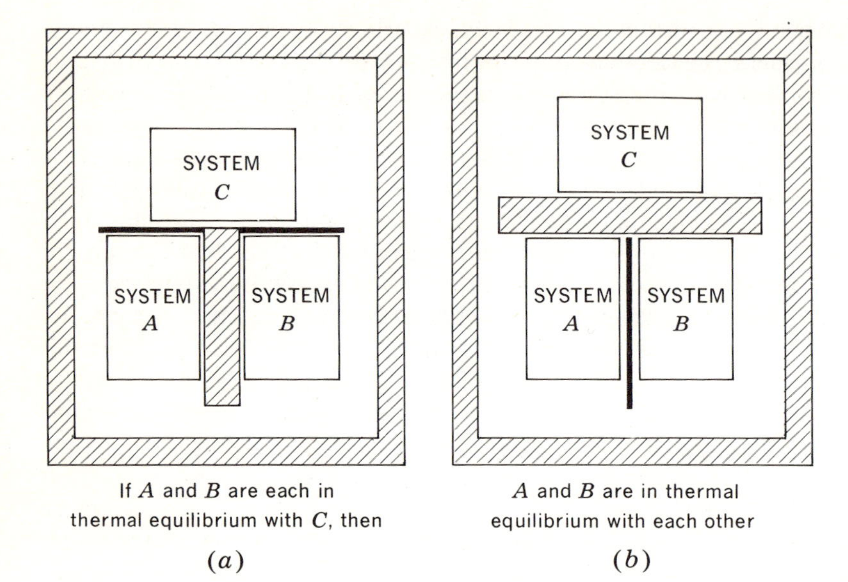

Fig. 1-2
The zeroth law of thermodynamics. (Adiabatic walls are designated by cross shading; diathermic walls, by heavy lines.)

insulators against heat transfer. If the two systems are separated by a *diathermic* wall (see Fig. 1-1*b*), the values of Y, X and Y', X' will change spontaneously until an equilibrium state of the combined system is attained. The two systems are then said to be in *thermal equilibrium* with each other. The commonest diathermic wall is a thin metallic sheet, which serves as an excellent conductor of heat. *Thermal equilibrium is the state achieved by two (or more) systems, characterized by restricted values of the coordinates of the systems, after they have been in communication with one another through a diathermic wall.*

Imagine two systems A and B separated from each other by an adiabatic wall but each in contact with a third system C through diathermic walls, the whole assembly being surrounded by an adiabatic wall as shown in Fig. 1-2*a*. Experiment shows that the two systems will come to thermal equilibrium with the third and that no further change will occur if the adiabatic wall separating A and B is then replaced by a diathermic wall (Fig. 1-2*b*). If, instead of allowing both systems A and B to come to equilibrium with C at the same time, we first have equilibrium between A and C and then equilibrium between B and C (the state of system C being the same in both cases), then, when A and B are brought into communication through a diathermic wall, they will be found to be in thermal equilibrium. We shall use the expression "two systems are in thermal equilibrium" to mean that the two systems are in states such that if the two *were* connected through a diathermic wall, the combined system *would be* in thermal equilibrium.

These experimental facts may then be stated concisely as follows: *Two systems in thermal equilibrium with a third are in thermal equilibrium with each other.* This postulate is known as the *zeroth law of thermodynamics.*

1-6 Temperature Concept

Consider a system A in the state Y_1, X_1 in thermal equilibrium with a system B in the state Y'_1, X'_1. If system A is removed and its state changed, there will be found another state Y_2, X_2 in which it is in thermal equilibrium with the *original* state Y'_1, X'_1 of system B. Experiment shows that there exists a whole set of states Y_1, X_1; Y_2, X_2; Y_3, X_3; . . . , every one of which is in thermal equilibrium with this *same* state Y'_1, X'_1 of system B and which, by the zeroth law, are in thermal equilibrium with one another. We shall suppose that *all* such states, when plotted on a YX diagram, lie on a curve such as I in Fig. 1-3, which we shall call an *isotherm. An isotherm is the locus of all points representing states at which a system is in thermal equilibrium with one state of another system.* We make no assumption as to the continuity of the isotherm, although experiments on simple systems indicate usually that at least a portion of an isotherm is a continuous curve.

Similarly, with regard to system B, we find a set of states Y'_1, X'_1; Y'_2, X'_2; . . . all of which are in thermal equilibrium with one state (Y_1, X_1) of

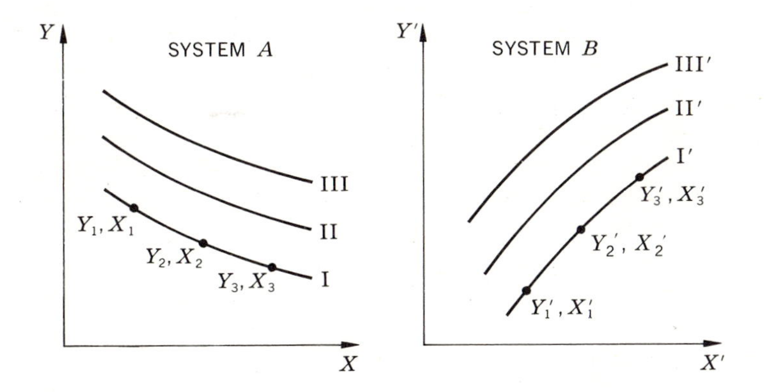

Fig. 1-3
Isotherms of two different systems.

system A, and therefore in thermal equilibrium with one another. These states are plotted on the $Y'X'$ diagram of Fig. 1-3 and lie on the isotherm I′. From the zeroth law, it follows that all the states on isotherm I of system A are in thermal equilibrium with all the states on isotherm I′ of system B. We shall call curves I and I′ *corresponding isotherms* of the two systems.

If the experiments outlined above are repeated with different starting conditions, another set of states of system A lying on curve II may be found, every one of which is in thermal equilibrium with every state of system B lying on curve II′. In this way, a family of isotherms, I, II, III, etc., of system A and a corresponding family I′, II′, III′, etc., of system B may be found. Furthermore, by repeated applications of the zeroth law, corresponding isotherms of still other systems C, D, etc., may be obtained.

All states of corresponding isotherms of all systems have something in common, namely, that they are in thermal equilibrium with one another. The systems themselves, in these states, may be said to possess a property that ensures their being in thermal equilibrium with one another. We call this property *temperature. The temperature of a system is a property that determines whether or not a system is in thermal equilibrium with other systems.*

The temperature of all systems in thermal equilibrium may be represented by a number. A temperature scale is established by the adoption of a set of rules by which one number is assigned to one set of corresponding isotherms and a different number to a different set of corresponding isotherms. Once this is done, the necessary and sufficient condition for thermal equilibrium between two systems is that they have the same temperature. Also, when the temperatures are different, we may be sure that the systems are not in thermal equilibrium.

The preceding operational treatment of the concept of temperature merely expresses the fundamental idea that the temperature of a system is a property which eventually attains the same value as that of other systems when all these systems are put in contact or separated by thin metallic walls within an enclosure of thick asbestos walls. This concept is identical with the everyday idea of temperature as a measure of the hotness or coldness of a system, since, so far as our senses may be relied upon, the hotness of all objects becomes the same after they have been together long enough. However, the expression of this simple idea in technical language establishes a rational set of rules for temperature measurement and also provides a solid foundation for the study of thermodynamics and statistical mechanics.

1-7 Measurement of Temperature

To establish an empirical temperature scale, we select some system with coordinates Y and X as a standard, which we call a *thermometer*, and adopt a set of rules for assigning a numerical value to the temperature associated with each of its isotherms. To every other system in thermal equilibrium with the thermometer, we assign the same number for the temperature. The simplest procedure is to choose any convenient path in the YX plane such as that shown in Fig. 1-4 by the dashed line $Y = Y_1$ which intersects the isotherms at points each of which has the same Y coordinate but a different X coordinate. The temperature associated with each isotherm is then taken to be a convenient function of the X at this intersection point. The coordinate X is called the *thermometric property*, and the form of the *thermometric function* $\theta(X)$ determines the temperature scale. Four important kinds of thermometers, each with their own thermometric property, are listed in Table 1-1.

Let X stand for any one of the thermometric properties listed in Table 1-1, and let us decide arbitrarily to define the temperature scale so that the temperature θ is directly proportional to X. Thus, the temperature common to the thermometer *and to all systems in thermal equilibrium with it* is given by

$$\theta(X) = aX \qquad (\text{const } Y). \tag{1-1}$$

It should be noted that *different* temperature scales usually result when this arbitrary relation is applied to different kinds of thermometers and even when it is applied to different systems of the same kind. One must thus ultimately select, either arbitrarily or in some rational way, one kind of thermometer and one particular system (or type of system) to serve as the standard thermometric device. But regardless of what standard is chosen,

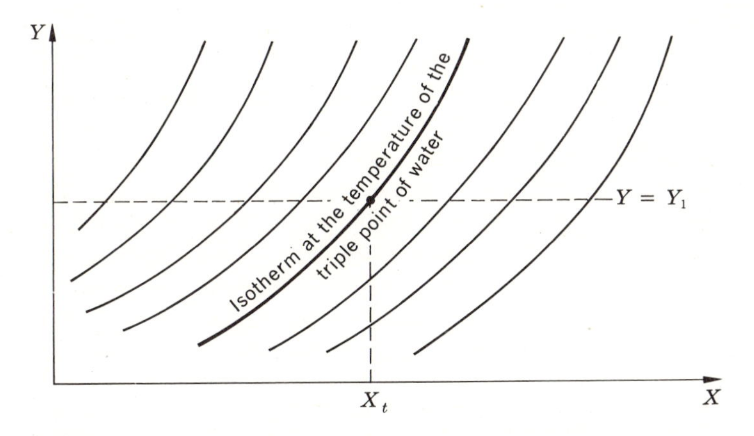

Fig. 1-4
Setting up a temperature scale involves assignment of numerical values to the isotherms of an arbitrarily chosen standard system, or thermometer.

Table 1-1 Thermometers and Thermometric Properties

Thermometer	Thermometric property	Symbol
Gas kept at constant volume	Pressure	P
Electric resistor (under constant pressure and tension)	Electric resistance	R
Thermocouple (under constant pressure and tension)	Thermal emf	e
Liquid column in a glass capillary	Length	L

the value of a in Eq. (1-1) must be established; only then does one have a numerical relation between the temperature $\theta(X)$ and the thermometric property X.

Equation (1-1) applies generally to a thermometer placed in contact with a system whose temperature $\theta(X)$ is to be measured. It therefore applies when the thermometer is placed in contact with an arbitrarily chosen standard system in an *easily reproducible state*; such a state of *an arbitrarily chosen standard system* is called *a fixed point*. Since 1954, only one standard fixed point has been in use, the *triple point of water*, the state of pure water existing as an equilibrium mixture of ice, liquid, and vapor. The temperature at which this state exists is arbitrarily assigned the value of 273.16 kelvins, abbreviated 273.16(K). Equation (1-1) solved for a now becomes

$$a = \frac{273.16}{X_t}, \tag{1-2}$$

where the subscript t identifies the property value X_t explicitly with the triple-point temperature. In view of Eq. (1-2), the general Eq. (1-1) may be written

$$\boxed{\theta(X) = 273.16\frac{X}{X_t} \qquad (\text{const } Y).} \tag{1-3}$$

The temperature of the triple point of water is the *standard fixed point* of thermometery. To achieve the triple point, one distills water of the highest purity into a vessel, depicted schematically in Fig. 1-5. When all air has been removed, the vessel is sealed off. With the aid of a freezing mixture in the inner well, a layer of ice is formed around the well. When the freezing mixture is replaced by a thermometer bulb, a thin layer of ice is melted nearby, as shown in Fig. 1-5. So long as the solid, liquid, and vapor phases coexist in equilibrium, the system is at the triple point. The actual shape of the apparatus used by the U.S. National Bureau of Standards is shown in Fig. 1-6.

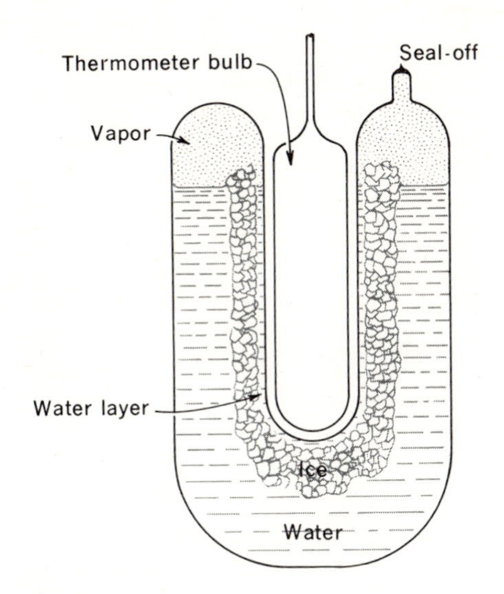

Fig. 1-5
Triple-point cell.

1-8 Comparison of Thermometers

Applying the principles outlined in the preceeding paragraphs to the four thermometers listed in Table 1-1, we have four different ways of measuring temperature. Thus, for a gas at constant volume,

$$\theta(P) = 273.16 \frac{P}{P_t} \qquad \text{(const } V\text{)};$$

for an electric resistor,

$$\theta(R) = 273.16 \frac{R}{R_t};$$

for a thermocouple,

$$\theta(e) = 273.16 \frac{e}{e_t};$$

and for a liquid-in-glass thermometer,

$$\theta(L) = 273.16 \frac{L}{L_t}.$$

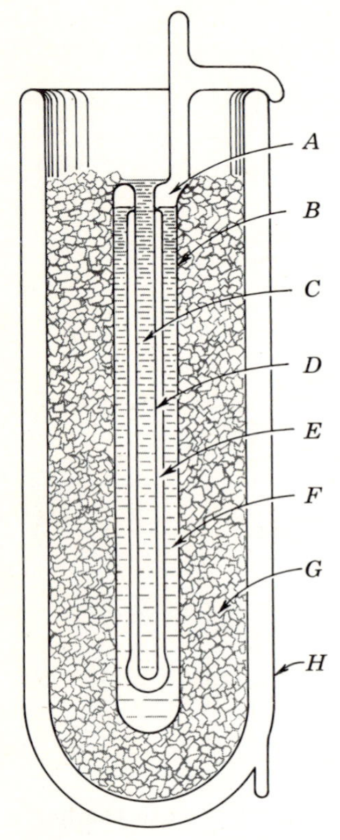

Fig. 1-6

Diagram of the NBS triple-point cell (B, D) in use in an ice bath (G) within a Dewar flask (H). A, water vapor; C, thermometer well; E, ice mantle; F, liquid water.

Now imagine a series of tests in which the temperature of a given system is measured simultaneously with each of the four thermometers. Results of such a test would show that there is considerable difference among the readings of the various thermometers. Further tests would show that different varieties of the same kind of thermometer yield different results. The smallest variation, however, is found among different constant-volume gas thermometers. In particular, the constant-volume hydrogen thermometer and the constant-volume helium thermometer agree more closely than do the others. For this reason a gas is chosen as the standard thermometric substance.

Example 1-1

At the triple point of water, the pressure in a certain constant-volume helium gas thermometer is determined to be 1.0000(atm); at the same fixed point, the resistance

Table 1-2

Bath	P(atm)	$R(\Omega)$	$\theta(P)$	$\theta(R)$
A	0.7322	18.391	200.0	196.0
B	1.0983	28.242	300.0	301.1
C	1.4643	38.181	400.0	407.0
D	1.8303	47.820	500.0	509.8

of a platinum resistance thermometer is 25.625(Ω). The two thermometers are successively immersed in four thermostated baths, each maintained at a different temperature level, and thermometer readings are taken for each bath. The measured values of P and R are listed in the second and third columns of Table 1-2. Compute the corresponding temperatures $\theta(P)$ and $\theta(R)$.

We use the first and second equations presented above. Since $P_t = 1.0000$(atm) and $R_t = 25.625(\Omega)$, these equations become, respectively,

$$\theta(P) = 273.16 \frac{P}{1.0000} \qquad \text{and} \qquad \theta(R) = 273.16 \frac{R}{25.625}.$$

Reduction of the P and R data by these expressions yields the values of $\theta(P)$ and $\theta(R)$ given in the last two columns of Table 1-2. Clearly, the two temperature scales are not identical.

1-9 Gas Thermometer

A schematic diagram of a constant-volume gas thermometer is shown in Fig. 1-7. The materials, construction, and dimensions differ in the various bureaus and institutes throughout the world where these instruments are used and depend on the nature of the gas and the temperature range for which the thermometer is intended. The gas is contained in the bulb B, which communicates with the mercury column M through a capillary. The volume of the gas must be kept constant; raising or lowering the reservoir adjusts the amount of mercury in column M so as to bring its surface just into contact with the tip of a small pointer in the space above M, known as the *dead space*, or *nuisance volume*. The difference in height h between the two mercury columns M and M' is measured when the bulb is surrounded by the system whose temperature is to be measured and when it is surrounded by water at the triple point.

The pressure P of the gas in bulb B is equal to atmospheric pressure plus the pressure equivalent to the mercury height h. This pressure may require

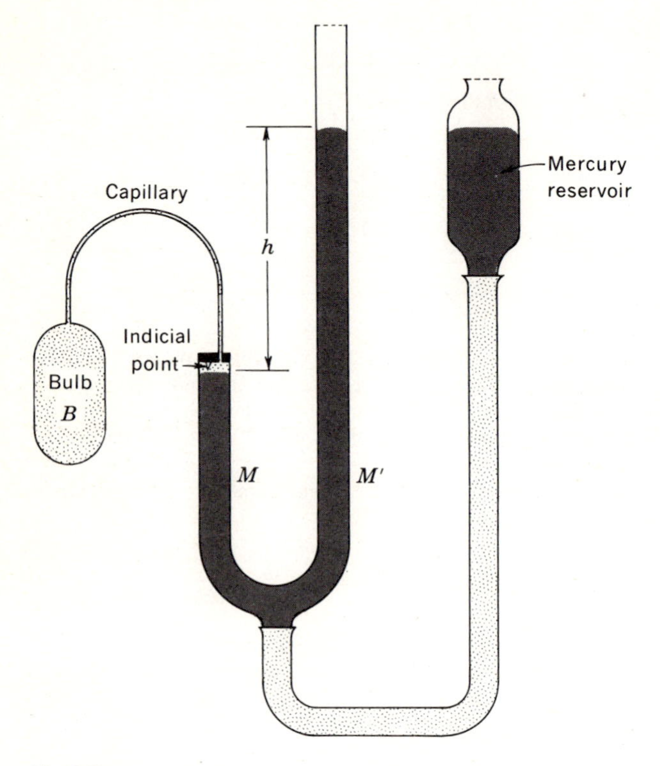

Fig. 1-7
Simplified constant-volume gas thermometer. Mercury reservoir is raised or lowered so that meniscus at left always touches indicial point. Bulb pressure equals h *plus atmospheric pressure.*

correction for certain known sources of error; but once such corrections are made, the temperature is calculated from the equation

$$\theta(P) = 273.16 \frac{P}{P_t} \qquad \text{(const } V\text{)}.$$

1-10 Ideal-Gas Temperature

Suppose that an amount of gas is introduced into the bulb of a constant-volume gas thermometer so that the pressure P_t, when the bulb is surrounded by water at its triple point, is equal to 1,000(mm Hg). With the volume V kept constant, suppose that the following procedures are carried out:

1 Surround the bulb with steam condensing at 1 (atm) pressure, determine the gas pressure P_s, and calculate

$$\theta(P_s) = 273.16 \frac{P_s}{1{,}000}.$$

2 Remove some of the gas so that P_t has a smaller value, say 500 (mm Hg). Determine the new value of P_s and calculate a new value:

$$\theta(P_s) = 273.16 \frac{P_s}{500}.$$

3 Continue reducing the amount of gas in the bulb so that P_t and P_s have smaller and smaller values, P_t having values of, say, 250 (mm Hg), 100 (mm Hg), etc. At *each* value of P_t, calculate the corresponding $\theta(P_s)$.

4 Plot $\theta(P_s)$ against P_t and extrapolate the resulting curve to the axis where $P_t = 0$. Read from the graph

$$\theta_s = \lim_{P_t \to 0} \theta(P_s).$$

The results of a series of tests of this sort are plotted in Fig. 1-8 for four different gases. The graph conveys the information that the readings of a constant-volume gas thermometer depend in general on the nature of the gas; however, as P_t (and P_s) approach zero, *the readings become identical,* yielding a value $\theta_s = 373.15$(K) for condensing steam at 1 (atm). Results of the same nature are obtained for other fixed points, such as those listed in Table 1-4. Since the behavior of gases at a pressure approaching zero is

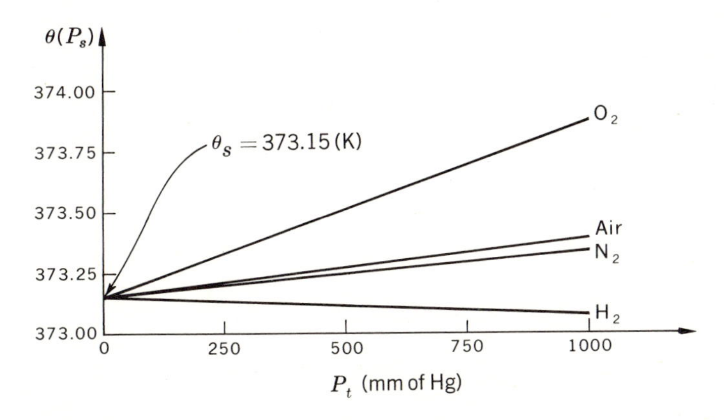

Fig. 1-8
Readings of a constant-volume gas thermometer for the temperature of condensing steam, when different gases are used at various values of P_t.

unique, we consider such behavior to be "ideal," and define the *ideal-gas temperature* T by the equation

$$T = 273.16 \lim_{P_t \to 0} \frac{P}{P_t} \qquad \text{(const } V\text{)}, \tag{1-4}$$

where $\theta(P)$ has been replaced by T to denote this particular temperature scale.

Although the ideal-gas temperature scale is independent of the properties of any one particular gas, it still depends on the properties of gases in general. Measurements at low temperatures can be made only with gases that do not condense at the temperature to be measured. The practical lower limit is about 1(K), obtained with low-pressure helium. *The temperature $T = 0$ remains as yet undefined.*

In Chap. 7 the Kelvin temperature scale, which is independent of the properties of any particular substance, will be developed. It will be shown that in the temperature region in which a gas thermometer may be used, the ideal-gas scale and the Kelvin scale are identical. In anticipation of this result we write (K) after an ideal-gas temperature. It will also be shown in Chap. 7 how the absolute zero of temperature is defined on the Kelvin scale. Until then, the phrase "absolute zero" will have no meaning.

1-11 Celsius Temperature Scale

The Celsius temperature scale employs a degree of the same magnitude as that of the ideal-gas scale, but its zero point is shifted so that *the Celsius temperature of the triple point of water is 0.01 degree Celsius,* abbreviated 0.01(°C). Thus, if t denotes the Celsius temperature,

$$t(°\text{C}) = T(\text{K}) - 273.15. \tag{1-5}$$

The Celsius temperature t_s at which steam condenses at 1(atm) pressure is therefore

$$t_s = T_s - 273.15.$$

Figure 1-8 provides the value of T_s (given as θ_s); thus

$$t_s = 373.15 - 273.15,$$

or

$$t_s = 100.00(°\text{C}).$$

Similar measurements for the ice point [the temperature at which ice and liquid water saturated with air at a pressure of 1(atm) are in equilibrium] show this temperature on the Celsius scale to be 0.00(°C). It should be noted, however, that these two temperatures are subject to the experimental uncertainty attending the determination of intercepts by extrapolation, as illustrated in Fig. 1-8. The only Celsius temperature which is fixed *by definition* is that of the triple point.

1-12 Liquid-in-glass Thermometer

The most familiar device for temperature measurement is the liquid-in-glass thermometer, which consists of a capillary tube of constant cross-sectional area connected to a reservoir (bulb) filled with a suitable liquid. Most fluids expand and therefore rise in the capillary when the temperature increases. The equilibrium length L of liquid in the capillary depends on the temperature, and hence L serves as a thermometric property (Table 1-1).[1]

The calibration of these instruments is conventionally based on *two* fixed points rather than on the single fixed point described in Secs. 1-7 and 1-8; an empirical temperature scale is *defined* such that t^* is linear in L:

$$t^* = aL + b,$$

where a and b are constants of the thermometer. These constants are determined once arbitrary values t_1^* and t_2^* are associated with the two liquid lengths L_1 and L_2 that result when the thermometer is maintained at two fixed points 1 and 2. At these points the linear relation between t^* and L gives

$$t_2^* = aL_2 + b \qquad \text{and} \qquad t_1^* = aL_1 + b.$$

The difference between these equations is

$$t_2^* - t_1^* = a(L_2 - L_1).$$

Similarly, the difference between an unknown temperature t^* and t_2^* is

$$t^* - t_2^* = a(L - L_2).$$

Division of this equation by the preceding one gives

$$\frac{t^* - t_2^*}{t_2^* - t_1^*} = \frac{L - L_2}{L_2 - L_1},$$

[1] For an interesting nontechnical account of the historical development of these devices, see W. E. K. Middleton, "A History of the Thermometer and Its Use in Meteorology," Johns Hopkins, Baltimore, 1966.

where the constants a and b have now been eliminated in favor of t_1^* and t_2^*. Rearrangement of this equation yields

$$t^* = \frac{t_2^*(L - L_1) + t_1^*(L_2 - L)}{L_2 - L_1}.$$

It is generally required that the empirical temperature scale approximate one of the conventional temperature scales. The extent to which this is possible in practice depends upon the liquid used and upon the required accuracy and range of application of the thermometer.

Example 1-2

We wish to construct a liquid-in-glass thermometer for measurement of Celsius temperatures in the range of −25 to 100(°C). Determine the relative suitability of the following liquids for this application: water, isopropyl alcohol, and mercury.

We choose as our fixed points the ice and steam points, and correspondingly set $t_1^* = 0$ and $t_2^* = 100$. The equation for t^* then reduces to

$$t^* = \frac{100(L - L_1)}{L_2 - L_1}.$$

Thus *by definition* $t^* = t(°C) = 0$ for $L = L_1$, and $t^* = t(°C) = 100$ for $L = L_2$. The question now arises whether t^* can be identified with $t(°C)$ over the temperature range of interest; if not, we wish to determine which of the three proposed fluids yields the best approximation.

If the dimensions of the thermometer glass remain constant, the ΔL resulting from a temperature change is proportional to ΔV, the corresponding change in the specific volume of the fluid. Thus, the last equation becomes

$$t^* = \frac{100(V - V_1)}{V_2 - V_1}.$$

Table 1-3 contains values for V of the three liquids at several Celsius temperatures, and the corresponding values of t^* computed from the above formula. Each thermometer reading agrees *by definition* with the Celsius scale at 0 and 100(°C); at other temper-

Table 1-3

	Water		Isopropyl alcohol		Mercury	
t(°C)	V(cm)³/(g)	t^*	V(cm)³/(g)	t^*	V(cm)³/(g)	t^*
−25	...	...	1.2167	−18.8	0.073220	−25.0
0	1.0002	0	1.2475	0	0.073556	0
25	1.0029	6.2	1.2800	19.8	0.073890	24.9
50	1.0121	27.5	1.3170	42.4	0.074225	49.9
75	1.0259	59.4	1.3604	68.8	0.074561	74.9
100	1.0435	100	1.4116	100	0.074898	100

atures agreement is more or less approximate. Water is clearly the worst of the three fluids; besides yielding very poor comparisons between t^* and t, it freezes at 0(°C) and cannot be used below this temperature. Isopropyl alcohol, although better than water, is also unacceptable. Mercury is best, which accounts for its use in commercial thermometers.

1-13 Electric-resistance Thermometry

Very accurate temperature measurements are possible with carefully calibrated resistance thermometers. The platinum resistance thermometer serves as a secondary temperature standard over a wide temperature range in laboratories throughout the world. It is a delicate instrument, made of a long fine wire, usually wound on a thin frame in such a way as to minimize changes in stress with temperature change.

To determine the resistance, one maintains a known constant current in the thermometer and measures the potential difference across it with the aid of a very sensitive potentiometer. The same instrument measures the potential difference across a standard resistor in series with the thermometer; this determines the current, which is held to the prescribed value by an appropriate rheostat setting. A typical circuit is shown in Fig. 1-9.

The platinum resistance thermometer may be used for very accurate work within the range of about −250 to 1200(°C). The calibration of the instrument involves the measurement of its resistance R at various known

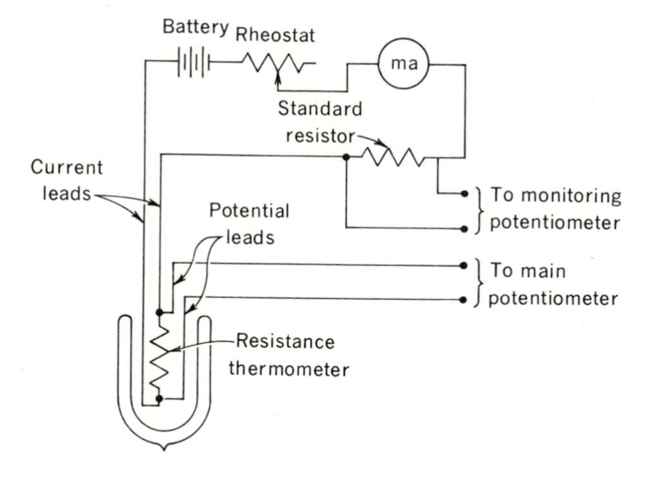

Fig. 1-9
Circuit for measuring the resistance of a resistance thermometer through which a constant current is maintained.

temperatures and the representation of the results by an empirical formula. In a restricted range, the following quadratic equation is often used:

$$R = R_0(1 + At + Bt^2),$$

where A and B are constants, and R_0 is the resistance of the platinum wire when it is surrounded by water at the ice point.

It should be noted that once the ideal-gas scale is adopted as the standard temperature scale, the relation between the resistance and temperature is not linear. This is clear from the results of Example 1-1. It is also true of other thermometric properties. Earlier in this chapter we noted the possibility of defining linear temperature scales based on several thermometric properties. However, once a particular property is chosen to provide a linear scale, the others will in general not be linear.

The thermistor, a resistance thermometer made of a semiconductor, has now come into widespread use. Its resistance *decreases* rapidly with increasing temperature, and when placed in a Wheatstone-bridge circuit, it can be used as a very sensitive thermometer. Thermistors are available in many shapes and sizes, and are thus suitable for a great variety of applications.

Example 1-3

The different temperature characteristics of metallic resistance thermometers and thermistors are best illustrated by example. The resistance ratio R/R_0 of a particular platinum resistance thermometer is given by

$$\frac{R}{R_0} = 1 + 3.984 \times 10^{-3}t - 5.852 \times 10^{-7}t^2.$$

For a typical ceramic thermistor, the corresponding formula is

$$\frac{R}{R_0} = \exp \frac{-14.52t}{t + 273.15}.$$

In both equations, R_0 is the resistance at 0(°C). In Fig. 1-10 are shown plots of R/R_0 vs. t computed from the two equations. For the temperature range 0 to 300(°C), the resistance of the platinum thermometer increases by a factor of 2, while R for the thermistor decreases by a factor of 2,000. This remarkable sensitivity of thermistors makes them suitable for applications requiring very precise temperature measurement and control.

1-14 Thermocouple

The correct use of a thermocouple is shown in Fig. 1-11. The thermal emf is measured with a potentiometer, which, as a rule, must be placed at some distance from the system whose temperature is to be measured. The reference junction, therefore, is placed near the test junction, and consists of two connections with copper wire, maintained at the ice-point temperature or at the triple-point temperature of water. This arrangement allows the use of copper wires for connection to the potentiometer. The binding posts of the potentiometer are usually made of brass; therefore, at the potentiometer there are two copper-brass thermocouples. If the two binding posts are at the same temperature, these two copper-brass thermocouples introduce no error.

The range and emf of a thermocouple depend upon the materials of which it is composed. A platinum–platinum-rhodium couple has a range of 0 to 1600(°C). The advantage of a thermocouple is that it comes to thermal equilibrium quite rapidly with the system whose temperature is to be measured, because its mass is small. It therefore follows temperature changes easily but is not so accurate as a platinum resistance thermometer.

1-15 International Practical Temperature Scale of 1968 (IPTS-68)

The use of an ideal-gas thermometer for routine calibrations or for the usual measurement of thermodynamic temperature is impractical. At the Seventh

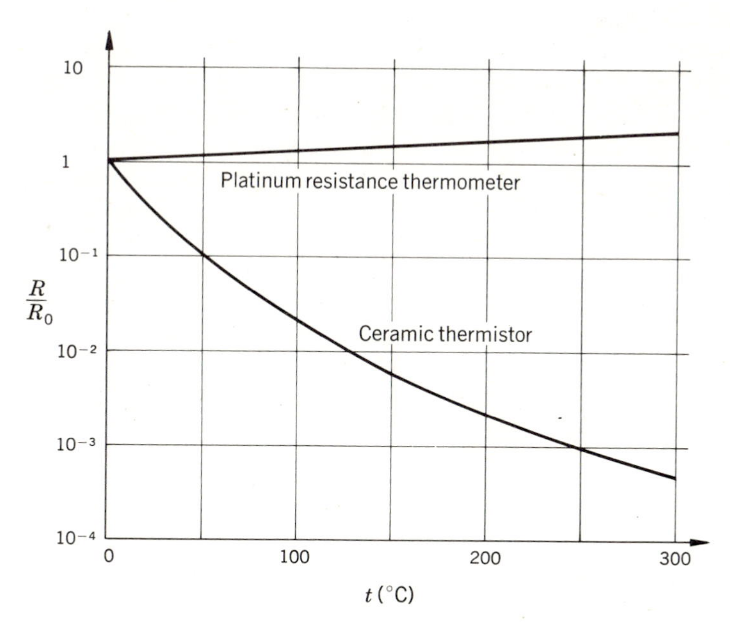

Fig. 1-10
Variation of R *with* t *for resistance thermometers.*

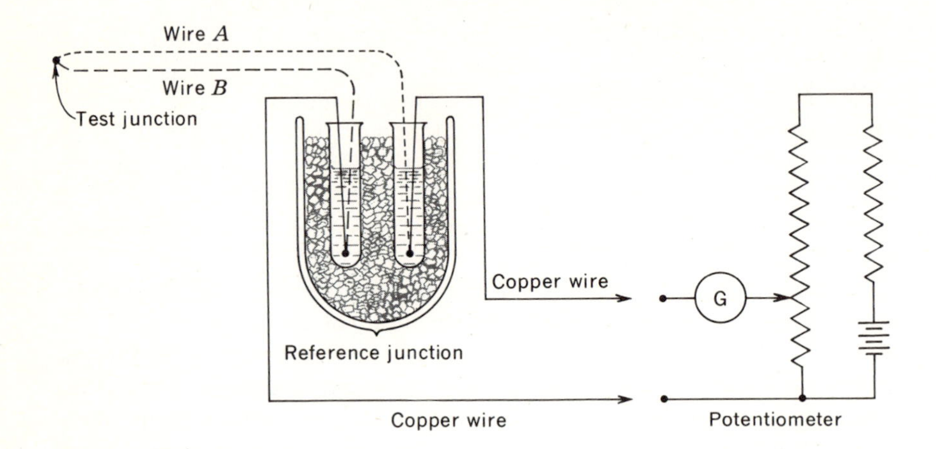

Fig. 1-11
Thermocouple of wires A and B with a reference junction consisting of two junctions with copper, connected to a potentiometer.

General Conference of Weights and Measures in 1927, an international practical temperature scale was adopted to provide the means for easy and rapid calibration of scientific and industrial instruments. Refinements and adjustments were incorporated into the scale in revisions adopted in 1948, 1960, and 1968.

The International Practical Temperature Scale of 1968 (IPTS-68) consists of a set of fixed points measured with a constant-volume gas thermometer (Table 1-4) and a set of procedures for interpolation between the fixed points. Although IPTS-68 is not intended to supplant the ideal-gas or Celsius scales, it is constructed so as to provide a very close approximation to them; the differences between the scales are within the limits of accuracy of measurement attainable in 1968.

The lower temperature limit of IPTS-68 is 13.81(K), the triple point of equilibrium hydrogen. Below this temperature the scale is undefined. Above the freezing point of gold [1337.58(K)], an optical method is used in conjunction with the Planck radiation formula. The interval between 13.81 and 1337.58(K) is divided into three main parts, as follows:

1 *From 13.81 to 273.15(K).* A strain-free annealed platinum resistance thermometer is used. The temperature range is divided into four subintervals, and within each interval the resistance is measured at specified fixed points selected from Table 1-4. The differences between the measured resistances and a tabulated reference function are fitted to polynomial equations in T or t; these equations serve as interpolation formulas for conversion of measured values of R to temperature.

2 *From 273.15 to 903.89(K)*. The same platinum resistance thermometer is used as in part 1. A polynomial equation for t as a function of R is employed, with the constants in the equation determined from resistance measurements at the triple point of water, the normal boiling point of water, and the normal freezing point of zinc.

3 *From 903.89 to 1337.58(K)*. A thermocouple, one wire of which is made of platinum of a specified purity and the other of an alloy of 90 percent platinum and 10 percent rhodium, has one junction maintained at 0(°C). The electromotive force e is represented by the formula

$$e = a + bt + ct^2,$$

where a, b, and c are calculated from measurements of e at 903.89 ± 0.2(K), as determined by a platinum resistance thermometer, and at the normal freezing points of silver and gold.

The text accompanying the official definition of IPTS-68 also describes recommended apparatus and experimental technique and contains a list of secondary reference points to supplement the fixed points given in Table 1-4.

1-16 The Rankine and Fahrenheit Scales

Two temperature scales commonly used by engineers are based on a degree five-ninths the size of the Kelvin and Celsius degree. By definition, the

Table 1-4 Defining Fixed Points for IPTS-68*

Defining fixed points	T(K)	t(°C)
Triple point of equilibrium hydrogen	13.81	−259.34
Boiling point of equilibrium hydrogen at 25/76 (atm)	17.042	−256.108
Normal boiling point of equilibrium hydrogen	20.28	−252.87
Normal boiling point of neon	27.102	−246.048
Triple point of oxygen	54.361	−218.789
Normal boiling point of oxygen	90.188	−182.962
Triple point of water	273.16	0.01
Normal boiling point of water	373.15	100.00
Normal freezing point of zinc	692.73	419.58
Normal freezing point of silver	1235.08	961.93
Normal freezing point of gold	1337.58	1064.43

* *Metrologia,* **5**:35 (1969)

Rankine scale is given by

$$\boxed{T(\mathrm{R}) = 1.8 \times T(\mathrm{K}).} \tag{1-6}$$

The Fahrenheit scale is defined in relation to the Rankine scale by the equation

$$\boxed{t(^\circ\mathrm{F}) = T(\mathrm{R}) - 459.67.} \tag{1-7}$$

Thus, at the ice point, where the Kelvin temperature is 273.15, the Rankine temperature is (1.8) (273.15) = 491.67. Hence the Fahrenheit temperature is

$$491.67 - 459.67 = 32.00(^\circ\mathrm{F}).$$

Similarly, at the steam point, the Rankine temperature is

$$(1.8)(373.15) = 671.67(\mathrm{R}),$$

and the Fahrenheit temperature is

$$671.67 - 459.67 = 212.00(^\circ\mathrm{F}).$$

The Fahrenheit temperature is related to the Celsius temperature by

$$\boxed{t(^\circ\mathrm{F}) = 1.8 \times t(^\circ\mathrm{C}) + 32.00.} \tag{1-8}$$

The relationships among the four temperature scales are shown schematically in Fig. 1-12.

Problems

1-1 The limiting value of the ratio of the pressures of a gas at the steam point and at the triple point of water when the gas is kept at constant volume is found to be 1.36605. What is the ideal-gas temperature of the steam point?

1-2 The resistance of a platinum wire is found to be 11.000(Ω) at the ice point, 15.247(Ω) at the steam point, and 27.949(Ω) at the zinc point. Find the constants A and B in the equation

$$R = R_0(1 + At + Bt^2),$$

and plot R against t in the range from 0 to 660(°C).

1-3 When the reference junction of a thermocouple is kept at the ice point and the test junction is at the Celsius temperature t, the emf e of the thermocouple is given by the equation

$$e = \alpha t + \beta t^2,$$

where $\alpha = 0.20(\mathrm{mV})/(^\circ\mathrm{C})$, and $\beta = -5.0 \times 10^{-4}(\mathrm{mV})/(^\circ\mathrm{C})^2$.

(*a*) Compute the emf when $t = -100$, 200, 400, and 500(°C), and draw a graph of e against t in this range.

(*b*) Suppose that the emf e is taken as a thermometric property, that a temperature scale t^* is defined by the linear equation

$$t^* = ae + b,$$

and that $t^* = 0°$ at the ice point and $t^* = 100°$ at the steam point. Find the numerical values of a and b, and draw a graph of e against t^*.

(*c*) Find the values of t^* when $t = -100$, 200, 400, and 500(°C), and draw a graph of t^* against t.

(*d*) Compare the Celsius scale with the t^* scale.

1-4 It will be shown in Chap. 5 that for the same number of moles of gas at the same temperature,

$$\lim_{P \to 0} (PV) = \text{a universal constant}$$

for all gases. Show how this fact may be used to set up a temperature scale, by

(*a*) Using the triple point of water as the only fixed point.

(*b*) Using both the ice point and the steam point as fixed points.

1-5 When the ice point and the steam point are chosen as fixed points with 100 degrees between them, the ideal-gas temperature of the ice point may be written

$$T_i = \frac{100}{r_s - 1},$$

where $r_s = \lim (P_s/P_i)$ at const V.

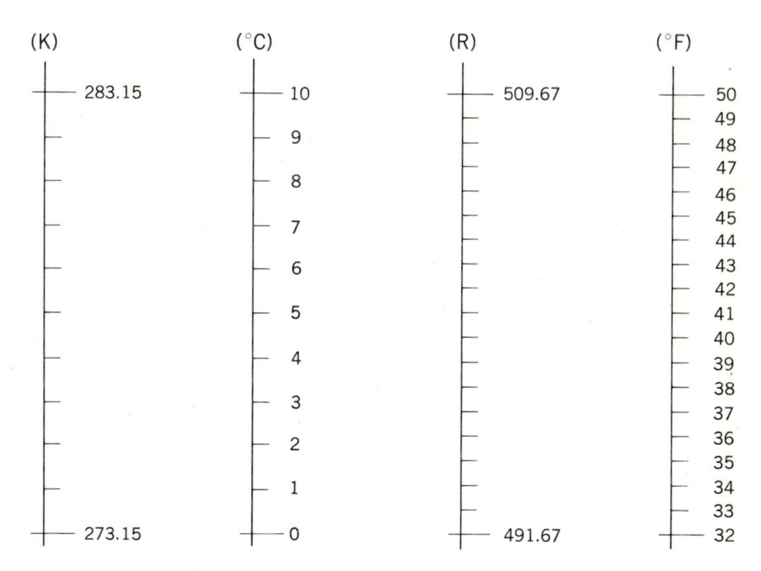

Fig. 1-12
Comparison of temperature scales.

(a) Show that the fractional error in T_i produced by an error in r_s is very nearly 3.73 times the fractional error in r_s, or

$$\frac{dT_i}{T_i} = 3.73 \frac{dr_s}{r_s}.$$

(b) Any ideal-gas temperature may be written

$$T = T_i r,$$

where $r = \lim(P/P_i)$ at const V. Show that the fractional error in T is

$$\frac{dT}{T} = \frac{dr}{r} + 3.73 \frac{dr_s}{r_s}.$$

(c) Now that there is only one fixed point at which the ideal-gas temperature is a universal constant, show that the fractional error in T is

$$\frac{dT}{T} = \frac{dr}{r},$$

where $r = \lim(P/P_t)$ at const V.

1-6 At what absolute temperature do the Celsius and Fahrenheit temperature scales give the same numerical value?

1-7 In the table below, a number in the top row represents the pressure of a gas in the bulb of a constant-volume gas thermometer (corrected for dead space, thermal expansion of bulb, etc.) when the bulb is immersed in a water triple-point cell. The bottom row represents the corresponding readings of pressure when the bulb is surrounded by a material at a constant unknown temperature. Calculate the ideal-gas temperature T of this material. (Use five significant figures.)

P_t(mm Hg)	1000.0	750.00	500.00	250.00
P(mm Hg)	1535.3	1151.6	767.82	383.95

2

THERMODYNAMIC SYSTEMS

2-1 Thermodynamic Equilibrium

Suppose that experiments have been performed on a thermodynamic system and that the coordinates necessary and sufficient for a macroscopic description have been determined. When these coordinates change in any way whatsoever, either spontaneously or by virtue of outside influence, the system is said to undergo a *change of state*.[1] When a system is not influenced in any way by its surroundings, it is said to be isolated. In practical applications of thermodynamics, isolated systems are of little importance. We usually have to deal with a system that is influenced in some way by its surroundings. In general, the surroundings may exert forces on the system or provide contact between the system and a body at some definite temperature. When the state of a system changes, interactions usually take place between the system and its surroundings.

When there is no unbalanced force in the interior of a system and also none between a system and its surroundings, the system is said to be in a state of *mechanical equilibrium*. When these conditions are not satisfied, either the system alone or both the system and its surroundings will undergo a change of state, which will cease only when mechanical equilibrium is restored.

When a system in mechanical equilibrium does not tend to undergo a spontaneous change of internal structure, such as a chemical reaction, or a transfer of matter from one part of the system to another, such as diffusion or solution, however slow, it is said to be in a state of *chemical equilibrium*. A system not in chemical equilibrium undergoes a change of state that may

[1] This must not be confused with the terminology of elementary physics, where the expression "change of state" is often used to signify a transition from solid to liquid or liquid to gas, etc. Such a change in the language of thermodynamics is called a "change of phase."

in some cases be exceedingly slow. The change ceases when chemical equilibrium is reached.

Thermal equilibrium exists where there is no spontaneous change in the coordinates of a system in mechanical and chemical equilibrium when it is separated from its surroundings by a diathermic wall. In thermal equilibrium, all parts of a system are at the same temperature, and this temperature is the same as that of the surroundings. When these conditions are not satisfied, a change of state will take place until thermal equilibrium is reached.

When the conditions for all three types of equilibrium are satisfied, the system is said to be in a state of *thermodynamic equilibrium*; in this condition, it is apparent that there will be no tendency whatever for any change of state of either the system or the surroundings to occur. *States of thermodynamic equilibrium can be described in terms of macroscopic coordinates that do not involve the time, i.e., in terms of thermodynamic coordinates.* Thermodynamics does not attempt to deal with any problem involving the rate at which a process takes place. The investigation of such problems is carried out in other branches of science and engineering, such as fluid mechanics, transport phenomena, and chemical kinetics.

When the conditions for any one of the three types of equilibrium that constitute thermodynamic equilibrium are not satisfied, the system is said to be in a *nonequilibrium state.* Thus, when there is an unbalanced force in the interior of a system or between a system and its surroundings, there may occur acceleration, turbulence, eddies, waves, etc. While such phenomena are in progress, a system passes through nonequilibrium states. If an attempt is made to give a macroscopic description of any one of these nonequilibrium states, it is found that the pressure varies from one part of a system to another. There is no single pressure that refers to the system as a whole. Similarly, in the case of a system at a different temperature from its surroundings, a nonuniform temperature distribution is set up and there is no single temperature that refers to the system as a whole. We therefore conclude that *when the conditions for mechanical and thermal equilibrium are not satisfied, the states traversed by a system cannot be described in terms of thermodynamic coordinates referring to the system as a whole.*

It must not be concluded, however, that we are entirely helpless in dealing with such nonequilibrium states. If we divide the system into a large number of small mass elements, then thermodynamic coordinates may be found in terms of which a macroscopic description of each mass element may be approximated. There are also special methods for dealing with systems in mechanical and thermal equilibrium but not in chemical equilibrium. At present we shall deal exclusively with systems in thermodynamic equilibrium.

2-2 Equation of State

Imagine for the sake of simplicity a constant mass of gas in a vessel so equipped that the pressure, volume, and temperature may be easily measured. If we fix the volume at some arbitrary value and cause the temperature to assume an arbitrarily chosen value, we shall not be able to vary the pressure at all. Once V and T are chosen by us, the value of P at equilibrium is determined by nature. Similarly, if P and T are chosen arbitrarily, the value of V at equilibrium is fixed. That is, of the three thermodynamic coordinates P, V, and T, only two are independent variables. This implies that there exists an equation of equilibrium which connects the thermodynamic coordinates and which robs one of them of its independence. Such an equation is called an *equation of state*. Every thermodynamic system has its own equation of state, although in some cases the relation may be so complicated that it cannot be expressed in terms of simple mathematical functions.

An equation of state expresses the individual peculiarities of one system in contradistinction to another, and must therefore be determined either by experiment or by molecular theory. A general theory like thermodynamics, based on general laws of nature, is incapable of predicting the specific behavior of any particular material. An equation of state is therefore not a theoretical deduction from thermodynamics but is usually an experimental addition to thermodynamics. It expresses the results of experiments in which the thermodynamic coordinates of a system were measured as accurately as possible, within a limited range of values. An equation of state is therefore only as accurate as the experiments that led to its formulation and holds only within the range of values measured. As soon as this range is exceeded, a different form of equation of state may be valid.

For example, a system consisting of exactly 1(mol) of gas [1(g mol) = M(g), or 1(lb mol) = M(lb), where M is the molecular weight] and at low pressure has the equation of state

$$PV = RT,$$

where R is a constant, and V is the molar volume of the gas. At higher pressures, its equation of state is more complicated, and is often represented by

$$\frac{PV}{RT} = 1 + \frac{B(T)}{V} + \frac{C(T)}{V^2} + \cdots,$$

where R is a constant, and $B(T)$ and $C(T)$ are functions of T. As far as thermodynamics is concerned, the important thing is that an equation of state exists, not that we can write it down mathematically.

It is obvious that no equation of state exists for the states traversed by a system which is not in mechanical and thermal equilibrium, since such states cannot be described in terms of thermodynamic coordinates referring to the system as a whole. For example, if a gas in a cylinder were to expand and to impart to a piston an accelerated motion, the gas might have, at any moment, a definite volume and temperature, but the corresponding pressure, calculated from an equation of state, would not apply to the system as a whole. This pressure would not be a thermodynamic coordinate because it would depend not only on the velocity and acceleration of the piston but also on the point of measurement.

The example of a gas at equilibrium is a special case of a general class of systems which can be described in terms of the three thermodynamic coordinates P, V, and T. The next section is devoted to a consideration of such systems. There are many other types of systems which may be described at least to a first approximation by means of three thermodynamic coordinates, which may be given the generalized designations X, Y, and Z. Such "XYZ systems" will be called *simple systems*. PVT systems represent an important type of simple system. Another type is represented by structural members in tension or compression; yet another, by surfaces. These and others will be considered later in this chapter and in subsequent chapters.

2-3 PVT Systems

Any isotropic system of constant mass and constant composition that exerts on the surroundings a uniform hydrostatic pressure, in the absence of surface, gravitational, electrical, and magnetic effects, we shall call a *PVT system*. PVT systems are divided into the following categories:

1. *A pure substance*, i.e., a single chemical species, in the form of a solid, a liquid, or a gas.
2. *A homogeneous mixture of different chemical species*, such as a mixture of gases, a mixture of liquids, or a solid solution.

Experiment shows that the states of equilibrium[1] of a PVT system can be described with the aid of three coordinates, namely, the pressure P exerted by the system on the surroundings, the volume V, and the absolute temperature T.

Every PVT system has an equation of state expressing a relation among these three coordinates that is valid for equilibrium states. If the system

[1] In the remainder of this book the word "equilibrium," unmodified by any adjective, will refer to thermodynamic equilibrium.

undergoes a small change of state whereby it passes from an initial state of equilibrium to another state of equilibrium very near the initial one, then all three coordinates, in general, undergo slight changes. If the change of, say, V is very small in comparison with V and very large in comparison with the space occupied by a few molecules, this change of V may be written as a differential dV. If V were a geometrical quantity referring to the volume of space, dV could be used to denote a portion of that space arbitrarily small. Since, however, V is a macroscopic coordinate denoting the volume of *matter*, then, for dV to have a meaning, it must be large enough to include enough molecules to warrant the use of the macroscopic point of view.

Similarly, if the change of P is very small in comparison with P and very large in comparison with local fluctuations of pressure caused by momentary variations in molecular concentration, it also may be represented by the differential dP. *Every infinitesimal in thermodynamics must satisfy the requirement that it represents a change in a quantity which is small with respect to the quantity itself and large in comparison with the effect produced by the behavior of a few molecules.* The reason for this is that thermodynamic coordinates such as volume, pressure, and temperature have no meaning when applied to a few molecules. This is another way of saying that thermodynamic coordinates are macroscopic coordinates.

We may imagine the equation of state solved for any coordinate in terms of the other two. Thus

$$V = \text{function of } (T,P), \qquad \text{or} \qquad V = V(T,P).$$

An infinitesimal change from one state of equilibrium to another state of equilibrium involves a dV, a dT, and a dP, all of which we shall assume satisfy the condition laid down in the previous paragraph. As a result of this functional relation, a fundamental theorem in partial differential calculus enables us to write

$$dV = \left(\frac{\partial V}{\partial T}\right)_P dT + \left(\frac{\partial V}{\partial P}\right)_T dP, \tag{2-1}$$

where each partial derivative is itself a function of T and P. Both have an important physical meaning. Recalling from elementary physics a quantity called the average coefficient of volume expansion, or volume expansivity, we can write

$$\text{Average volume expansion} = \frac{\text{change of volume per unit volume}}{\text{change of temperature}},$$

where the changes occur at constant pressure.

If the change of temperature is made smaller and smaller until it becomes infinitesimal, the change in volume also becomes infinitesimal and we have what is known as the instantaneous volume expansivity, or just volume expansivity, which is denoted by β. Thus

$$\boxed{\beta \equiv \frac{1}{V}\left(\frac{\partial V}{\partial T}\right)_P.} \tag{2-2}$$

A similar quantity called the *isothermal compressibility*, denoted by κ, shows the influence of pressure on volume at constant temperature. Thus

$$\boxed{\kappa \equiv -\frac{1}{V}\left(\frac{\partial V}{\partial P}\right)_T.} \tag{2-3}$$

Since a positive change (increase) of pressure produces a negative change (decrease) of volume, the minus sign is introduced to make the isothermal compressibility a positive number.

Both β and κ are properties and, like volume, are functions of T and P. Numerical values of β and κ obtained by experiment are often tabulated for liquids and solids in compilations of data. A list of values for some common liquids is given in Table 2-1. The temperature and pressure dependence of β and κ for liquids and solids is treated in Chap. 9. However, it is often found that β and κ vary only slowly with T and P, and are therefore sometimes treated as though constant.

If the equation of state is solved for P, then

$$P = \text{function of } (T,V), \qquad \text{or} \qquad P = P(T,V)$$

and

$$dP = \left(\frac{\partial P}{\partial T}\right)_V dT + \left(\frac{\partial P}{\partial V}\right)_T dV. \tag{2-4}$$

Similarly, if T is taken to be a function of P and V, i.e., if $T = T(P,V)$,

Table 2-1 Values of β and κ at 20(°C) for Some Liquids

Substance	$\beta \times 10^3 (\mathrm{K})^{-1}$	$\kappa \times 10^6 (\mathrm{bar})^{-1}$
Diethyl ether	1.66	187
Benzene	1.24	95
n-Heptane	1.23	144
Ethyl alcohol	1.12	111
Glycerine	0.505	21
Water	0.207	45.9
Mercury	0.182	4.02

then

$$dT = \left(\frac{\partial T}{\partial P}\right)_V dP + \left(\frac{\partial T}{\partial V}\right)_P dV. \tag{2-5}$$

In all the equations above, the system was assumed to undergo an infinitesimal change from an initial state of equilibrium to another. This enabled us to use an equation of equilibrium (equation of state) and to solve it for any coordinate in terms of the other two. The differentials dP, dV, and dT are therefore differentials of actual functions, and are called *exact differentials*. In general, if z is a function of x and y, i.e., if $z = z(x,y)$, then dz is given by the exact differential expression

$$dz = \left(\frac{\partial z}{\partial x}\right)_y dx + \left(\frac{\partial z}{\partial y}\right)_x dy.$$

An infinitesimal that is not the differential of an actual function is called an *inexact differential* and cannot be expressed by an equation of the type above. Other distinctions between exact and inexact differentials will be made clear later.

2-4 Mathematical Theorems

Three simple theorems in partial differential calculus are used very often in this subject. The proofs are as follows. Suppose that there exists a functional relationship among the three coordinates x, y, and z:

$$f(x,y,z) = 0.$$

Then x may be considered a function of y and z, and

$$dx = \left(\frac{\partial x}{\partial y}\right)_z dy + \left(\frac{\partial x}{\partial z}\right)_y dz. \tag{2-6}$$

Alternatively, y may be considered a function of x and z, and

$$dy = \left(\frac{\partial y}{\partial z}\right)_x dz + \left(\frac{\partial y}{\partial x}\right)_z dx. \tag{2-7}$$

Substituting Eq. (2-7) into Eq. (2-6) and rearranging, we have

$$\left[\left(\frac{\partial x}{\partial y}\right)_z \left(\frac{\partial y}{\partial x}\right)_z - 1\right] dx + \left[\left(\frac{\partial x}{\partial y}\right)_z \left(\frac{\partial y}{\partial z}\right)_x + \left(\frac{\partial x}{\partial z}\right)_y\right] dz = 0.$$

Now of the three coordinates, two are independent. If x and z are chosen as independent coordinates, then the equation above must hold for all possible values of dx and dz, and the coefficients of dx and dz must be identically zero. Thus,

$$\left(\frac{\partial x}{\partial y}\right)_z \left(\frac{\partial y}{\partial x}\right)_z - 1 = 0,$$

from which

$$\boxed{\left(\frac{\partial x}{\partial y}\right)_z = \left(\frac{\partial y}{\partial x}\right)_z^{-1}.} \tag{2-8}$$

Similarly,

$$\left(\frac{\partial x}{\partial y}\right)_z \left(\frac{\partial y}{\partial z}\right)_x + \left(\frac{\partial x}{\partial z}\right)_y = 0,$$

from which

$$\boxed{\left(\frac{\partial x}{\partial y}\right)_z = -\left(\frac{\partial x}{\partial z}\right)_y \left(\frac{\partial z}{\partial y}\right)_x.} \tag{2-9}$$

The third theorem is developed by introduction of a fourth state variable w. Dividing Eq. (2-6) by the differential dw, we find

$$\frac{dx}{dw} = \left(\frac{\partial x}{\partial y}\right)_z \frac{dy}{dw} + \left(\frac{\partial x}{\partial z}\right)_y \frac{dz}{dw}.$$

At constant z, this becomes

$$\left(\frac{\partial x}{\partial w}\right)_z = \left(\frac{\partial x}{\partial y}\right)_z \left(\frac{\partial y}{\partial w}\right)_z,$$

and rearrangement yields

$$\boxed{\left(\frac{\partial x}{\partial y}\right)_z = \left(\frac{\partial x}{\partial w}\right)_z \left(\frac{\partial w}{\partial y}\right)_z.} \tag{2-10}$$

Example 2-1

In a PVT system, differential changes in P are related to those in T and V by Eq. (2-4):

$$dP = \left(\frac{\partial P}{\partial T}\right)_V dT + \left(\frac{\partial P}{\partial V}\right)_T dV. \tag{2-4}$$

We wish to eliminate the partial derivatives in favor of the volume expansivity β and the isothermal compressibility κ.

From the definitions, Eqs. (2-2) and (2-3), we find

$$\left(\frac{\partial P}{\partial V}\right)_T = -\frac{1}{\kappa V} \quad \text{and} \quad \left(\frac{\partial V}{\partial T}\right)_P = \beta V.$$

An expression is required for $(\partial P/\partial T)_V$. Application of Eq. (2-9), with $x = P$, $y = T$, and $z = V$, yields

$$\left(\frac{\partial P}{\partial T}\right)_V = -\left(\frac{\partial P}{\partial V}\right)_T\left(\frac{\partial V}{\partial T}\right)_P. \tag{2-11}$$

Combination of the last three equations gives

$$\left(\frac{\partial P}{\partial T}\right)_V = \frac{\beta}{\kappa},$$

and thus Eq. (2-4) becomes

$$dP = \frac{\beta}{\kappa}\,dT - \frac{1}{\kappa V}\,dV. \tag{2-12}$$

Actually, Eq. (2-12) can be obtained by a more direct method, not requiring use of Eq. (2-9). If V is treated as a function of T and P, then Eq. (2-1) applies:

$$dV = \left(\frac{\partial V}{\partial T}\right)_P dT + \left(\frac{\partial V}{\partial P}\right)_T dP. \tag{2-1}$$

Direct application of the definitions of β and κ then gives

$$dV = \beta V\,dT - \kappa V\,dP. \tag{2-13}$$

Solution for dP yields Eq. (2-12)

Example 2-2

A mass of mercury at an initial pressure of 1 (bar) and a temperature of 0 (°C) is heated to 10 (°C) under constant-volume conditions. What is the final pressure? It may be assumed that β and κ remain constant and have the values

$$\beta = 181 \times 10^{-6}(\text{K})^{-1}$$

$$\kappa = 3.88 \times 10^{-6}(\text{bar})^{-1}.$$

Restriction of Eq. (2-12) to constant volume gives

$$dP = \frac{\beta}{\kappa}\,dT \qquad (\text{const } V).$$

If we cause the temperature to change a finite amount from T_1 to T_2 at constant V, the pressure will change from P_1 to P_2. Integrating between these two states, we get

$$P_2 - P_1 = \int_{T_1}^{T_2} \frac{\beta}{\kappa}\,dT$$

or

$$P_2 = P_1 + \frac{\beta}{\kappa}(T_2 - T_1),$$

where the second expression follows from the assumptions that β and κ remain constant. Substitution of numerical values into this expression gives

$$P_2 = 1(\text{bar}) + \frac{181 \times 10^{-6}(\text{K})^{-1}}{3.88 \times 10^{-6}(\text{bar})^{-1}} \times 10(\text{K}),$$

or

$$P_2 = 467(\text{bar}).$$

2-5 Dimensions and Units

The same fundamental quantities needed in the science of mechanics, i.e., length, time, and mass (or force), are used in thermodynamics. Additional quantities essential to thermodynamics are the temperature and, for chemical applications, a measure of the amount of a substance. Quantitative application of any of these dimensions requires the use of arbitrary scales of measure, divided into specific *units* of size. Primary units have been set by international agreement and form the basis for the SI (Système International), or International System, of Units.

The SI employs the following dimensions: length, time, mass, temperature, amount of substance, electric current, and luminous intensity. The primary standard of length is the wavelength in vacuum of orange-red light from krypton-86 atoms, and the SI unit of length is the *meter* (m), defined as 1,650,763.73 standard wavelengths. The *second* (s) is the SI unit of time, defined as the duration of 9,192,631,770 cycles of radiation associated with a specified transition of the cesium atom. The *kilogram* (kg) is the only SI base unit defined by a physical artifact. It is equal to the mass of a platinum-iridium cylinder preserved at the International Bureau of Weights and Measures at Sèvres, France.

The thermodynamic temperature scale has its origin at absolute zero and a single fixed point at the triple point of water. The *kelvin* (K) is the fraction 1/273.16 of the thermodynamic temperature of the triple point of water. A measure of the amount of a substance is provided by the *mole* (mol). It is defined as the amount of substance in a system containing as many elementary entities (e.g., molecules) as there are atoms in 0.012(kg) of carbon-12.[1]

[1] This mole is equivalent to the *gram mole* commonly used by chemists. At the time of the writing of this text, the mole has not been officially adopted by the General Conference of Weights and Measures.

Table 2-2 SI Base Units

Physical quantity	Unit	Symbol
Length	meter	(m)
Time	second	(s)
Mass	kilogram	(kg)
Temperature	kelvin	(K)
Amount of substance	mole	(mol)
Electric current	ampere	(A)
Luminous intensity	candela	(cd)

The SI base units for electric current and luminous intensity are the ampere (A) and the candela (cd), respectively; the SI similarly provides precise definitions for them.

Other units are defined in terms of these seven base units. Some of the secondary or derived units occur so frequently that they are given special names and symbols. Thus the SI unit for force is the *newton* (N), derived from Newton's second law, $F = ma$, and equal to 1(kg)(m)/(s)^2. Similarly, the SI unit for energy is the *joule* (J), derived from the expression for work, $W = F_s \Delta s$, and equal to $1(\text{N})(\text{m}) \equiv 1(\text{kg})(\text{m})^2/(\text{s})^2$. The seven basic units are listed in Table 2-2, and some common SI derived units are given in Table 2-3.

Decimal multiples and fractions of SI units can be designated by prefixes. Some commonly used prefixes are listed in Table 2-4. In addition, certain decimal multiples and fractions of particular basic and derived units are given special names and symbols. These include the *angstrom* (Å), equal to

Table 2-3 Some SI Derived Units

Physical quantity	Unit	Symbol	Definition
Force	newton	(N)	$1(\text{kg})(\text{m})/(\text{s})^2$
Energy	joule	(J)	$1(\text{kg})(\text{m})^2/(\text{s})^2$ [$= 1(\text{N})(\text{m})$]
Pressure	pascal	(Pa)	$1(\text{kg})/(\text{m})(\text{s})^2$ [$= 1(\text{N})/(\text{m})^2$]
Power	watt	(W)	$1(\text{kg})(\text{m})^2/(\text{s})^3$ [$= 1(\text{J})/(\text{s})$]
Electric charge	coulomb	(C)	$1(\text{A})(\text{s})$
Electric potential difference	volt	(V)	$1(\text{kg})(\text{m})^2/(\text{A})(\text{s})^3$ [$= 1(\text{A})(\Omega)$]
Electric resistance	ohm	(Ω)	$1(\text{kg})(\text{m})^2/(\text{A})^2(\text{s})^3$ [$= 1(\text{V})/(\text{A})$]
Electric capacitance	farad	(F)	$1(\text{A})^2(\text{s})^4/(\text{kg})(\text{m})^2$ [$= 1(\text{C})/(\text{V})$]
Magnetic flux	weber	(Wb)	$1(\text{kg})(\text{m})^2/(\text{A})(\text{s})^2$ [$= 1(\text{V})(\text{s})$]
Inductance	henry	(H)	$1(\text{kg})(\text{m})^2/(\text{s})^2(\text{A})^2$ [$= 1(\text{Wb})/(\text{A})$]

Table 2-4 Some Prefixes for SI Units

Fraction or multiple	Prefix	Symbol
10^{-9}	nano	n
10^{-6}	micro	μ
10^{-3}	milli	m
10^{-2}	centi	c
10^{3}	kilo	k
10^{6}	mega	M
10^{9}	giga	G

10^{-10}(m); the *dyne* (dyn), equal to 10^{-5}(N); the *erg* (erg), equal to 10^{-7}(J); and the *bar* (bar), a unit of pressure equal to $10^{5}(\text{N})/(\text{m})^2$.

One of the important advantages of the SI is the *coherence* of its derived units with the base units, viz., no numerical factor is required for the conversion of a combination of base units into a derived unit. The elimination of numerical conversion factors removes a source of computational error.

Despite the advantages of the SI, many engineers in the English-speaking countries continue to use a system of units based upon the pound, the foot, and the second. The dual role played by the pound in the English engineering system to designate both force and mass leads to an unfortunate ambiguity. Since force and mass are different concepts, the pound *force* must be carefully distinguished from the pound *mass*. In this book the mass unit will be designated by (lb_m) and the force unit by (lb_f).

Force and mass are related by Newton's second law of motion, which states that the force on a body is proportional to its mass and acceleration. Most rational systems of units are contrived to make the proportionality constant dimensionless and equal to unity. Thus we normally write

$$F = ma.$$

In systems of units compatible with this form of Newton's law, the dimensions and units of force, mass, length, and time are not independent of one another. Thus in the SI the newton is equal to $1(\text{kg})(\text{m})/(\text{s})^2$. Similarly, in the English absolute system the *poundal* is equal to $1(\text{lb}_m)(\text{ft})/(\text{s})^2$ In the English gravitational system, which treats force rather than mass as a fundamental dimension, the unit of mass is the *slug*, equal to $1(\text{lb}_f)(\text{s})^2/(\text{ft})$.

In the English engineering system, on the other hand, force, mass, length, and time are each treated as fundamental dimensions. The pound *force* is defined, independently of Newton's second law, as that force which accelerates 1 pound *mass* 32.174 feet per second per second. Substitution into the expression $F = ma$ gives

$$1(\text{lb}_f) = 1(\text{lb}_m) \times 32.174(\text{ft})/(\text{s})^2.$$

Although this expression could serve as a *definition* for the pound *force*, this would be inconsistent with the concept that the four dimensions are *independent.* The difficulty is resolved when a dimensional proportionality factor g_c^{-1} is included in Newton's law:

$$F = \frac{1}{g_c} ma.$$

In the English engineering system, g_c therefore has the value 32.174 and units of $(\text{lb}_m)(\text{ft})/(\text{lb}_f)(\text{s})^2$. For any other system in which force, mass, length, and time are considered independent fundamental dimensions, Newton's law must also contain a dimensional proportionality constant. As already noted, $g_c = 1$(dimensionless) for the SI, but since both the SI and the English engineering system are used in this book, all formulas derived from Newton's law will be written to include g_c.

English engineering units for pressure and for energy are, respectively, the $(\text{lb}_f)/(\text{ft})^2$ [or $(\text{lb}_f)/(\text{in})^2$] and the $(\text{ft})(\text{lb}_f)$. Additional units belonging neither to the SI nor to the English engineering system are often used by scientists and engineers. These include the *millimeter of mercury* (mm Hg) and the standard *atmosphere* (atm), which are units of pressure, and the *calorie* (cal) and *British thermal unit* (Btu), both units of energy. A list of commonly used conversion factors is given in Appendix A.

Example 2-3

The *weight* of a body is that *force* which, when applied to the body, gives it an acceleration equal to the local acceleration of gravity g. An astronaut weighs 670(N) in Houston, Texas, where $g = 9.792(\text{m})/(\text{s})^2$. What is the mass of the astronaut, and what does he weigh on the surface of the moon, where $g = 1.67(\text{m})/(\text{s})^2$?

Letting $a = g$, we write Newton's second law as

$$F = \frac{mg}{g_c}$$

where, for SI units, $g_c = 1$(dimensionless). Thus

$$m = \frac{g_c F}{g} = \frac{1 \times 670(\text{N})}{9.792(\text{m})/(\text{s})^2} = 68.4 \frac{(\text{N})(\text{s})^2}{(\text{m})},$$

or

$$m = 68.4(\text{kg}).$$

The *mass* of the astronaut is the same regardless of his location; however, his *weight* is smaller on the moon than in Houston, because g is smaller:

$$F_{\text{moon}} = \frac{68.4(\text{kg}) \times 1.67(\text{m})/(\text{s})^2}{1} = 114\frac{(\text{kg})(\text{m})}{(\text{s})^2},$$

or

$$F_{\text{moon}} = 114(\text{N}).$$

The layman does not usually distinguish between weight and mass, and would say that the astronaut "weighs" 68.4(kg) on earth but that he "weighs" only (1.67/9.792) × 68.4 = 11.7(kg) on the surface of the moon.

It is instructive to rework this example in English engineering units. One newton is equivalent to 0.224809(lb_f) and 1 meter is equivalent to 3.28084(ft), so that the corresponding values for the problem statement are: weight of astronaut in Houston = 150.6(lb_f), g in Houston = 32.13(ft)/(s)^2, and g on the moon = 5.48(ft)/(s)^2. Application of Newton's law gives

$$m = \frac{32.174(\text{lb}_m)(\text{ft})/(\text{lb}_f)(\text{s})^2 \times 150.6(\text{lb}_f)}{32.13(\text{ft})/(\text{s})^2},$$

or

$$m = 150.8(\text{lb}_m).$$

Thus in this system of units the astronaut's mass and weight in Houston are *numerically* almost identical. On the moon, however, the numerical values are quite different; his mass is still 150.8(lb_m) but his weight is now 150.8 × 5.48/32.174 = 25.7(lb_f).

Example 2-4

Pressure is sometimes measured and reported in units of an equivalent height ("head") of a specified liquid. We wish to establish a basis for the definition of such units.

Consider a column of fluid of height h and cross-sectional area A. The weight of the column is determined from Newton's second law:

$$F = \frac{mg}{g_c} = \frac{\rho A h g}{g_c},$$

where ρ is the mass density of the fluid. The hydrostatic pressure exerted by the column is just F/A, given by the above equation as

$$P = \frac{\rho g h}{g_c}.$$

Precise determination of the pressure equivalent to a given height of fluid thus requires accurate knowledge of both the fluid density and the local acceleration of gravity.

The most common unit of this type is the millimeter of mercury (also called the *torr*), defined as the pressure exerted by a 1(mm) column of liquid mercury in the standard gravitational field at 0(°C). By definition, g = 9.80665(m)/(s)^2 in the standard gravitational field, and precise determinations of the density of liquid mercury give ρ = 13.5951

$\times\ 10^3(\text{kg})/(\text{m})^3$ at $0(^\circ\text{C})$. Substitution into the last equation yields

$$P = \frac{13.5951 \times 10^3(\text{kg})/(\text{m})^3 \times 9.80665(\text{m})/(\text{s})^2 \times 10^{-3}(\text{m})}{1}$$

$$= 133.322 \frac{(\text{kg})}{(\text{m})\,(\text{s})^2},$$

or

$$P = 133.322(\text{N})/(\text{m})^2.$$

Thus,

$$1\,(\text{mm Hg}) = 133.322(\text{N})/(\text{m})^2 = 133.322(\text{Pa})$$

$$= 1.33322 \times 10^{-3}(\text{bar}).$$

2-6 Bars in Tension and Compression

Structural members in tension and compression usually exist under conditions of atmospheric pressure. Furthermore, their volume changes are usually quite negligible, and for most practical purposes it is therefore unnecessary to include pressure and volume as thermodynamic coordinates. A sufficiently complete thermodynamic description of bars in tension or compression is given in terms of just three coordinates:

1 The *stress* in the member σ, defined as the load per unit of cross-sectional area, and usually measured in $(\text{kN})/(\text{m})^2$ or $(\text{lb}_f)/(\text{in})^2$. Stress is taken as + for tension and − for compression.

2 The *strain* ϵ, a dimensionless measure of the relative length of the member. By definition,

$$d\epsilon \equiv \frac{dL}{L}. \tag{2-14}$$

It is seen that $d\epsilon$ is positive for tension and negative for compression.

3 The ideal-gas temperature T.

States of thermodynamic equilibrium are connected by an equation of state, which frequently is by no means simple. In the absence of a specific equation, we can still write a functional relationship connecting the thermodynamic coordinates, in this instance,

$$\epsilon = \epsilon(T,\sigma).$$

From this we have immediately, for a bar which undergoes an infinitesimal change from one equilibrium state to another,

$$d\epsilon = \left(\frac{\partial \epsilon}{\partial T}\right)_\sigma dT + \left(\frac{\partial \epsilon}{\partial \sigma}\right)_T d\sigma.$$

where both partial derivatives represent important physical quantities. The first is called the linear expansivity, or coefficient of linear (thermal) expansion,

$$\boxed{\alpha \equiv \left(\frac{\partial \epsilon}{\partial T}\right)_\sigma,} \tag{2-15}$$

and the second is the reciprocal of Young's modulus E, where

$$\boxed{E \equiv \left(\frac{\partial \sigma}{\partial \epsilon}\right)_T.} \tag{2-16}$$

Thus we may write

$$d\epsilon = \alpha\, dT + \frac{1}{E}\, d\sigma.$$

If ϵ is constant, this equation gives

$$\left(\frac{\partial \sigma}{\partial T}\right)_\epsilon = -\alpha E,$$

a result also obtained from Eq. (2-9).

The SI units of α are $(K)^{-1}$ or $(°C)^{-1}$. For metals, α is nearly independent of σ and is a weak function of T. Thus, for small temperature changes, it may be regarded as essentially constant. Typically, α for a pure metal is positive and of the order of 10 to $30 \times 10^{-6}(K)^{-1}$ at normal temperatures. For rubber and elastomers, α is positive in the unstressed state but becomes negative when the material is stretched. Fuzed quartz has unusually low values of α [$\approx 0.5 \times 10^{-6}(K)^{-1}$], making it ideal for applications requiring dimensional integrity.

Young's modulus has the same dimensions as stress and is reported in the same units. Solids with no permanent strain ("set") exhibit a region, for small to moderate σ, where σ is proportional to ϵ:

$$\sigma = k\epsilon \qquad \text{(const } T\text{)}.$$

This expression is a statement of Hooke's law; it implies that $E = k$ and that E depends upon T only. Actually E is often only a weak function of T, so that for small temperature changes it may be taken as approximately constant. Young's modulus is always positive; for most metals it is of the order of 5 to $25 \times 10^7 (\mathrm{kN})/(\mathrm{m})^2$. Extreme values are found for lead [$\approx 1.5 \times 10^7 (\mathrm{kN})/(\mathrm{m})^2$] and for tungsten [$\approx 35 \times 10^7 (\mathrm{kN})/(\mathrm{m})^2$].

Example 2-5

An exposed vertical steel girder 5(m) in length supports a constant load. What is the maximum variation in the length of the girder if the ambient temperature varies between -35 and $+35$(°C)? For steel, take $\alpha = \text{constant} = 13 \times 10^{-6} (\mathrm{K})^{-1}$.

As a result of Eq. (2-14) we can write Eq. (2-15) as

$$\frac{1}{L}\left(\frac{\partial L}{\partial T}\right)_\sigma = \alpha.$$

Integration from an initial state (L_1, T_1) to a final state (L_2, T_2) gives

$$\ln \frac{L_2}{L_1} = \alpha(T_2 - T_1).$$

Writing the argument of the logarithm as $1 + (L_2 - L_1)/L_1$ and noting that $\ln(1 + x) \approx x$ for small x, we find an approximate form of this equation, valid for small strain:

$$L_2 - L_1 = \alpha L_1 (T_2 - T_1).$$

The initial length L_1 is taken as 5(m), the nominal length of the girder. Substitution of given values into the equation then yields

$$L_2 - L_1 = 13 \times 10^{-6} (\mathrm{K})^{-1} \times 5(\mathrm{m}) \times [35 - (-35)](\mathrm{K}),$$

or

$$L_2 - L_1 = 4.55 \times 10^{-3} (\mathrm{m}).$$

Thus seasonal temperature changes cause a maximum variation of 4.55(mm) in the length of the girder. This variation is small but by no means insignificant; such effects are always considered in a good structural design.

2-7 Surfaces

The surfaces of liquids and solids have properties which are distinctly different from the bulk properties of the underlying material. The simplest experimental demonstration of the influence of surfaces is given by a film of liquid (usually a soap solution) formed on a wire frame as shown in Fig. 2-1.

This film consists of two surfaces and a layer of liquid of small but finite thickness between them. It is found that a force F applied to the movable wire of the frame is necessary to hold the film in equilibrium. It is further observed that the force F is directly proportional to the length of the wire l but independent of x and therefore independent of the film thickness. This leads immediately to the conclusion that the origin of the force F is in the

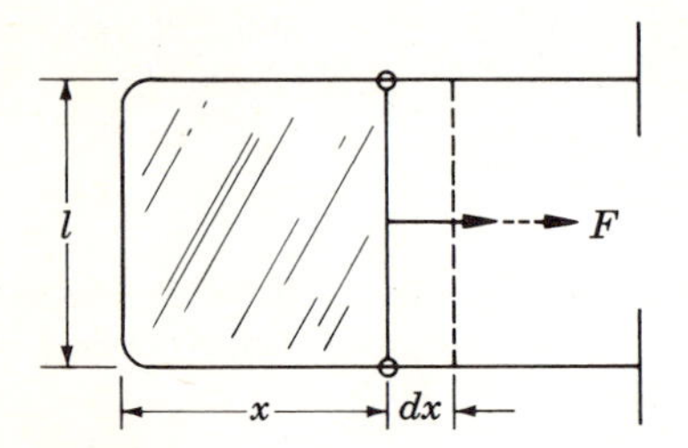

Fig. 2-1
Liquid film formed on a wire frame.

surfaces of the film and not in its interior. Thus the applied force F is a direct measure of the surface tension of the two surfaces which confine the bulk of the film:

$$\gamma = \frac{F}{2l}.$$

The *surface tension* γ by definition is the force attributed to the surface acting perpendicular to any boundary, real or imaginary, per unit length of boundary. It is usually expressed in (dyn)/(cm). A list of values for some common liquids is given in Table 2-5.

When the film of Fig. 2-1 is extended, material moves from the bulk of the liquid to the surfaces so as to form additional surface. The film does not *stretch* in the usual sense of the word. Only when the film is extended sufficiently to decrease its thickness to the point where no interior remains, so that it consists of just the two surfaces but a few molecules thick, does the film tend to stretch, and then it ruptures.

There is no sharp line of demarcation between a surface region (which is what is meant by *surface*) and the bulk of the underlying liquid. By its very nature the surface cannot exist separately. Thus liquid surfaces cannot be considered without including the accompanying liquid as part of the system. We merely attribute to the surface that part of a "total system property" which results from the abnormality of a surface region. For a pure liquid, the surface tension and the influence of the surface on the properties of the entire system are found to depend solely on temperature. A number of empirical equations have been proposed to express γ as a function of T.

The phenomenon of capillarity depends upon the existence of surface tension in liquids, and certain properties of small drops and bubbles are significantly affected by their surfaces. One cannot proceed very far with a study of this topic without finding it to be exceedingly complex. We do not

Table 2-5 Surface Tension γ at 20(°C) for Some Liquids

Substance	γ(dyn)/(cm)
Diethyl ether	17.1
n-Hexane	18.4
Ethyl alcohol	22.8
Benzene	28.9
Water	72.8
Mercury	476

pretend here to give more than the most elementary introduction. For our purposes it is appropriate to treat a surface as a simple system characterized by the three coordinates surface tension γ, surface area A, and temperature T.

Example 2-6

Determine an expression for the difference in pressure between the inside and outside of a spherical liquid droplet.

If only hydrostatic forces were present, the pressure inside the droplet would be virtually the same as on the outside. However, the force resulting from surface tension must be considered, as illustrated in Fig. 2-2. The left half of the spherical droplet of radius r is acted on by three forces:

1 The surface force, directed right, is equal to the product of the surface tension and the circumference of the shaded cross-section, and is given by

$$(\gamma)\,(2\pi r).$$

2 The hydrostatic force on the shaded cross-section, directed left, resulting from the

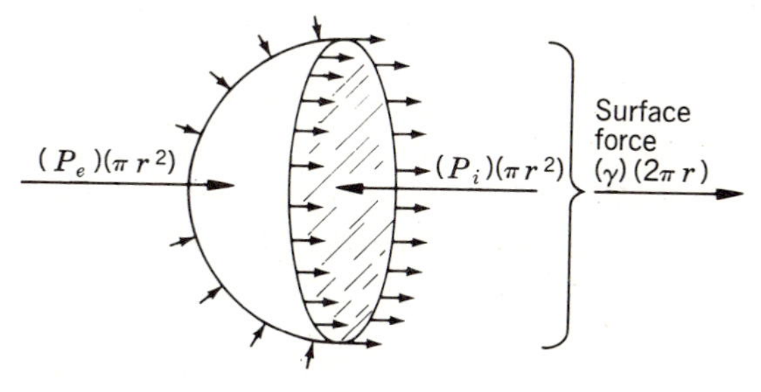

Fig. 2-2
Force balance on half-droplet.

Table 2-6

r(m)	$P_i - P_e$(bar)
10^{-2}	0.000952
10^{-4}	0.0952
10^{-6}	9.52
0	∞

internal pressure acting over the area, is equal to

$$(P_i)(\pi r^2).$$

3 The force, directed right, resulting from the external pressure on the surface of the half-droplet is given as the product of the external pressure and the area obtained when the surface of the half-droplet is projected on the plane dividing the droplet. Since this is the shaded area, the force is

$$(P_e)(\pi r^2)$$

For equilibrium these forces must be in balance; thus

$$(P_e)(\pi r^2) + (\gamma)(2\pi r) = (P_i)(\pi r^2),$$

from which we obtain

$$P_i - P_e = \frac{2\gamma}{r}.$$

Thus for a droplet of finite radius the internal pressure is *greater* than the external pressure. Extreme examples of magnitudes of the pressure difference are provided by values calculated for mercury at 20(°C). These are summarized in Table 2-6.

To illustrate the calculation, we take $r = 10^{-4}$(m) and note from Table 2-5 that $\gamma = 476$(dyn)/(cm):

$$P_i - P_e = \frac{2 \times 476(\text{dyn})/(\text{cm})}{10^{-4}(\text{m}) \times 10^2(\text{cm})/(\text{m}) \times 10^6(\text{dyn})/(\text{cm})^2(\text{bar})}$$

$$= 0.0952(\text{bar}).$$

2-8 Reversible Cell

There are several common types of reversible electrochemical cells, but all consist of one or more electrolytes in which two electrodes are immersed. The electromotive force e of the cell depends on the nature of the materials, on the concentrations of the electrolytes, and on the temperature. The Daniell cell, which is representative of one type of reversible cell, is shown schematically in Fig. 2-3. A zinc electrode immersed in a saturated $ZnSO_4$ solution is separated by a porous wall from a copper electrode immersed in a saturated solution of $CuSO_4$.

Experiment shows that the copper electrode is positive with respect to the zinc. Thus, if the electrodes of Fig. 2-3 are connected externally by a purely resistive circuit, the current in the external circuit may be described conventionally as a transfer of positive electricity from the copper to the zinc electrode. The electrons flow in the opposite directions as indicated in Fig. 2-3.

The changes which occur in the cell when the electrodes are connected externally by a purely resistive circuit are represented by the chemical equation

$$Zn + CuSO_4 \rightarrow Cu + ZnSO_4.$$

Thus zinc goes into solution, zinc sulfate is formed, copper sulfate is used up, and copper is deposited on the positive electrode. The reaction may be divided into a number of ionic reactions, as indicated in Fig. 2-3. The sum of these ionic reactions gives the overall reaction of the cell.

Suppose that the cell is connected to a potentiometer whose potential difference opposes but is slightly smaller than the emf of the cell. Under these conditions the changes described above occur very slowly. If the imposed potential difference is slightly larger than the emf of the cell, the transfer of electricity in the cell will be reversed and the chemical changes will proceed slowly in the opposite direction. The essential feature of a reversible cell is that the chemical changes accompanying the transfer of electricity in one direction take place to the same extent in the reverse direction when the same quantity of electricity is transferred in the reverse

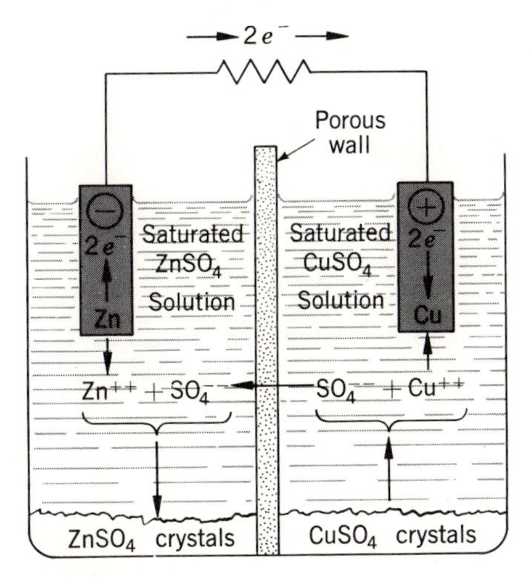

Fig. 2-3
The Daniell cell.

direction. Furthermore, according to one of Faraday's laws of electrolysis, the simultaneous solution of 1(mol) of zinc and deposition of 1(mol) of copper, or the reverse, are accompanied by the transfer of exactly $j\mathfrak{F}$(C) of electricity, where j is the number of moles of electrons transferred per mole of advancement of the reaction, and $\mathfrak{F}$ is Faraday's constant, equal to 96,487(C)/(mol).

We may define a quantity q, called the *charge* of the cell, as a number whose absolute magnitude is of no consequence but whose change is numerically equal to the quantity of electricity that is transferred during the chemical reaction. Thus, if dn(mol) of material react, the charge of the cell changes by an amount dq, where

$$dq = -j\mathfrak{F}\, dn. \tag{2-17}$$

The minus sign arises because we adopt the convention that dn is positive when the chemical reaction proceeds in the forward direction, corresponding to *discharge* of the cell, and negative when the chemical reaction is reversed so as to *charge* the cell. For a finite change,

$$q_2 - q_1 = -j\mathfrak{F}\, \Delta n.$$

Now, if we limit ourselves to reversible cells in which no gases are liberated and which operate at constant pressure (usually atmospheric), we may ignore the pressure and volume and consider the cell a simple system characterized by the following three coordinates:

1 The emf of the cell e in (V).
2 The charge q in (C).
3 The ideal-gas temperature T.

A reversible cell is in equilibrium *only* when it is connected to a potentiometer and the circuit is adjusted until there is no current. The emf of the cell is then balanced, and the cell is in mechanical and chemical equilibrium. When thermal equilibrium is also established, the cell is in thermodynamic equilibrium. For saturated cells, such as the Daniell cell, e is a function of T only. Some values of e for representative cells are given in Table 11-5.

Problems

2-1 A metal whose volume expansivity is $5.0 \times 10^{-5}(\text{K})^{-1}$ and isothermal compressibility is $1.2 \times 10^{-6}(\text{atm})^{-1}$ is at a pressure of 1(atm) and a temperature of 20(°C). A thick surrounding cover of Invar, of negligible expansivity and compressibility, fits it very snugly.

(*a*) What will be the final pressure if the temperature is raised to 32(°C)?
(*b*) If the surrounding cover can withstand a maximum pressure of 1,200(atm), what is the highest temperature to which the system may be raised?

2-2 A block of the same metal as in Prob. 2-1 at a pressure of 1(atm), a volume of 5(l), and a temperature of 20(°C) undergoes a temperature rise of 12(°C) and an increase in volume of 0.5(cm)^3. Calculate the final pressure.

2-3 Making use of the fact that

$$\frac{\partial^2 V}{\partial T\,\partial P} = \frac{\partial^2 V}{\partial P\,\partial T},$$

prove that

$$\left(\frac{\partial \beta}{\partial P}\right)_T = -\left(\frac{\partial \kappa}{\partial T}\right)_P.$$

2-4 For water at 25(°C) and 1(atm), $\kappa = 4.5 \times 10^{-5}(atm)^{-1}$. To what pressure must water be compressed at 25(°C) to change its density by 1 percent? Assume κ to be independent of P.

2-5 A hypothetical substance has the following volume expansivity and isothermal compressibility:

$$\beta = \frac{3aT^3}{V}, \qquad \kappa = \frac{b}{V},$$

where a and b are constants. Find the equation of state.

2-6 The volume expansivity and the isothermal compressibility of a certain gas are

$$\beta = \frac{nR}{PV}, \qquad \kappa = \frac{1}{P} + \frac{a}{V},$$

where n, R, and a are constants. Find the equation of state.

2-7 Express the volume expansivity and the isothermal compressibility in terms of the density ρ and its partial derivatives.

2-8 An instrument to measure the acceleration of gravity on Mars is constructed of a spring from which is suspended a mass of 0.24(kg). At a place on earth where the local acceleration of gravity is 9.80(m)/(s)^2, the spring extends 0.61(cm). When the instrument package is landed on Mars, it radios back the information that the spring is extended 0.20(cm). What is the Martian acceleration of gravity?

2-9 A group of scientists have landed on the moon and would like to determine the mass of several unusual rocks. They have a spring scale calibrated to read pounds *mass* at a location where the local acceleration of gravity is 32.2(ft)/(s)^2. One of the moon rocks gives a reading of 25 on the scale. What is its mass? What is its weight on the moon? The value of g on the surface of the moon is 5.47(ft)/(s)^2.

2-10 The equation of state of an ideal elastic substance under tension is

$$\sigma = KT\left(\lambda - \frac{1}{\lambda^2}\right),$$

where σ = nominal stress (force divided by the no-load area),
K = a constant,
λ = extension ratio, defined by $\lambda = L/L_0$, where L_0 (the no-load value of L) is a function of temperature only.

(a) Show that the isothermal Young's modulus is given by

$$E = \lambda \left(\frac{\partial \sigma}{\partial \lambda}\right)_T = \sigma + \frac{3KT}{\lambda^2}.$$

(b) Show that the isothermal Young's modulus at zero tension is given by

$$E_0 = 3KT.$$

(c) Show that the linear expansivity is given by

$$\alpha = \alpha_0 - \frac{\sigma}{ET},$$

where α_0 is the value of α at zero tension, or

$$\alpha_0 = \left(\frac{\partial \epsilon}{\partial T}\right)_{\sigma=0}.$$

(d) Assume the following values for a sample of rubber: $T = 300$(K), $K = 1.86 \times 10^3$(N)/(m)^2(K), and $\alpha_0 = 5 \times 10^{-4}(K)^{-1}$. Calculate σ, E, and α for the following values of λ: 0.5, 1.0, 1.5, 2.0. Show on a graph how σ, E, and α depend on λ.

2-11 (a) A steel rod 0.5(in)^2 in cross-sectional area, under a tensile stress of 8,000 (lb$_f$)/(in)^2 at 80(°F), is held between two fixed supports 10(ft) apart. If the temperature falls to 20(°F), what is the tensile stress in the rod? Take $E = 30.0 \times 10^6$(lb$_f$)/(in)^2 and $\alpha = 6.5 \times 10^{-6}(R)^{-1}$.

(b) If, in addition, the supports approach each other by 0.02(in), what is the final tensile stress in the rod?

2-12 Show that if Young's modulus E is independent of temperature, the linear expansivity α is independent of stress.

2-13 For a PVT system we wrote in Sec. 2-3

$$V = V(T,P),$$

and for a stressed bar we wrote in Sec. 2-6

$$\epsilon = \epsilon(T,\sigma).$$

In Secs. 2-7 and 2-8 on surfaces and reversible cells, we did not write

$$A = A(T,\gamma), \qquad q = q(T,e).$$

Why not? How are these systems different?

2-14 The *natural strain* ϵ is defined in terms of the length L of an axially stressed bar by the differential equation $d\epsilon = dL/L$. Similarly, the *engineering strain* ϵ' can be defined by the equation $d\epsilon' = dL/L_0$, where L_0 is the initial (unstrained) length of the bar. Show that ϵ and ϵ' are related by the equation $\epsilon = \ln(1 + \epsilon')$ and that the two measures of strain differ by less than 1 percent for ϵ' less than about 0.02.

2-15 What are the *basic* SI units for each of the following physical quantities?

(a) Avogadro's constant,
(b) Absolute viscosity,
(c) Electrical capacitance,
(d) Electrical charge,
(e) Frequency,
(f) Moment of inertia,

(*g*) Momentum,
(*h*) Thermal conductivity,
(*i*) Torque.

2-16 Is the pressure difference across a liquid droplet greater or less than the pressure difference across an equal-sized, air-filled bubble of the same liquid floating in the air?

2-17 Verify the following equivalencies:

(*a*) $1(\mathrm{J}) = 1(\mathrm{V})^2(\mathrm{F}) = 1(\mathrm{V})(\mathrm{C}) = 1(\mathrm{H})(\mathrm{A})^2$,

(*b*) $1(\mathrm{W}) = 1(\mathrm{V})(\mathrm{A}) = 1(\Omega)(\mathrm{A})^2 = 1(\mathrm{V})^2/(\Omega)$.

3

WORK

3-1 Work

If a system undergoes a displacement under the action of a force, *work* is said to be done, the amount of work being equal to the product of the force and the component of the displacement parallel to the force. If a system as a *whole* exerts a force on its surroundings and a displacement takes place, the work that is done either by or on the system is called *external work*. Thus, a gas, confined in a cylinder and at uniform pressure, while expanding and imparting motion to a piston, does external work on its surroundings. The work done, however, by part of a system on another part is called *internal work*. The interactions of molecules or electrons on one another constitute internal work.

Internal work has no place in thermodynamics. Only the work that involves an interaction between a system and its surroundings is significant. When a system does external work, the changes that take place can be described by means of macroscopic quantities referring to the system as a whole, in which case the changes may be imagined to accompany the raising or lowering of a suspended body, the winding or unwinding of a spring, or, in general, the alteration of the position or configuration of some external mechanical device. This may be regarded as the ultimate criterion as to whether external work is done or not. *Unless otherwise indicated, the word work, unmodified by any adjective, will mean external work.*

A few examples will be found helpful. If an electric cell is on open circuit, changes that take place in the cell (such as diffusion) are not accompanied by the performance of work. If, however, the cell is connected to an external circuit through which electricity is transferred, the current may be imagined to produce rotation of the armature of a motor, thereby lifting a weight or winding a spring. Therefore, *for an electric cell to do work it must be connected to an external circuit.* As another example, consider a magnet far removed from any external electric conductor. A change of magnetization within the magnet is not accompanied by the performance of work. If, however, the magnet undergoes a change of magnetization while it is surrounded by an electric conductor, eddy currents are set up in the conductor, constituting an external transfer of electricity. Hence, *for a magnetic system to do work it must interact with an electric conductor or with other magnets.*

In mechanics, we are concerned with the behavior of systems acted on by external forces. When the resultant force exerted *on* a mechanical system is in the same direction as the displacement of the system, the work of the force is positive, work is said to be done on the system, and the energy of the system increases.

In this text, we adopt the same sign convention for work that is employed in mechanics. Thus, when the displacement of the point of application of an external force acting on a thermodynamic system is in the *same* direction as the force, work is done *on* the system, and the work is regarded as *positive.* Conversely, when the displacement is in the direction *opposite* to the force, work is done *by* the system, and is regarded as *negative.*[1]

3-2 Quasi-static Process

A system in thermodynamic equilibrium satisfies the following stringent requirements:

1 *Mechanical equilibrium.* There are no unbalanced forces acting on any part of the system or on the system as a whole.
2 *Thermal equilibrium.* There are no temperature differences between parts of the system or between the system and its surroundings.
3 *Chemical equilibrium.* There are no chemical reactions within the system and no motion of any chemical constituent from one part of a system to another part.

Once a system is in thermodynamic equilibrium and the surroundings are kept unchanged, no motion will take place and no work will be done. If, however, the sum of the external forces is changed so that there is a finite unbalanced force acting on the system, then the condition for mechanical equilibrium is no longer satisfied and the following situations may arise:

[1] This is the convention recommended by the International Union of Pure and Applied Chemistry and commonly used by workers in physics and fluid mechanics. It is opposite to that used in most American engineering thermodynamics texts, where work is treated as positive if done by a system and negative if done on it.

1 Unbalanced forces may be created within the system; as a result, turbulence, waves, etc., may be set up. Also, the system as a whole may execute some sort of accelerated motion.
2 As a result of this turbulence, acceleration, etc., a nonuniform temperature distribution may be brought about, as well as a finite difference of temperature between the system and its surroundings.
3 The sudden change in the forces and in the temperature may produce a chemical reaction or the motion of a chemical constituent.

It follows that a finite unbalanced force may cause the system to pass through nonequilibrium states. If it is desired during a process to describe every state of a system by means of thermodynamic coordinates referring to the system as a whole, the process must *not* be brought about by a finite unbalanced force. We are led, therefore, to conceive of an ideal situation in which the external forces acting on a system are varied only slightly so that the unbalanced force is infinitesimal. A process performed in this ideal way is said to be *quasi-static. During a quasi-static process, the system is at all times infinitesimally near a state of thermodynamic equilibrium*, and all the states through which the system passes can be described by means of thermodynamic coordinates referring to the system as a whole. An equation of state is valid, therefore, for all these states. A quasi-static process is an idealization that is applicable to all thermodynamic systems, including electric and magnetic systems. The conditions for such a process can never be rigorously satisfied in the laboratory, but they can be approached with almost any degree of accuracy. In the next few sections we will treat quasi-static processes for several types of simple thermodynamic systems.

3-3 Work in Changing the Volume of a *PVT* System

Imagine a *PVT* system contained in a cylinder equipped with a movable piston on which the system and the surroundings may act. Suppose that the cylinder has a cross-sectional area A, that the pressure exerted *by the system* at the piston face is P, and that the force is PA. The surroundings also exert an opposing force on the piston. The origin of this opposing force is irrelevant; it might be due to friction or a combination of friction and the push of a spring. The system within the cylinder does not have to know how the opposing force originated. The important condition that must be satisfied is that the opposing force must differ only slightly from the force PA. If, under these conditions, the piston moves a distance dx, then an infinitesimal amount of work

$$\delta W = PA\,dx$$

is done on the system. (The use of δ as a symbol for a differential will be explained later.) But

$$A\,dx = -dV,$$

where V is the *total* volume of the system, and hence

$$\boxed{\delta W = -P\,dV.} \tag{3-1}$$

During this process, we might have a chemical reaction or a transport of a constituent from one point to another taking place slowly enough to keep the system near mechanical equilibrium; or we might have some dissipative process such as friction taking place—or even all these processes. The lack of chemical equilibrium (and therefore of complete thermodynamic equilibrium) and the presence of dissipation *do not preclude* writing $\delta W = -P\,dV$. A lack of mechanical equilibrium, however, such as exists when there is *no* opposing force, definitely precludes the expression of δW as $-P\,dV$. We note that the validity of Eq. (3-1) does not depend upon the piston-and-cylinder device used in its derivation; it can be applied to an expanding or contracting PVT system of arbitrary shape.

Example 3-1

Explain the presence of the minus sign in Eq. (3-1).

The thermodynamic pressure P is a positive quantity, so that the sign of δW is determined by the sign of dV. If the volume of a PVT system decreases by a differential amount, then dV is negative, and hence δW is *positive* by Eq. (3-1). This is consistent with our adopted sign convention for work, as work clearly must be done *on* a PVT system to decrease its volume. If the volume of a PVT system differentially increases, then dV is positive, and δW is *negative* by Eq. (3-1). Again this agrees with our convention, for an expanding PVT system (e.g., the mixture of gases in the cylinder of an internal combustion engine) does work *against* external forces.

In a *finite* quasi-static process in which the volume changes from V_1 to V_2 the work is

$$W = -\int_{V_1}^{V_2} P\,dV. \tag{3-2}$$

Since the change in volume is performed quasi-statically, P is at all times a thermodynamic coordinate and can be expressed as a function of T and V by means of an equation of state. The evaluation of the integral can be accomplished once the behavior of T is specified, because then P can be expressed as a function of V only. If P is expressed as a function of V, the *path* of integration is defined. Along a particular quasi-static path, the work done in going from a volume V_1 to a volume V_2 is expressed as

$$W_{12} = -\int_{V_1}^{V_2} P\,dV;$$

whereas from 2 to 1, along the same path but in the opposite direction, the work is

$$W_{21} = -\int_{V_2}^{V_1} P\,dV.$$

When the path is quasi-static,

$$W_{12} = -W_{21}.$$

Sufficient approximation to a quasi-static process is achieved in practice when the external pressure differs from that exerted by the system by only a small finite amount.

The coherent SI unit for P is the $(\mathrm{N})/(\mathrm{m})^2$, and that of V is $(\mathrm{m})^3$. The SI unit for work is therefore $[(\mathrm{N})/(\mathrm{m})^2] \times (\mathrm{m})^3 = (\mathrm{N})(\mathrm{m})$, or the joule (J). The usual engineering unit of pressure is the pound *force* per square inch (psi). Absolute pressure must be employed, and this is often stated explicitly as pounds *force* per square inch absolute, or (psia). Pressure gauges normally indicate the difference between absolute pressure and atmospheric pressure, and the gauge reading is given in pounds *force* per square inch gauge, or (psig). Addition of atmospheric pressure to the gauge pressure gives the absolute pressure. If volume is expressed in $(\mathrm{ft})^3$, we multiply the absolute pressure in (psia) by $144(\mathrm{in})^2/(\mathrm{ft})^2$ to convert pressure to pounds *force* per square foot $(\mathrm{lb}_f)/(\mathrm{ft})^2$. The resulting unit of work is then the foot-pound *force* $(\mathrm{ft})(\mathrm{lb}_f)$.

3-4 *PV* Diagram

As the volume of a *PVT* system changes by virtue of the motion of a piston in a cylinder, the position of the piston at any moment is proportional to the volume. A pen whose motion along the X axis of a diagram follows exactly the motion of the piston will trace out a line every point of which represents an instantaneous value of the volume. If, at the same time, this pen is given a motion along the Y axis such that the Y coordinate is proportional to the pressure, then the pressure and volume changes of the system during expansion or compression are indicated simultaneously on the same diagram. Such a device is called an *indicator*. The diagram in which pressure

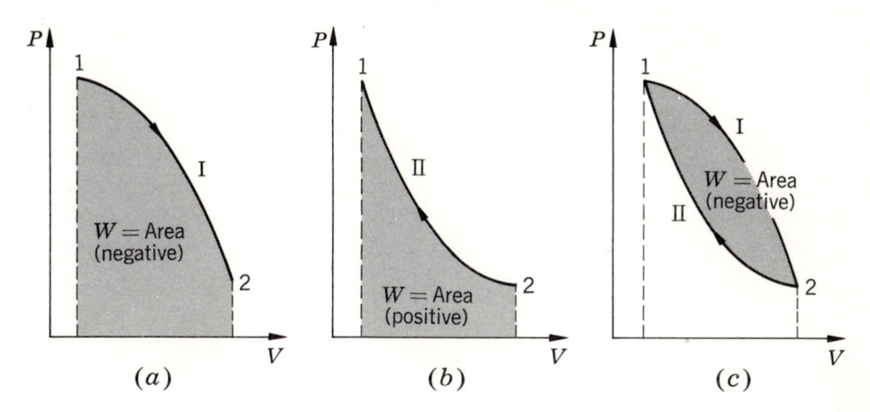

Fig. 3-1
PV diagram. (a) Curve I, expansion; (b) Curve II, compression; (c) Curves I and II together constitute a cycle.

is plotted along the Y axis and volume along the X axis is called an *indicator diagram*, or *PV diagram.*

On the indicator diagram shown in Fig. 3-1*a*, the states traversed by a gas during an expansion process are indicated by curve I. Since V increases during the process, the work as given by Eq. (3-2) is negative. The integral $\int P\,dV$ can be interpreted as the area under curve I; and to identify this area directly with the expansion work, we assign the area a *negative* value. Similarly, the work for a compression process can be represented by the shaded area under curve II in Fig. 3-1*b*; to maintain conformity with our sign convention for work, we take this area as *positive*. In Fig. 3-1*c*, curves I and II are superimposed, and we see that they represent the net effect of two processes, by means of which the gas is brought back to its initial state. Such a series of processes, represented by a closed figure, is called a *cycle*. The area within the closed figure is the sum of the areas under curves I and II and therefore represents the *net* work done in the cycle. For this particular case, where the direction of the cycle is clockwise, the area between curves I and II is *negative* and net work is done *by* the system. If the direction were reversed, the area would be *positive*, indicating net work done *on* the system.

3-5 Work Depends on the Path

On the PV diagram depicted in Fig. 3-2, an initial equilibrium state and a final equilibrium state of a PVT system are represented by the two points

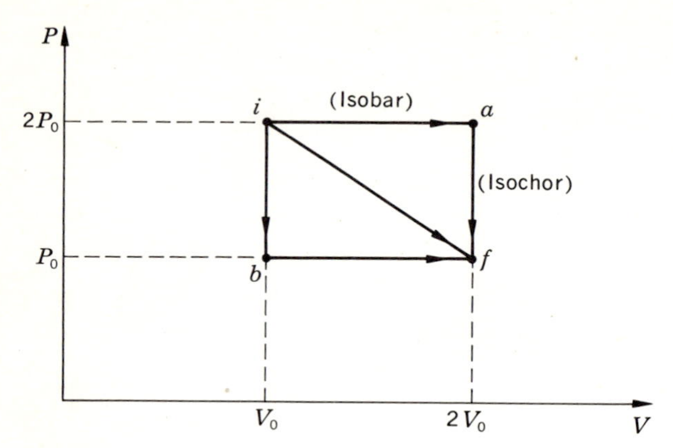

Fig. 3-2
Work depends on the path.

i and f, respectively. There are many ways in which the system may be taken from i to f. For example, the pressure may be kept constant from i to a (*isobaric process*) and then the volume kept constant from a to f (*isochoric process*), in which case the work is equal to the area under the line ia, which is equal to $-2P_0V_0$. Another possibility is the path, ibf, in which case the work is the area under the line bf, or $-P_0V_0$. The straight line from i to f represents another path, where the work is $-\frac{3}{2}P_0V_0$. We can see, therefore, that the *work depends not only on the initial and final states but also on the intermediate states*, i.e., *on the path*. This is merely another way of saying that for a quasi-static process, the expression

$$W = -\int_{V_1}^{V_2} P\,dV$$

cannot be integrated until P is specified as a function of V.

The expression $-P\,dV$ is an infinitesimal amount of work and has been represented by the symbol δW. There is, however, an important distinction between an infinitesimal amount of work and the other infinitesimals we have considered up to now. An infinitesimal amount of work is an *inexact differential*; that is, it is *not* the differential of an actual function of the thermodynamic coordinates. There is no function of the thermodynamic coordinates representing the work in a body. The phrase "work in a body" has no meaning. The use of δ as a differential sign indicates that an infinitesimal amount of work δW is *not* the differential of a mathematical function and emphasizes that it is an inexact differential.[1]

[1] The symbol δ does not represent a "virtual" change, as it does in mechanics; δW is *not* "virtual work."

Example 3-2

We wish to determine an expression for the work of quasi-static isothermal expansion or compression of an ideal gas. Let the gas change from an initial state (P_1, V_1) to a final state (P_2, V_2). Then, by Eq. (3-2),

$$W = -\int_{V_1}^{V_2} P\,dV.$$

The equation of state of an ideal gas is

$$PV = nRT,$$

where V is the *total* volume occupied by the gas, n is the number of moles of gas, and R is the universal gas constant. Substituting for P, we get

$$W = -\int_{V_1}^{V_2} \frac{nRT}{V}\,dV = -nRT \int_{V_1}^{V_2} \frac{dV}{V},$$

or

$$W = nRT \ln \frac{V_1}{V_2}. \tag{3-3}$$

Equation (3-3) is the desired result. An alternative expression can be found. For constant T, the ideal-gas equation gives

$$\frac{V_1}{V_2} = \frac{P_2}{P_1}.$$

Substitution in Eq. (3-3) yields

$$W = nRT \ln \frac{P_2}{P_1}. \tag{3-4}$$

As an illustration, consider the expansion of 2(mol) of an ideal gas from 5 to 3(bar) at 300(K). Substitution of numerical values into Eq. (3-4) gives

$$W = 2(\text{mol}) \times 8.314(\text{J})/(\text{mol})(\text{K}) \times 300(\text{K}) \times \ln \tfrac{3}{5},$$

or

$$W = -2.55 \times 10^3(\text{J}) = -2.55(\text{kJ}),$$

where the minus sign indicates that work is done *by* the expanding gas. The numerical value of 8.314 for R is taken from Appendix B.

Example 3-3

The pressure on 5(kg) of solid copper is increased quasi-statically and isothermally from 1 to 1,000(bar) at 0(°C). Determine the work done. For copper at 0(°C), $\kappa = 0.770 \times 10^{-6}(\text{bar})^{-1}$ and the mass density $\rho = 8.93 \times 10^3(\text{kg})/(\text{m})^3$.

We start with the expression

$$W = -\int P\,dV.$$

The volumetric behavior of metals is most conveniently described by the quantities β

and κ, introduced in Chap. 2. For constant T, Eq. (2-13) becomes

$$dV = -\kappa V\, dP \qquad (\text{const } T).$$

Combination of the two equations gives

$$W = \int_{P_1}^{P_2} \kappa VP\, dP.$$

At constant T, variations in κ and V may safely be neglected in the integration over P. The last equation then becomes

$$W = \frac{\kappa V}{2}(P_2^2 - P_1^2).$$

Substitution of given numerical values yields

$$W = 0.770 \times 10^{-6}(\text{bar})^{-1} \times \frac{5(\text{kg})}{8.93 \times 10^3(\text{kg})/(\text{m})^3} \times \frac{1}{2}$$

$$\times [(1{,}000)^2 - (1)^2](\text{bar})^2 \times 10^5\ (\text{N})/(\text{m})^2(\text{bar}),$$

or $$W = 21.6(\text{N})(\text{m}) = 21.6(\text{J}).$$

The positive value of W indicates that work is done *on* the solid copper.

3-6 Work in Straining a Bar

If the length of a bar in tension or compression is changed from L to $L + dL$, the differential work δW is

$$\delta W = F\, dL,$$

where F is the total axial load on the bar. In agreement with the conventions of Sec. 2-6, F and dL are taken as positive for tension and negative for compression.

Since $F = \sigma A$, where σ is stress and A is cross-sectional area, and since $dL = L\, d\epsilon$, where ϵ is the strain, the above equation may be written

$$\boxed{\delta W = V\sigma\, d\epsilon,} \tag{3-5}$$

where $V = AL$ is the volume of the bar. For a finite change in strain, we have

$$W = \int_{\epsilon_1}^{\epsilon_2} V\sigma\, d\epsilon,$$

or, if the volume remains constant,

$$\frac{W}{V} = \int_{\epsilon_1}^{\epsilon_2} \sigma\, d\epsilon. \tag{3-6}$$

When the external load is applied slowly, the process is essentially quasi-static, and an equation of state may be used to relate σ and ϵ so that the integration can be carried out.

A typical stress-strain curve for a metal in tension is shown in Fig. 3-3. The work per unit volume of a constant-volume bar stressed in tension from state 1 to state 2 is represented by the shaded area under the curve between points 1 and 2. For this constant-temperature process the curve is a straight line, representative of Hooke's law.

Example 3-4

How much work is required to stretch a 1 (m) length of 0.5 (mm) piano wire isothermally and quasi-statically by 1 (cm)? Young's modulus E for the wire is 20×10^7(kN)/(m)^2.

We assume the validity of Hooke's law:

$$\sigma = E\epsilon.$$

Substitution in Eq. (3-6) and integration gives

$$W = \frac{VE}{2} (\epsilon_2^2 - \epsilon_1^2).$$

Further, we take $\epsilon_1 = 0$, and let V = constant = $\pi r_1^2 L_1$, where r_1 and L_1 are the initial radius [0.25 (mm)] and length of the wire. Thus

$$W = \frac{\pi r_1^2 L_1 E}{2} \epsilon_2^2.$$

All that remains is to relate ϵ_2 to the elongation. Integration of the defining equation

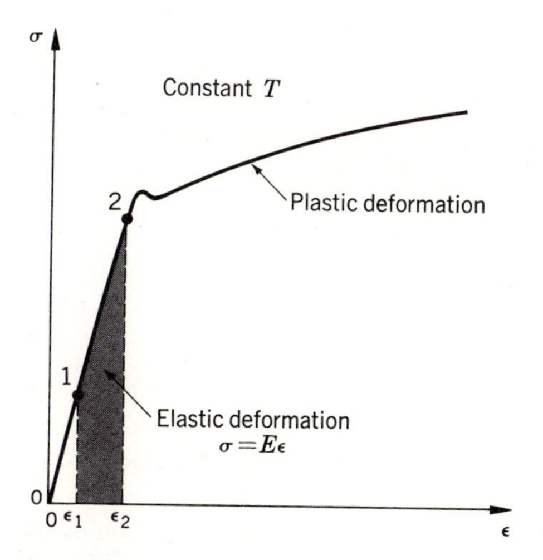

Fig. 3-3
Conventional stress-strain diagram.

for ϵ, Eq. (2-14), provides the relationship:

$$\epsilon_2 = \ln \frac{L_2}{L_1},$$

which becomes, for small elongation (see Example 2-5),

$$\epsilon_2 = \frac{L_2 - L_1}{L_1}.$$

The final expression for W is then

$$W = \frac{\pi r_1^2 E}{2L_1} (L_2 - L_1)^2.$$

Substitution of numerical values into this equation gives finally

$$W = \frac{3.142 \times (0.25 \times 10^{-3})^2 (\mathrm{m})^2 \times 20 \times 10^{10} (\mathrm{N})/(\mathrm{m})^2}{2 \times 1 (\mathrm{m})} \times (10^{-2})^2 (\mathrm{m})^2,$$

or $\quad W = 1.96 (\mathrm{N}) (\mathrm{m}) = 1.96 (\mathrm{J}).$

3-7 Work in Changing the Area of a Surface

Consider a double surface with liquid in between stretched across a wire framework one side of which is movable, as in Fig. 2-1. As stated in Sec. 2-7, the force exerted by both films is $2\gamma l$. For an infinitesimal displacement dx, the work is

$$\delta W = 2\gamma l \, dx.$$

But

$$2l \, dx = dA;$$

therefore,

$$\boxed{\delta W = \gamma \, dA.} \tag{3-7}$$

For a finite change of area from A_1 to A_2,

$$W = \int_{A_1}^{A_2} \gamma \, dA. \tag{3-8}$$

A quasi-static process may be approximated if one maintains on the slider wire a constant external force only slightly different from that exerted by the film. During a quasi-static process, γ is a function of thermodynamic coordinates only, and the integral can be evaluated once the path is known. The area under a curve on a γA diagram represents the work done on or by a surface film. When γ is expressed in (dyn)/(cm) and A in (cm^2), W is in (erg) $\equiv 10^{-7}$(J).

Example 3-5

Equation (3-8) is valid not only for the case of a planar surface but also for the formation of spherical surfaces. Estimate the work required to form the surface associated with a water mist having a droplet radius of 10(μm), if 1(l) ≡ 10^{-3}(m)^3 of liquid water at 20(°C) is to be made into mist at 20(°C).

The surface-to-volume ratio of a spherical drop is

$$\frac{A}{V} = \frac{4\pi r^2}{\frac{4}{3}\pi r^3} = \frac{3}{r},$$

where r is the radius of the sphere. For a volume of 10^{-3}(m)^3 of liquid and a radius of 10(μm),

$$A = \frac{3V}{r} = \frac{3 \times 10^{-3}(\text{m})^3}{10 \times 10^{-6}(\text{m})} = 3 \times 10^2(\text{m})^2 = 3 \times 10^6(\text{cm})^2.$$

We assume that the process is carried out quasi-statically. By Eq. (3-8),

$$W = \int_0^A \gamma\, dA = \gamma A.$$

From Table 2-5, γ for water at 20(°C) is 72.8(dyn)/(cm). Thus,

$$W = 72.8(\text{dyn})/(\text{cm}) \times 3 \times 10^6(\text{cm})^2 = 218.4 \times 10^6(\text{dyn})(\text{cm}),$$

or

$$W = 21.84 \times 10^7(\text{erg}) = 21.84(\text{J}).$$

3-8 Work in Varying the Charge of a Reversible Cell

We recall from the discussion of Sec. 2-8 that an equilibrium condition for a reversible cell requires the exact balancing of its emf by a potentiometer. If the external potential difference applied by the potentiometer is made infinitesimally smaller or larger than the cell emf e, the cell can be discharged or charged by a quasi-static process. For such a process, during which a quantity of electricity dq is transferred, the work is given by

$$\delta W = e\, dq. \tag{3-9}$$

It was shown in Sec. 2-8 that

$$dq = -j\mathfrak{F}\, dn.$$

Thus Eq. (3-9) may be written in the alternative form

$$\delta W = -ej\mathfrak{F}\, dn, \tag{3-10}$$

where $j\mathfrak{F}$ is the charge transferred per mole of reacting material, and dn is the differential number of moles reacted. For a finite change,

$$W = -j\mathfrak{F} \int_{n_1}^{n_2} e \, dn.$$

If e is constant, the last equation becomes

$$W = -ej\mathfrak{F} \, \Delta n. \tag{3-11}$$

The emf of a Daniell cell is a function of temperature only. The reason for this is that the electrolytes exist in contact with excess crystals and are therefore saturated solutions of constant concentration so long as the temperature is constant. Thus a finite amount of reaction can easily be obtained at constant e. This is not the case for all cells. However, one can always imagine a cell of infinite size for which a finite reaction and alteration of charge would not change the emf. The only measurement necessary to allow the calculation of the work that *would* occur in a quasi-static process for a cell is its emf, and this can readily be made with a cell of finite size and a potentiometer.

Example 3-6

A Daniell cell at 25(°C) is discharged quasi-statically through an external circuit. During the discharging process, 10(g) of the zinc electrode dissolves. How much electrical work is done? At 25(°C), e for the Daniell cell is 1.0821(V).

The emf e remains constant during discharge, so that Eq. (3-11) is used. The overall cell reaction during discharge is (see Sec. 2-8)

$$Zn + CuSO_4 \rightarrow Cu + ZnSO_4.$$

According to Fig. 2-3, 2(mol) of electrons are transferred through the external circuit for each mole of reactant (Zn or $CuSO_4$) used up or, equivalently, for each mole of product (Cu or $ZnSO_4$) formed. Thus $j = 2$. The molecular weight of zinc is 65.37, and the loss of 10(g) from the zinc electrode therefore corresponds to

$$\Delta n = 10(\text{g}) \frac{1(\text{mol})}{65.37(\text{g})} = 0.153(\text{mol}).$$

Note that Δn is *positive* by the conventions described in Sec. 2-8.

Substitution of numerical values into Eq. (3-11) gives

$$W = -ej\mathfrak{F} \, \Delta n = -1.0821(\text{V}) \times 2 \times 96{,}487 \, (\text{C})/(\text{mol}) \times 0.153(\text{mol}),$$

or

$$W = -31{,}900(\text{V})(\text{C}) = -31{,}900(\text{J}).$$

3-9 Work in Changing the Polarization of a Dielectric in a Parallel-Plate Capacitor

Consider a capacitor consisting of two parallel conducting plates of area A, whose linear dimensions are large compared with their separation l, filled with an isotropic solid or fluid dielectric. If a potential difference v is established across the plates, one plate is given a charge $+q$ and the other a charge $-q$. If the charge of the capacitor is changed by an amount dq, the total work done is

$$\delta W_t = v\, dq.$$

Work supplied by an outside source accomplishes two objectives: (1) an increase in the *electric field strength* $\mathcal{E}$ in the space between the two plates, and (2) an increase in the *total polarization* $\mathcal{P}$ of the dielectric. We seek to find the contribution of each of these effects to the total work.

If $\mathcal{E}$ is the electric field strength in the dielectric, then the potential difference v between the capacitor plates is given by

$$v = \mathcal{E}\, l.$$

The charge q on the plates is equal to

$$q = \mathcal{D}A,$$

where $\mathcal{D}$ is the *electric displacement*. Combination of the three equations gives

$$\delta W_t = V\mathcal{E}\, d\mathcal{D}, \tag{3-12}$$

where $V = Al$ is the volume of the capacitor. The electric displacement can be written as the sum of two terms:

$$\mathcal{D} = \epsilon_0 \mathcal{E} + \frac{\mathcal{P}}{V}, \tag{3-13}$$

where ϵ_0 is the *permittivity of vacuum* (a constant), and $\mathcal{P}$ is the total polarization of the dielectric, or its total dipole moment. Differentiation of Eq. (3-13) and substitution into Eq. (3-12) yields

$$\delta W_t = V\epsilon_0 \mathcal{E}\, d\mathcal{E} + \mathcal{E}\, d\mathcal{P}. \tag{3-14}$$

The first term on the right-hand side of Eq. (3-14) is the work required to increase the electric field strength by $d\mathcal{E}$, and would be present even if a vacuum existed between the plates. The second term is the work required to increase the polarization of the dielectric; it is zero when no matter is present in the capacitor. Therefore the *net work done on the dielectric* is just this second term:

$$\boxed{\delta W = \mathcal{E}\, d\mathcal{P}.} \tag{3-15}$$

Although our derivation has been specific for the case of a dielectric in a parallel-plate capacitor, the result is general for a dielectric in a uniform electric field.

Table 3-1 Values of χ_e for Some Dielectric Substances

Substance	t(°C)	χ_e
Air [at 1(atm)]	20	0.000536
n-Hexane	20	0.890
Benzene	20	1.28
Plexiglas	27	2.40
Neoprene	24	5.70
Acetone	25	19.7
Glycerine	25	41.5
Water	25	77.5

The SI units for $\mathscr{D}$ are (C)/(m)^2; for $\mathscr{E}$, (V)/(m); and for $\mathscr{P}$, (C)(m). The permittivity of vacuum ϵ_0 has the value 8.85419×10^{-12}(F)/(m), where the *farad* (F) is a derived SI unit equal to 1(C)/(V) $\equiv$ 1(A)^2(s)^4/(kg)(m)^2.

The thermodynamic coordinates of a solid or liquid dielectric are $\mathscr{E}$, $\mathscr{P}$, and T. The dielectric properties of gases (actually of all materials, to some extent) depend upon pressure as well. The equation of state for a dielectric substance is commonly written

$$\frac{\mathscr{P}}{V} = \chi_e \epsilon_0 \mathscr{E}, \tag{3-16}$$

where χ_e is the *electric susceptibility,* a positive dimensionless number. For solids and liquids, χ_e is to a good approximation a function of T only. Values of χ_e for some dielectrics are listed in Table 3-1.

Example 3-7

A parallel-plate air capacitor is quasi-statically charged at room temperature to a potential difference of 100(V). What is the *total* work required, and how much of the work is used in polarizing the air? The capacitor plates are 5(cm) square, and their separation is 1(mm).

The differential expression for *total* work δW_t is given by Eq. (3-14). Taking Eq. (3-16) as the dielectric equation of state for air, we find

$$\delta W_t = V\epsilon_0 \mathscr{E}\, d\mathscr{E} + V\chi_e\epsilon_0 \mathscr{E}\, d\mathscr{E} = V\epsilon_0(\chi_e + 1)\mathscr{E}\, d\mathscr{E}.$$

The work required to increase the electric field strength from 0 to $\mathscr{E}$ is obtained on integration of this equation. Thus,

$$W_t = \tfrac{1}{2}V\epsilon_0(\chi_e + 1)\mathscr{E}^2.$$

From the given data we have

$$V = Al = (5 \times 10^{-2})^2(\text{m})^2 \times 10^{-3}(\text{m}) = 25 \times 10^{-7}(\text{m})^3$$

$$\mathcal{E} = \frac{v}{l} = \frac{100(\text{V})}{10^{-3}(\text{m})} = 10^5(\text{V})/(\text{m});$$

from Table 3-1, we find $\chi_e = 5.36 \times 10^{-4}$ for the air. Substitution into the expression for W_t gives

$$W_t = \tfrac{1}{2} \times 25 \times 10^{-7}(\text{m})^3 \times 8.85419 \times 10^{-12}\,(\text{C})/(\text{V})\,(\text{m}) \times 1.000536$$
$$\times\ (10^5)^2\ (\text{V})^2/(\text{m})^2,$$

or $$W_t = 1.11 \times 10^{-7}(\text{V})\,(\text{C}) = 1.11 \times 10^{-7}(\text{J}) = 1.11\,(\text{erg}).$$

This is the *total* work. The differential work δW of polarizing the air is given by

$$\delta W = V\chi_e\epsilon_0\,\mathcal{E}\,d\mathcal{E},$$

from which

$$W = \tfrac{1}{2}V\chi_e\epsilon_0\mathcal{E}^2.$$

Substitution of numerical values gives finally

$$W = 5.92 \times 10^{-11}(\text{J}) = 5.92 \times 10^{-4}(\text{erg}).$$

Thus, only 0.05 percent of the total work is used in polarizing the dielectric; the rest is required to set up the electric field. This is an extreme example, because gases have very small electric susceptibilities. For other dielectrics, W may well constitute the greater part of the required work.

3-10 Work in Changing the Magnetization of a Magnetic Solid

Consider a sample of magnetic material in the form of a ring of cross-sectional area A and mean circumference l. Suppose that an insulated wire is wound on top of the sample, forming a toroidal winding of N closely spaced turns, as shown in Fig. 3-4. A current may be maintained in the winding by a battery, and by moving the sliding contactor of a rheostat we can change the current.

The effect of a current in the winding is to set up a magnetic field with *magnetic induction* $\mathcal{B}$. If the relative dimensions are as shown in the figure, $\mathcal{B}$ will be nearly uniform over the cross section of the toroid. Suppose that the current is changed and that in time $d\tau$ the magnetic induction changes by an amount $d\mathcal{B}$. Then, by Faraday's principle of magnetic induction, there is induced in the winding a back emf e, where

$$e = -NA\frac{d\mathcal{B}}{d\tau}.$$

During the time interval $d\tau$, a quantity of electricity dq is transferred in the

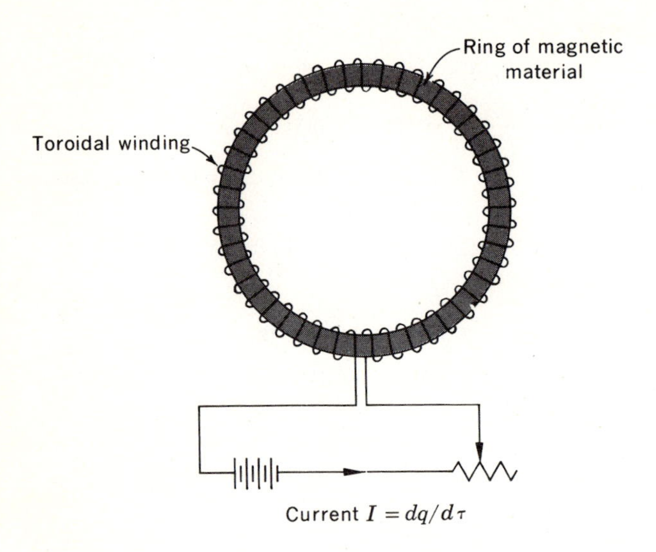

Fig. 3-4
Changing the magnetization of a magnetic solid.

circuit, and the *total* work done to maintain the current is

$$\delta W_t = -e\,dq = -eI\,d\tau,$$

where $I = dq/d\tau$ is the instantaneous current. The current in the winding is related to the *magnetic field strength* $\mathcal{H}$ by

$$I = \frac{\mathcal{H}V}{NA},$$

where $V = Al$ is the volume of magnetic material. Combination of the three equations gives

$$\delta W_t = V\mathcal{H}\,d\mathcal{B}. \qquad (3\text{-}17)$$

The magnetic induction can be written as the sum of two terms:

$$\mathcal{B} = \mu_0\mathcal{H} + \mu_0\frac{\mathcal{M}}{V}, \qquad (3\text{-}18)$$

where μ_0 is the *permeability of vacuum* (a constant), and $\mathcal{M}$ is the *total magnetization* of the magnetic material, or its total magnetic moment. Differentiation of Eq. (3-18) and substitution into Eq. (3-17) gives

$$\delta W_t = V\mu_0\mathcal{H}d\mathcal{H} + \mu_0\mathcal{H}d\mathcal{M}. \qquad (3\text{-}19)$$

The first term on the right-hand side of Eq. (3-19) is the work required to increase the magnetic field strength by $d\mathcal{H}$, and would be present even if a vacuum existed within the toroidal winding. The second term is the work required to increase the magnetization of the magnetic material; it is zero when no material is present within the winding. Therefore the *net work done on the magnetic material* is just this second term:

$$\boxed{\delta W = \mu_0 \mathcal{H} d\mathcal{M}.} \tag{3-20}$$

Modern experiments on paramagnetic materials are usually performed on samples in the form of cylinders or ellipsoids, not toroids. In these cases, the $\mathcal{H}$ field inside the material is somewhat smaller than the $\mathcal{H}$ field generated by the electric current in the surrounding winding because of the reverse field (demagnetizing field) set up by magnetic poles which form on the surfaces of the samples. In longitudinal magnetic fields, the demagnetizing effect may either be rendered negligible through use of cylinders whose length is much larger than the diameter or corrected for in a simple way. In transverse magnetic fields, a correction factor must be applied. We shall limit ourselves to toroids or to long thin cylinders in longitudinal fields where the internal and external $\mathcal{H}$ fields are the same.

The SI units for $\mathcal{B}$ are (Wb)/(m)^2, where the *weber* (Wb) is equal to 1(V)(s) $\equiv$ 1(kg)(m)^2/(s)^2(A). The units for $\mathcal{H}$ are (A)/(m), and for $\mathcal{M}$, (A)(m)^2. The permeability of vacuum μ_0 has the value $4\pi \times 10^{-7}$(H)/(m) = 1.25664×10^{-6}(H)/(m), where the *henry* (H) is equal to 1(Wb)/(A) $\equiv$ 1(kg) (m)^2/(s)^2(A)^2.

Most experiments on magnetic materials are performed at constant atmospheric pressure and involve only minute volume changes. Consequently, we may ignore the pressure and the volume and describe a magnetic material through the use of just three thermodynamic coordinates: the magnetic intensity $\mathcal{H}$, the magnetization $\mathcal{M}$, and the temperature T. This is not true for *ferromagnetic* materials, however, for they are subject to hysteresis. The state of such a material therefore depends not only on $\mathcal{H}$, $\mathcal{M}$, and T but also on its past history.

The equation of state for a *diamagnetic* or *paramagnetic* substance is usually written

$$\frac{\mathcal{M}}{V} = \chi_m \mathcal{H}, \tag{3-21}$$

where χ_m is the *magnetic susceptibility,* a dimensionless number which is commonly taken to be a function of T only. For diamagnetic materials, χ_m is a small negative number; for paramagnetic materials it is a small positive number. Values of χ_m for some substances are listed in Table 3-2.

Table 3-2 Values of χ_m for Some Substances

Diamagnetic			Paramagnetic		
Substance	$t(°C)$	$\chi_m \times 10^5$	Substance	$t(°C)$	$\chi_m \times 10^5$
Mercury	18	−3.24	Oxygen [at 1(atm)]	20	0.177
Silver	18	−2.64	Sodium	18	0.621
Quartz	25	−1.65	Aluminum	18	2.21
Ice	0	−0.805	Platinum	18	29.7
Acetone	25	−0.578	Copper sulfate	25	38.9
Nitrogen [at 1(atm)]	20	−0.000500	Uranium	18	61.0

Example 3-8

The current in the winding of an aluminum Rowland ring (the device shown in Fig. 3-4) is slowly increased at room temperature from 0 to 1(A). The ring has 500 turns of wire, its cross-sectional area is 0.5(cm)², and it has a mean circumference of 50(cm). How much work is done, and what fraction of the work is used in magnetizing the aluminum core?

We assume a quasi-static process. The differential expression for *total* work δW_t is given by Eq. (3-19). Taking Eq. (3-21) as the magnetic equation of state for aluminum, we find

$$\delta W_t = V\mu_0 \mathcal{H} d\mathcal{H} + V\chi_m\mu_0 \mathcal{H} d\mathcal{H} = V\mu_0(\chi_m + 1)\mathcal{H} d\mathcal{H}.$$

The work required to increase the magnetic field strength from 0 to $\mathcal{H}$ is obtained on integration of this equation. Thus,

$$W_t = \tfrac{1}{2} V\mu_0(\chi_m + 1)\mathcal{H}^2.$$

From the given data, we have

$$V = Al = 0.5 \times 10^{-4}(\mathrm{m})^2 \times 50 \times 10^{-2}(\mathrm{m}) = 2.5 \times 10^{-5}(\mathrm{m})^3$$

and

$$\mathcal{H} = \frac{NAI}{V} = \frac{NI}{l} = \frac{500 \times 1(\mathrm{A})}{50 \times 10^{-2}(\mathrm{m})} = 1{,}000(\mathrm{A})/(\mathrm{m});$$

from Table 3-2, we find $\chi_m = 2.21 \times 10^{-5}$ for aluminum. Substitution into the expression for W_t gives

$$W_t = \tfrac{1}{2} \times 2.5 \times 10^{-5}(\mathrm{m})^3 \times 1.25664 \times 10^{-6}(\mathrm{H})/(\mathrm{m}) \times 1.0000221 \times (10^3)^2\,(\mathrm{A})^2/(\mathrm{m})^2,$$

or

$$W_t = 1.57 \times 10^{-5}(\mathrm{H})(\mathrm{A})^2 = 1.57 \times 10^{-5}(\mathrm{J}) = 157(\mathrm{erg}).$$

This is the total work. The differential work δW of magnetizing the core is given by

$$\delta W = V\chi_m\mu_0 \mathcal{H} d\mathcal{H},$$

from which

$$W = \tfrac{1}{2} V \chi_m \mu_0 \mathscr{H}^2.$$

Substitution of numerical values gives

$$W = 3.47 \times 10^{-10}(\text{J}) = 3.47 \times 10^{-3}(\text{erg}).$$

Only 0.002 percent of the total work is used in magnetizing the core, the rest being required to set up the magnetic field. This result is typical, because χ_m is extremely small for a paramagnetic substance at normal temperatures. Only for very low temperatures of the order of tenths or hundredths of a kelvin is χ_m large enough so that the magnetization $\mathscr{M}$ is an appreciable fraction of $\mathscr{H}$.

3-11 Summary

The results of this chapter are summarized in Table 3-3. Each expression for work involves the product of a *generalized force* Y and a *generalized displacement* dX. A generalized work diagram is obtained if Y is plotted against X. The most common diagram of this type, the PV (indicator) diagram, was treated in Sec. 3-4; it was shown there that PV work is proportional to the area under a PV curve.

Similar interpretations obtain for other YX diagrams. For the last five systems in Table 3-3, however, the sign conventions are different from those for a PV diagram, because the expression for PV work contains a minus sign. Thus for these non-PVT systems the area under a YX curve is *positive* in the direction of increasing X, and *negative* in the reverse direction. As a result, the net area of a cycle traversed in the clockwise direction is positive, and in the counterclockwise direction, negative.

Table 3-3 Work in Simple Systems

System	Generalized force Y	Generalized displacement dX	Generalized work $W = k\int Y\,dX$
PVT System	P	dV	$-\int P\,dV$
Stressed bar	σ	$d\epsilon$	$V\int \sigma\,d\epsilon$
Surface	γ	dA	$\int \gamma\,dA$
Reversible cell	e	dq	$\int e\,dq$
Dielectric	$\mathscr{E}$	$d\mathscr{P}$	$\int \mathscr{E}\,d\mathscr{P}$
Magnetic material	$\mathscr{H}$	$d\mathscr{M}$	$\mu_0\int \mathscr{H}\,d\mathscr{M}$

Problems

3-1 (*a*) Steam is admitted to the cylinder of a steam engine at a constant pressure of 300(psia). The bore of the cylinder is 8(in), and the stroke of the piston is 12(in). How much work is done per stroke?

(*b*) If 10(lb_m) of water is evaporated at atmospheric pressure until a volume of 288.5$(ft)^3$ is occupied, how much work is done?

3-2 A thin-walled metal bomb of volume V_B contains n(mol) of gas at high pressure. Connected to the bomb is a capillary tube and stopcock. When the stopcock is opened slightly, the gas leaks slowly into a cylinder equipped with a nonleaking, frictionless piston where the pressure remains constant at the atmospheric value P_0.

(*a*) Show that after as much gas as possible has leaked out, an amount of work

$$W = P_0(V_B - nV_0)$$

has been done, where V_0 is the *molar* volume of the gas at atmospheric pressure and temperature.

(*b*) How much work would be done if the gas leaked directly into the atmosphere?

3-3 Calculate the work done by 1(mol) of a gas during a quasi-static, isothermal expansion from an initial volume V_1 to a final volume V_2 when the equation of state is

$$P(V - b) = RT,$$

where b is a positive constant. If the gas were ideal, would the same process produce more or less work?

3-4 A vertical cylinder, closed at the bottom, is placed on a spring scale. The cylinder contains a gas whose volume may be changed with the aid of a frictionless, nonleaking piston. The piston is pushed down.

(*a*) How much work is done by an outside agent in compressing the gas an amount dV while the spring scale goes down a distance dy?

(*b*) If this device is used only to produce effects in the gas—or, in other words, if the gas is the system—what expression for work is appropriate?

3-5 A stationary vertical cylinder, closed at the top, contains a gas whose volume may be changed with the aid of a heavy, frictionless, nonleaking piston of weight w.

(*a*) How much work is done by an outside agent in compressing the gas an amount dV by raising the piston a distance dy?

(*b*) If this device is used only to produce temperature changes of the gas, what expression for work would be appropriate?

(*c*) Compare this situation with that of Prob. 3-4, and also with that involved in increasing the magnetic induction of a ring of magnetic material.

3-6 Figure P3-6 shows the relation of pressure to volume for a closed PVT system during a quasi-static process. Calculate the work done by the system for each of the three steps 12, 23, and 31, and for the entire process 1231.

3-7 Compute the quasi-static work done in compressing 1$(ft)^3$ of mercury at a constant temperature of 32(°F) from a pressure of 1(atm) to a pressure of 3,000(atm). The isothermal compressibility of mercury at 32(°F) is given as a function of pressure by

$$\kappa = 3.9 \times 10^{-6} - 0.1 \times 10^{-9}P,$$

where P is in (atm) and κ is in $(\text{atm})^{-1}$. Justify any assumption you make.

3-8 (*a*) The tensile stress in a metal bar is increased quasi-statically and isothermally from σ_1 to σ_2. If the volume of the bar and Young's modulus remain essentially constant, show that the work is given by

$$W = \frac{V}{2E}(\sigma_2^2 - \sigma_1^2).$$

(*b*) The tensile stress in a steel bar 10(ft) long and $1(\text{in})^2$ in area is increased quasi-statically and isothermally from 10,000 to $40{,}000(\text{lb}_f)/(\text{in})^2$. What is the work in (ft) (lb_f)? $E = 30.0 \times 10^6(\text{lb}_f)/(\text{in})^2$ for steel.

3-9 The equation of state for an ideal elastic substance is

$$\sigma = \frac{A_0}{A}KT\left(\lambda - \frac{1}{\lambda^2}\right),$$

where K is a constant, λ is the extension ratio L/L_0, and the ratio of no-load area to area of the stressed specimen A_0/A converts the stress σ from the nominal value (based on the no-load area) to the actual value. L_0 is the no-load length and is a function of temperature only.

(*a*) Using the formula for work in the form

$$\delta W = V\sigma\, d\epsilon,$$

show that

$$\frac{\delta W}{V_0} = KT\left(\lambda - \frac{1}{\lambda^2}\right) d\lambda$$

for isothermal, quasi-static extension or compression.

(*b*) For a sample of rubber, $L_0 = 0.3(\text{m})$ and $A_0 = 0.65(\text{cm})^2$, and for which $K = 1.86 \times 10^3(\text{N})/(\text{m})^2(\text{K})$ at 300(K), calculate W for elongation of the sample from $L = L_0$ to $L = 2L_0$.

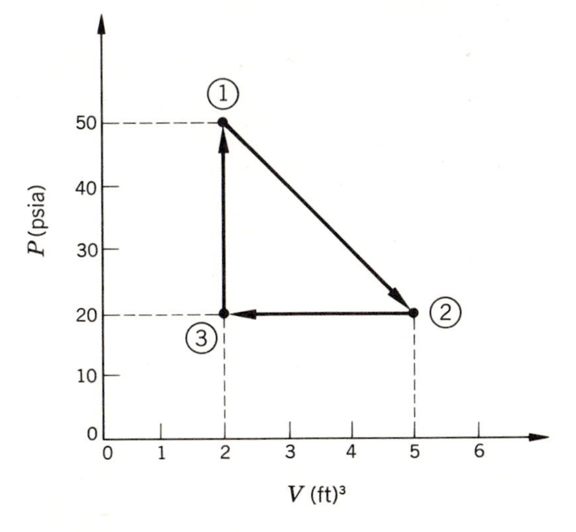

Fig. P3-6

3-10 For liquid water the isothermal compressibility is given by

$$\kappa = \frac{c}{V(P + b)}$$

where c and b are functions of temperature only. If 1 (kg) of water is compressed isothermally and quasi-statically from 1 to 500 (atm) at 60 (°C), how much work is required? At 60 (°C), $b = 2{,}700$ (atm), $c = 0.125\,(\text{cm})^3/(\text{g})$.

3-11 Show that the work done in forming the area of a spherical "soap bubble" of radius r by an isothermal quasi-static process is

$$W = 8\pi\gamma r^2.$$

3-12 The magnetic susceptibility of a certain paramagnetic substance is described by Curie's equation,

$$\chi_m = \frac{C}{T},$$

where C is constant. Show that the work done during a quasi-static isothermal change of state of the substance is

$$W = \frac{\mu_0 T}{2VC}(\mathcal{M}_2^2 - \mathcal{M}_1^2) = \frac{\mu_0 VC}{2T}(\mathcal{H}_2^2 - \mathcal{H}_1^2).$$

3-13 A paramagnetic substance occupying a volume of 200 (cm)³ is maintained at constant temperature. A magnetic field is increased quasi-statically and isothermally from 0 to 1.2×10^6 (A)/(m). Assuming Curie's equation (see Prob. 3-12) to hold and the Curie constant to be 0.163(K);

(a) How much work would be required if no material were present?

(b) How much work is done to change the magnetization of the material when the temperature is 300 (K) and when it is 1 (K)?

(c) How much work is done by the agent supplying the magnetic field at each temperature?

3-14 The electric susceptibility of a certain dielectric substance is described by the equation

$$\chi_e = a + \frac{b}{T},$$

where a and b are constants. Show that the work done during a quasi-static isothermal change of state of the substance is

$$W = \frac{T}{2\epsilon_0 V(b + aT)}(\mathcal{P}_2^2 - \mathcal{P}_1^2) = \frac{\epsilon_0 V}{2}\left(a + \frac{b}{T}\right)(\mathcal{E}_2^2 - \mathcal{E}_1^2).$$

4

HEAT AND THE FIRST LAW

4-1 Work and Heat

It was shown in Chap. 3 how a system could be transferred from an initial to a final state by means of a quasi-static process and how the work done during the process could be calculated. There are, however, other means of changing the state of a system that do not necessarily involve the performance of work. Consider for example the three processes depicted schematically in Fig. 4-1. In (*a*) a fluid undergoes an adiabatic expansion in a cylinder-piston combination that is coupled to the surroundings with a suspended body so that, as the expansion takes place, the body is lifted while the fluid remains always close to equilibrium.[1] In (*b*) a liquid in equilibrium with its vapor is in contact through a diathermic wall with the hot combustion products of a bunsen burner and undergoes vaporization, accompanied by a rise of temperature and of pressure, without the performance of work. In (*c*) a fluid is expanded while in contact with the flame of a bunsen burner.

What happens when two systems at different temperatures are placed together is one of the most familiar experiences of mankind. It is well known that the final temperature reached by both systems is intermediate between the two starting temperatures. Up to the beginning of the nineteenth century, such phenomena, which constitute the subject of *calorimetry*, were explained by the supposition that a substance or form of matter termed *caloric*, or heat, exists in every body. It was believed that a body at a high temperature contained much caloric and that one at a low temperature had only a little. When the two bodies were put together, the body rich in caloric lost some to the other, and thus the final temperature was intermediate.

[1] The clever coupling device in Fig. 4-1 is due to E. Schmidt, "Thermodynamics: Principles and Applications to Engineering," Dover, New York, 1966.

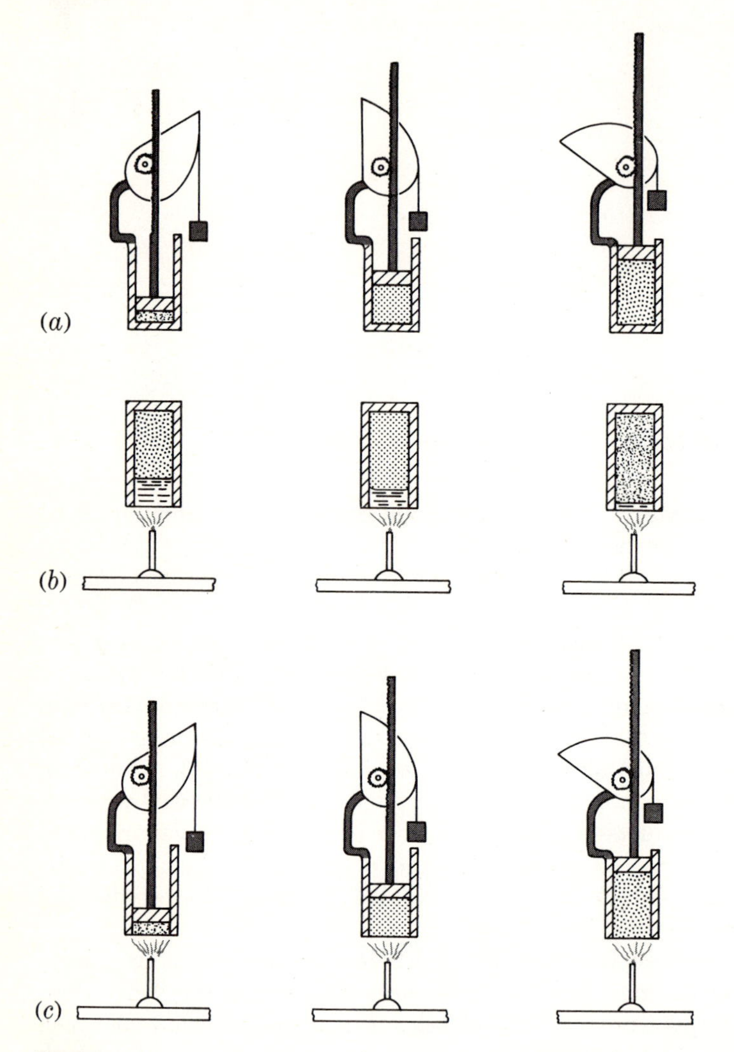

Fig. 4-1
Distinction between work and heat. (*a*) *Adiabatic work;* (*b*) *heat flow without work;* (*c*) *work and heat.*

Although we now know that heat is not a substance whose total amount remains constant, nevertheless we ascribe the changes that take place in Fig. 4-1*b* and *c* to the transfer of "something" from the body at the higher temperature to the one at the lower, and this something we call *heat.* We therefore adopt as a *calorimetric* definition of heat *that which is transferred* *between a system and its surroundings by virtue of a temperature difference only.*

It is important to observe that the decision about whether a particular change of state involves the performance of work or the transfer of heat requires first an unequivocal answer to the questions: What is the system, and what are the surroundings? For example, in Fig. 4-2, a resistor immersed in water carries a current provided by an electric generator that is rotated with the aid of a descending body. If we assume the absence of friction in the shafts of the pulleys and the absence of electrical resistance in the generator and the connecting wires, we have a device whereby the thermodynamic state of a system composed of *water and the resistor* is changed by purely mechanical means, i.e., by the performance of *work*. If, however, the resistor is regarded as the system and the water as the surroundings, then there is a *transfer of heat from the resistor* by virtue of the temperature difference between the resistor and the water. Also, if a small part of the water is regarded as the system, with the rest of the water considered the surroundings, then again there is a transfer of heat. However, the composite system comprising both the water and the resistor is contained in an adiabatic enclosure; therefore no heat is transferred between this composite system and its surroundings.

4-2 Adiabatic Work

Figure 4-1*a* and 4-2 show systems completely surrounded by an adiabatic envelope but coupled to the surroundings so that work may be done. Three

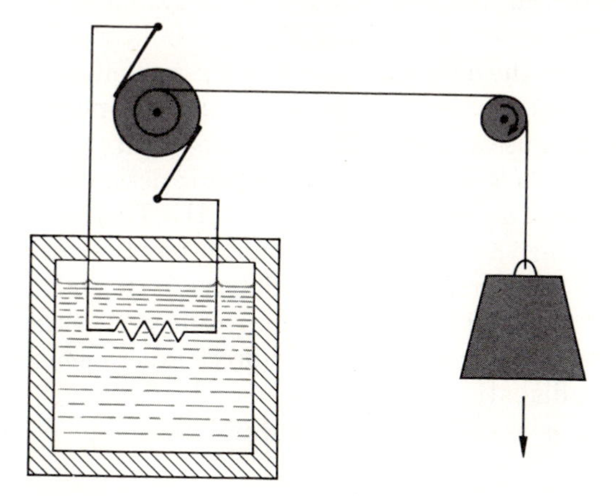

Fig. 4-2
Whether a process is designated as a work or a heat interaction depends on the choice of system.

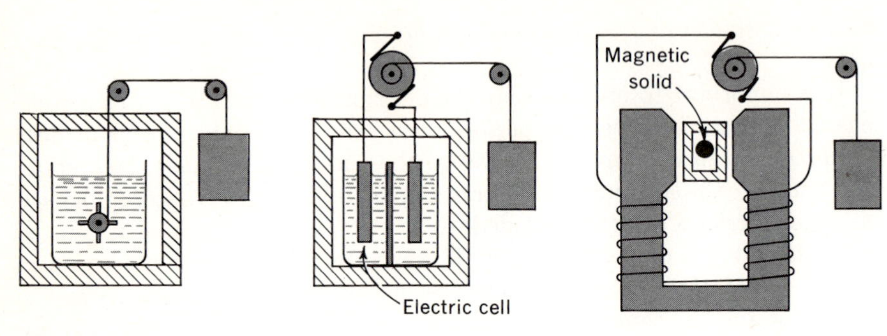

Fig. 4-3
Adiabatic work.

other simple examples of adiabatic work are shown in Fig. 4-3. It is an important fact of experience that the state of a system may be caused to change from a given initial state to a different final state by the performance of adiabatic work *only*.

Consider a composite system consisting of a mass of fluid in which is immersed an electrical resistor, all enclosed in a piston-and-cylinder device. We imagine the device to be made of a perfect thermal insulator, so that any process in the composite system is *adiabatic*. We define an initial state i of the system by the coordinates $P_i = 1(\text{bar})$ and $T_i = 295(\text{K})$, and a final state f by the coordinates $P_f = 1(\text{bar})$ and $T_f = 300(\text{K})$. These states are shown on Fig. 4-4, a PV diagram for the composite system.

To cause the system to proceed from state i to state f, we could maintain an electrical current in the resistor, adjusting the volume of the system so as to maintain P at the constant value of $1(\text{bar})$. This process is indicated in Fig. 4-4 as path I. But path I is not the only path by which the system may be changed from i to f by the performance of adiabatic work. We might compress the fluid from i to j, then use a current in the resistor from j to k, and then expand from k to f, the whole series of processes being designated by path II. Or we might make use of a similar adiabatic path III. There are an infinite number of paths by which a system may be transferred from an initial state to a final state by the performance of adiabatic work only. Although accurate measurements of adiabatic work along different paths between the same two states have never been made, indirect experiments nevertheless indicate that the adiabatic work is the same along all such paths. The generalization of this result is a fundamental postulate of thermodynamics:

> *If a system is caused to change from an initial state to a final state by adiabatic means only, the work done is the same for all adiabatic paths connecting the two states.*

When a quantity is found to depend only on the initial and final states of a system, and not upon the path connecting the states, an important conclusion can be drawn. We recall from mechanics that when an object is moved from one point to another in a gravitational field, the work done (in the absence of friction) depends only upon the positions of the two points and not on the path through which the body is moved. It is concluded from this that there exists a *potential-energy function* which depends upon the space coordinates, and whose final value minus its initial value equals the work done. Similarly, the work done when a quantity of electricity is transferred from one point to another in an electrostatic field is expressible as the change in an *electric-potential function*, evaluated between initial and final states. Analogously, we conclude from the postulate in italics that there exists for a thermodynamic system some function of the thermodynamic coordinates whose change in value equals the adiabatic work. This function is known as the *internal-energy function*, or simply the *internal energy* U. Thus we write

$$W(\text{adiabatic}) = \Delta U, \tag{4-1}$$

where ΔU is the *change* in the internal energy; for a change in state *from* state 1 *to* state 2, $\Delta U = U_2 - U_1$.

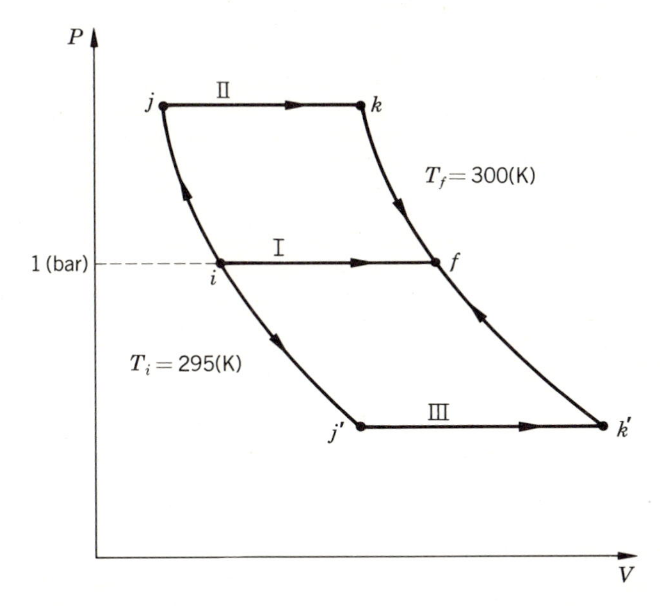

Fig. 4-4
The state of a system is changed from i *to* f *along three different paths.*

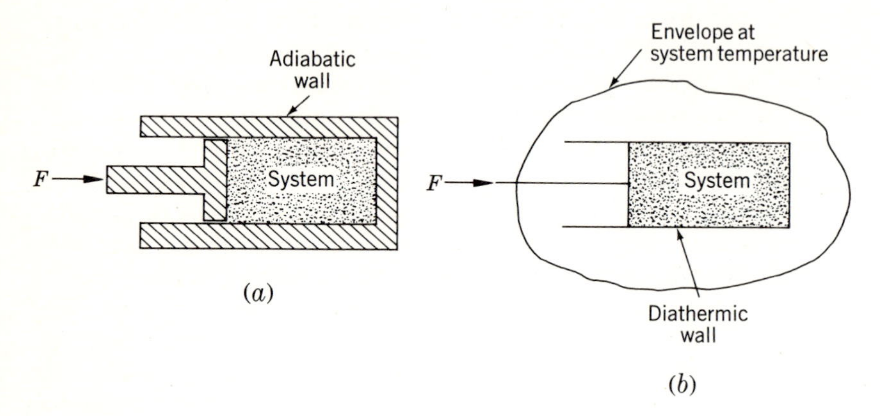

Fig. 4-5
Two methods for doing adiabatic work.

Example 4-1

Explain how adiabatic work may be accomplished with a piston-cylinder apparatus when the piston and cylinder do not form an adiabatic enclosure.

Figure 4-5*a* shows a piston and cylinder which form adiabatic walls within which the system (a gas) is enclosed. When the force F is large enough to cause the piston to move inward, work is done *on* the system; the gas is compressed; hence W is positive and, according to Eq. (4-1), ΔU is positive. Thus the internal energy increases, and we know from experience that the temperature of the gas (the system) increases.

If the piston and cylinder are made of a conductive metal like copper, then they are diathermic rather than adiabatic walls, and heat transfer is possible. However, if the piston and cylinder are surrounded by an envelope as shown in Fig. 4-5*b*, whose temperature is always maintained equal to that of the system by an automatic control device, then no heat flows to or from the system. When work is done and the temperature of the system changes, the temperature *difference* between the system and its immediate surroundings is always negligible. The work done is therefore adiabatic work, and Eq. (4-1) applies. This conclusion rests on the calorimetric definition of heat.

4-3 Internal Energy

The quantity ΔU as given by Eq. (4-1) is the net change in the internal energy of a system as the result of an adiabatic process. Equation (4-1) therefore expresses the principle of conservation of energy for a special kind of process. It should be emphasized, however, that this equation represents more than a conservation principle. It states that there *exists* a special energy function U which is a property of the system and a function of as many thermodynamic coordinates as are necessary to specify the state of a system.

The equilibrium states of a *PVT* system, for example, describable by means of three thermodynamic coordinates P, V, and T, are completely determined by only *two*, since the third is fixed by the equation of state. Therefore the internal energy may be thought of as a function of only two (any two) of the thermodynamic coordinates. This is true for each of the simple systems described in Chap. 2. It is not always possible to write this function in simple mathematical form. Usually the exact form of the function is unknown. It must be understood, however, that it is not necessary to know actually what the internal-energy function is, so long as we can be sure that it exists.

If the coordinates characterizing the two states differ from each other only infinitesimally, the change of internal energy is dU, where dU is an exact differential, since it is the differential of an actual function. In the case of a *PVT* system, if U is regarded as a function of T and V, then

$$dU = \left(\frac{\partial U}{\partial T}\right)_V dT + \left(\frac{\partial U}{\partial V}\right)_T dV; \tag{4-2}$$

similarly, if U is taken as a function of T and P,

$$dU = \left(\frac{\partial U}{\partial T}\right)_P dT + \left(\frac{\partial U}{\partial P}\right)_T dP. \tag{4-3}$$

Note that the two partial derivatives $(\partial U/\partial T)_V$ and $(\partial U/\partial T)_P$ are not equal. They are different mathematically and also have different physical meanings.

4-4 Formulation of the First Law

We have considered up to now processes by which a system undergoes a change of state through the performance of adiabatic work only. Many real processes are not of this type. In Fig. 4-6 two examples of nonadiabatic processes are depicted. In (*a*) an expanding gas is in contact with a flame whose temperature is higher than that of the gas. In (*b*) the magnetization of a paramagnetic solid is increased while it is in contact with liquid helium at a lower temperature. During the magnetization, some of the helium boils away.

Let us now imagine two different experiments performed on the same system. In one, we measure the adiabatic work required to change the system from state 1 to state 2; by Eq. (4-1), this work is equal to $U_2 - U_1$. In the other experiment we cause the system to undergo the same change of

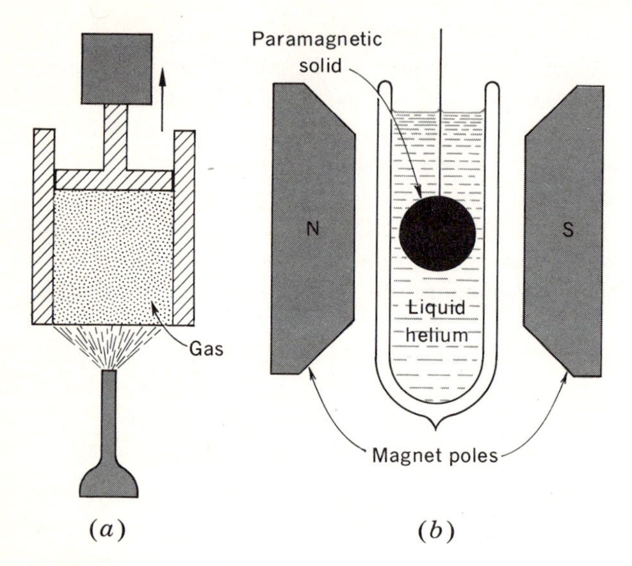

Fig. 4-6
Nonadiabatic processes.

state *non*adiabatically, and again we measure the work. The result of all such experiments is that the *non*adiabatic work is *not* equal to $U_2 - U_1$. If this result is to be consistent with the conservation principle, we must conclude that energy has been transferred by means other than the performance of work. This other mechanism, which operates by virtue of the temperature difference between system and surroundings, is the transfer of what we have previously called *heat*. We therefore adopt the following as a thermodynamic definition of heat: *When a system whose surroundings are at a different temperature undergoes a process during which work may be done, the energy transferred by nonmechanical means, equal to the difference between the internal-energy change and the work, is called heat.* Denoting this difference by Q, we have

$$Q = (U_2 - U_1) - W$$

or

$$\boxed{\Delta U = Q + W.} \tag{4-4}$$

In writing Eq. (4-4), we adopt the convention that heat is *positive* when transferred *to* a system and *negative* when transferred *from* it. This is the same as the sign convention adopted for work. Equation (4-4) is a mathematical statement of the first law of thermodynamics, valid for systems of constant mass.

It should be emphasized that the above formulation of the first law contains three related ideas: (1) the existence of an internal-energy function,

(2) the principle of the conservation of energy, and (3) the definition of heat as energy in transit.

4-5 Concept of Heat

Heat flows from one part of a system to another, or from one system to another, by virtue of only a temperature difference. When this flow has ceased, there is no longer any occasion to use the word "heat." It would be just as incorrect to refer to the "heat in a body" as it would be to speak of the "work in a body." The performance of work and the transfer of heat are methods by which the energy of a system is changed. It is impossible to separate or divide the internal energy into a mechanical and a thermal part.

We have seen that in general, i.e., for a nonadiabatic process, the work is not a function of the coordinates of a system but depends on the path by which a system is brought from an initial to a final state. The same is true for heat; Q is not in general a function of thermodynamic coordinates, but depends on the path. An infinitesimal amount of heat is therefore an *in*exact differential and is represented by the symbol δQ.

In general, when heat enters a system, a change of state occurs, involving a change in some or all of the thermodynamic coordinates. A definite change under specified conditions in any one convenient coordinate of an easily reproduced system may serve to indicate the absorption of a standard amount of heat. Thus it is not surprising that the unit of heat was, in early times, tied to the properties of water.[1] For many years the unit of heat was taken to be the *calorie* (cal), defined as that amount of heat whose absorption by 1(g) of water at constant atmospheric pressure is accompanied by a temperature rise from 14.5 to 15.5(°C). Similarly, the *British thermal unit* (Btu) was long defined as that quantity of heat which produces a rise in temperature from 63 to 64(°F) in 1(lb_m) of water.

Once heat is identified as a form of energy, there is no need for special calorimetric units. Thus there is now just one internationally recognized unit of heat, the joule. This is not to say, however, that units called the calorie and the Btu have passed out of use, desirable as that might be. Instead they are now used as units of energy and *defined* by common consent as multiples of the joule. Various conversion factors for energy units are tabulated in Appendix A.

[1] Readable accounts of the historical development of the first law and of the concept of heat are given by M. Mott-Smith, "The Concept of Energy Simply Explained," Dover, New York, 1964, and by D. Roller, "The Early Development of the Concepts of Temperature and Heat: The Rise and Decline of the Caloric Theory," Harvard University Press, Cambridge, 1950.

Example 4-2

The indicator diagram of Fig. 4-7 depicts a quasi-static cyclic process carried out on a mass of fluid. During each step of the cycle, heat is transferred to or from the fluid: $Q_{ab} = 800(\text{J})$, $Q_{bc} = -950(\text{J})$, and $Q_{ca} = 250(\text{J})$. Determine W and ΔU for each step of the cycle and for the complete cycle *abca*.

The work for each step is proportional to the area under the PV curve representing the step. Thus we can determine the work terms directly from the diagram; the corresponding values of ΔU then follow from W and Q by application of the first law, Eq. (4-4), to each step. For step *ab* we have

$$W_{ab} = \text{Area } abdeca = -3(\text{bar})(\text{l}) = -300(\text{J})$$

and

$$\Delta U_{ab} = Q_{ab} + W_{ab} = 800 - 300 = 500(\text{J}).$$

Similarly, for step *bc*,

$$W_{bc} = \text{Area } bdecb = +2(\text{bar})(\text{l}) = 200(\text{J})$$

and

$$\Delta U_{bc} = Q_{bc} + W_{bc} = -950 + 200 = -750(\text{J}).$$

For step *ca*, the volume of the system remains constant, and no work is done. Thus,

$$W_{ca} = 0$$

and

$$\Delta U_{ca} = Q_{ca} + W_{ca} = 250 + 0 = 250(\text{J}).$$

The values of W, Q, and ΔU for the cycle are the sums of the values for the individual steps. The results are summarized in Table 4-1. Note that $\Delta U_{\text{cycle}} = 0$, and that $W_{\text{cycle}} = -Q_{\text{cycle}}$, in accordance with Eq. (4-4) as applied to the cycle.

Table 4-1

Step	$\Delta U(\text{J})$	$Q(\text{J})$	$W(\text{J})$
ab	500	800	−300
bc	−750	−950	200
ca	250	250	0
Cycle *abca*	0	100	−100

Example 4-3

System A is in thermal contact with system B. The *composite* system undergoes an adiabatic process during which no work is exchanged between systems A and B. Determine the consequences of the first law.

The first law applies separately to each system. For system A alone,

$$\Delta U_A = Q_A + W_A,$$

and for system B alone,

$$\Delta U_B = Q_B + W_B.$$

Adding, we get

$$\Delta U_A + \Delta U_B = (Q_A + Q_B) + (W_A + W_B).$$

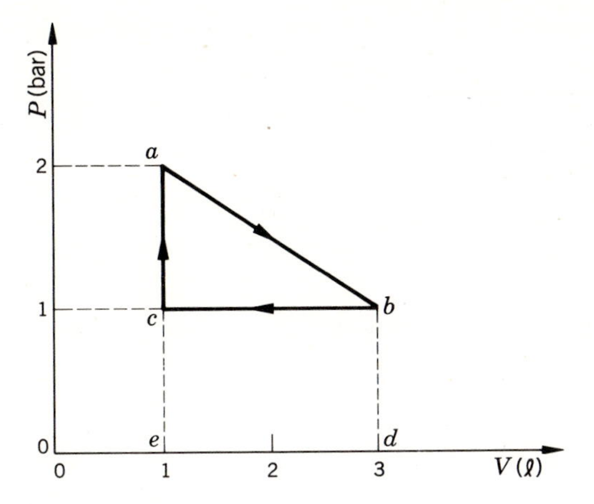

Fig. 4-7

But the composite system must also conform to the first law:

$$\Delta U = W,$$

where $Q = 0$ by the problem statement. Comparison of the last two equations gives

$$\Delta U_A + \Delta U_B = \Delta U,$$

$$W_A + W_B = W,$$

and

$$Q_A + Q_B = 0,$$

or

$$Q_A = -Q_B \qquad \text{(adiabatic)}.$$

Thus, under adiabatic conditions, the heat lost (or gained) by system A is equal to the heat gained (or lost) by system B. This equation finds frequent application to problems in calorimetry.

4-6 General Form of the First Law

To facilitate the introduction of the concept of internal energy, we implicitly limited consideration in the preceding sections to thermodynamic systems at rest. This allowed us to identify the adiabatic work with ΔU, and led logically to the introduction of heat via the statement of the first law, Eq. (4-4). We know from mechanics, however, that performance of work on a purely mechanical system (e.g., a mass point) may change both the *kinetic energy* E_K and the *gravitational potential energy* E_P of the system. A comprehensive energy-conservation principle must include in its statement these forms of energy, in addition to the internal energy. A more general mathematical statement of the first law for a constant-mass system is therefore given by the expression

$$\boxed{\Delta E = Q + W,} \tag{4-5}$$

where ΔE represents the sum of several terms:

$$\Delta E = \Delta E_K + \Delta E_P + \Delta U,$$

and hence Eq. (4-5) becomes

$$\Delta E_K + \Delta E_P + \Delta U = Q + W. \tag{4-6}$$

By definition, the kinetic energy is

$$E_K = \frac{mu^2}{2g_c}, \tag{4-7}$$

and the gravitational potential energy is

$$E_P = \frac{mzg}{g_c}, \tag{4-8}$$

where u is the velocity, z is the elevation above a datum level, and g is the acceleration of gravity. When the sum $E_K + E_P$ does not change, or when E_K and E_P are negligible, Eq. (4-4) is recovered from Eq. (4-6) as a special case. The treatment of systems encountered in engineering practice often requires application of Eq. (4-6). This equation is fully developed in Chap. 10.

Example 4-4

One kilogram of fluid is decelerated from $u = 1\text{(m)/(s)}$ to $u = 0\text{(m)/(s)}$ and undergoes a simultaneous decrease in elevation of 1(m). Determine ΔE_K and ΔE_P.

From Eq. (4-7),

$$\Delta E_K = \frac{m\Delta(u^2)}{2g_c} = \frac{1\text{(kg)} \times (0^2 - 1^2)\text{(m)}^2/\text{(s)}^2}{2 \times 1} = -0.5\text{(kg)(m)}^2/\text{(s)}^2,$$

or $\qquad \Delta E_K = -0.5\text{(N)(m)} = -0.5\text{(J)}.$

From Eq. (4-8), with $g = 9.807\text{(m)/(s)}^2$, the standard acceleration of gravity, we find

$$\Delta E_P = \frac{mg\,\Delta z}{g_c} = \frac{1\text{(kg)} \times 9.807\text{(m)/(s)}^2 \times [-1\text{(m)}]}{1} = -9.807\text{(kg)(m)}^2/\text{(s)}^2,$$

or $\Delta E_P = -9.807\text{(N)(m)} = -9.807\text{(J)}.$

For many processes, terms of this magnitude are entirely negligible with respect to the other quantities in the first-law expression.

4-7 Differential Form of the First Law

A process involving only infinitesimal changes in the thermodynamic coordinates of a system is known as an infinitesimal process. For such a process the first law, Eq. (4-4), becomes

$$\boxed{dU = \delta Q + \delta W.} \tag{4-9}$$

If the infinitesimal process is quasi-static, both dU and δW can be expressed in terms of thermodynamic coordinates only. An infinitesimal quasi-static process is one in which the system passes from an initial equilibrium state to a neighboring equilibrium state.

For an infinitesimal quasi-static process of a PVT system, the first law becomes

$$dU = \delta Q - P\,dV, \tag{4-10}$$

where U is a function of any two of the three thermodynamic coordinates, and P is, of course, a function of V and T. A similar equation may be written for each of the other simple systems, as shown in Table 4-2.

In the case of more complicated systems, it is merely necessary to replace δW in the first law by two or more expressions. For example, the operation of a reversible cell in which gases are liberated may involve work in changing not only its charge but also its volume. Hence

$$dU = \delta Q - P\,dV + e\,dq,$$

and U is a function of *three* of P, V, T, e, q. In the case of a magnetic gas,

$$dU = \delta Q - P\,dV + \mu_0 \mathcal{H}\,d\mathcal{M},$$

and U is a function of *three* of P, V, T, $\mathcal{H}$, $\mathcal{M}$. In the case of a liquid and

Table 4-2 The First Law for Simple Systems

System	First law	U is a function of *only two* of:
PVT system	$dU = \delta Q - P\,dV$	P, V, T
Stressed bar	$dU = \delta Q + V\sigma\,d\epsilon$	σ, ϵ, T
Surface	$dU = \delta Q + \gamma\,dA$	γ, A, T
Reversible cell	$dU = \delta Q + e\,dq$	e, q, T
Dielectric	$dU = \delta Q + \mathcal{E}\,d\mathcal{P}$	$\mathcal{E}$, $\mathcal{P}$, T
Magnetic material	$dU = \delta Q + \mu_0 \mathcal{H}\,d\mathcal{M}$	$\mathcal{H}$, $\mathcal{M}$, T

its surface,

$$dU = \delta Q - P\,dV + \gamma\,dA,$$

and U is a function of *three* of P, V, T, γ, A.

4-8 Enthalpy

It proves convenient for the analysis of certain types of processes to deal with an auxiliary thermodynamic function related to the internal energy, and called the *enthalpy* H.[1] For a PVT system the enthalpy is *by definition*

$$\boxed{H \equiv U + PV.} \tag{4-11}$$

Clearly, H has the same units as U, for example, joules. Differentiation of Eq. (4-11) yields

$$dH = dU + d(PV),$$

and combination with the differential form of the first law, Eq. (4-9), gives

$$dH = \delta Q + \delta W + d(PV). \tag{4-12}$$

Equation (4-12) is merely an *alternative form* of the first law and is as valid as Eq. (4-9), from which it was derived. The corresponding integral statements of the first law, analogous to Eqs. (4-4) and (4-6), are easily found.

The enthalpy, like the internal energy, is a *property* of a system, and it is a function of as many thermodynamic coordinates as are necessary to specify the state of a system. Thus, for a PVT system we may consider H a function of T and V and write for a differential change of state

$$dH = \left(\frac{\partial H}{\partial T}\right)_V dT + \left(\frac{\partial H}{\partial V}\right)_T dV. \tag{4-13}$$

Alternatively, we may regard H as a function of T and P, and write

$$dH = \left(\frac{\partial H}{\partial T}\right)_P dT + \left(\frac{\partial H}{\partial P}\right)_T dP. \tag{4-14}$$

Equations (4-13) and (4-14) are similar to Eqs. (4-2) and (4-3) for the internal energy.

Enthalpy functions may also be defined (see Sec. 11-1) for the other simple systems treated in Chap. 2.

[1] Pronounced *en thal′ pi.*

Example 4-5

The internal energy and enthalpy play analogous roles in the description of certain constant-volume and constant-pressure processes for PVT systems. We wish to develop this analogy.

For a *constant-volume* quasi-static process, Eq. (4-10) simplifies to

$$\delta Q = dU \qquad (\text{const } V). \tag{4-15a}$$

Integration yields

$$Q = \Delta U \qquad (\text{const } V). \tag{4-15b}$$

Thus, for a quasi-static constant-volume process the heat transferred is equal to the internal-energy change of the system.

For a quasi-static process, Eq. (4-12) becomes

$$dH = \delta Q - P\,dV + d(PV),$$

or

$$dH = \delta Q + V\,dP. \tag{4-16}$$

Equation (4-16) is equivalent to Eq. (4-10). Restriction to *constant pressure* gives

$$\delta Q = dH \qquad (\text{const } P), \tag{4-17a}$$

and integration yields

$$Q = \Delta H \qquad (\text{const } P). \tag{4-17b}$$

Thus, for a quasi-static constant-pressure process the heat transferred equals the enthalpy change of the system. Engineering processes often occur at constant pressure, and Eq. (4-17b) constitutes one justification for the introduction of H as a thermodynamic variable.

4-9 Heat Capacity

When heat is transferred to or from a system, a change of temperature may or may not take place, depending on the process. If a system undergoes a change of temperature from T_1 to T_2 during the transfer of Q units of heat, the *average heat capacity* of the system is defined as the ratio

$$\text{Average heat capacity} = \frac{Q}{T_2 - T_1}.$$

As both Q and $T_2 - T_1$ get smaller, this ratio approaches the *instantaneous heat capacity*, or heat capacity C, thus:

$$C = \lim_{T_2 \to T_1} \frac{Q}{T_2 - T_1},$$

or

$$\boxed{C = \frac{\delta Q}{dT}.} \tag{4-18}$$

The heat capacity may be negative, zero, positive, or infinite, depending on the process the system undergoes during the heat transfer. It has a

Table 4-3 Heat Capacities of Simple Systems

System	Heat capacity	Symbol
PVT system	At constant pressure	C_P
	At constant volume	C_V
Stressed bar	At constant stress	C_σ
	At constant strain	C_ϵ
Surface	At constant surface tension	C_γ
	At constant area	C_A
Reversible cell	At constant emf	C_e
	At constant charge	C_q
Dielectric system	At constant field strength	$C_\mathcal{E}$
	At constant polarization	$C_\mathcal{P}$
Magnetic system	At constant field strength	$C_\mathcal{H}$
	At constant magnetization	$C_\mathcal{M}$

definite value only for a definite process. In the case of a PVT system undergoing a quasi-static process, the ratio $\delta Q/dT$ has a unique value when the pressure is kept constant. Under these conditions, C is called the *heat capacity at constant pressure*, and is denoted by the symbol C_P, where

$$C_P = \left(\frac{\delta Q}{dT}\right)_P. \tag{4-19}$$

In general, C_P is a function of P and T (or V and T). Similarly, the heat capacity for a constant-volume quasi-static process is

$$C_V = \left(\frac{\delta Q}{dT}\right)_V, \tag{4-20}$$

and depends on both V and T (or P and T). In general, C_P and C_V are different. Both will be discussed thoroughly throughout the book. Each system has its own heat capacities, as shown in Table 4-3.

Each heat capacity is a function of two variables. Within a small range of variation of these coordinates, however, the heat capacity may sometimes be regarded as practically constant. Occasionally, one heat capacity can be set equal to another without much error. Thus the $C_\mathcal{H}$ of a paramagnetic solid is at times very nearly equal to C_P.

4-10 Notation

In this and earlier chapters we have not always stated specifically the amount of material in the system to which our symbols refer. For example, the first law as written for an infinitesimal quasi-static process of a PVT system was given as

$$dU = \delta Q - P\,dV. \tag{4-10}$$

In this equation U, Q, and V are *extensive quantities*, dependent upon the amount of the constant-mass system under consideration. The first-law equation as written could apply to a system of any amount, where U, Q, and V would represent total quantities for the entire system. However, our usual practice hereafter with respect to the thermodynamic *properties*, such as U and V, will be to let the capital-letter symbols represent *intensive quantities*, i.e., properties of a *unit mass* or a *mole* of the material making up the system. Intensive properties referred to a unit mass are called *specific* properties; when referred to a mole, they are called *molar* properties. If we wish to deal with a system of m(kg) or m(lb$_m$) or a system of n(mol), we will usually express the amount of material explicitly by writing mU or nU, mV, or nV, etc. In certain circumstances we will use a superscript t to denote extensive properties; for example, $U^t \equiv mU$ or nU, $V^t \equiv mV$ or nV, etc. Thus, for a PVT system of n(mol), the first law becomes, for a quasi-static process,

$$d(nU) = \delta Q - P\,d(nV),$$

or

$$dU^t = \delta Q - P\,dV^t,$$

where Q always represents the *total* heat transferred to the entire system, whatever its amount. The work W is treated in the same way as Q.

The symbol C for heat capacity will always mean either the molar heat capacity or the specific heat capacity ("specific heat"). By this convention, an expression relating C and Q will contain an explicit indication of the amount of substance. For example, if C_P is the *molar* heat capacity, then Eq. (4-19) is written

$$nC_P = \left(\frac{\delta Q}{dT}\right)_P,$$

where C_P has the units of energy per *mole* per unit temperature change. Similarly, if C_P is the *specific* heat capacity, then

$$mC_P = \left(\frac{\delta Q}{dT}\right)_P,$$

where C_P has the units of energy per *unit mass* per unit temperature change.

Example 4-6

A 2(kg) chunk of iron, initially at 75(°C), is immersed in 5(kg) of liquid water, initially at 25(°C). What is the final temperature of the composite system? For water, take C_P = constant = 4,180(J)/(kg)(K), and for iron, C_P = constant = 460(J)/(kg)(K).

Heat will be transferred from the iron to the water until the temperatures become equal. We assume the *overall* process to be adiabatic, so that the result of Example 4-3 applies:

$$Q_A = -Q_B, \tag{A}$$

where subscript A refers to the water and subscript B to the iron. We assume further that the process is isobaric and equivalent to a quasi-static process. Thus, by Eq. (4-19),

$$\delta Q_A = m_A C_{PA}\, dT_A$$

and

$$\delta Q_B = m_B C_{PB}\, dT_B,$$

where the m's are masses of the materials and the C_P's are *specific* heat capacities. Integration under the assumption of constant C_P's gives

$$Q_A = m_A C_{PA}(T_f - T_{Ai}) \tag{B}$$

and

$$Q_B = m_B C_{PB}(T_f - T_{Bi}), \tag{C}$$

where T_f is the final temperature of the composite system and T_{Ai} and T_{Bi} are the initial temperatures of A and B. Combination of Eqs. (A), (B), and (C), and solution for T_f yields

$$T_f = \frac{m_A C_{PA} T_{Ai} + m_B C_{PB} T_{Bi}}{m_A C_{PA} + m_B C_{PB}}.$$

Substitution of numerical values gives

$$T_f = \frac{(5)(4{,}180)(298.15) + (2)(460)(348.15)}{(5)(4{,}180) + (2)(460)},$$

or

$$T_f = 300.3(\text{K}); \qquad t_f = 27.1(°\text{C}).$$

Example 4-7

Derive expressions for C_V and C_P as temperature derivatives of U and H.

For an infinitesimal constant-volume quasi-static process, δQ is given by Eq. (4-15a). Division by dT in accordance with Eq. (4-20) yields

$$\boxed{C_V = \left(\frac{\partial U}{\partial T}\right)_V.} \tag{4-21}$$

For an infinitesimal isobaric quasi-static process, δQ is given by Eq. (4-17a). Division by dT gives, by Eq. (4-19),

$$\boxed{C_P = \left(\frac{\partial H}{\partial T}\right)_P.} \tag{4-22}$$

Equations (4-21) and (4-22) are the preferred working definitions for C_V and C_P, because they contain no reference to particular *processes*; they involve *properties* only. Moreover, these definitions permit more concise expression of formulas containing temperature derivatives of U and H. Thus the expression for dU given by Eq. (4-2) can be written in the equivalent form

$$dU = C_V\, dT + \left(\frac{\partial U}{\partial V}\right)_T dV. \tag{4-23}$$

Similarly, Eq. (4-14) for dH can be written

$$dH = C_P\, dT + \left(\frac{\partial H}{\partial P}\right)_T dP. \tag{4-24}$$

For quasi-static processes, we can therefore write, equivalent to Eqs. (4-15) and (4-16),

$$\delta Q = C_V\, dT \qquad (\text{const } V), \tag{4-25a}$$

$$Q = \int C_V\, dT \qquad (\text{const } V), \tag{4-25b}$$

$$\delta Q = C_P\, dT \qquad (\text{const } P), \tag{4-26a}$$

and

$$Q = \int C_P\, dT \qquad (\text{const } P). \tag{4-26b}$$

The last four equations also follow, of course, from the definitions given by Eqs. (4-19) and (4-20).

Example 4-8

Figure 4-8 depicts an idealized *throttling process.* Two nonconducting pistons, separated by a rigid porous plug, are situated in a thermally insulated cylinder. Initially (Fig. 4-8*a*), with the right-hand piston held flush against the plug, the space between the left-hand piston and the plug contains a total volume $V_1{}^t$ of gas at pressure P_1. Both pistons are now moved simultaneously and slowly in such a way that the constant pressure P_1 is maintained on the left side of the plug, and a constant (lower) pressure P_2 is maintained on the right side. As the pistons move, gas streams through the porous plug until the system reaches the final configuration of Fig. 4-8b, in which all of the

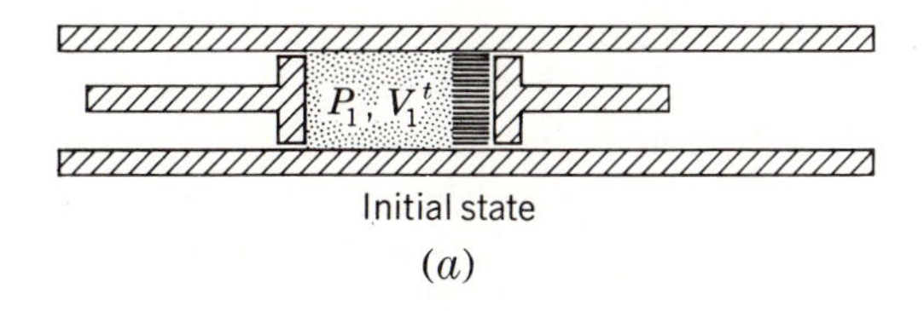

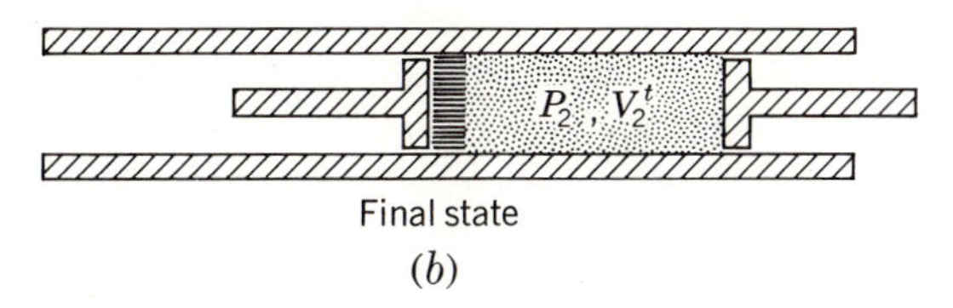

Fig. 4-8
Throttling process.

gas is at pressure P_2 and occupies a total volume V_2^t. We wish to determine the consequences of the first law for this process.

We use the integrated form of Eq. (4-12), applied to the gas in its initial and final states:

$$\Delta H^t = Q + W + \Delta(PV^t),$$

or

$$H_2^t - H_1^t = Q + W + P_2V_2^t - P_1V_1^t. \tag{A}$$

But $Q = 0$, and the work is given by

$$W = -\int_{V_1^t}^{0} P_1\,dV^t - \int_0^{V_2^t} P_2\,dV^t,$$

or

$$W = -(P_2V_2^t - P_1V_1^t).$$

Thus Eq. (A) reduces to

$$H_2^t - H_1^t = 0 - (P_2V_2^t - P_1V_1^t) + (P_2V_2^t - P_1V_1^t) = 0,$$

or

$$H_2^t = H_1^t.$$

The quantity of gas is the same in both states, so that we have, equivalently,

$$H_2 = H_1.$$

For a throttling process, therefore, *the initial and final enthalpies are the same.*

In practice, a partially opened valve or other constriction can serve as a throttling device. A continuous throttling process may be achieved by a pump that maintains a constant high pressure on one side of a constriction. The throttling process is considered again in Chap. 10.

4-11 Quasi-static Flow of Heat. Heat Reservoir

It was shown in Chap. 3 that a process caused by a finite unbalanced force is attended by phenomena such as turbulence and acceleration which cannot be handled by means of thermodynamic coordinates that refer to the system as a whole. A similar situation exists when there is a finite difference between the temperature of a system and that of its surroundings. A nonuniform temperature distribution is set up in the system, and the calculation of this distribution and its variation with time is in most cases an elaborate mathematical problem. During a quasi-static process, however, the difference between the temperature of a system and that of its surroundings is infinitesimal. As a result, the temperature of the system is at any moment uniform throughout, and its changes are infinitely slow. The flow of heat is also infinitely slow, and may be calculated in a simple manner in terms of thermodynamic coordinates referring to the system as a whole.

Suppose that a system is in good thermal contact with a body of extremely large mass and that a quasi-static process is performed. A finite amount of heat that flows during this process will not bring about an appreciable change in the temperature of the surrounding body if the mass of the surrounding body is large enough. For example, a cake of ice of ordinary size, thrown into the ocean, will not produce a drop in temperature of the ocean. No ordinary flow of heat into the outside air will produce an appreciable rise of temperature of the air. The ocean and the outside air are approximate examples of an ideal body called a heat reservoir. *A heat reservoir is a body of such a large mass that it may absorb or reject an unlimited quantity of heat without suffering an appreciable change in temperature or in any other thermodynamic coordinate.* It is not to be understood that there is *no* change in the thermodynamic coordinates of a heat reservoir when a finite amount of heat flows in or out. There is a change, but an extremely small one, too small to be measured.

Any quasi-static process of a system in contact with a given heat reservoir is bound to be isothermal. To describe a quasi-static flow of heat involving a change of temperature, one could conceive of a system placed in contact successively with a series of reservoirs. Thus, if we imagine a series of reservoirs ranging in temperature from T_1 to T_2 placed successively in contact with a system at constant pressure and with heat capacity $C_P{}^t$, in such a way that the difference in temperature between the system and the reservoir with which it is in contact is infinitesimal, the flow of heat will be quasi-static, and can be calculated as follows: By definition,

$$C_P{}^t = \left(\frac{\delta Q}{dT}\right)_P,$$

and therefore

$$Q = \int_{T_1}^{T_2} C_P{}^t \, dT.$$

For example, the heat that is absorbed by 1(g) of water from a series of reservoirs varying in temperature from T_1 to T_2 during a quasi-static isobaric process is given by this equation, and if C_P is assumed to remain practically constant,

$$Q = C_P(T_2 - T_1).$$

For a quasi-static isochoric process,

$$Q = \int_{T_1}^{T_2} C_V \, dT.$$

Similar considerations hold for other systems and other quasi-static processes.

Problems

4-1 A combustion experiment is performed by burning a mixture of fuel and oxygen in a constant-volume "bomb" surrounded by a water bath. During the experiment the temperature of the water is observed to rise. Regarding the mixture of fuel and oxygen as the system:

(*a*) Has heat been transferred?
(*b*) Has work been done?
(*c*) What is the sign of ΔU?

4-2 A liquid is irregularly stirred in a well-insulated container and undergoes a rise in temperature. Regarding the liquid as the system:

(*a*) Has heat been transferred?
(*b*) Has work been done?
(*c*) What is the sign of ΔU?

4-3 A vessel with rigid walls and covered with asbestos is divided into two parts by a partition. One part contains a gas, and the other is evacuated. If the partition is suddenly broken, show that the initial and final internal energies of the gas are equal.

4-4 A vessel with rigid walls and covered with asbestos is divided into two parts by an insulating partition. One part contains a gas at temperature T and pressure P. The other part contains a gas at temperature T' and pressure P'. The partition is removed. What conclusion may be drawn by applying the first law of thermodynamics?

4-5 A mixture of hydrogen and oxygen is enclosed in a rigid insulating container and exploded by a spark. The temperature and pressure both increase considerably. Neglecting the small amount of energy provided by the spark itself, what conclusion may be drawn by applying the first law of thermodynamics?

4-6 When a system is taken from state a to state b, in Fig. P4-6, along the path acb, 80(Btu) of heat flow into the system, and the system does 30(Btu) of work.

(*a*) How much heat flows into the system along path adb if the work done is 10(Btu)?
(*b*) When the system is returned from b to a along the curved path, the work done on the system is 20(Btu). Does the system absorb or liberate heat and how much?
(*c*) If $U_a = 0$ and $U_d = 40$(Btu), find the heat absorbed in the processes ad and db.

4-7 A scientist proposes to determine the heat capacities of liquids by use of a Joule calorimeter. In this device work is done by a paddle wheel on a liquid in an insulated container. The heat capacity is calculated from the measured temperature rise of the liquid and the measured work done by the paddle wheel. It is assumed that there is no heat exchanged between the liquid and its surroundings. To check this assumption, the scientist performs a preliminary experiment on 10(mol) of benzene, for which C_P is 31.8(cal)/(mol)(K). His data are as follows: work done by the paddle wheel = 1,500(cal); temperature rise of the liquid = 4(°C). If both C_P and the pressure on the liquid remain constant during the experiment, show that these results are not consistent with the stated assumptions, and offer an explanation for the inconsistency.

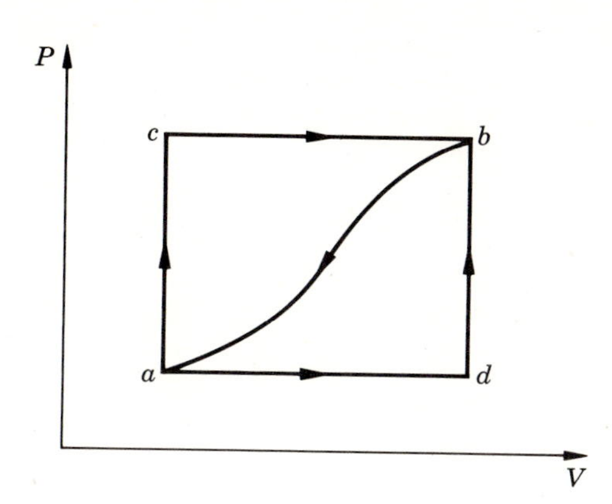

Fig. P4-6

4-8 An exhausted chamber with nonconducting walls is connected through a valve to the atmosphere, where the pressure is P_0. The valve is opened, and air flows into the chamber until the pressure within the chamber is P_0. Prove that $H_0 = U_f$, where H_0 is the molar enthalpy of the air at the temperature and pressure of the atmosphere, and U_f is the molar internal energy of the air in the chamber. (*Hint:* Connect to the chamber a cylinder equipped with a frictionless nonleaking piston. Suppose the cylinder to contain exactly the amount of air that will enter the chamber when the valve is opened. As soon as the first small quantity of air enters the chamber, the pressure in the cylinder is reduced a small amount below atmospheric pressure, and the outside air forces the piston in.)

4-9 A bomb of volume V^t contains n(mol) of gas at high pressure. Connected to the bomb is a capillary tube through which the gas may slowly leak out into the atmosphere, where the pressure is P_0. Surrounding the bomb and capillary is a water bath in which is immersed an electric resistor. The gas is allowed to leak slowly through the capillary into the atmosphere while, at the same time, electric energy is dissipated in the resistor at such a rate that the temperature of the gas, the bomb, the capillary, and the water is kept equal to that of the outside air. Show that after as much gas as possible has leaked out during time τ, the change of internal energy is

$$\Delta U^t = vI\tau - P_0(nV_0 - V^t),$$

where V_0 = molar volume of the gas at atmospheric pressure,
v = electric potential difference across the resistor,
I = current in the resistor.

4-10 A thick-walled insulated metal chamber contains n_1(mol) of helium at high pressure P_1. It is connected through a valve with a large, almost empty gasholder in which the pressure is maintained at a constant value P', very nearly atmospheric. The valve is opened slightly, and the helium flows slowly and adiabatically into the gasholder until the pressures on the two sides of the valve are equalized. Prove that

$$\frac{n_2}{n_1} = \frac{U_1 - H'}{U_2 - H'},$$

where n_2 = number of moles of helium left in the chamber,
U_1 = initial molar internal energy of helium in the chamber,
U_2 = final molar internal energy of helium in the chamber,
H' = molar enthalpy of helium in the gasholder.

4-11 Derive the following equations:

(a) For a PVT system,

$$C_P = \left(\frac{\partial U}{\partial T}\right)_P + PV\beta$$

(b) For a stressed bar,

$$C_\sigma = \left(\frac{\partial U}{\partial T}\right)_\sigma - V\sigma\alpha$$

(c) For a paramagnetic solid obeying Curie's equation, $\chi_m = C/T$,

$$C_{\mathcal{H}} = \left(\frac{\partial U}{\partial T}\right)_{\mathcal{H}} + \frac{\mu_0 \mathcal{M}^2}{VC}.$$

4-12 It has been suggested that the kitchen in your house could be cooled in the summer by closing the kitchen from the rest of the house and opening the door to the electric refrigerator. Comment on this. State clearly and concisely the basis for your conclusions.

4-13 One mole of a gas obeys van der Waals' equation of state

$$\left(P + \frac{a}{V^2}\right)(V - b) = RT,$$

and its internal energy is given by

$$U = cT - \frac{a}{V},$$

where a, b, c, and R are constants. Calculate the heat capacities C_V and C_P.

4-14 The equation of state of a monatomic solid is

$$PV + f(V) = \Gamma(U - U_0)$$

where V is the molar volume, and Γ and U_0 are constants. Prove that

$$\Gamma = \frac{\beta V}{C_V \kappa}$$

where κ is the isothermal compressibility. This relation, first derived by Gruneisen, plays a role in the theory of the solid state.

4-15 Starting with the first law for 1 (mol) of a PVT system, derive the equations:

(a) $$\delta Q = C_V\, dT + \left[\left(\frac{\partial U}{\partial V}\right)_T + P\right] dV \qquad \text{(quasi-static)}.$$

(b) $$C_P = C_V + \left[\left(\frac{\partial U}{\partial V}\right)_T + P\right] V\beta.$$

(c) $$\delta Q = C_V\, dT + \frac{C_P - C_V}{V\beta}\, dV \qquad \text{(quasi-static)}.$$

4-16 (*a*) Energy is supplied electrically at a constant rate to a thermally insulated substance. The heating curve has the form shown in Fig. P4-16*a*. Draw a rough graph showing the dependence of heat capacity on temperature.

(*b*) Repeat for a heating curve of the form shown in Fig. P4-16*b*.

4-17 The molar heat capacity at constant pressure of a gas varies with the temperature according to the equation

$$C_P = a + bT - \frac{c}{T^2},$$

where a, b, and c are constants. How much heat is transferred during an isobaric process in which n(mol) of gas undergo a temperature rise from T_1 to T_2?

4-18 The molar heat capacity at constant magnetic field of a paramagnetic solid at

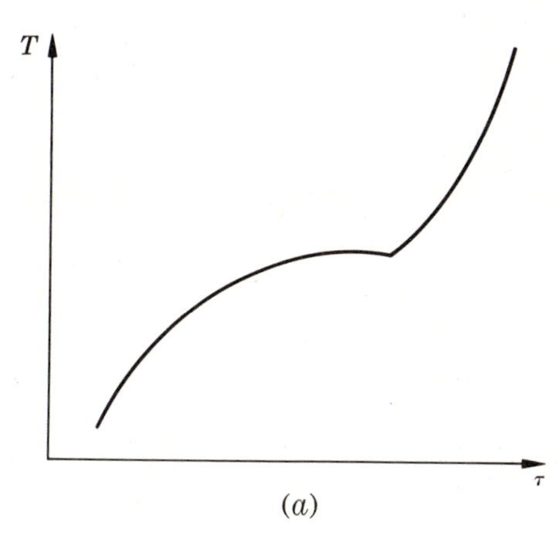

(*a*)

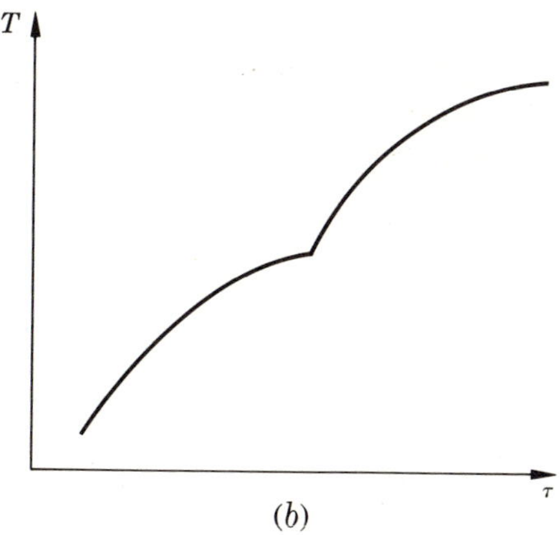

(*b*)

Fig. P4-16

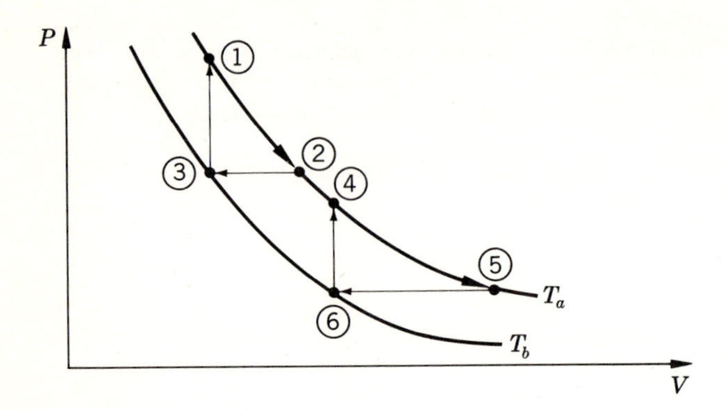

Fig. P4-21

low temperatures varies with temperature and field according to the relation

$$C_{\mathscr{H}} = \frac{b + c\mathscr{H}^2}{T^2} + DT^3,$$

where b, c, and D are constants. How much heat is transferred during a process in which $\mathscr{H}$ remains constant at the value of $\mathscr{H}_0$ and the temperature of n(mol) of material changes from T_1 to T_2?

4-19 A vapor expands quasi-statically in a piston and cylinder assembly. During the process the pressure and total volume are related as follows:

$P(\text{lb}_f)/(\text{in})^2$	$V^t(\text{ft})^3$
90	6.00
75	7.06
60	8.57
45	10.95
30	15.5
15	27.8

The internal energy change of the vapor is known to be $\Delta U^t = -75.2$(Btu). Calculate W in (ft)(lb$_f$) and Q in (Btu) for the process.

4-20 Starting with the first law for 1(mol) of a PVT system, derive the equations:

(a) $\delta Q = C_P\,dT + \left[\left(\frac{\partial H}{\partial P}\right)_T - V\right]dP$ (quasi-static).

(b) $C_P = C_V - \left[\left(\frac{\partial H}{\partial P}\right)_T - V\right]\frac{\beta}{\kappa}$.

(c) $\delta Q = C_P\,dT - \frac{\kappa}{\beta}(C_P - C_V)\,dP$ (quasi-static).

4-21 Figure P4-21 depicts two quasi-static processes undergone by 1 (mol) of an ideal gas. Curves T_a and T_b are isotherms, paths 23 and 56 are isobars, and paths 31 and 64 are isochores (paths of constant volume). Show that W and Q are the same for processes 1231 and 4564.

4-22 One mole of a gas undergoes a quasi-static process from state 1 to state 2. Show that if the process can be represented as a straight line on a PV diagram, then the work is given as

$$W = -\tfrac{1}{2}(P_1 + P_2)(V_2 - V_1).$$

5

THERMAL PROPERTIES OF GASES

The gas phase is only one of the many states of aggregation assumed by material substances. Preliminary to our discussion of thermal properties of gases, we consider in the first four sections of this chapter the phase behavior of pure substances as represented qualitatively by *phase diagrams.*

5-1 *PV* Diagram for a Pure Substance

If 1(g) of water at about 94(°C) is introduced into a vessel about 2(l) in volume from which all the air has been exhausted, the water will evaporate completely and the system will be in the condition known as *superheated vapor.* On the PV diagram shown in Fig. 5-1, this state is represented by a point such as A. If the vapor is then compressed isothermally, the pressure will rise until there is *saturated vapor* at the point B. If the compression is continued, condensation takes place, with the pressure remaining constant as long as the temperature remains constant. The straight line BD represents the isothermal isobaric condensation of water vapor; the constant pressure is called the *vapor pressure.* At any point between B and D, water and steam are in equilibrium; at point D, there is only *saturated liquid.* Since a very large increase of pressure is needed to compress liquid water, the line DE, which represents states for *compressed liquid,* is almost vertical. At any point on the line DE, the water is in the *liquid phase*; at any point on AB, in the *vapor phase*; and at any point on BD, there is equilibrium between the liquid and the vapor phases. $ABDE$ is a typical isotherm of a pure substance on a PV diagram.

At other temperatures the isotherms are of similar character, as shown by dashed lines in Fig. 5-1. It is seen that the lines representing equilibrium between liquid and vapor phases, or *vaporization lines,* get shorter as the temperature rises until a certain temperature is reached—the *critical*

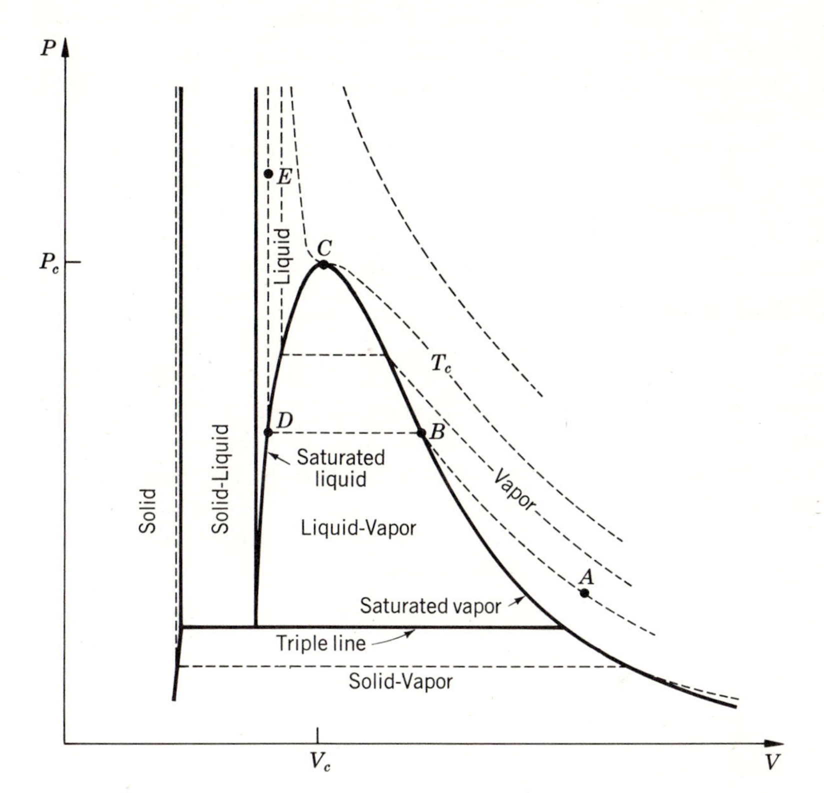

Fig. 5-1
PV diagram for a pure substance.

temperature T_c—above which there is no longer any distinction between a liquid and a vapor. The isotherm at the critical temperature is called the *critical isotherm*, and the point C that represents the limit of the vaporization lines is called the *critical point*. It is seen that the critical isotherm exhibits a horizontal inflection at the critical point. The pressure and volume at the critical point are known as the *critical pressure* P_c and the *critical volume* V_c. All points at which the liquid is saturated lie on the *liquid saturation curve*, and all points representing saturated vapor lie on the *vapor saturation curve*. The two saturation curves meet at the critical point C. Above the critical point the isotherms are continuous curves that at large volumes and low pressures approach the isotherms of an ideal gas.

The PV diagram of Fig. 5-1 also shows regions of solid-liquid and solid-vapor equilibrium. The isotherm shown traversing the solid-vapor region is similar in character to those at higher temperatures. The horizontal portion of this isotherm represents the transition from saturated solid to saturated vapor, or *sublimation.* There is obviously one such line that is the boundary between the liquid-vapor region and the solid-vapor region. This *line* is associated with the *triple point.* In the case of ordinary water, the triple point is at a pressure of 4.58(mm Hg) and a temperature of 0.01(°C), and the line extends from a volume of 1.00(cm)^3/(g) (saturated liquid) to a volume of 206,000(cm)^3/(g) (saturated vapor).

5-2 Critical State

The critical point is the state at which liquid and vapor phases in equilibrium become indistinguishable from one another, i.e., at which all of the properties of the two phases (density, enthalpy, etc.) become identical. For a pure substance the critical point is the highest temperature and pressure at which liquid and vapor phases can coexist in equilibrium. The horizontal inflection exhibited by the critical isotherm at the critical point implies the two mathematical conditions

$$\left(\frac{\partial P}{\partial V}\right)_{T,\text{cr}} = 0 \tag{5-1a}$$

and

$$\left(\frac{\partial^2 P}{\partial V^2}\right)_{T,\text{cr}} = 0, \tag{5-1b}$$

where subscript cr denotes the critical state.

To measure precise values of T_c, P_c, and V_c, one must determine exactly *when* the critical state is reached. For this purpose, a substance is raised in temperature slowly and uniformly in a sealed tube of constant volume. Depending upon the relative amounts of liquid and vapor initially in the tube, three types of behavior are observed. These are discussed with reference to the three constant-volume paths shown on Fig. 5-2. Point 1 represents an initial state of vapor-liquid equilibrium for which the meniscus separating the phases is initially near the top of the tube. As T is increased along path 12, the meniscus rises to the top of the tube as the last of the vapor condenses; the final state 2 is that of saturated liquid. If the initial loading is such that the meniscus is near the bottom of the tube (point 4), the system follows path 45, during which the meniscus falls to the bottom of the tube as the liquid evaporates; the final state 5 is saturated vapor. For an initial loading such that $V = V_c$, the specific or molar critical volume (point 3), the level of the meniscus changes very little when the

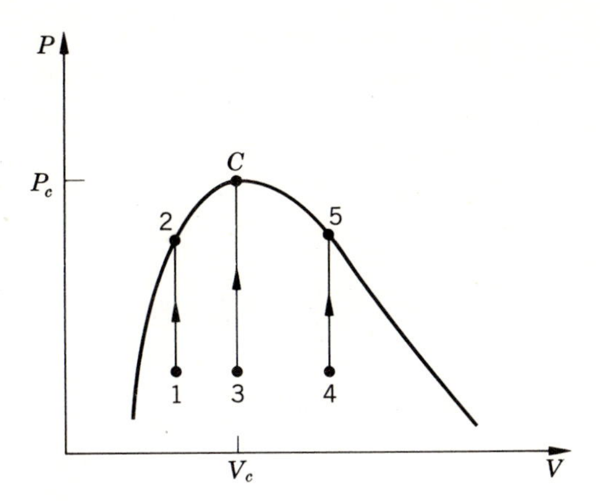

Fig. 5-2

temperature is increased along path $3C$. As the critical point is approached, the meniscus becomes less distinct, then hazy, and finally disappears. The temperature and pressure at which the meniscus vanishes are T_c and P_c.

At all points within the two-phase vapor-liquid region, including the limiting case of the critical state, the three physical properties, β, κ, and C_P, are *infinite*. These factors conspire to make difficult the actual observation of the critical point:

1 Since κ is infinite, the gravitational field of the earth causes large density gradients from the top to the bottom of the tube. This effect may be overcome somewhat if the tube is placed in a horizontal position.
2 Since C_P is infinite, thermal equilibrium is difficult to achieve. One must keep the system at a constant temperature [within $\pm 10^{-2}$ or even $\pm 10^{-3}$(K)] for a long time, and stir constantly.
3 Since β is infinite, small local temperature variations within the system produce large density fluctuations, which give rise to light scattering, and cause the material to become almost opaque. This phenomenon is called *critical opalescence.*

In spite of the difficulty of experimental determination of the critical state, critical data have been measured for many materials. Values of T_c, P_c, and V_c for selected substances are given in Appendix C.

5-3 *PT* Diagram for a Pure Substance

If the vapor pressure of a solid is measured at various temperatures until the triple point is reached and then that of the liquid is measured until the

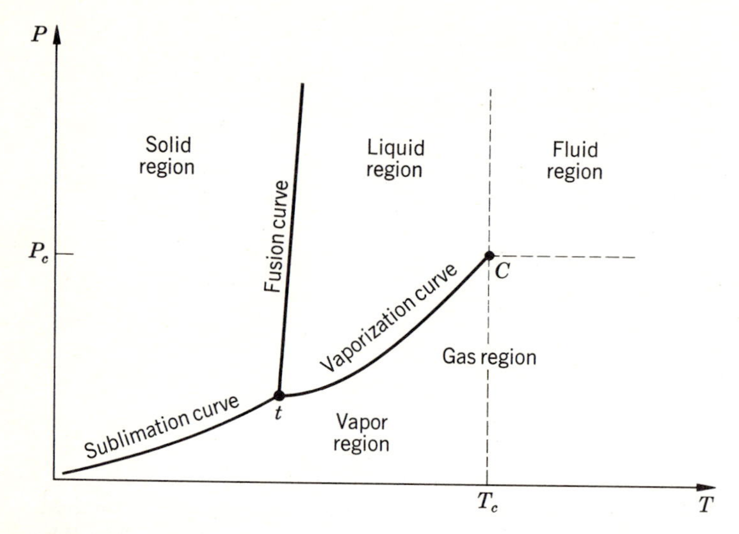

Fig. 5-3
PT diagram for a substance which expands on melting.

critical point is reached, the results when plotted on a PT diagram appear as the two lower lines on Fig. 5-3. Point t is the triple point and point C is the critical point. If the substance at the triple point is compressed until there is no vapor left and the pressure on the resulting mixture of liquid and solid is increased, the temperature must be changed if equilibrium is to continue to exist between the solid and the liquid. Measurements of these pressures and temperatures give rise to a third curve on the PT diagram, starting at the triple point and continuing indefinitely. The experimental data define three curves:

1. The points representing the coexistence of solid and vapor lie on the *sublimation curve.*
2. Those representing the coexistence of liquid and vapor lie on the *vaporization curve.*
3. Those representing the coexistence of liquid and solid lie on the *fusion curve.*

In the particular case of water, the sublimation curve is called *the frost line,* the vaporization curve is called the *steam line,* and the fusion curve is called the *ice line.*

The slopes of the sublimation and the vaporization curves for all substances are positive. The slope of the fusion curve, however, may be positive or negative, but for most substances it is positive. Water is one of the important exceptions. Any substance, such as water, which contracts upon

Table 5-1 Triple-point Data

Substance	T(K)	P(mbar)
Helium 4 (λ point)	2.172	50.40
Hydrogen (normal)	13.84	70.4
Deuterium (normal)	18.63	171.
Neon	24.57	432.
Oxygen	54.36	1.52
Nitrogen	63.18	125.
Ammonia	195.40	60.75
Sulfur dioxide	197.68	1.675
Carbon dioxide	216.55	5,170.
Water	273.16	6.105

melting has a fusion curve with a negative slope, whereas the opposite is true for a substance which expands upon melting. The triple point t is the point of intersection of the sublimation, fusion, and vaporization curves. Only on a PT diagram is the triple point represented by a point. On a PV diagram it is a line. Triple-point data for a number of substances are given in Table 5-1.

The states of *single-phase* equilibrium define *areas* on a PT diagram. These areas are bordered or bounded by the various saturation curves which are the loci of states representing two-phase equilibrium. Thus, the solid region on Fig. 5-3 is the area above the sublimation curve and to the left of the fusion curve. The classification of regions is somewhat arbitrary for phases bordered by the vaporization curve, which terminates at the critical point C. Conventionally, the liquid region is defined as that area bordered by the fusion curve, the vaporization curve, and the critical isotherm (the vertical dashed line on Fig. 5-3). The vapor region is bordered by the sublimation curve, the vaporization curve, and the critical isotherm; thus a vapor can be condensed either by a decrease of temperature or an increase of pressure. The gas region includes, in addition to the vapor region, those states for which $T > T_c$ but for which $P < P_c$. A gas may be condensed by a decrease in temperature but not necessarily by an increase in pressure. For $T > T_c$ *and* $P > P_c$, no clear distinction can be drawn between gas and liquid, and we call this region the *fluid* region.[1]

Example 5-1

Show how it is possible to change a vapor into a liquid without condensation.

When a vapor condenses it undergoes an abrupt change of phase, characterized by

[1] The word *fluid* is commonly used to designate collectively *all* nonsolid phases.

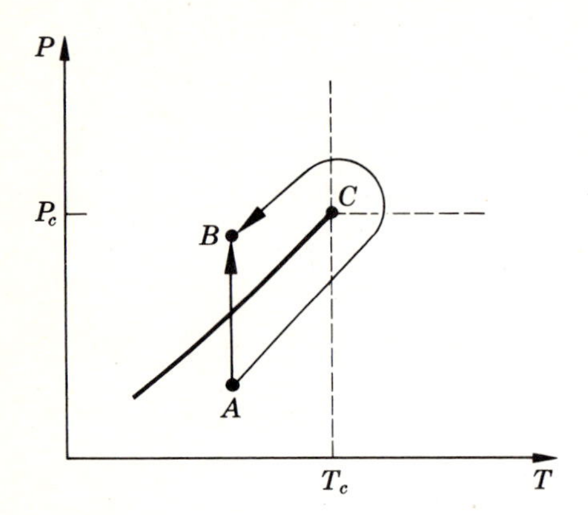

Fig. 5-4

the appearance of a meniscus. This can only occur when the vaporization curve of Fig. 5-3 is crossed. Figure 5-4 shows two points A and B on either side of the vaporization curve. Point A represents a vapor state and point B a liquid state. If the pressure on the vapor at A is raised at constant T, then the vertical path from A to B is followed and this path crosses the vaporization curve, at which point condensation occurs at a fixed pressure. However, it is also possible to follow a path from A to B that goes around the critical point C and which then does not cross the vaporization curve, as can be seen in Fig. 5-4. When this path is followed, the transition from vapor to liquid is gradual, and at no point is an abrupt change in properties observed.

Example 5-2

At ambient conditions [about 1(atm) and 295(K)], solid carbon dioxide (dry ice) does not melt but sublimes to give CO_2 vapor. Explain this behavior.

From Table 5-1, we see that the triple-point pressure of CO_2 is 5.17(bar) = 5.10(atm), which is higher than ambient pressure. Reference to the PT diagram of Fig. 5-3 shows that the relevant phase transition below the triple-point pressure is sublimation. The sublimation temperature at 1(atm) is lower than the triple-point temperature of 216.55(K), which is lower than ambient temperature. Therefore dry ice sublimes when exposed to ambient conditions. It can be made to *melt* only if compressed to a pressure equal to or greater than its triple-point pressure.

5-4 *PVT* Surface

All the information contained on both the PV and the PT diagrams can be shown on a single figure if the three coordinates, P, V, and T, are plotted in three-dimensional space. The result is a PVT *surface*. Such a surface is shown in Fig. 5-5 for a substance like carbon dioxide which expands upon melting.

If the PVT surface is projected on the PV plane, a PV diagram, similar

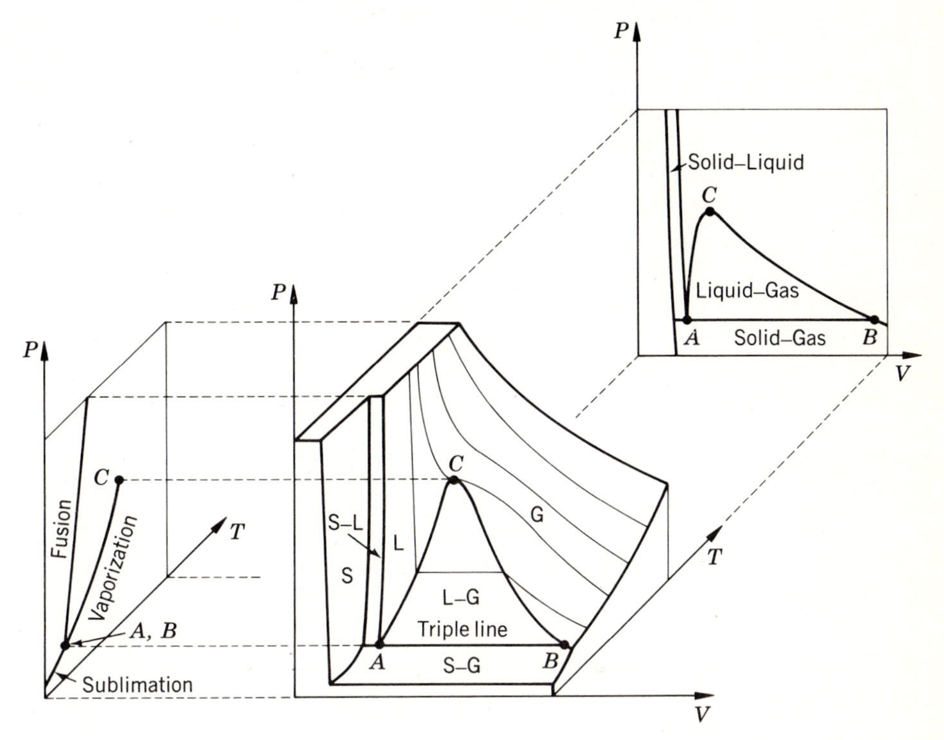

Fig. 5-5
PVT surface for a substance which expands on melting.

to Fig. 5-1, is obtained as shown. When projected onto the *PT* plane, the surface generates a *PT* diagram like Fig. 5-3. In the latter construction, the whole solid-vapor region projects into the sublimation curve, the whole liquid-vapor region projects into the vaporization curve, the whole solid-liquid region projects into the fusion curve, and the *triple line* projects into the triple point.

5-5 Equation of State of a Gas

It was emphasized in Chap. 1 that a gas is the most satisfactory standard thermometric substance because of the fact that the ratio of the pressure P of a gas at any temperature to the pressure P_t of the same gas at the triple point, as both P and P_t approach zero at the same constant volume, approaches a value independent of the nature of the gas. The limiting value of this ratio, multiplied by 273.16, was defined to be the ideal-gas temperature T. We now consider the behavior of the product PV for a gas as a function of P at constant T.

Suppose that the pressure P and the molar volume V of a gas held at any constant temperature are measured over a wide range of values of the pressure, and the product PV is plotted as a function of P. A set of such isotherms for nitrogen is shown in Fig. 5-6. The relation between PV and P for a given isotherm may be expressed by means of a power series of the form

$$PV = A(1 + B'P + C'P^2 + D'P^3 + \cdots), \tag{5-2}$$

where A, B', C', etc., depend on the temperature and on the nature of the gas. It should be noticed in Fig. 5-6 that for nitrogen in the pressure range from 0 to about 40(atm), the relation between PV and P is practically linear; thus only the first two terms in the series are significant. In general, the greater the pressure range, the larger the number of terms required.

The remarkable property of gases that makes them so valuable in

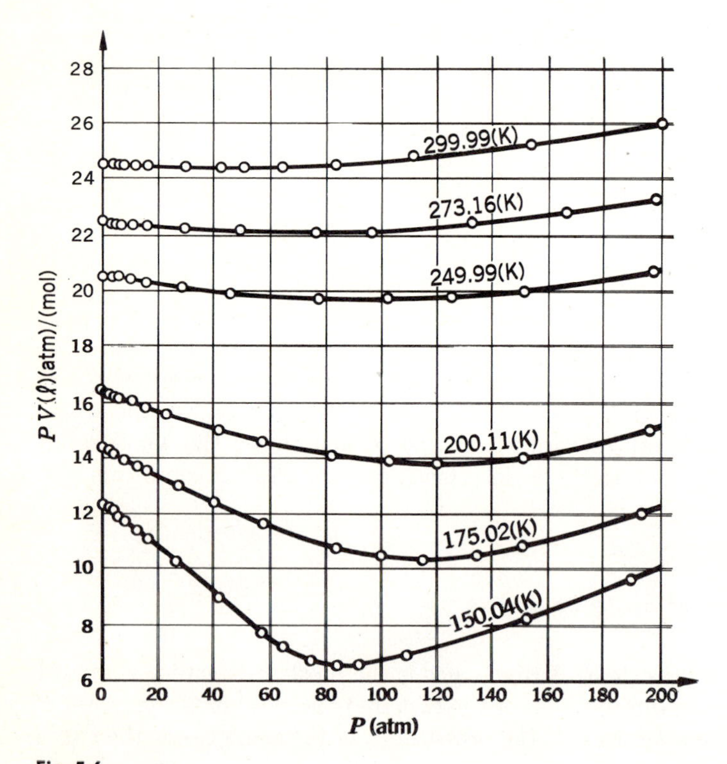

Fig. 5-6

Variation of PV of nitrogen with pressure at constant temperature. (A. S. Friedman et al., Ohio State University Cryogenic Laboratory, 1950.)

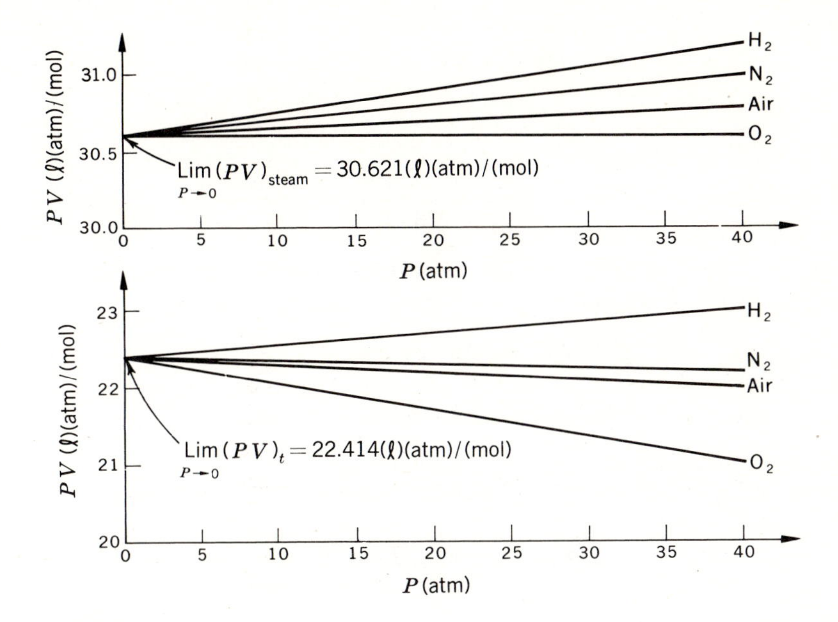

Fig. 5-7
Fundamental property of gases: $\lim(PV)_T$ *is independent of the nature of the gas and depends only on* T.

thermometry is displayed in Fig. 5-7, where the product PV is plotted against P for four different gases at two different temperature levels. The upper graph is for the normal boiling point of water, and the lower graph is for the triple point of water. In both cases it is seen that as P approaches zero the PV product approaches a single value, dependent only upon temperature, for all gases. This behavior is observed at all temperature levels. It therefore follows from Eq. (5-2) that

$$\lim_{P\to 0}\,(PV) = A = \text{function of } T \text{ only, independent of gas.} \qquad (5\text{-}3)$$

The ideal-gas temperature T is defined by Eq. (1-4):

$$T = 273.16 \lim \frac{P}{P_t} \qquad (\text{const } V). \qquad (1\text{-}4)$$

where the indication of a limit (lim) signifies that *both* P and P_t approach zero. Thus we can write

$$T = 273.16 \lim \frac{PV}{P_t V} = 273.16\, \frac{\lim\,(PV)}{\lim\,(PV)_t},$$

from which

$$\lim_{P \to 0} (PV) = RT, \tag{5-4}$$

where

$$R \equiv \frac{\lim (PV)_t}{273.16} \tag{5-5}$$

The quantity R is called the *universal gas constant.* The currently accepted value for $\lim (PV)_t$ is 22.4144(l) (atm)/(mol); then by Eq. (5-5)

$$R = \begin{cases} 0.08206(\text{l})\,(\text{atm})/(\text{mol})\,(\text{K}), \\ 8.314(\text{J})/(\text{mol})\,(\text{K}). \end{cases}$$

Values of R in other units are given in Appendix B.

According to Eqs. (5-3) and (5-4), $A = RT$; therefore Eq. (5-2) becomes,

$$\frac{PV}{RT} = 1 + B'P + C'P^2 + D'P^3 + \cdots$$

The ratio PV/RT is called the *compressibility factor* and is denoted by Z:

$$Z \equiv \frac{PV}{RT}. \tag{5-6}$$

Thus, the equation of state may be written

$$\boxed{Z = 1 + B'P + C'P^2 + D'P^3 + \cdots} \tag{5-7}$$

An alternative expression for Z which is also in common use is

$$\boxed{Z = 1 + \frac{B}{V} + \frac{C}{V^2} + \frac{D}{V^3} + \cdots.} \tag{5-8}$$

Both these equations are known as *virial expansions,* and the coefficients $B', C', D', \ldots$ and $B, C, D, \ldots$ are called virial coefficients. B' and B are termed second virial coefficients; C' and C, third virial coefficients; etc. For a pure substance or a constant-composition mixture these coefficients are functions of temperature only.

Many other equations of state have been proposed for gases, but the virial equations are the only ones having a firm basis in theory. The methods of statistical mechanics allow the derivation of the virial equations and provide physical significance to the virial coefficients. Thus, for the expansion in $1/V$, the term B/V arises on account of interactions between pairs of molecules; the C/V^2 term, on account of three-body interactions; etc. Since two-body interactions are many times more common than three-body interactions, and three-body interactions are many times more common than four-body interactions, etc., the contributions to Z of successively higher-ordered terms fall off rapidly.

The coefficients of the expansion in pressure are related to the coefficients of the expansion in density $(1/V)$ as follows:

$$B' = \frac{B}{RT},$$

$$C' = \frac{C - B^2}{(RT)^2}, \tag{5-9}$$

$$D' = \frac{D - 3BC + 2B^3}{(RT)^3},$$

etc.

These relations are obtained by elimination of P on the right-hand side of the expansion in pressure through use of the expansion in $1/V$. The resulting equation represents a power series in $1/V$ which may be compared term by term with the original virial expansion in $1/V$. This comparison provides equations which may be reduced to the above set.

Although it is the expansion in density, or $1/V$, which results directly as a consequence of statistical mechanics, nevertheless the expansion in pressure finds widespread use. As we have already seen, isotherms of PV, and hence of Z, are nearly linear at low pressure. Thus, for low pressures, a very satisfactory equation of state is

$$Z = 1 + B'P,$$

or

$$\boxed{Z = 1 + \frac{BP}{RT}}. \tag{5-10}$$

At elevated temperatures this linear relationship is a good approximation even to high pressures. This is illustrated in Fig. 5-8, which shows the compressibility factor for hydrogen plotted against pressure at a number of temperatures. When a pressure is reached above which Eq. (5-10) becomes invalid, more terms must be retained in the virial equations. The preferred three-term virial equation is the truncated version of Eq. (5-8):

$$\boxed{Z = 1 + \frac{B}{V} + \frac{C}{V^2}}. \tag{5-11}$$

The experimental determination of accurate values for virial coefficients is

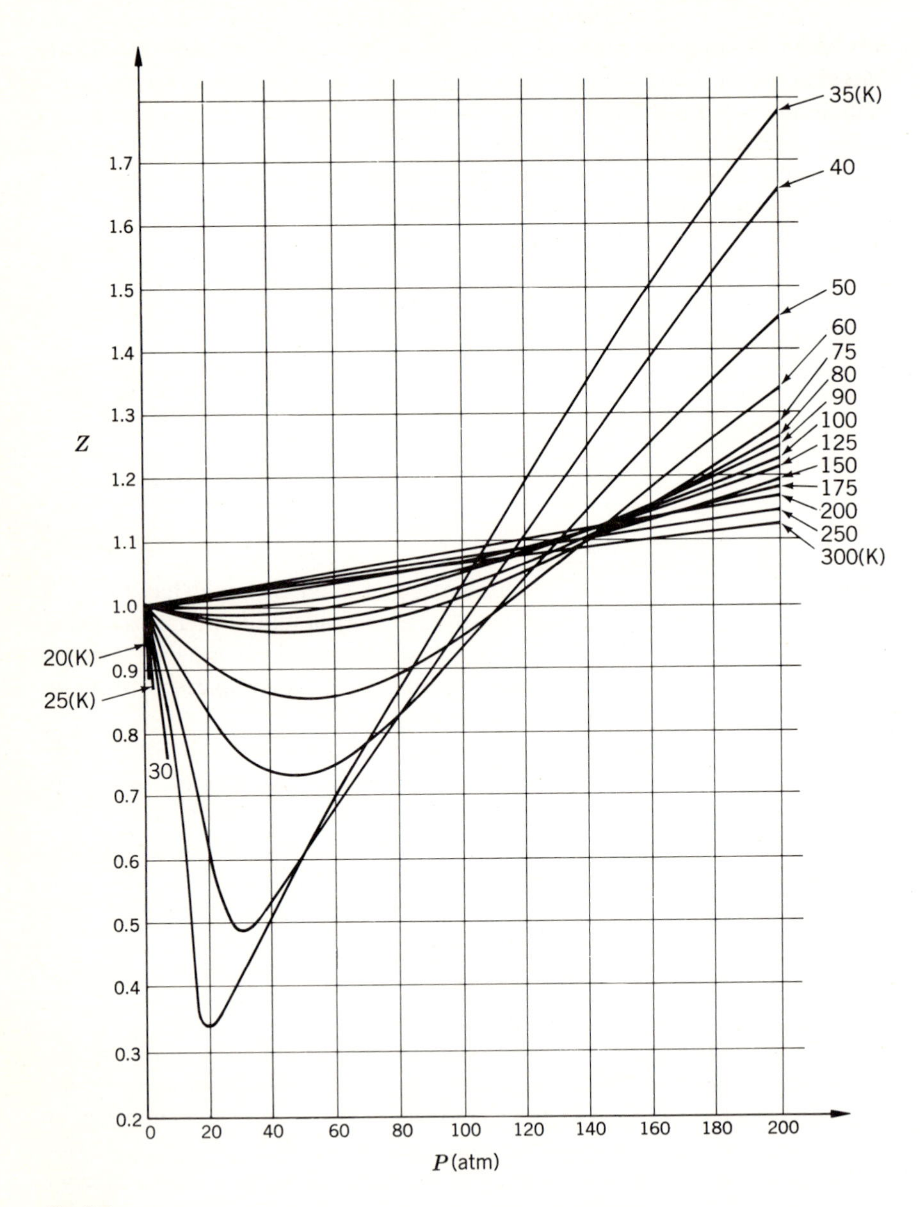

Fig. 5-8

Variation of Z of hydrogen with pressure at constant temperature. (A. S. Friedman et al., Ohio State University Cryogenic Laboratory, 1950.)

very difficult, and reliable data on coefficients beyond the third are rare. Thus Eq. (5-11) represents the usual limit to use of the virial expansion in engineering practice.

The nth virial coefficient in the pressure series, Eq. (5-7), is usually expressed in units of $(\text{bar})^{1-n}$ or $(\text{atm})^{1-n}$; the nth virial coefficient in the volume series, Eq. (5-8), is almost always reported in units of $[(\text{cm})^3/(\text{mol})]^{n-1}$. In this text, unless noted otherwise, we mean the coefficients in Eq. (5-8) when we speak of virial coefficients.

Virial coefficient data for many gases have been compiled by J. H. Dymond and E. B. Smith, "The Virial Coefficients of Gases: A Critical Compilation," Oxford University Press, New York, 1968. Values may also be found in handbooks, e.g., "American Institute of Physics Handbook", 3d ed., Sec. 4, McGraw-Hill, New York, 1972. In Sec. 9-6 we describe a simple method for estimation of second virial coefficients.

Example 5-3

Estimate the volume occupied by 2(mol) of methane gas at 0(°C) and 10(atm) pressure. At 0(°C), the second virial coefficient B for methane is $-53.4(\text{cm})^3/(\text{mol})$.

Since a value for B only is given, we use the simplest form of the virial equation, Eq. (5-10). This equation may be written

$$Z = \frac{PV}{RT} = \frac{P(nV)}{nRT} = \frac{PV^t}{nRT} = 1 + \frac{BP}{RT},$$

where V^t is the *total* volume. Solution for V^t gives

$$V^t = \frac{nRT}{P} + nB,$$

and substitution of numerical values yields

$$V^t = \frac{2(\text{mol}) \times 82.06(\text{cm})^3(\text{atm})/(\text{mol})(\text{K}) \times 273.15(\text{K})}{10(\text{atm})} - 2(\text{mol}) \times 53.4(\text{cm})^3/(\text{mol}),$$

or $$V^t = 4{,}483 - 106.8 = 4{,}376(\text{cm})^3 = 4.376(\text{l}).$$

Example 5-4

We wish to determine the relative contributions of successive terms in the virial expansion in volume, Eq. (5-8). For purposes of illustration, we choose nitrogen vapor at 120(K), and apply Eq. (5-8) at four pressures: 0.1, 1, 10, and 20(bar). Virial coefficients for nitrogen at 120(K) are: $B = -114.6(\text{cm})^3/(\text{mol})$, $C = 4.8 \times 10^3(\text{cm})^6/(\text{mol})^2$, and $D = -2.7 \times 10^5(\text{cm})^9/(\text{mol})^3$.

Comparison of the terms in Eq. (5-8) requires values for the molar volume V; these must be found by solution of the equation

$$\frac{PV}{RT} = 1 + \frac{B}{V} + \frac{C}{V^2} + \frac{D}{V^3}, \tag{A}$$

Table 5-2

P(bar)	V(cm)^3/(mol)	Z	$= 1 +$	B/V	$+$	C/V^2	$+$	D/V^3
0	∞	1.0000	$= 1 -$	0	$+$	0	$-$	0
0.1	99,650	0.9989	$= 1 -$	0.0011	$+$	0.0000	$-$	0.0000
1.0	9861	0.9884	$= 1 -$	0.0116	$+$	0.0000	$-$	0.0000
10.0	872.5	0.8746	$= 1 -$	0.1313	$+$	0.0063	$-$	0.0004
20.0	353.1	0.7078	$= 1 -$	0.3246	$+$	0.0385	$-$	0.0061

which is quartic in V. An iterative solution is therefore required at each pressure. Initial estimates for V can be found from the equation (see Example 5-3):

$$V = \frac{RT}{P} + B.$$

Thus, for $P = 10$(bar), we find

$$V \approx \frac{83.14(\text{cm})^3(\text{bar})/(\text{mol})(\text{K}) \times 120(\text{K})}{10(\text{bar})} - 114.6(\text{cm})^3/(\text{mol}),$$

or $\qquad V \approx 883.1(\text{cm})^3/(\text{mol}).$

Substitution into Eq. (A) yields, with the given values of B, C, and D,

$$\text{Left-hand side of Eq. } (A) = 0.8852;$$

$$\text{Right-hand side of Eq. } (A) = 0.8760.$$

The initial value of V is evidently too large. Choosing successively smaller values of V, we find finally that Eq. (A) is satisfied when $V = 872.5(\text{cm})^3/(\text{mol})$, from which $Z = PV/RT = 0.8746$.

Repeating the calculation for the other pressures, we obtain the values shown in the second and third columns of Table 5-2. The last three columns list the contributions of the separate terms for all four pressures, and also for the limiting zero-pressure case. Up to a pressure of 10(bar), only the term containing B is significant; at higher pressures, more terms are needed. The numerical values of the virial coefficients for successive terms increase rapidly; however, they are coefficients of increasing powers of $1/V$. Since V is a large number, the contributions of the successive terms decrease.

The results of this example are typical; however, the precise pressure below which a given degree of truncation of Eq. (5-8) is acceptable depends in general both on the temperature level and on the gas being considered.

5-6 Internal Energy of a Gas

Imagine a thermally insulated vessel with rigid walls divided into two compartments by a partition. Suppose that there is a gas in one compartment and that the other is empty. If the partition is removed, the gas will undergo what is known as a *free expansion* in which no work is done and no heat is transferred. From the first law, since both Q and W are zero, it follows that *the internal energy remains unchanged during a free expansion.* The question as to whether or not the temperature of a gas changes during a free expansion, and, if it does, the magnitude of the temperature change, has engaged the attention of scientists for about a hundred years. Starting with Joule in 1843, many attempts have been made to measure either the quantity $(\partial T/\partial V)_U$, which is called the *Joule coefficient,* or related quantities, all of which are a measure in one way or another of the effect of a free expansion, or as it is often called, the *Joule effect.*

In general, the energy of a gas is a function of any two of the coordinates P, V, and T. Considering U as a function of T and V, we have by Eq. (4-2)

$$dU = \left(\frac{\partial U}{\partial T}\right)_V dT + \left(\frac{\partial U}{\partial V}\right)_T dV. \tag{4-2}$$

If no temperature change $(dT = 0)$ takes place in a free expansion $(dU = 0)$, it follows that

$$\left(\frac{\partial U}{\partial V}\right)_T = 0,$$

or in other words, U does not depend on V. Considering U to be a function of T and P, we have by Eq. (4-3)

$$dU = \left(\frac{\partial U}{\partial T}\right)_P dT + \left(\frac{\partial U}{\partial P}\right)_T dP. \tag{4-3}$$

If no temperature change $(dT = 0)$ takes place in a free expansion $(dU = 0)$, it follows that

$$\left(\frac{\partial U}{\partial P}\right)_T = 0,$$

or in other words, U does not depend on P. It is apparent then that if no temperature change takes place in a free expansion, U is independent of V and of P, and therefore *U is a function of T only.*

Because of inherent experimental difficulties, no satisfactory direct measurement of the temperature change associated with a free expansion has ever been made. However, alternative measurements which involve the *isothermal* expansion of a gas and which yield values of $(\partial U/\partial P)_T$ have been successfully carried out. The most extensive series of measurements of this kind were reported by Rossini and Frandsen.[1] They were made at the National Bureau of Standards with an apparatus described by Washburn,[2]

[1] F. D. Rossini and M. Frandsen, *J. Research Nat. Bur. Standards,* **9**:733 (1932).
[2] E. W. Washburn, *J. Research Nat. Bur. Standards,* **9**:521 (1932).

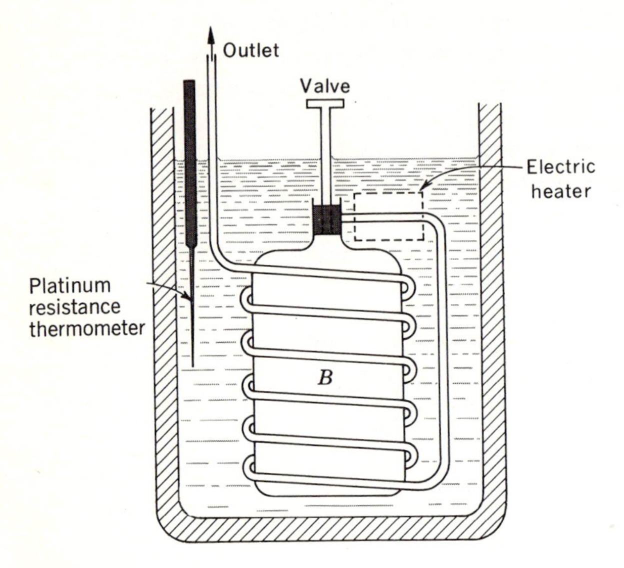

Fig. 5-9
Apparatus of Rossini and Frandsen for measuring $(\partial U/\partial P)_T$ of a gas.

which is shown in Fig. 5-9. A vessel B contains n(mol) of gas at a pressure P and communicates with the atmosphere through a long coil wrapped around the vessel. The whole apparatus is immersed in a water bath whose temperature can be maintained constant at exactly the same value as that of the surrounding atmosphere.

The experiment is performed as follows: When the valve is opened slightly, the gas flows slowly through the long coil and out into the air until the pressures equalize. At the same time, the temperature of the gas, the vessel, the coils, and the water is maintained constant by an electric heating coil immersed in the water. The electric energy supplied to the water is therefore the heat Q absorbed by the gas during the expansion. The work done by the gas during the entire process which reduces the pressure in the vessel to P_0 is

$$W = -P_0(nV_0 - V_B),$$

where P_0 = atmospheric pressure,
V_0 = molar volume at atmospheric temperature and pressure,
V_B = total volume of the vessel.

If $U(P,T)$ is the molar energy at pressure P and temperature T and $U(P_0,T)$ is the molar energy at atmospheric pressure and the same temperature, then, from the first law,

$$U(P,T) - U(P_0,T) = \frac{-(Q + W)}{n},$$

provided that corrections are made to take account of the energy changes due to the contraction of the walls of the vessel. In this way, the energy change was measured for various values of the initial pressure and was plotted against this pressure, as shown in Fig. 5-10. Since $U(P_0,T)$ is constant, it follows that the slope of the resulting curve at any value of P is equal to $(\partial U/\partial P)_T$. Within the pressure range of 1 to 40(atm), it is seen that $(\partial U/\partial P)_T$ *is independent of the pressure,* depending only on the temperature. Thus

$$\left(\frac{\partial U}{\partial P}\right)_T = f(T)$$

and

$$U = f(T)P + F(T),$$

where $F(T)$ is another function of the temperature only.

Rossini and Frandsen's experiments with air, oxygen, and mixtures of oxygen and carbon dioxide led to the conclusion that the internal energy of a gas is a function of both temperature and pressure. They found no pressure or temperature range in which the quantity $(\partial U/\partial P)_T$ was equal to zero.

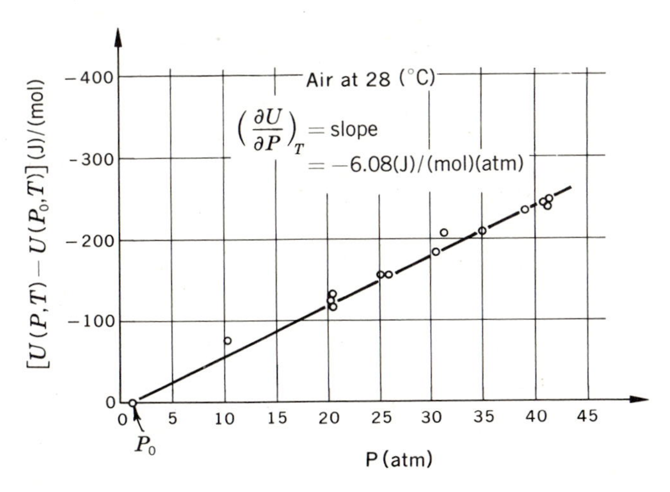

Fig. 5-10
Pressure dependence of U for air.

The principal source of experimental error in these measurements arises from the fact that the heat capacity of the gas is much smaller than that of the calorimeter and water bath. Thus, if Q is to represent with reasonable accuracy the heat transferred *to the gas*, then the water temperature must be held constant to within less than a thousandth of a degree Celsius. The final precision of Rossini and Frandsen's measurements was estimated to be $2\frac{1}{2}$ percent. The few additional measurements made since have substantiated their results.

Example 5-5

Show how the Joule coefficient $(\partial T/\partial V)_U$ is related to the pressure derivative $(\partial U/\partial P)_T$.

Application of Eq. (2-9) gives

$$\left(\frac{\partial T}{\partial V}\right)_U = -\left(\frac{\partial T}{\partial U}\right)_V\left(\frac{\partial U}{\partial V}\right)_T.$$

But, by Eq. (4-21),

$$\left(\frac{\partial T}{\partial U}\right)_V = \frac{1}{C_V},$$

and, by Eq. (2-10),

$$\left(\frac{\partial U}{\partial V}\right)_T = \left(\frac{\partial U}{\partial P}\right)_T\left(\frac{\partial P}{\partial V}\right)_T.$$

Thus we obtain finally

$$\left(\frac{\partial T}{\partial V}\right)_U = -\frac{1}{C_V}\left(\frac{\partial P}{\partial V}\right)_T\left(\frac{\partial U}{\partial P}\right)_T. \qquad (A)$$

The derivative $(\partial P/\partial V)_T$ is always negative for a real fluid and C_V is always positive; therefore according to Eq. (A) the sign of the temperature change for a free expansion is determined by the sign of $(\partial U/\partial P)_T$. For air at 28(°C), the data of Fig. 5-10 indicate that $(\partial U/\partial P)_T$ is negative. Thus a free expansion of air at this temperature level is accompanied by a *decrease* in the temperature of the air.

If an equation of state is available for the gas, then the derivative $(\partial P/\partial V)_T$ may be computed. For the conditions of the experiments for which data are given on Fig. 5-10, this derivative may be approximated by the ideal-gas value:

$$\left(\frac{\partial P}{\partial V}\right)_T \approx -\frac{P}{V} = -\frac{P^2}{RT}.$$

Thus,

$$\left(\frac{\partial T}{\partial V}\right)_U \approx \frac{P^2}{C_V RT}\left(\frac{\partial U}{\partial P}\right)_T. \qquad (B)$$

For air at 28(°C), C_V is about 20.8(J)/(mol)(K), and from the data in Fig. 5-10, $(\partial U/\partial P)_T$ is -6.08(J)/(mol)(atm). Substitution of numerical values into Eq. (B) yields, for air at 28(°C),

$$\left(\frac{\partial T}{\partial V}\right)_U \approx -0.012P^2\text{(K)(mol)/(l)},$$

where P is expressed in (atm). Thus for an infinitesimal free expansion of air at an average pressure of 10(atm), we would expect the temperature to drop at a rate of about 1.2(°C) for each (l)/(mol) increase in V.

5-7 Ideal Gas

We have already pointed out that the terms B/V, C/V^2, etc., of the virial expansion arise on account of various kinds of molecular interactions. If no such interactions existed, the virial coefficients B, C, etc., would be identically zero and the virial expansion, Eq. (5-8), would reduce to

$$Z = 1, \qquad \text{or} \qquad PV = RT.$$

For a real gas, molecular interactions *do* exist and exert an influence on the observed behavior of the gas. As the pressure of a real gas is reduced at constant temperature, V increases and the contributions of the terms B/V, C/V^2, etc., decrease. For a pressure approaching zero, Z approaches unity, not because of any change in the virial coefficients but because V becomes infinite. Thus, in the limit as the pressure approaches zero, the equation of state assumes the simple form

$$Z = 1, \qquad \text{or} \qquad PV = RT.$$

We have also found that the internal energy of a real gas is a function of pressure as well as of temperature. This pressure dependency arises as a result of molecular interactions. If such interactions did not exist, no energy would be required to alter the average intermolecular distance. Since this is all that is accomplished when volume (and hence pressure) changes occur at constant temperature, we conclude that in the absence of molecular interactions, the internal energy of a gas would depend on temperature only. These considerations of the behavior of a hypothetical gas in which no molecular interactions exist and of a real gas at the limiting pressure of zero lead us at this point to define an *ideal gas* whose properties, although not corresponding to those of any existing gas, are approximately those of a real gas at low pressures. By definition, an ideal gas satisfies the following equations at all temperatures and pressures:

$$\boxed{PV = RT} \qquad \text{(ideal gas)}, \tag{5-12}$$

and

$$\boxed{\left(\frac{\partial U}{\partial V}\right)_T = 0} \qquad \text{(ideal gas)}. \tag{5-13}$$

The requirement that $(\partial U/\partial V)_T = 0$ may be written in other ways.

Application of Eq. (2-10) gives

$$\left(\frac{\partial U}{\partial P}\right)_T = \left(\frac{\partial U}{\partial V}\right)_T \left(\frac{\partial V}{\partial P}\right)_T,$$

and since $(\partial V/\partial P)_T = -RT/P^2 = -V/P$, and therefore is not zero, whereas $(\partial U/\partial V)_T$ is zero, it follows that, for an ideal gas,

$$\left(\frac{\partial U}{\partial P}\right)_T = 0 \qquad \text{(ideal gas).} \tag{5-14}$$

Finally, since both $(\partial U/\partial P)_T$ and $(\partial U/\partial V)_T$ are zero,

$$U = U(T) \text{ only} \qquad \text{(ideal gas).} \tag{5-15}$$

It may similarly be shown that the *enthalpy* of an ideal gas depends upon temperature only. Thus,

$$\left(\frac{\partial H}{\partial P}\right)_T = 0 \qquad \text{(ideal gas),} \tag{5-16}$$

$$\left(\frac{\partial H}{\partial V}\right)_T = 0 \qquad \text{(ideal gas),} \tag{5-17}$$

and

$$H = H(T) \text{ only} \qquad \text{(ideal gas).} \tag{5-18}$$

The first law for a PVT system undergoing an infinitesimal quasi-static process and consisting of 1(mol) of material is given by Eq. (4-10),

$$\delta Q = dU + P\,dV.$$

Replacing dU by Eq. (4-23), we find an equivalent expression:

$$\delta Q = C_V\,dT + \left(\frac{\partial U}{\partial V}\right)_T dV + P\,dV.$$

But $(\partial U/\partial V)_T = 0$ for an ideal gas. Thus,

$$\delta Q = C_V\,dT + P\,dV \qquad \text{(ideal gas).} \tag{5-19}$$

An alternative expression is based on Eq. (4-16):

$$\delta Q = dH - V\,dP.$$

Replacing dH by Eq. (4-24), we get

$$\delta Q = C_P\,dT + \left(\frac{\partial H}{\partial P}\right)_T dP - V\,dP.$$

But $(\partial H/\partial P)_T = 0$ for an ideal gas. Thus,

$$\delta Q = C_P\,dT - V\,dP \qquad \text{(ideal gas).} \qquad (5\text{-}20)$$

Equations (5-19) and (5-20) are entirely equivalent and apply to any infinitesimal quasi-static process for an ideal gas.

Since U for an ideal gas is a function of T only, it follows from the definition of C_V, Eq. (4-21), that

$$C_V = C_V(T) \text{ only} \qquad \text{(ideal gas).} \qquad (5\text{-}21)$$

Similarly, since H for an ideal gas is a function of T only, we have from the definition of C_P, Eq. (4-22), that

$$C_P = C_P(T) \text{ only} \qquad \text{(ideal gas).} \qquad (5\text{-}22)$$

Since Eqs. (5-19) and (5-20) are equivalent, we may eliminate δQ between them, giving

$$C_P\,dT - C_V\,dT = P\,dV + V\,dP.$$

For an ideal gas $P\,dV + V\,dP \equiv d(PV) = R\,dT$. Therefore,

$$C_P\,dT - C_V\,dT = R\,dT,$$

and

$$C_P - C_V = R \qquad \text{(ideal gas).} \qquad (5\text{-}23)$$

Thus, *the heat capacities of an ideal gas are functions of temperature only, and their difference has a constant value, equal to the universal gas constant.*

It is often convenient to deal with the *ratio* C_P/C_V, which is given the special symbol γ:

$$\gamma \equiv \frac{C_P}{C_V}. \qquad (5\text{-}24)^1$$

For an ideal gas we have, from Eqs. (5-24), (5-21), and (5-22),

$$\gamma = \gamma(T) \text{ only} \qquad \text{(ideal gas).} \qquad (5\text{-}25)$$

Whether an actual gas may be treated as an ideal gas depends upon the error that may be tolerated in a given calculation. Errors greater than a few percent are rarely introduced when gases at pressures up to 2 or 3(atm) are treated as ideal gases. Even in the case of a saturated vapor in equilibrium with its liquid, the ideal-gas equation of state may be used with only a small error if the vapor pressure is low.

[1] The symbol γ as used here is not to be confused with the surface tension.

Example 5-6

An ideal gas undergoes an arbitrary process between two equilibrium states. Determine expressions for ΔU and ΔH.

In general, we may write

$$dU = C_V\,dT + \left(\frac{\partial U}{\partial V}\right)_T dV. \tag{4-23}$$

However, by Eq. (5-13), $(\partial U/\partial V)_T = 0$ for an ideal gas. Thus,

$$dU = C_V\,dT,$$

from which

$$\Delta U = \int_{T_1}^{T_2} C_V\,dT \qquad \text{(ideal gas)}. \tag{5-26}$$

Similarly, we have in general

$$dH = C_P\,dT + \left(\frac{\partial H}{\partial P}\right)_T dP. \tag{4-24}$$

However, by Eq. (5-16), $(\partial H/\partial P)_T = 0$ for an ideal gas. Thus,

$$\Delta H = \int_{T_1}^{T_2} C_P\,dT \qquad \text{(ideal gas)}. \tag{5-27}$$

Equations (5-26) and (5-27) apply to *any* process which takes an ideal gas from one equilibrium state to another.

5-8 Ideal-Gas Heat Capacities

The experimental determination of heat capacities of gases through use of the defining equations of the preceding chapter is a most difficult task. Such measurements are presently almost never made. Rather, such information is obtained by calculation through the methods of statistical mechanics from spectral data. This procedure yields values of C_V for gases at the limiting pressure of zero. Since the equations which define an ideal gas are also valid for the zero-pressure state and since the heat capacity of an ideal gas is independent of pressure, it is presumed that the zero-pressure heat capacity is identical with the ideal-gas heat capacity. To put it another way, if a gas at zero pressure is imagined to obey the ideal-gas equations as it is compressed isothermally to finite pressures, this imaginary ideal gas would always have a heat capacity identical with the zero-pressure value. Thus zero-pressure heat capacities are commonly referred to as ideal-gas heat capacities, or heat capacities for the ideal-gas state. It is these heat capacities that are tabulated in compilations of thermodynamic data. They are functions of temperature but not of pressure, and they depend on the gas.

The heat capacities of *real* gases are functions of pressure as well as of temperature, and are not often used. Other methods, to be considered later, are employed to take into account the effect of pressure on the thermodynamic properties of gases.

The general characteristics of molar-heat-capacity data for the *ideal-gas state* can be given quite simply as follows:

1 *All gases:*

(*a*) C_V is a function of T only.
(*b*) C_P is a function of T only, and $> C_V$.
(*c*) $C_P - C_V = \text{const} = R$.
(*d*) $\gamma = C_P/C_V$ = a function of T only, and > 1.

2 *Monatomic gases,* such as He, Ne, and Ar, and most metallic vapors, such as the vapors of Na, Cd, and Hg:

(*a*) C_V is constant over a wide temperature range and is very nearly equal to $\frac{3}{2}R$.
(*b*) C_P is constant over a wide temperature range and is very nearly equal to $\frac{5}{2}R$.
(*c*) γ is constant over a wide temperature range and is very nearly equal to $\frac{5}{3}$.

3 *So-called permanent diatomic gases,* namely H_2, D_2, O_2, N_2, NO, and CO:

(*a*) C_V is nearly constant at ordinary temperatures, being equal to about $\frac{5}{2}R$, and increases slowly at higher temperatures.
(*b*) C_P is nearly constant at ordinary temperatures, being equal to about $\frac{7}{2}R$, and increases slowly at higher temperatures.
(*c*) γ is nearly constant at ordinary temperatures, being equal to about $\frac{7}{5}$, and decreases slowly at higher temperatures.

4 *Polyatomic gases and gases that are chemically active,* such as CO_2, NH_3, CH_4, Cl_2, and Br_2: C_P, C_V, and C_P/C_V vary with the temperature, the variation being different for each gas.

Molar heat capacities have the same units as the universal gas constant R, and this suggests use of the dimensionless ratios C_V/R and C_P/R in correlations. Ideal-gas values of C_P/R can be represented over a substantial temperature range by the empirical equation

$$\frac{C_P}{R} = a + bT + cT^2, \tag{5-28}$$

where a, b, and c are constants, different for each gas. As a result of Eq.

Table 5-3 C_P/R for the Ideal-Gas State: Constants in Eq. (5-28) for 300 to 1500(K)

Gas	Formula	a	$b \times 10^3 (K)^{-1}$	$c \times 10^6 (K)^{-2}$
Acetylene	C_2H_2	3.689	6.352	−1.957
Ammonia	NH_3	3.063	4.435	−0.758
Benzene	C_6H_6	−0.206	39.064	−13.301
Carbon dioxide	CO_2	3.127	5.232	−1.784
Carbon monoxide	CO	3.231	0.838	−0.099
Chlorine	Cl_2	3.813	1.220	−0.486
Ethyl alcohol	C_2H_6O	3.518	20.001	−6.002
Hydrogen	H_2	3.496	−0.101	0.242
Hydrogen chloride	HCl	3.388	0.218	0.186
Hydrogen sulfide	H_2S	3.353	2.584	−0.430
Methane	CH_4	1.702	9.081	−2.164
Nitrogen	N_2	3.283	0.629	−0.001
Oxygen	O_2	3.094	1.561	−0.465
Sulfur dioxide	SO_2	3.581	4.787	1.767
Water	H_2O	3.652	1.157	0.142

(5-23), C_V/R is given by

$$\frac{C_V}{R} = \frac{C_P}{R} - 1.$$

Table 5-3 gives values of a, b, and c for several gases.

Example 5-7

The temperature of a stream of pure methane gas at 1(atm) pressure is to be increased from 100 to 1000(°C). How much heat is required per (mol) of methane?

The pressure is sufficiently low and the temperature sufficiently high so that we can assume ideal-gas behavior. For an isobaric quasi-static process, Eq. (5-20) becomes

$$\delta Q = C_P \, dT.$$

Integration between T_1 and T_2 gives

$$Q = \int_{T_1}^{T_2} C_P \, dT. \tag{A}$$

If C_P is represented by Eq. (5-28), Eq. (A) becomes

$$Q = \int_{T_1}^{T_2} R(a + bT + cT^2) \, dT,$$

or

$$Q = R\left[a(T_2 - T_1) + \frac{b}{2}(T_2^2 - T_1^2) + \frac{c}{3}(T_2^3 - T_1^3)\right]. \tag{B}$$

Values of a, b, and c for methane are listed in Table 5-3: $a = 1.702$, $b = 9.081 \times$

$10^{-3}(\text{K})^{-1}$, and $c = -2.164 \times 10^{-6}(\text{K})^{-2}$. Substitution into Eq. (B), with $R = 8.314(\text{J})/(\text{mol})(\text{K})$, $T_1 = 100 + 273.15 \approx 373(\text{K})$, and $T_2 = 1000 + 273.15 \approx 1273(\text{K})$, gives

$$Q = 8.314 \times \left[1.702 \times (1{,}273 - 373) + \frac{9.081}{2} \times 10^{-3} \times (1{,}273^2 - 373^2) - \frac{2.164}{3} \times 10^{-6} \times (1{,}273^3 - 373^3)\right],$$

or $\quad Q = 56{,}610(\text{J})/(\text{mol}).$

This problem could also be solved by application of the results of Examples 4-5 and 5-6. Thus, from Example 4-5, we have for an isobaric quasi-static process

$$Q = \Delta H, \tag{4-17b}$$

and from Example 5-6, for any change in state undergone by an ideal gas,

$$\Delta H = \int_{T_1}^{T_2} C_P \, dT. \tag{5-27}$$

Combination of these two expressions gives Eq. (A).

5-9 Quasi-static Adiabatic Process for an Ideal Gas

When an ideal gas undergoes a quasi-static adiabatic process, the pressure, volume, and temperature change in a manner that is described by a relation between P and V, T and V, or P and T. In order to derive the relation between P and V, we start with Eqs. (5-19) and (5-20). Thus, for 1(mol) of an ideal gas,

$$\delta Q = C_V \, dT + P \, dV, \tag{5-19}$$

and

$$\delta Q = C_P \, dT - V \, dP. \tag{5-20}$$

Since, in an adiabatic process, $\delta Q = 0$,

$$P \, dV = -C_V \, dT$$

and

$$V \, dP = C_P \, dT.$$

Dividing the second equation by the first and rearranging, we find

$$\frac{dP}{P} = -\frac{C_P}{C_V}\frac{dV}{V} = -\gamma \frac{dV}{V}.$$

This equation cannot be integrated until we know something about the behavior of γ. We have seen that for monatomic gases γ is constant, whereas for diatomic and polyatomic gases it may vary with the temperature. It requires, however, a large change of temperature to produce an appreciable change in γ. For example, for carbon monoxide, a temperature rise from 0 to 2000(°C) produces a decrease in γ from 1.4 to 1.3. Most adiabatic processes that we deal with do not involve such a large temperature change. We are therefore entitled, in an adiabatic process that involves only a moderate temperature change, to neglect the small accompanying change in γ. Regarding γ, therefore, as constant and integrating, we obtain

$$\ln P = -\gamma \ln V + \ln \text{const},$$

or

$$PV^{\gamma} = k, \tag{5-29}$$

where k is a constant, different in general for each particular process. Equation (5-29) holds for all equilibrium states through which an ideal gas with constant heat capacities passes during a quasi-static adiabatic process. We emphasize that a *free* expansion is an adiabatic process but is *not* quasi-static. It is therefore incorrect to apply Eq. (5-29) to the states traversed by an ideal gas during a free expansion.

A family of curves representing quasi-static adiabatic processes may be plotted on a PV diagram, each curve corresponding to a different value of the constant k. The slope of such a curve is

$$\left(\frac{\partial P}{\partial V}\right)_S = -\gamma k V^{-\gamma-1} = -\gamma \frac{P}{V},$$

where the subscript S denotes a quasi-static adiabatic process.

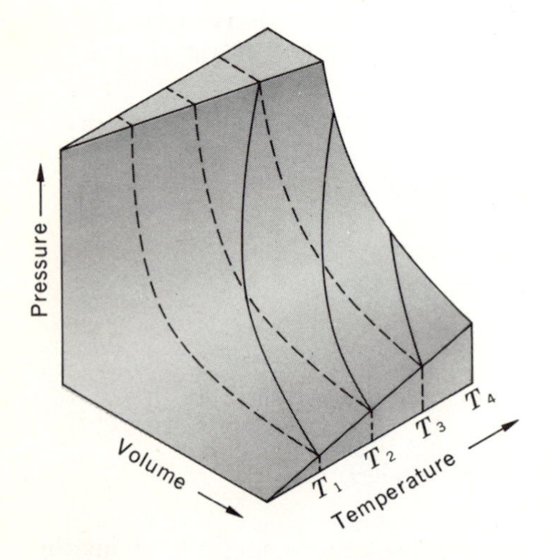

Fig. 5-11
Surface for an ideal gas. (Isotherms are represented by dashed curves, and adiabatics by full curves.)

Quasi-static *isothermal* processes are represented by a family of equilateral hyperbolas obtained when different values are assigned to T in the equation $PV = RT$. Since

$$\left(\frac{\partial P}{\partial V}\right)_T = -\frac{P}{V},$$

it follows that an adiabatic curve has a steeper negative slope than does an isothermal curve at the same point.

The isothermal curves and adiabatic curves of an ideal gas may be shown in a revealing way on a PVT surface. This surface is shown in Fig. 5-11, where it is seen that the adiabatic curves cut across the isotherms.

Example 5-8

An ideal gas with constant heat capacities undergoes an adiabatic quasi-static process. Determine expressions which relate T to V and T to P.

We start with Eq. (5-29). Since $P = RT/V$,

$$PV^\gamma = \left(\frac{RT}{V}\right)V^\gamma = RT\,V^{\gamma-1} = k,$$

from which

$$TV^{\gamma-1} = k', \tag{5-30}$$

where $k' \equiv k/R$. The equation relating T and P is found similarly. Since $V = RT/P$, we have, by Eq. (5-29),

$$P\left(\frac{RT}{P}\right)^\gamma = R^\gamma P^{1-\gamma}T^\gamma = k,$$

from which

$$TP^{-[(\gamma-1)/\gamma]} = k'', \tag{5-31}$$

where $k'' \equiv k^{1/\gamma}/R$.

Suppose that a mass of monatomic gas initially at 300(K) and 1(atm) is compressed quasi-statically and adiabatically to 2(atm). The final temperature is determined by application of Eq. (5-31) separately to the initial state and to the final state. Since k'' is a constant for the process, we have

$$T_1P_1^{-[(\gamma-1)/\gamma]} = T_2P_2^{-[(\gamma-1)/\gamma]},$$

or

$$T_2 = T_1\left(\frac{P_2}{P_1}\right)^{[(\gamma-1)/\gamma]}.$$

From Sec. 5-8 we find that γ for an ideal monatomic gas is $\frac{5}{3}$. Thus $(\gamma - 1)/\gamma = \frac{2}{5}$, and the above equation yields

$$T_2 = 300(\text{K})\left[\frac{2(\text{atm})}{1(\text{atm})}\right]^{2/5},$$

or

$$T_2 = 396(\text{K}).$$

Example 5-9

Two moles of an ideal diatomic gas with constant heat capacities, initially at 300(K) and 5(bar), expand adiabatically and quasi-statically to 3(bar). Determine the work done.

The first law for an adiabatic process is $W = \Delta U$. For an ideal gas with constant heat capacities, Eq. (5-26) becomes $\Delta U = C_V(T_2 - T_1)$. We can therefore write

$$W = C_V(T_2 - T_1). \tag{A}$$

The definition of γ in combination with Eq. (5-23) gives

$$\gamma = \frac{C_P}{C_V} = \frac{C_V + R}{C_V} = 1 + \frac{R}{C_V},$$

from which
$$C_V = \frac{R}{\gamma - 1}.$$

Substitution into (A) gives

$$W = \frac{RT_2 - RT_1}{\gamma - 1} = \frac{P_2V_2 - P_1V_1}{\gamma - 1}. \tag{B}$$

Equation (5-29) allows us to write

$$P_1V_1^{\gamma} = P_2V_2^{\gamma},$$

from which
$$V_2 = V_1\left(\frac{P_1}{P_2}\right)^{1/\gamma} = V_1\left(\frac{P_2}{P_1}\right)^{-1/\gamma}.$$

We now eliminate V_2 from (B):

$$W = \frac{P_2V_1(P_2/P_1)^{-1/\gamma} - P_1V_1}{\gamma - 1} = \frac{P_1V_1}{\gamma - 1}\left[\left(\frac{P_2}{P_1}\right)\left(\frac{P_2}{P_1}\right)^{-1/\gamma} - 1\right],$$

or
$$W = \frac{P_1V_1}{\gamma - 1}\left[\left(\frac{P_2}{P_1}\right)^{(\gamma-1)/\gamma} - 1\right] = \frac{RT_1}{\gamma - 1}\left[\left(\frac{P_2}{P_1}\right)^{(\gamma-1)/\gamma} - 1\right]. \tag{C}$$

For n(mol) of gas, (C) becomes

$$W = \frac{nRT_1}{\gamma - 1}\left[\left(\frac{P_2}{P_1}\right)^{[(\gamma-1)/\gamma]} - 1\right]. \tag{5-32}$$

This is our working equation.

From Sec. 5-8 we find $\gamma = \frac{7}{5}$ for an inert diatomic ideal gas, from which $\gamma - 1 = \frac{2}{5}$ and $(\gamma - 1)/\gamma = \frac{2}{7}$. Thus, for the given conditions,

$$W = \frac{2 \times 8.314 \times 300}{\frac{2}{5}}\left[\left(\frac{3}{5}\right)^{2/7} - 1\right],$$

or
$$W = -1694(\mathrm{J}).$$

This value is numerically much smaller than the work obtained for the corresponding isothermal expansion of Example 3-2.

5-10 Speed of a Longitudinal Wave

Let c' be the velocity of propagation of a pressure increase in a fluid at rest. To show the relation of c' to the properties of the fluid, we consider the propagation of a pressure increase in a tube, as shown in Fig. 5-12.

The piston on the left is moved into the tube with constant velocity u by exertion of a force $A(P + \Delta P)$. The effect of this is to increase the pressure in front of the piston from P to $P + \Delta P$. This pressure increase propagates into the fluid with the constant velocity c'. The entire region affected by the pressure increase at time τ moves with velocity u, and the effect of the process is to increase the momentum of this region from zero to a finite value. The impulse-momentum principle may be applied to the entire tube.

At time τ the mass of fluid already set in motion is

$$m = \frac{c'\tau A}{V},$$

where V is the *specific* volume of undisturbed fluid. The impulse-momentum theorem gives

$$F\tau = \frac{\Delta(\Sigma m_i u_i)}{g_c},$$

where F, the resultant force acting on the entire fluid column, is given by $F = A\,\Delta P$. The quantity $\Delta(\Sigma m_i u_i)$ indicates the difference between the total momentum of the fluid at time τ and at zero time. The Σ indicates that we are considering all parts of the fluid. However, the only part of the fluid having momentum is that mass m which has been given the velocity

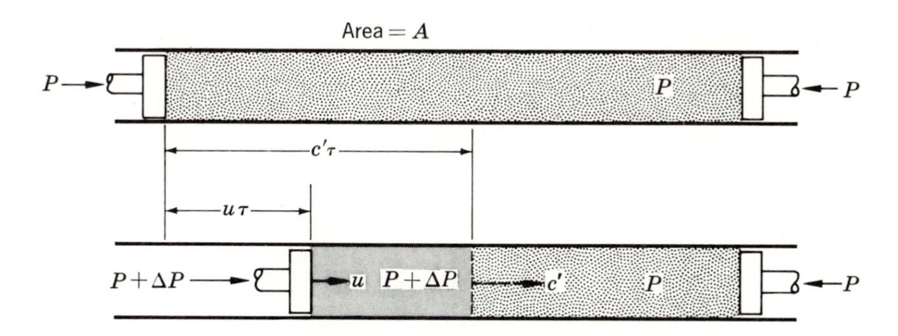

Fig. 5-12
Propagation of a compression with constant velocity c' by motion of a piston with constant velocity u. Upper diagram at the start; lower diagram after time τ.

u at time τ. Hence

$$\Delta(\Sigma m_i u_i) = mu,$$

where

$$m = \frac{c'\tau A}{V}.$$

Substitution of these elements into the impulse-momentum equation gives

$$A\ \Delta P_T = \frac{c'\tau A u}{V g_c},$$

or

$$\Delta P = \frac{c' u}{V g_c}.$$

The change in total volume of the fluid column is $-u\tau A$, and the change in the specific volume of the fluid set in motion is

$$\Delta V = \frac{-u\tau A}{m} = \frac{-u\tau A}{c'\tau A/V} = -\frac{uV}{c'}.$$

Therefore,

$$\frac{\Delta P}{\Delta V} = \frac{-c'u/Vg_c}{uV/c'} = \frac{-(c')^2}{V^2 g_c}.$$

In the limit as ΔP approaches zero, the propagation velocity approaches the *sonic velocity* or *soundspeed* c, which is the velocity of propagation of an infinitesimal pressure pulse. Thus

$$\lim_{\Delta P \to 0} \frac{\Delta P}{\Delta V} = \frac{dP}{dV} = -\frac{c^2}{V^2 g_c},$$

or

$$c^2 = -g_c V^2 \frac{dP}{dV}.$$

This formula for the sonic velocity was first obtained by Newton, who regarded the propagation process as being isothermal. Laplace later showed that it is really adiabatic. Actually, these considerations apply not so much to a single pulse as to a succession of waves composed of alternate compressions and rarefactions such as transmit sound. The temperature of a fluid rises with compression and falls with expansion unless heat is transferred. The very existence of adjacent regions of compression and rarefaction with their associated differences in temperature provides the driving force for heat transfer between these regions. However, at audible frequencies it is found that both the thermal conductivity and the temperature gradient are too small for any appreciable heat transfer to occur during the brief time that the temperature gradient exists at any one location. At very high frequencies, of the order of 10^9(cps), the temperature gradients become appreciable because of the very close spacing of the regions of compression and rarefaction, and the adiabatic condition is less closely approached.

Returning now to the expression for the sonic velocity, we identify the changes represented by dP/dV as being adiabatic by use of the subscript S:

$$c^2 = -g_c V^2 \left(\frac{\partial P}{\partial V}\right)_S. \tag{5-33}$$

Thus the soundspeed c is a thermodynamic property, a function in general of any two of P, V, and T. Some values of c for gases and liquids are listed in Table 5-4.

For the special case of an ideal gas, we showed in Sec. 5-9 that

$$\left(\frac{\partial P}{\partial V}\right)_S = -\frac{\gamma P}{V}.$$

Hence

$$c^2 = g_c \gamma P V,$$

where V is still the specific volume. If V is to represent the *molar* volume, we must divide by the molecular weight M. In this event,

$$c^2 = \frac{g_c \gamma P V}{M},$$

where V is now the *molar* rather than specific volume. Replacing PV by RT, we have, finally,

$$c = \sqrt{\frac{g_c \gamma R T}{M}} \qquad \text{(ideal gas)}. \tag{5-34}$$

Table 5-4 Soundspeed c for Some Gases and Liquids

Substance	t(°C)	c(m)/(s)
Nitrogen (g)	27, 1(atm)	353
	27, 100(atm)	379
Diethyl ether (l)	20	1,006
n-Heptane (l)	20	1,150
Hydrogen (g)	0, 1(atm)	1,200
	0, 100(atm)	1,281
Mercury (l)	20	1,451
Water (l)	20	1,482
Glycerine (l)	20	1,895

Equation (5-34) enables us to calculate γ from experimental low-pressure measurements of c and T. The speed of a sound wave in a gas can be measured with fair accuracy by means of Kundt's tube. The gas is admitted to a cylindrical tube closed at one end and supplied at the other end with a movable piston capable of being set in vibration parallel to the axis of the tube. In the tube is a small amount of light powder. For a given frequency, a position of the piston can be found at which standing waves are set up. Under these conditions small heaps of powder pile up at the nodes. The distance between any two adjacent nodes is one-half a wavelength, and the speed of the waves is the product of the frequency and the wavelength. Values of γ obtained by this method are in good agreement with those obtained from both calorimetric measurements and statistical calculations.

Example 5-10

We illustrate the use of Eq. (5-34) for the determination of γ from soundspeed measurements. According to Table 5-4, c is 353(m)/(s) for nitrogen gas at 300(K) and 1(atm). The molecular weight of nitrogen is 28.01(g)/(mol) $\equiv$ 0.02801(kg)/(mol). Solution of Eq. (5-34) for γ gives

$$\gamma = \frac{Mc^2}{g_c RT},$$

and substitution of numerical values yields

$$\gamma = \frac{0.02801\,(\mathrm{kg})/(\mathrm{mol}) \times (353)^2(\mathrm{m})^2/(\mathrm{s})^2}{1 \times 8.314\,(\mathrm{kg})\,(\mathrm{m})^2/(\mathrm{s})^2(\mathrm{mol})\,(\mathrm{K}) \times 300\,(\mathrm{K})},$$

or

$$\gamma = 1.40.$$

This value is identical with that determined by statistical calculations.

Problems

5-1 For an ideal gas show that:

(a) $\beta = 1/T$.
(b) $\kappa = 1/P$.

5-2 Using the virial expansion in the form

$$Z = 1 + B'P + C'P^2 + \cdots,$$

show that

(a) $$\beta = \frac{1}{T} + \frac{(dB'/dT)P + (dC'/dT)P^2 + \cdots}{1 + B'P + C'P^2 + \cdots}.$$

(b) $$\kappa = \frac{1}{P} - \frac{B' + 2C'P + \cdots}{1 + B'P + C'P^2 + \cdots}.$$

What are the limiting values of β and κ as pressure approaches zero at constant temperature? Do these equations give the ideal-gas values of β and κ when B', C', etc., are set equal to zero? When $P = 0$?

5-3 Using the virial expansion truncated to two terms,

$$Z = 1 + \frac{B}{V},$$

show that

(a) $$\beta = \frac{1 + B/V + (T/V)(dB/dT)}{T(1 + 2B/V)}.$$

(b) $$\kappa = \frac{1}{P + BRT/V^2}.$$

5-4 Prove

$$B' = \lim_{P\to 0} \left[\frac{Z-1}{P}\right]_T \quad \text{and} \quad C' = \lim_{P\to 0} \left\{\frac{\partial[(Z-1)/P]}{\partial P}\right\}_T$$

$$= \lim_{P\to 0} \left\{\frac{[(Z-1)/P] - B'}{P}\right\}_T.$$

5-5 Derive expressions for the volume expansivity β and the isothermal compressibility κ in terms of T, P, and Z and its derivatives.

5-6 Estimate the mass of the air in a $20 \times 15 \times 8$(ft) room. Assume that room temperature is 70(°F).

5-7 (a) Show that the heat transferred during an infinitesimal quasi-static process for 1(mol) of an ideal gas can be written

$$\delta Q = \frac{C_V}{R} V\,dP + \frac{C_P}{R} P\,dV.$$

Applying this equation to an adiabatic process, show that $PV^\gamma = \text{const.}$

(b) An ideal gas of volume 1,400(cm)^3 and pressure 8(bar) undergoes a quasi-static adiabatic expansion until the pressure drops to 1(bar). Assuming γ to remain constant at the value 1.4: (i) What is the final volume? (ii) How much work is done?

5-8 (a) Helium ($\gamma = \frac{5}{3}$) at 300(K) and 1(atm) pressure is compressed quasi-statically and adiabatically to a pressure of 5(atm). Assuming that the helium behaves as an ideal gas, what is the final temperature?

(b) At about 100(ms) after detonation of a uranium fission bomb, the "ball of fire" consists of a sphere of gas with a radius of about 15(m) and a temperature of 300,000(K). Making very rough assumptions, estimate at what radius its temperature would be 3000(K).

5-9 A cylindrical highball glass 6(in) high and 3(in) in diameter contains water up to the 4(in) level. A card is placed over the top and held there as the glass is inverted. When the support is removed, what mass of water must leave the glass in order that the rest of the water remain in the glass, neglecting the weight of the card? (*Caution*: Try this over a sink.)

5-10 The temperature of an ideal gas in a capillary of constant cross-sectional area varies exponentially from one end ($x = 0$) to the other ($x = L$) according to the equation

$$T = T_0 e^{-kx}.$$

If the volume of the capillary is V^t and the pressure P is uniform throughout, show that the number of moles of gas n is given by

$$PV^t = nR \frac{kLT_0}{e^{kL} - 1}.$$

Show that, as $k \to 0$, $PV^t \to nRT_0$, as it should.

5-11 A horizontal insulated cylinder contains a frictionless nonconducting piston. On each side of the piston is $0.05(\mathrm{m})^3$ of an inert monatomic ideal gas at 1(atm) and 273(K). Heat is slowly supplied to the gas on the left side until the piston has compressed the gas on the right side to 7(atm).

(*a*) How much work is done on the gas on the right side?
(*b*) What is the final temperature of the gas on the right side?
(*c*) What is the final temperature of the gas on the left side?
(*d*) How much heat is added to the gas on the left side?

5-12 Sketch a VT diagram for a pure substance. Include on your sketch the region of two-phase vapor-liquid equilibrium.

5-13 (*a*) If y is the height above sea level, show that the decrease of atmospheric pressure due to a rise dy is given by

$$\frac{dP}{P} = \frac{-Mg}{g_c RT} dy,$$

where M = molecular weight of the air,
g = acceleration of gravity,
T = absolute temperature at height y.

(*b*) If the change in pressure in (*a*) is due to an adiabatic expansion, show that

$$\frac{dP}{P} = \frac{\gamma}{\gamma - 1} \frac{dT}{T}.$$

(*c*) From (*a*) and (*b*), taking $\gamma = 1.4$, calculate dT/dy in (K)/(km).

5-14 A rigid nonconducting tank with a volume of $2(\mathrm{m})^3$ is divided in half by a partition. An ideal monatomic gas is contained in both halves of the tank. In one half the temperature and pressure are 170(°C) and 0.3(bar), and in the other they are 50(°C) and 1(bar). The partition is removed, allowing complete mixing. What are the final T and P?

5-15 Using the data of Table 5-3, calculate the value of $\int C_P\, dT$ in (J)/(mol) for sulfur dioxide for a temperature change from 25 to 425(°C). Calculate the mean value of the heat capacity for this temperature interval. How does it compare with the heat capacity evaluated at the arithmetic average of T_1 and T_2?

5-16 An empirical equation, PV^δ = const, where δ is a constant, is often used to relate P and V for *any* quasi-static process. (If there is no restraint on the process, it is often called *polytropic*.) Assuming this equation to be valid for an ideal gas, show that

$$W = \frac{RT_1}{\delta - 1}\left[\left(\frac{P_2}{P_1}\right)^{(\delta-1)/\delta} - 1\right].$$

If the process is isothermal, $\delta = 1$. Show that the above equation reduces in this case to the isothermal work equation

$$W = RT \ln \frac{P_2}{P_1}.$$

5-17 For a gas at a pressure low enough so that the virial expansion may be truncated to

$$Z = 1 + B'P,$$

show that the equation for isothermal work in a quasi-static compression is

$$W = RT \ln \frac{P_2}{P_1},$$

the same as for an ideal gas.

5-18 For SO_2 at 157.5(°C) the virial coefficients are

$$B = -159(\text{cm})^3/(\text{mol}),$$

$$C = 9{,}000(\text{cm})^6/(\text{mol})^2.$$

Calculate the work of quasi-static isothermal compression of 1(lb mol) of SO_2 from 1(atm) to 75(atm) at 157.5(°C). Use the following forms of the virial equation:

(a) $Z = 1 + \frac{B}{V} + \frac{C}{V^2}$,

(b) $Z = 1 + B'P + C'P^2$,

where $B' = \frac{B}{RT}$ and $C' = \frac{C - B^2}{(RT)^2}$.

Why do not both equations give exactly the same result?

5-19 One pound mole of an ideal gas, initially at 70(°F) and 1(atm), undergoes the following quasi-static changes: It is first compressed isothermally to a point such that when it is heated at constant volume to 200(°F), its final pressure is 10(atm). Taking $C_V = 5$(Btu)/(lb mol)(R), calculate Q, W, and ΔU for the process.

5-20 From the data of Table 5-3, write equations for C_P/R for ammonia as a function of temperature in (a) °C, (b) R, and (c) °F.

5-21 In the equation

$$\int_{T_1}^{T_2} C_P \, dT = C_{P_m}(T_2 - T_1),$$

where C_{P_m} is the mean heat capacity, show that when C_P is linear with T, C_{P_m} is the heat capacity evaluated at the arithmetic average of T_1 and T_2.

5-22 Equation (5-8) when truncated to four terms accurately represents the volumetric behavior of methane gas at 0(°C). Values for B, C, and D for methane at 0(°C) are: $B = -53.4(\text{cm})^3/(\text{mol})$, $C = 2{,}620(\text{cm})^6/(\text{mol})^2$, and $D = 5{,}000(\text{cm})^9/(\text{mol})^3$.

(*a*) Using the above data, prepare a plot of Z vs. P for methane at 0(°C) from 0 to 400(atm).
(*b*) To what pressure does Eq. (5-11) provide a good approximation to the curve? To what pressure does Eq. (5-8) truncated to two terms provide a good approximation?
(*c*) If B', C', and D' are determined by Eq. (5-9) and are used in Eq. (5-7), what is the error in Z at 400(atm)? Why should there be any error at all?

5-23 Consider the following series of quasi-static processes to be carried out with 1(lb mol) of an ideal gas having a constant heat capacity, $C_V = 5(\text{Btu})/(\text{lb mol})(\text{R})$:

(*a*) The gas at 70(°F) and 1(atm) is heated at constant volume to a temperature of 400(°F).
(*b*) The gas is then expanded adiabatically to the initial temperature of 70(°F).
(*c*) Finally, the gas is compressed isothermally to the initial pressure of 1(atm). Calculate Q, W, and ΔU for each step and for the cycle.

5-24 Air is compressed quasi-statically from an initial state of 1(bar) and 20(°C) to a final state of 5(bar) and 20(°C) by three different processes:

(*a*) Heating at constant volume followed by cooling at constant pressure.
(*b*) Isothermal compression.
(*c*) Adiabatic compression followed by cooling at constant volume.

At these conditions air may be considered an ideal gas with a heat capacity $C_P = 29.3(\text{J})/(\text{mol})(\text{K})$. Calculate Q, W, ΔU for each process. Sketch the paths followed in each process on a single PV diagram.

5-25 Table P5-25 contains data for the known triple points of water. Using these data, prepare a schematic PT diagram for water. (Note: There is no "ice IV.")

Table P5-25 Triple Points of Water

Phases in equilibrium	P(bar)	t(°C)
Ice I, liquid, vapor	0.006105	+0.01
Ice I, liquid, ice III	2,074	−22.0
Ice I, ice II, ice III	2,130	−34.7
Ice II, ice III, ice V	3,440	−24.3
Ice III, liquid, ice V	3,460	−17.0
Ice V, liquid, ice VI	6,260	+0.16
Ice VI, liquid, ice VII	22,000	+81.6

5-26 Using data from Table 5-3, calculate the heat required in (Btu) to raise the temperature of 1(lb mol) of carbon dioxide from 200 to 1500(°F) at a constant low pressure where CO_2 is essentially an ideal gas.

5-27 The heat capacity of a gas sample is to be calculated from the following data: The sample was put in a flask and came to the initial equilibrium conditions, 25(°C) and 910(mm Hg). A stopcock was then opened for a short time to allow the pressure inside the flask to drop to 760(mm Hg). With the stopcock again closed, the flask was warmed; and when it was again at 25(°C), its pressure was found to be 780(mm Hg). Determine C_P in (J)/(mol)(K). It may be assumed that the gas is ideal, that C_P is constant, and that expansion of the gas remaining in the flask was quasi-static and adiabatic.

5-28 One mole of solid NH_4Cl is contained in a cylinder at 350(°C). At this temperature NH_4Cl decomposes slowly to give a gaseous mixture of NH_3 and HCl. If the cylinder is fitted with a frictionless piston which exerts a constant pressure of 1(atm), the entire charge of NH_4Cl can be decomposed by supplying the heat necessary to hold the temperature at 350(°C). This heat requirement is 175.7(kJ) for the decomposition of 1(mol) of NH_4Cl. Assuming the gases ideal and the volume of NH_4Cl negligible, calculate ΔU for the process.

5-29 The speed of a longitudinal wave in a mixture of helium and neon at 300(K) was found to be 758(m)/(s). What is the composition of the mixture?

5-30 The atomic weight of iodine is 127. A standing wave in iodine vapor at 400(K) produces nodes that are 6.77(cm) apart when the frequency is 1,000$(s)^{-1}$. Is iodine vapor monatomic or diatomic?

5-31 It has been proposed that a high-speed rapid transit system could be constructed to operate on the principle of a pneumatic tube. The vehicle would consist of a train of circular cross section running on rails in a closely fitting round tube. The tube in front of the train would be evacuated to a low pressure, and the train would be given a start by admitting atmospheric air behind it as it traveled a specified distance into the tube. The inlet port would then be closed and the train would be acted upon by the expanding air behind it and by the air being compressed in front of it. The train accelerates as long as the pressure behind it exceeds the pressure in front of it. Eventually, the reverse becomes true, and the train decelerates. When the pressure in front of the train builds up to atmospheric, the port at the far end of the tube opens to hold the front pressure at atmospheric, and the train continues to decelerate until it comes to a halt at the end of the tube solely as a result of the pressure forces. Thus the train is treated as a "floating" piston in a tube. Figure P5-31 gives a schematic picture of the process, showing roughly how the pressures and train velocity vary during a trip. In working the problem, make the following assumptions:

(i) The train moves without friction.
(ii) There is no leakage of air past the train.
(iii) Air is an ideal gas.
(iv) The train length is negligible compared with L.
(v) The compression and expansion of air in the tube are quasi-static, and P and V are related by the equation

$$PV^{\delta} = k,$$

where δ and k are constants.
(vi) The train must stop at the final station without mechanical braking.

Consider now a transit tube of diameter D between two stations i and f traversed

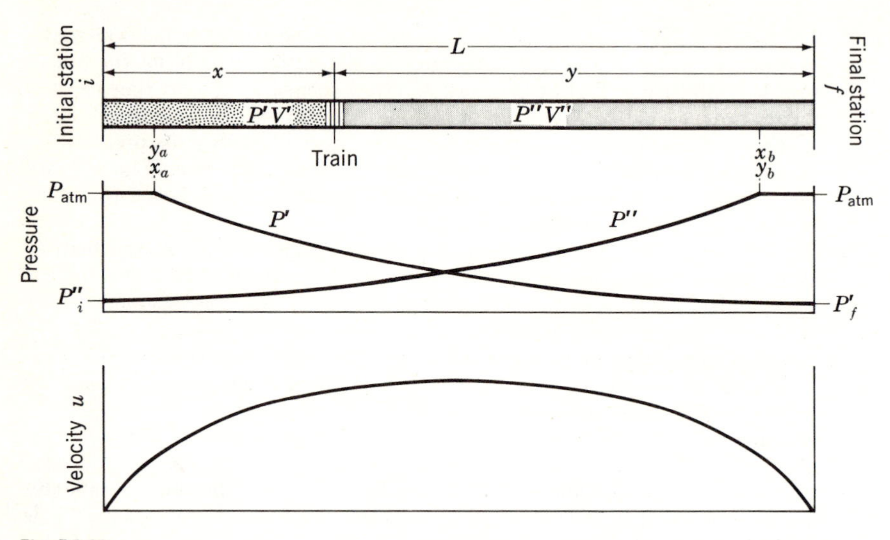

Fig. P5-31

by a train of mass m. The following parameters are specified:

$$L = 100(\text{mi}),$$
$$D = 10(\text{ft}),$$
$$m = 10^6(\text{lb}_m),$$
$$x_a = 4(\text{mi}),$$
$$P_{\text{atm}} = 14.7(\text{psia}),$$
$$\delta = 1.1.$$

Determine the time required for the trip from i to f and plot:

(*a*) P' versus x.
(*b*) P'' versus x.
(*c*) u versus x.

(*Note*: This problem is suitable for a group project. For different groups, one could vary such parameters as x_a and δ.)

5-32 A piston-and-cylinder assembly contains an ideal gas with constant heat capacities. The initial volume of the gas is 2,000(cm)³, the initial pressure is 5(bar), and the initial temperature is 150(°C). During a quasi-static process, carried out at constant pressure, 1,000(J) of heat is transferred to the gas, and its temperature rises to 255(°C). Calculate:

(*a*) The *total* heat capacity of the gas at constant *volume* $C_V{}^t$.
(*b*) The molar heat capacity of the gas at constant *volume* C_V.

5-33 One mole of an ideal gas with constant heat capacities undergoes an arbitrary process. Show that

$$\Delta U = \frac{1}{\gamma - 1}\Delta(PV).$$

6

ENGINES, REFRIGERATORS, AND THE SECOND LAW

6-1 Conversion of Work into Heat, and Vice Versa

When two stones are rubbed together under water, the work done against friction is transformed into internal energy tending to produce a rise of temperature of the stones. As soon as the temperature of the stones rises above that of the surrounding water, however, there is a flow of heat into the water. If the mass of water is large enough or if the water is continually flowing, there will be no appreciable rise of temperature, and the water can be regarded as a heat reservoir. Since the state of the stones is the same at the end of the process as at the beginning, the net result of the process is merely the conversion of mechanical work into heat. Similarly, when an electric current is maintained in a resistor immersed either in running water or in a very large mass of water, there is also a conversion of electrical work into heat, without any change in the thermodynamic coordinates of the wire. In general, work of any kind W may be done upon a system in contact with a reservoir, giving rise to a flow of heat Q without altering the state of the system. The system acts merely as an intermediary. It is apparent from the first law that the work W is equal in magnitude to the heat Q; or in other words, the transformation of work into heat is accomplished with 100 percent efficiency. Moreover, this transformation can be continued indefinitely.

To study the converse process, namely, the conversion of heat into work, we must also have at hand a process or series of processes by means of which such a conversion may continue indefinitely without involving any resultant changes in the state of any system. At first, one might think the isothermal expansion of an ideal gas a suitable process for the conversion of heat into work. In this case there is no internal-energy change since the temperature remains constant; therefore $Q + W = 0$ and heat has been converted completely into work. This process, however, involves a change of state of the gas. The volume increases and the pressure decreases until atmospheric pressure is reached, at which point the process stops. It therefore cannot be used indefinitely.

What is needed is a series of processes in which a system is brought back to its initial state, i.e., a *cycle*. Each of the processes that constitute a cycle may involve a flow of heat to or from the system and the performance of work by or on the system. For one complete cycle, let

$$|Q_H| = \text{amount of heat absorbed by the system,}$$

$$|Q_C| = \text{amount of heat rejected by the system,}$$

$$W = \text{net work.}$$

Both $|Q_H|$ and $|Q_C|$ are absolute values. If $|Q_H|$ is larger than $|Q_C|$, then W is negative and work is done *by* the system on the environment; the mechanical device by whose agency the system is caused to undergo the cycle is called a *heat engine*. It is convenient to speak of the system as the *working substance* and to say that the engine operates in a cycle. The purpose of a heat engine is to deliver work continuously to the outside by performing the same cycle over and over again. The net work of the cycle is the output, and the heat absorbed by the working substance is the input. The *thermal efficiency* of the engine η is defined as

$$\text{Thermal efficiency} = \frac{\textit{net} \text{ work output, in any energy units}}{\text{heat input, in the same energy units}},$$

or

$$\eta \equiv \frac{|W|}{|Q_H|}. \tag{6-1}$$

Applying the first law to one complete cycle and remembering that there is no net change of internal energy, we get

$$W = -Q = -(|Q_H| - |Q_C|),$$

and since W is negative

$$|W| = |Q_H| - |Q_C|. \tag{6-2}$$

Therefore

$$\eta = \frac{|Q_H| - |Q_C|}{|Q_H|},$$

or

$$\eta = 1 - \frac{|Q_C|}{|Q_H|}. \tag{6-3}$$

It is seen from this equation that η will be unity (efficiency of 100 percent) when $|Q_C|$ is zero. In other words, if an engine can be built to operate in a cycle in which there is no outflow of heat from the system, there will be 100 percent conversion of the absorbed heat into work. We shall see later under what conditions this is possible in principle and why it is not possible in practice.

Since the purpose of a heat engine is to deliver work continuously, the heat and work terms are often expressed as *rates*. Equations (6-1) through (6-3) are then understood to apply on a unit-time basis, and $|W|$, $|Q_C|$, and $|Q_H|$ all have dimensions of *power*.

Example 6-1

A particular heat engine generates 1 (kW) of power and discards heat to the atmosphere at the rate of 600 (J)/(s). Determine the rate of heat transfer to the engine, and the thermal efficiency of the engine.

The rate of heat transfer to the engine $|Q_H|$ is found from the first law, Eq. (6-2):

$$|Q_H| = |W| + |Q_C| = 1{,}000\,(\mathrm{J})/(\mathrm{s}) + 600\,(\mathrm{J})/(\mathrm{s}),$$

or

$$|Q_H| = 1{,}600\,(\mathrm{J})/(\mathrm{s}).$$

We compute the thermal efficiency η from Eq. (6-3):

$$\eta = 1 - \frac{|Q_C|}{|Q_H|} = 1 - \frac{600\,(\mathrm{J})/(\mathrm{s})}{1{,}600\,(\mathrm{J})/(\mathrm{s})} = 1 - 0.375,$$

or

$$\eta = 0.625.$$

The transformation of heat into work is usually accomplished in practice by two general types of engine: the *external combustion engine*, such as the Stirling engine and the steam engine, and the *internal-combustion engine*, such as the gasoline engine and the Diesel engine. In both types, a confined gas or mixture of gases undergoes a cycle, causing a piston or turbine to impart to a shaft a motion of rotation against an opposing force. It is necessary in both engines that at some time in the cycle the gas be raised to a high temperature and pressure. In the Stirling and steam engines this is accomplished by an outside furnace. The high temperature and pressure achieved in the internal-combustion engine, however, are produced by a chemical reaction between a fuel and air that takes place in the engine itself. In the gasoline engine, the combustion of the gasoline and air takes place explosively through the agency of an electric spark. In the Diesel engine, however, oil is sprayed into the engine at a convenient rate, and combustion is accomplished more slowly.

6-2 The Stirling Engine

In 1816, before the science of thermodynamics had even been begun, a minister of the Church of Scotland named Robert Stirling designed and patented a hot-air engine that could convert some of the energy liberated by a burning fuel into work. The Stirling engine remained useful and popular for many years but, with the development of steam engines and internal-combustion engines, finally became obsolete. In the 1940s the Stirling engine was revived by the engineers of the Philips company in Eindhoven, Holland, and has once again become the subject of a great amount of interest and research.

The steps in the operation of a somewhat idealized Stirling engine are shown schematically in Fig. 6-1*a*. Two pistons, an expansion piston on the left and a compression piston on the right, are connected to the same shaft. As the shaft rotates, these pistons move in different phase, with the aid of suitable connecting linkages. The space between the two pistons is filled with gas, and the left-hand portion of the space is kept in contact with a hot reservoir (burning fuel), while the right-hand portion is in contact with a cold reservoir. Between the two portions of gas is a device R, called a *regenerator*, consisting of a packing of steel wool or a series of metal baffles, whose thermal conductivity is low enough to support the temperature difference between the hot and cold ends without appreciable heat conduction. The Stirling cycle consists of four processes depicted schematically in Fig. 6-1*a*. The corresponding *idealized* Stirling cycle is plotted on the PV diagram of Fig. 6-1*b*.

$1 \rightarrow 2$ While the left piston remains at the top, the right piston moves halfway up, compressing cold gas while in contact with the cold reservoir and therefore causing heat $|Q_C|$ to leave. This is an approximately isothermal compression and is depicted as a strictly isothermal process at the temperature T_C in Fig. 6-1*b*.

$2 \rightarrow 3$ The left piston moves down and the right piston up, so that there is no change in volume, but gas is forced through the regenerator from the cold side to the hot side and enters the left-hand side at the higher temperature T_H. During the transfer process, the regenerator supplies heat $|Q_R|$ to the gas. This step is depicted as a constant-volume process in Fig. 6-1*b*.

$3 \rightarrow 4$ The right piston now remains stationary as the left piston continues moving down while in contact with the hot reservoir, causing the gas to undergo an approximately isothermal expansion, during which heat $|Q_H|$ is absorbed at the temperature T_H, as shown in Fig. 6-1*b*.

$4 \rightarrow 1$ Both pistons move in opposite directions, forcing gas from the hot to the cold side. In passing through the regenerator, the gas gives up approximately the same amount of heat $|Q_R|$ to the regenerator that it absorbed in the process $2 \rightarrow 3$. This process takes place at practically constant volume.

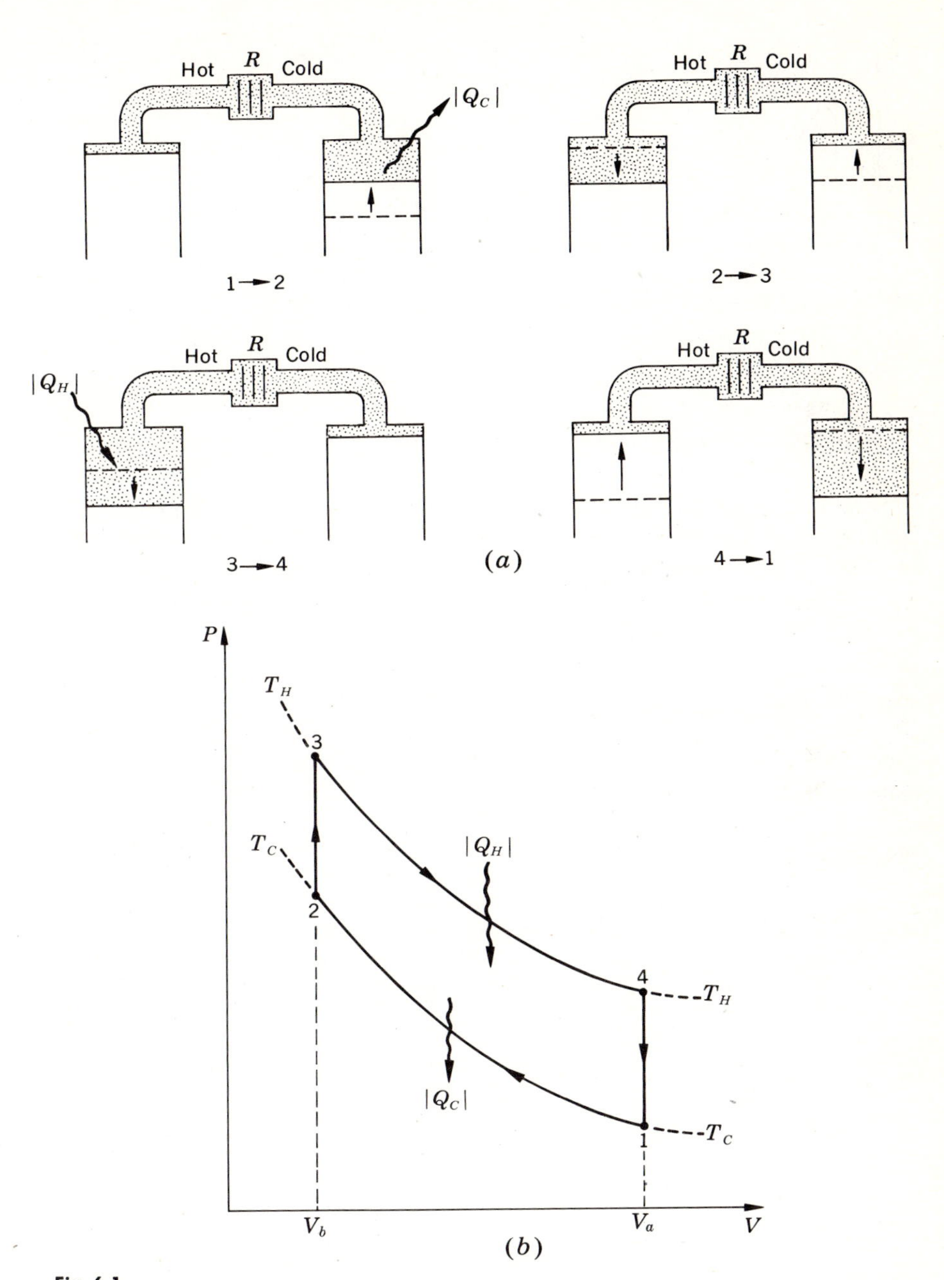

Fig. 6-1

(a) Schematic diagram of steps in the operation of an idealized Stirling engine. (The numbers under each diagram refer to the processes shown on PV diagram in Fig. 6-1b.) (b) Idealized Stirling engine cycle on PV diagram.

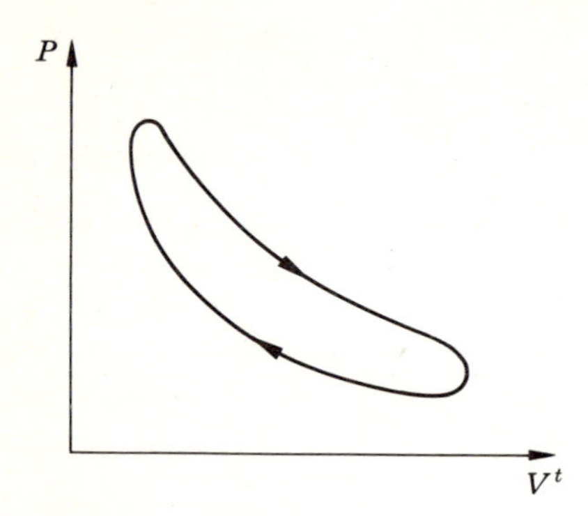

Fig. 6-2
Indicator diagram of an actual Stirling engine.

The net result of the cycle is the absorption of heat $|Q_H|$ at the high temperature T_H, the rejection of heat $|Q_C|$ at the low temperature T_C, and the delivery of work $|W| = |Q_H| - |Q_C|$ to the surroundings, with no net heat transfer resulting from the two constant-volume processes. It must be emphasized that Fig. 6-1*b* represents an *idealized* Stirling cycle. An indicator diagram of an actual Stirling cycle would look much more like the diagram in Fig. 6-2. Even if the idealized cycle could be realized in practice, there would still be some heat $|Q_C|$ rejected at the lower temperature, and therefore all the heat input $|Q_H|$ could not be converted into work.

Example 6-2

Determine an expression for the thermal efficiency of an idealized Stirling cycle in which the working substance is an ideal gas. Refer to Fig. 6-1*b*.

We apply the first law to each step, assuming quasi-static, frictionless operation. For an isothermal change of state of an ideal gas, $\Delta U = 0$ by Eq. (5-15). Thus $\Delta U_{12} = 0$, and the first law becomes

$$Q_{12} = -W_{12}.$$

Taking as a basis for our derivation 1(mol) of gas, we find from Eq. (3-3) that

$$W_{12} = RT_C \ln \frac{V_a}{V_b}$$

and therefore

$$Q_{12} = RT_C \ln \frac{V_b}{V_a},$$

or, since $V_b < V_a$,

$$|Q_C| = |Q_{12}| = RT_C \ln \frac{V_a}{V_b}. \tag{A}$$

Step $2 \to 3$ is one of constant volume; therefore $W_{23} = 0$, and the first law becomes

$$Q_{23} = \Delta U_{23}.$$

But we have from Eq. (5-26) that

$$\Delta U_{23} = \int_{T_C}^{T_H} C_V \, dT,$$

and therefore

$$Q_{23} = \int_{T_C}^{T_H} C_V \, dT. \qquad (B)$$

This is the quantity of heat supplied to the gas by the regenerator. Since $T_H > T_C$, it is positive in sign, as it must be for heat addition.

Steps $3 \to 4$ and $4 \to 1$ are treated similarly to steps $1 \to 2$ and $2 \to 3$. The expressions for the heat terms are

$$|Q_H| = |Q_{34}| = RT_H \ln \frac{V_a}{V_b} \qquad (C)$$

and

$$Q_{41} = \int_{T_H}^{T_C} C_V \, dT. \qquad (D)$$

The term Q_{41} represents the heat transferred from the gas to the regenerator during the last step of the cycle. Comparison of Eqs. (B) and (D) shows that

$$Q_{41} = -Q_{23},$$

or that

$$|Q_{41}| = |Q_{23}| = |Q_R|,$$

in accord with the idealization that there is no *net* exchange of heat between the gas and the regenerator during operation of the cycle.

The thermal efficiency of the cycle is found by combination of Eqs. (A) and (C) with Eq. (6-3):

$$\eta = 1 - \frac{|Q_C|}{|Q_H|} = 1 - \frac{RT_C \ln(V_a/V_b)}{RT_H \ln(V_a/V_b)},$$

or

$$\eta = 1 - \frac{T_C}{T_H}.$$

Thus the efficiency of the ideal-gas Stirling cycle can be increased either by raising T_H or by lowering T_C.

6-3 Steam Power Plant

A schematic diagram of an elementary steam power plant is shown in Fig. 6-3*a*. To understand how such a plant works, one must follow the pressure and volume changes of a unit mass of water as it is conveyed from the condenser, through the boiler, into an expansion device (turbine), and back to the condenser. The water in the condenser is at a pressure less than

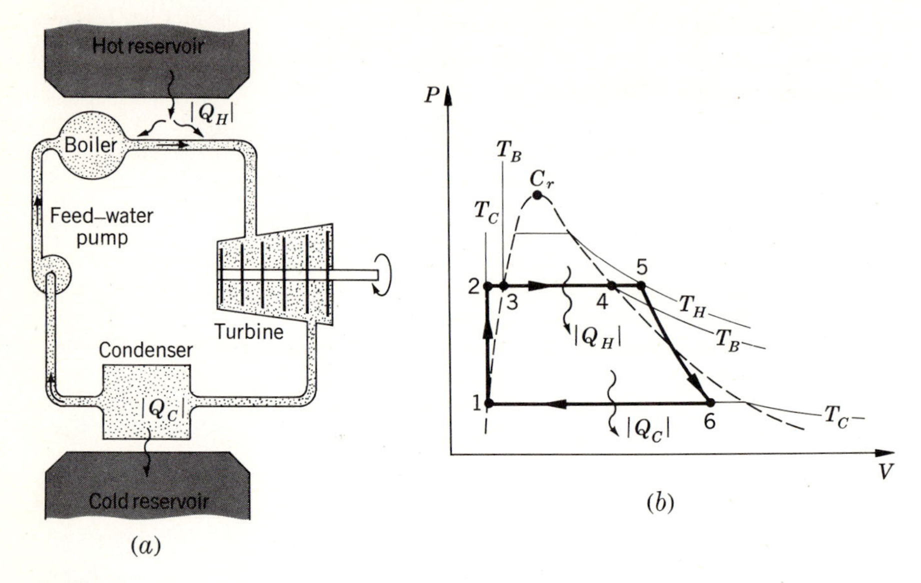

Fig. 6-3
(a) Elementary steam power plant. (b) PV diagram of Rankine cycle.

atmospheric and at a temperature less than the normal boiling point. By means of a pump it is introduced into the boiler, which is at a much higher pressure and temperature. In the boiler the water is first heated to its boiling point and then vaporized, both processes taking place approximately at constant pressure. The steam is then superheated at the same pressure. It is then allowed to flow into a turbine, where it expands approximately adiabatically against a set of rotating blades, until its pressure and temperature drop to that of the condenser. In the condenser, finally, the steam condenses into water at the same temperature and pressure as at the beginning, and the cycle is complete.

In the actual operation of the steam plant there are several processes that render an exact analysis difficult. These are: (1) acceleration and turbulence caused by the pressure difference required to cause flow of the steam from one part of the apparatus to another, (2) friction, and (3) conduction of heat through the walls during expansion of the steam.

For a first approximation to the modeling of a steam plant we introduce some simplifying assumptions which, although not strictly realizable in practice, provide at least an upper limit to the efficiency of such a plant and which define a cycle called the *Rankine cycle*, in terms of which the actual behavior of a steam plant may be discussed.

In Fig. 6-3*b* three isotherms of water are shown on a *PV* diagram: one at

T_C corresponding to the temperature of the condenser, another at T_B for the temperature of the boiler, and a third at a still higher temperature T_H. The dashed curves are the liquid and the vapor saturation curves. In the Rankine cycle all processes are assumed to be well-behaved; complications that arise from acceleration, turbulence, friction, and heat losses are disregarded. Starting at point 1, representing the state of a unit mass of saturated liquid water at the temperature and pressure of the condenser, the Rankine cycle comprises the following six processes:

$1 \rightarrow 2$ Adiabatic compression of water to the pressure of the boiler (only a very small change of temperature takes place during this process).
$2 \rightarrow 3$ Isobaric heating of water to the boiling point.
$3 \rightarrow 4$ Isobaric, isothermal vaporization of water into saturated steam.
$4 \rightarrow 5$ Isobaric superheating of steam into superheated steam at temperature T_H.
$5 \rightarrow 6$ Adiabatic expansion of steam into wet steam.
$6 \rightarrow 1$ Isobaric, isothermal condensation of steam into saturated water at the temperature T_C.

During the processes $2 \rightarrow 3$, $3 \rightarrow 4$, and $4 \rightarrow 5$, heat $|Q_H|$ enters the system from a hot reservoir; whereas during the condensation process $6 \rightarrow 1$, heat $|Q_C|$ is rejected by the system to a reservoir at T_C. This condensation process *must* occur in order to bring the system back to its initial state 1. Since heat is always rejected during the condensation of water, $|Q_C|$ cannot be made equal to zero, and therefore the heat input $|Q_H|$ cannot be converted completely into work.

6-4 Internal Combustion Engines: Otto Cycle and Diesel Cycle

In the gasoline engine, the cycle involves the performance of six processes, four of which require motion of the piston and are called *strokes*:

1 *Intake stroke.* A mixture of gasoline vapor and air is drawn into the cylinder by the suction stroke of the piston. The pressure of the outside is greater than that of the mixture by an amount sufficient to cause acceleration and to overcome friction.
2 *Compression stroke.* The mixture of gasoline vapor and air is compressed until its pressure and temperature rise considerably. This is accomplished by the compression stroke of the piston, in which friction, acceleration, and heat loss by conduction are present.
3 *Explosion.* Combustion of the hot mixture is caused to take place very

rapidly by an electric spark. The resulting combustion products attain a very high pressure and temperature, but the volume remains unchanged. The piston does not move during this process.

4 *Power stroke.* The hot combustion products expand and push the piston out, thus suffering a drop in pressure and temperature. This is the power stroke of the piston and is also accompanied by friction, acceleration, and heat conduction.

5 *Valve exhaust.* The combustion products at the end of the power stroke are still at a higher pressure and temperature than the outside. An exhaust valve allows gas to escape until the pressure drops to that of the atmosphere. The piston does not move during this process.

6 *Exhaust stroke.* The piston pushes almost all the remaining combustion products out of the cylinder by exerting a pressure sufficiently larger than that of the outside to cause acceleration and overcome friction. This is the exhaust stroke.

In the processes above there are several phenomena that render an exact mathematical analysis almost impossible. Among these are friction, acceleration, loss of heat by conduction, and the chemical reaction between gasoline vapor and air. A drastic but useful simplification results if we ignore these troublesome effects. When this is done, we have a sort of idealized gasoline engine that performs a cycle known as an *Otto cycle.*

An idealization of the behavior of a gasoline engine is based on the following assumptions: (1) The working substance is 1(mol) or a unit mass of an ideal gas with constant heat capacities. (2) All processes are quasi-static. (3) There is no friction. On the basis of these assumptions the ideal-gas Otto cycle (or *air-standard* Otto cycle) is composed of six simple processes of an ideal gas, which are plotted on a PV diagram in Fig. 6-4a.

Step $5 \rightarrow 1$ represents a quasi-static isobaric intake at a low pressure P_0. There is no friction and no acceleration. The volume of gas in the cylinder increases from V_5 to V_1.

Step $1 \rightarrow 2$ represents a quasi-static, adiabatic compression. There is no friction and no loss of heat through the cylinder wall. The temperature rises from T_1 to T_2 according to Eq. (5-30):

$$T_1 V_1^{\gamma-1} = T_2 V_2^{\gamma-1}.$$

Step $2 \rightarrow 3$ represents a quasi-static isochoric increase of temperature and pressure, brought about by an absorption of heat $|Q_H|$ from a series of external reservoirs whose temperatures range from T_2 to T_3. If there were only one reservoir at temperature T_3, the flow of heat would not be quasi-static. This process is meant to approximate the effect of the explosion in a gasoline engine.

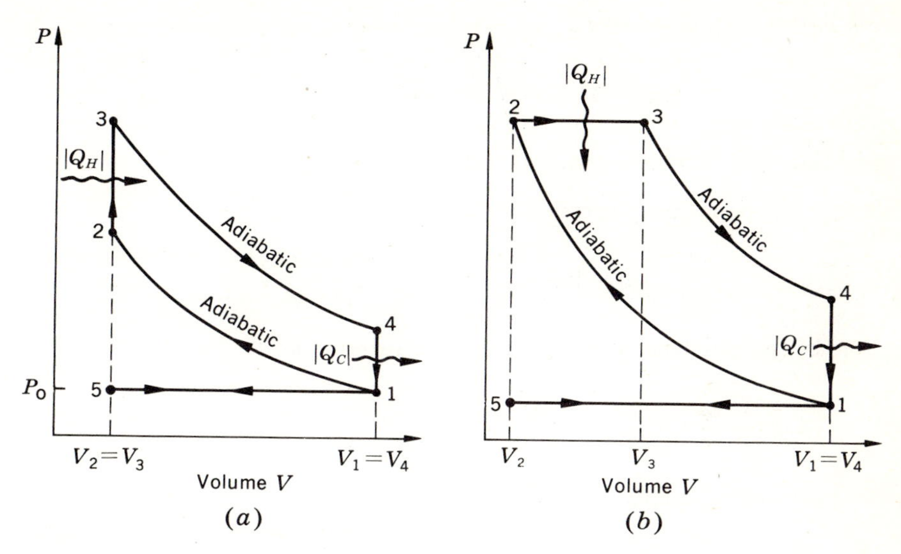

Fig. 6-4
(a) Otto cycle. (b) Diesel cycle.

Step $3 \rightarrow 4$ represents a quasi-static adiabatic expansion, involving a drop in temperature from T_3 to T_4 according to the equation

$$T_3 V_2^{\gamma-1} = T_4 V_1^{\gamma-1}.$$

Step $4 \rightarrow 1$ represents a quasi-static isochoric drop in temperature and pressure, brought about by a rejection of heat $|Q_C|$ to a series of external reservoirs ranging in temperature from T_4 to T_1. This process is meant to approximate the drop to pressure P_0 when the exhaust valve opens.

Step $1 \rightarrow 5$ represents a quasi-static isobaric exhaust at pressure P_0. The volume decreases from V_1 to V_5.

The two isobaric processes $5 \rightarrow 1$ and $1 \rightarrow 5$ obviously cancel each other and need not be considered further. Of the four remaining processes, only two involve a flow of heat. There occur an absorption of $|Q_H|$ units of heat at high temperatures from $2 \rightarrow 3$ and a rejection of $|Q_C|$ units of heat at lower temperatures from $4 \rightarrow 1$, as indicated in Fig. 6-4*a*.

Assuming C_V to be constant along the line $2 \rightarrow 3$, we get, since $W = 0$ (isochoric process),

$$Q_{23} = \Delta U_{23} = \int_{T_2}^{T_3} C_V \, dT = C_V(T_3 - T_2),$$

and thus $$|Q_H| = |Q_{23}| = C_V(T_3 - T_2).$$

Similarly, for process $4 \rightarrow 1$,

$$Q_{41} = \Delta U_{41} = \int_{T_4}^{T_1} C_V \, dT = C_V(T_1 - T_4),$$

and thus $\quad |Q_C| = |Q_{41}| = C_V(T_4 - T_1).$

The thermal efficiency is therefore

$$\eta = 1 - \frac{|Q_C|}{|Q_H|} = 1 - \frac{T_4 - T_1}{T_3 - T_2}.$$

For the two adiabatic processes we have the equations

$$T_4 V_1^{\gamma-1} = T_3 V_2^{\gamma-1}$$

and

$$T_1 V_1^{\gamma-1} = T_2 V_2^{\gamma-1},$$

which, after subtraction, yield

$$(T_4 - T_1) V_1^{\gamma-1} = (T_3 - T_2) V_2^{\gamma-1},$$

or

$$\frac{T_4 - T_1}{T_3 - T_2} = \left(\frac{V_2}{V_1}\right)^{\gamma-1}.$$

Denoting the ratio V_1/V_2 by r, where r is called the *compression ratio*, we have finally

$$\eta = 1 - \frac{1}{(V_1/V_2)^{\gamma-1}} = 1 - \frac{1}{r^{\gamma-1}} \tag{6-4}$$

In an actual gasoline engine, r cannot be made greater than about 10, because if r is larger, the rise of temperature upon compression of the mixture of gasoline and air is great enough to cause an explosion before the advent of the spark. This is called *preignition.* Taking r equal to 10 and γ equal to $\frac{7}{5}$ (the diatomic gas value),

$$\eta = 1 - \frac{1}{10^{2/5}}$$

$$\approx 0.60 = 60 \text{ percent.}$$

All the troublesome effects present in an actual gasoline engine, such as acceleration, turbulence, and heat conduction by virtue of finite temperature differences, are such as to make the efficiency much lower than that of the ideal-gas Otto cycle.

In the Diesel engine, only air is admitted on the intake. The air is compressed adiabatically until the temperature is high enough to ignite oil that is sprayed into the cylinder after the compression. The rate of supply of oil is adjusted so that combustion takes place approximately isobarically, the piston moving out during combustion. The rest of the cycle—namely, power stroke, valve exhaust, and exhaust stroke—is exactly the same as in the gasoline engine. The usual troublesome effects, such as chemical reaction, friction, acceleration, and heat losses, are present in the Diesel engine as in the gasoline engine. Ignoring these effects by making the same assumptions as before, we are left with an idealized engine that performs an *ideal-gas Diesel cycle* (or *air-standard Diesel cycle*). If the line $2 \rightarrow 3$ in Fig. 6-4*a* is imagined horizontal instead of vertical, the resulting cycle will be the ideal-gas Diesel cycle. This is shown in Fig. 6-4*b*.

It is a simple matter to show that the efficiency of an engine operating in an idealized Diesel cycle is given by

$$\eta = 1 - \frac{1}{\gamma} \frac{(1/r_E)^\gamma - (1/r)^\gamma}{(1/r_E) - (1/r)}, \tag{6-5}$$

where $$r_E = \frac{V_1}{V_3} = \text{expansion ratio},$$

and $$r = \frac{V_1}{V_2} = \text{compression ratio}.$$

In practice, the compression ratio of a Diesel engine can be made much larger than that of a gasoline engine, because there is no chance of preignition since only air is compressed. Taking, for example, $r = 15$, $r_E = 5$, and $\gamma = \frac{7}{5}$,

$$\eta = 1 - \frac{5}{7} \frac{(\frac{1}{5})^{7/5} - (\frac{1}{15})^{7/5}}{\frac{1}{5} - \frac{1}{15}}$$

$$\approx 0.56 = 56 \text{ percent}.$$

The efficiencies of actual Diesel engines are still lower, of course, for the reasons mentioned in connection with the gasoline engine.

In the Diesel engine just considered, four strokes of the piston are needed for the execution of a cycle, and only one of the four is a power stroke. Since only air is compressed in the Diesel engine, it is possible to do away with the exhaust and intake strokes and thus complete the cycle in two strokes. In the two-stroke-cycle Diesel engine, every other stroke is a power stroke, and thus the power is doubled. The principle is very simple: At the conclusion of the power stroke, when the cylinder is full of combustion products, the valve opens, exhaust takes place until the combustion products are at atmospheric pressure, and then instead of using the piston itself to exhaust the remaining gases, fresh air is blown into the cylinder, replacing the combustion products. A blower, operated by the engine itself, is used for this purpose, and thus it accomplishes in one simple operation what formerly required two separate piston strokes.

6-5 Kelvin-Planck Statement of the Second Law

In the preceding pages, four different heat engines have been briefly and somewhat superficially described. There are, of course, more types of engines and a tremendous number of structural details, methods of increasing thermal efficiency, mathematical analyses, etc., included within the subject matter of engineering thermodynamics. Thermodynamics owes its origin to the attempt to convert heat into work and to develop the theory of operation of devices for this purpose. It is therefore fitting that one of the fundamental laws of thermodynamics is based upon the operation of heat engines. Reduced to their simplest terms, the important characteristics of heat-engine cycles may be summed up as follows:

1 There is some process or series of processes during which there is an absorption of heat from an external reservoir at a high temperature (called simply the *hot reservoir*).
2 There is some process or series of processes during which heat is rejected to an external reservoir at a lower temperature (called simply the *cold reservoir*).
3 There is as a result of the complete series of processes a net *output* of work.

This is represented schematically in Fig. 6-5. No engine has ever been developed that converts the heat extracted from one reservoir totally into work; some heat is always rejected to a reservoir at a lower temperature. This statement, which is the result of engineering experience, constitutes the *second law of thermodynamics* and has been formulated in several ways. The original statement of Kelvin[1] is: "It is impossible by means of inanimate material agency to derive mechanical effect from any portion of matter by cooling it below the temperature of the coldest of the surrounding objects." In the words of Planck,[2] "It is impossible to construct an engine which,

[1] Lord Kelvin, or William Thomson (1824–1907), an English physicist and engineer. A founder of thermodynamics and a pioneer in the application of thermodynamics to practical processes.

[2] Max Planck (1858–1947), a German physicist. Planck's contributions to the theory of radiation formed the basis for the development of the quantum theory.

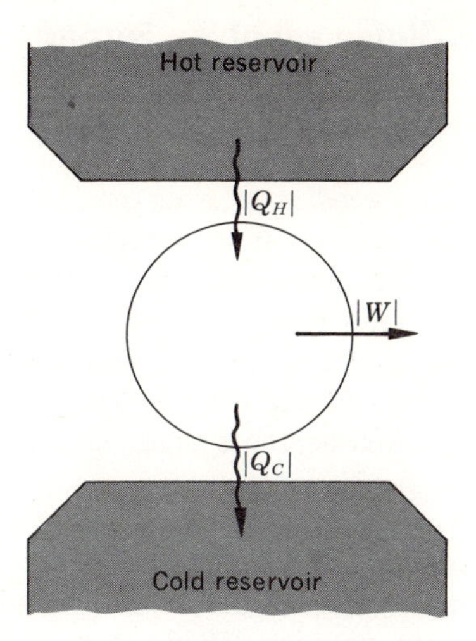

Fig. 6-5
Symbolic representation of heat engine.

working in a complete cycle, will produce no effect other than the raising of a weight and the cooling of a heat reservoir." We may combine these statements into one equivalent statement, to which we shall refer hereafter as the *Kelvin-Planck statement of the second law,* thus:

> *No process is possible whose sole result is the absorption of heat from a reservoir and the conversion of this heat into work.*

If the second law were not true, a steamship could be propelled by the extraction of heat from the ocean and a power plant could operate by extraction of heat from the surrounding air. Neither of these "impossibilities" violates the first law of thermodynamics. After all, both the ocean and the surrounding air contain an enormous store of internal energy that, in principle, may be extracted in the form of a flow of heat. Nothing in the first law precludes the possibility of conversion of this heat completely into work. The second law, therefore, is not a deduction from the first but stands by itself as a separate law of nature, referring to an aspect of nature different from that contemplated by the first law. The first law denies the possibility of creating or destroying energy; the second denies the possibility of using energy in a particular way. The continual operation of a machine that creates its own energy and thus violates the first law is called *perpetual motion of the first kind.* The operation of a machine that exchanges heat with only one heat reservoir, thus violating the second law, is called *perpetual motion of the second kind.*

6-6 The Refrigerator. The Clausius Statement of the Second Law

We have seen that a heat engine is a device by which a system is taken through a cycle in such a direction that some heat is absorbed while the temperature is high, a smaller amount is rejected at a lower temperature, and a net amount of work is done on the outside. If we imagine a cycle performed in a direction opposite to that of an engine, the net result would be the absorption of some heat at a low temperature, the rejection of a *larger* amount at a higher temperature, and a net amount of work done *on* the system. A device that performs a cycle in this direction is called a *refrigerator*, and the working substance, taken as the system, is called a *refrigerant*.

The Stirling cycle is capable of being reversed, and when reversed, it gives rise to one of the most useful types of refrigerator. The operation of an ideal Stirling refrigerator may best be understood with the aid of the schematic diagrams shown in Fig. 6-6*a*, and the accompanying PV diagram of Fig. 6-6*b*.

$1 \rightarrow 2$ While the right piston remains stationary, the left piston moves up, compressing the gas isothermally at the temperature T_H and rejecting heat $|Q_H|$ to the hot reservoir.

$2 \rightarrow 3$ Both pistons move the same amount simultaneously, forcing gas through the regenerator, where some heat $|Q_R|$ is given up. The cold gas emerges in the right-hand space. This process takes place at constant volume.

$3 \rightarrow 4$ While the left piston remains stationary, the right piston moves down and causes an isothermal expansion at the low temperature T_C, during which heat $|Q_C|$ is absorbed by the gas from the cold reservoir.

$4 \rightarrow 1$ Both pistons move and force gas at constant volume from the cold to the hot end through the regenerator, where the gas absorbs approximately the same heat $|Q_R|$ that was supplied to the regenerator in process $2 \rightarrow 3$.

In order to gain a little more insight into the working of a refrigerator, let us consider some of the details of a commercial refrigeration plant that are reflected in most electric home refrigerators. The schematic diagram in Fig. 6-7*a* shows the path of a constant mass of refrigerant as it is conveyed from the liquid storage, where it is at the temperature and pressure of the condenser, through the throttling valve, through the evaporator, into the compressor, and finally back to the condenser.

In the condenser the refrigerant is at a high pressure; it is cooled by air

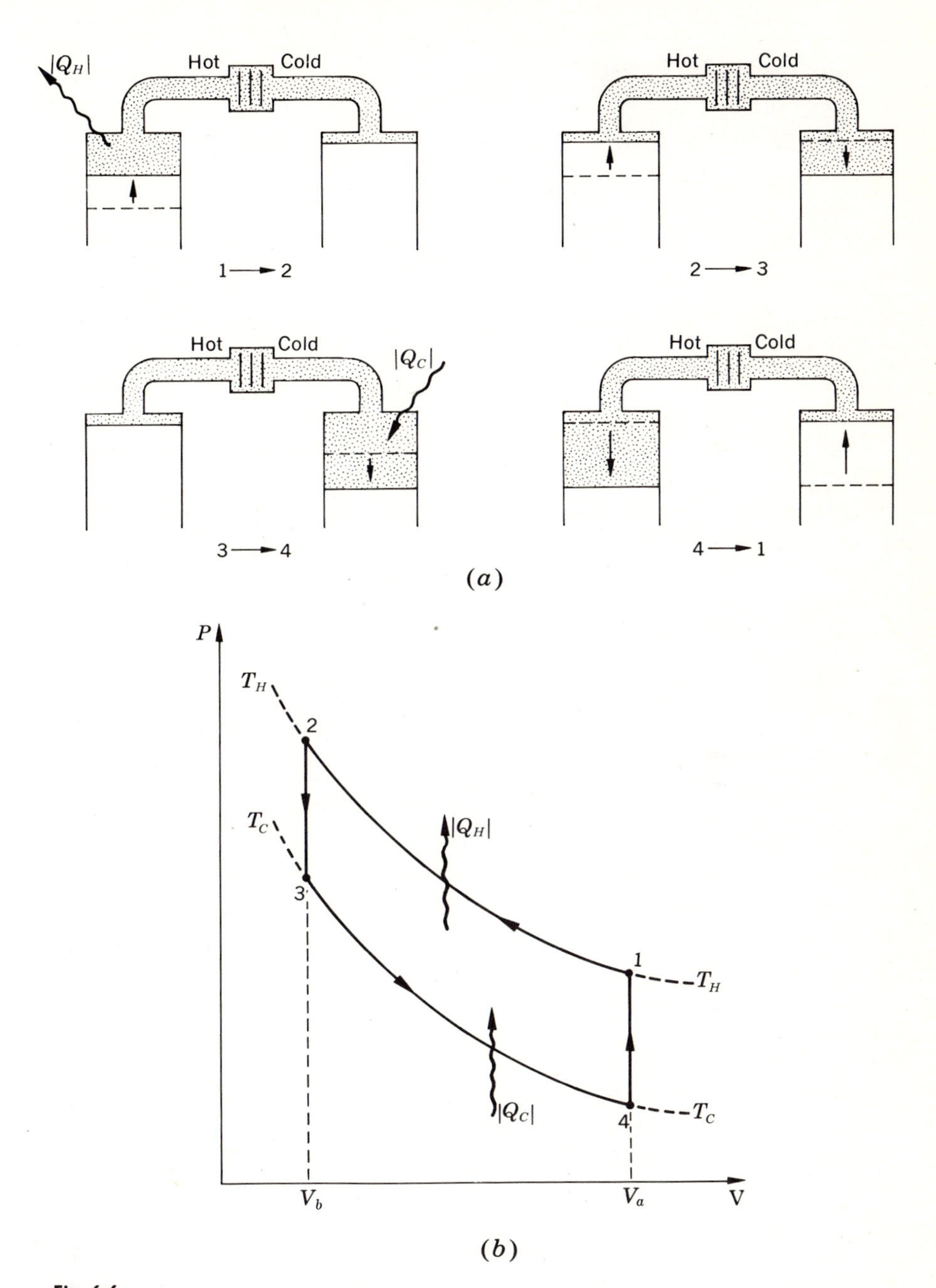

Fig. 6-6

(a) Schematic diagram of steps in the operation of idealized Stirling refrigerator. (b) Idealized Stirling refrigeration cycle on PV diagram.

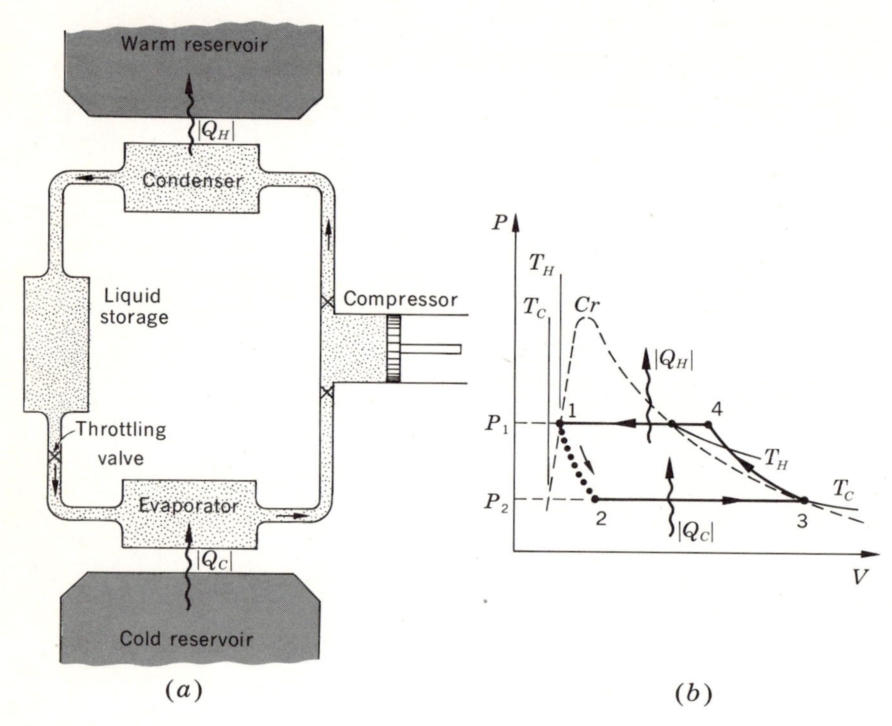

Fig. 6-7
(a) Elementary refrigeration plant. (b) PV diagram of commercial refrigerator.

or water to as low a temperature as practical, and leaves as a liquid. When a fluid passes through a narrow opening (a needle valve) from a region of constant high pressure to a region of constant lower pressure adiabatically, it is said to undergo a *throttling process* (see Example 4-8), or a *Joule-Thomson* or *Joule-Kelvin expansion.* This process will be considered in some detail in Chap. 10. It is a property of saturated liquids (not of gases) that a throttling process always produces cooling and partial vaporization. In the evaporator the fluid is completely vaporized, with the heat of vaporization being supplied by the cold reservoir (in which the refrigerated materials are embedded). The vapor is then compressed adiabatically, thereby increasing in temperature. In the condenser, the vapor is cooled until it condenses and becomes completely liquefied.

The ideal refrigeration cycle depicted on a *PV* diagram in Fig. 6-7*b* results when we ignore the usual difficulties due to turbulence, friction, heat losses, etc. In Fig. 6-7*b*, two isotherms of a fluid such as ammonia or Freon are shown on a *PV* diagram—one at T_H, the temperature of the condenser, and the other at T_C, the temperature of the evaporator. Starting at the point 1, representing the state of a unit mass of saturated liquid at the temperature and pressure of the condenser, the commercial refrigeration cycle comprises the following processes:

$1 \rightarrow 2$ Throttling process involving a drop of pressure and temperature. The states between the initial and final states of a fluid during a throttling process cannot be described with the aid of thermodynamic coordinates referring to the system as a whole and, therefore, cannot be represented by points on a PV diagram. Hence the series of dots between 1 and 2.

$2 \rightarrow 3$ Isothermal, isobaric vaporization in which heat $|Q_C|$ is absorbed by the refrigerant at the low temperature T_C, thereby cooling the materials of the cold reservoir.

$3 \rightarrow 4$ Adiabatic compression of the vapor to a temperature higher than that of the condenser T_H.

$4 \rightarrow 1$ Isobaric cooling and condensation at T_H.

The purpose of any refrigerator is to extract as much heat as possible from a cold reservoir with the expenditure of as little work as possible. The "output," so to speak, is the heat extracted from the cold reservoir, and the "input" is work. A convenient measure, therefore, of the performance of a refrigerator is expressed by the *coefficient of performance* ω, where

$$\omega = \frac{\text{heat extracted from cold reservoir}}{\text{work done on refrigerant}}.$$

If in one cycle heat $|Q_C|$ is absorbed by the refrigerant from the cold reservoir and work $|W|$ is done by the electric motor that operates the refrigerator, then

$$\omega \equiv \frac{|Q_C|}{|W|} \tag{6-6}$$

Example 6-3

An ideal-gas Stirling refrigerator absorbs heat $|Q_C|$ at 0(°C) and rejects heat $|Q_H|$ at 30(°C). What is the coefficient of performance of the refrigerator?

We assume quasi-static, frictionless operation. With reference to Fig. 6-6*b* we find, by the same methods used in Example 6-2, that

$$W_{12} = RT_H \ln \frac{V_a}{V_b},$$

$$W_{34} = -RT_C \ln \frac{V_a}{V_b},$$

and $$|Q_C| = |Q_{34}| = |W_{34}| = RT_C \ln \frac{V_a}{V_b}.$$

The net work $|W|$ is

$$|W| = |W_{12} + W_{34}| = R(T_H - T_C) \ln \frac{V_a}{V_b}.$$

Substitution into the definition of ω, Eq. (6-6), gives

$$\omega = \frac{T_C}{T_H - T_C}.$$

For the given values of T_C and T_H, we find

$$\omega = \frac{273}{303 - 273} = 9.1$$

The coefficient of performance of a *real* refrigerator may also be considerably larger than unity. If, for the sake of argument, one assumes the value 5,

$$\omega = \frac{|Q_C|}{|W|} = 5;$$

since $$|Q_C| = |Q_H| - |W|,$$

we have $$\frac{|Q_H| - |W|}{|W|} = 5,$$

or $$\frac{|Q_H|}{|W|} = 6.$$

Thus the *heat liberated at the higher temperature is equal to six times the work done.* If the work is supplied by an electric motor, for every joule of electrical energy supplied, 6(J) of heat will be liberated; whereas if 1(J) of electrical energy were dissipated in a resistor, one could obtain at most 1(J) of heat. Consequently, it would seem to be highly advantageous to heat a house by refrigerating the outdoors.

This was first proposed in 1852 by Lord Kelvin, who designed a machine for the purpose. The device was never built. It remained for Haldane, about 75 years later, to use the principle and heat his house in Scotland by refrigerating the outdoor air.

In recent years, many devices known as "heat pumps" have appeared on the market for warming the house in winter by refrigerating either the ground, the outside air, or available water. By turning a valve and reversing the flow of refrigerant, one may use the heat pump to cool the house in summer, as shown in Fig. 6-8. Various commercial units have coefficients of

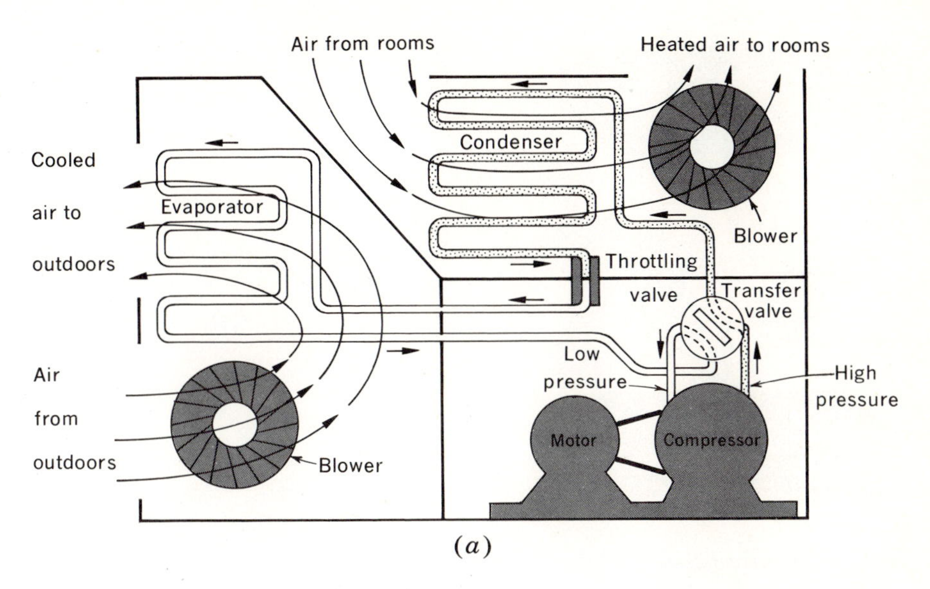

(*a*)

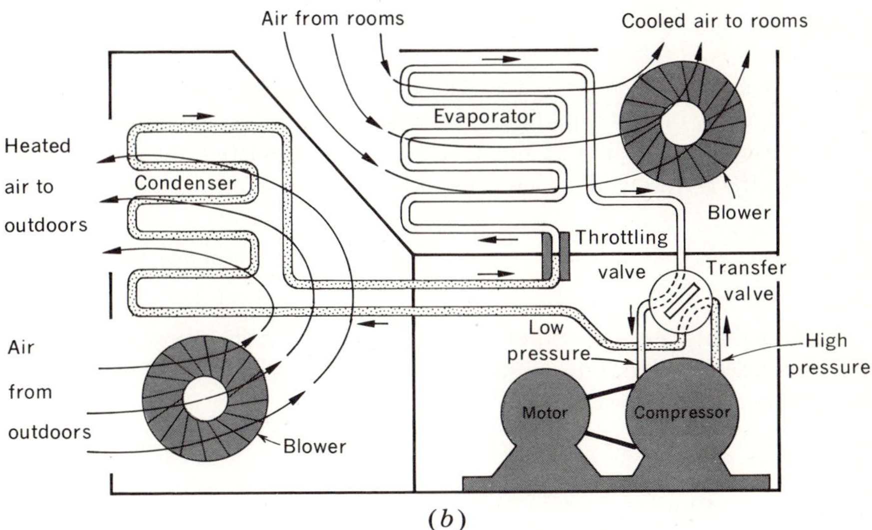

(*b*)

Fig. 6-8
(*a*) *Heating the house by refrigerating the outside air.* (*b*) *Cooling the house by heating the outside air.* (*J. Partington, Jr., G. E. Educational Service News, December, 1951.*)

performance ranging from about 2 to 7. The design, installation, and operation of such units represent an important branch of engineering.

The operation of a refrigerator may be symbolized by the schematic diagram shown in Fig. 6-9, which should be compared with the corre-

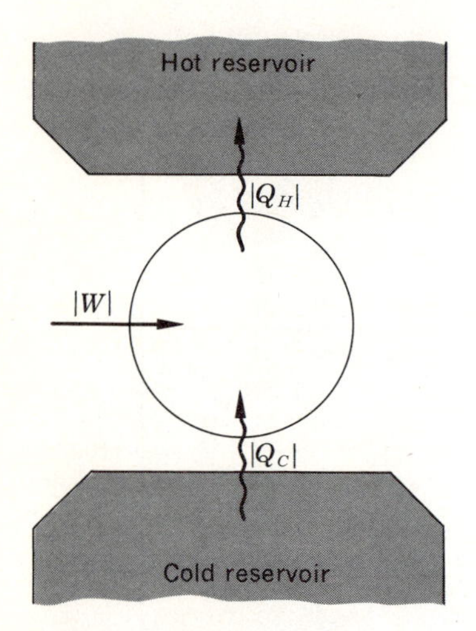

Fig. 6-9
Symbolic representation of refrigerator.

sponding engine diagram of Fig. 6-5. Work is always required to transfer heat from a cold to a hot reservoir. In household refrigerators, this work is usually done by an electric motor, whose cost of operation appears regularly on the monthly bill. It would be a boon to mankind if no external supply of energy were needed, but it must certainly be admitted that experience indicates the contrary. This leads us to the *Clausius*[1] *statement of the second law*:

> *No process is possible whose* **sole** *result is the transfer of heat from a cooler to a hotter body.*

At first sight, the Kelvin-Planck and the Clausius statements appear to be quite unconnected, but we show in the following example that they are in all respects equivalent.

Example 6-4

We wish to demonstrate the equivalence of the Kelvin-Planck and the Clausius statements of the second law. We adopt the following notation:

K = truth of the Kelvin-Planck statement,
$-K$ = falsity of the Kelvin-Planck statement,
C = truth of the Clausius statement,
$-C$ = falsity of the Clausius statement.

[1] R. J. Clausius (1822–1888), a German physicist. Clausius was one of the major contributors to the development of the second law and the kinetic theory of gases; he coined the word *entropy*.

Two propositions or statements are said to be equivalent when the truth of one implies the truth of the second and the truth of the second implies the truth of the first. Using the symbol $\supset$ to mean "implies" and the symbol $\equiv$ to denote equivalence, we have, by definition,

$$K \equiv C$$

when $\quad K \supset C \quad$ and $\quad C \supset K.$

Now, it may easily be shown that

$$K \equiv C$$

also when $\quad -K \supset -C \quad$ and $\quad -C \supset -K.$

Thus, in order to demonstrate the equivalence of K and C, we have to show that a violation of one statement implies a violation of the second, and vice versa.

1 To prove that $-C \supset -K$, consider a refrigerator, shown in the left-hand side of Fig. 6-10, which requires *no work* to transfer Q_2 units of heat from a cold reservoir to a hot reservoir and which therefore violates the Clausius statement. Suppose that a heat engine (on the right) also operates between the same two reservoirs in such a way that heat Q_2 is delivered to the cold reservoir. The engine, of course, does not violate any law, but the refrigerator and engine *together* constitute a self-acting device whose sole effect is to take heat $|Q_1| - |Q_2|$ from the hot reservoir and to convert *all* this heat into work. Therefore the refrigerator and engine together constitute a violation of the Kelvin-Planck statement.
2 To prove that $-K \supset -C$, consider a heat engine, shown on the left-hand side of Fig. 6-11, which rejects no heat to the cold reservoir and which therefore violates the Kelvin-Planck statement. Suppose that a refrigerator (on the right) also operates between the same two reservoirs and uses up all the work liberated by the engine. The refrigerator violates no law, but the engine and refrigerator *together* constitute

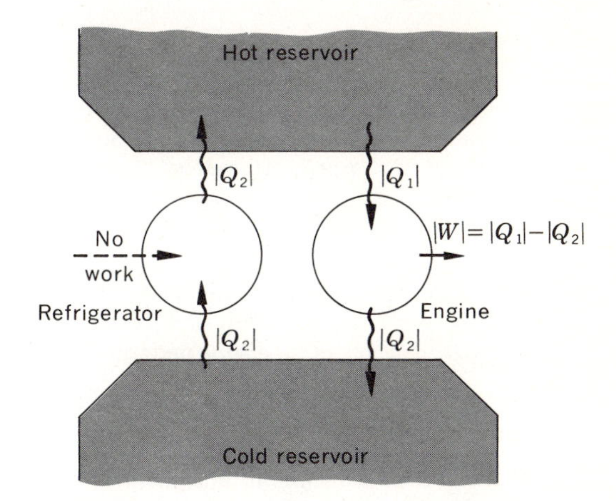

Fig. 6-10

Proof that $-C \supset -K$. The refrigerator on the left is a violation of C; the refrigerator and engine acting together constitute a violation of K.

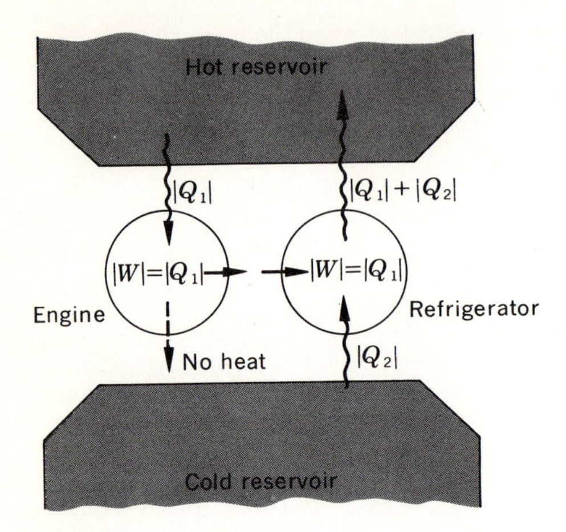

Fig. 6-11
Proof that $-K \supset -C$. *The engine on the left is a violation of K; the engine and refrigerator acting together constitute a violation of C.*

a self-acting device whose sole effect is to transfer heat Q_2 from the cold to the hot reservoir. Therefore the engine and refrigerator together constitute a violation of the Clausius statement.

We therefore arrive at the conclusion that both statements of the second law are equivalent. It is a matter of indifference which one is used in a particular argument.

Problems

6-1 Figure P6-1 represents a simplified PV diagram of the Joule ideal-gas cycle. All processes are quasi-static, and C_P is constant. Prove that the thermal efficiency of an engine performing this cycle is

$$\eta = 1 - \left(\frac{P_1}{P_2}\right)^{(\gamma-1)/\gamma}.$$

6-2 Figure P6-2 represents a simplified PV diagram of the Sargent ideal-gas cycle. All processes are quasi-static, and the heat capacities are constant. Prove that the thermal efficiency of an engine performing this cycle is

$$\eta = 1 - \gamma \frac{T_4 - T_1}{T_3 - T_2}.$$

6-3 Figure P6-3 represents an imaginary ideal-gas engine cycle. Assuming constant heat capacities, show that the thermal efficiency is

$$\eta = 1 - \gamma \frac{(V_1/V_2) - 1}{(P_3/P_2) - 1}.$$

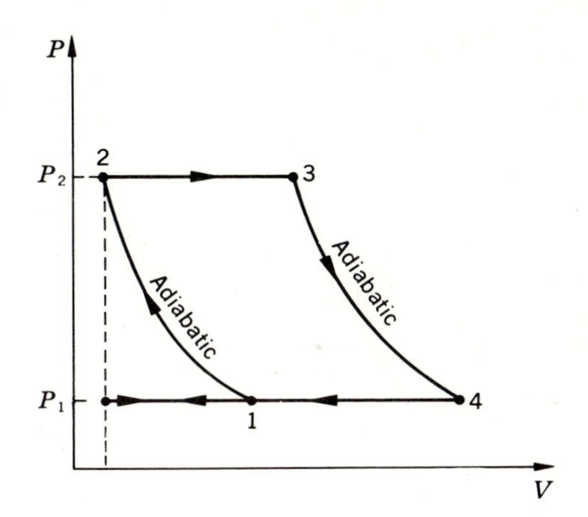

Fig. P6-1
Joule ideal-gas cycle.

6-4 An ideal-gas engine operates in a cycle which, when represented on a PV diagram, is a rectangle. Call P_1, P_2 the lower and higher pressures, respectively; call V_1, V_2 the lower and higher volumes, respectively.

(*a*) Calculate the work done in one cycle.
(*b*) Indicate which parts of the cycle involve heat flow into the gas, and calculate the amount of heat flowing into the gas in one cycle. (Assume constant heat capacities.)

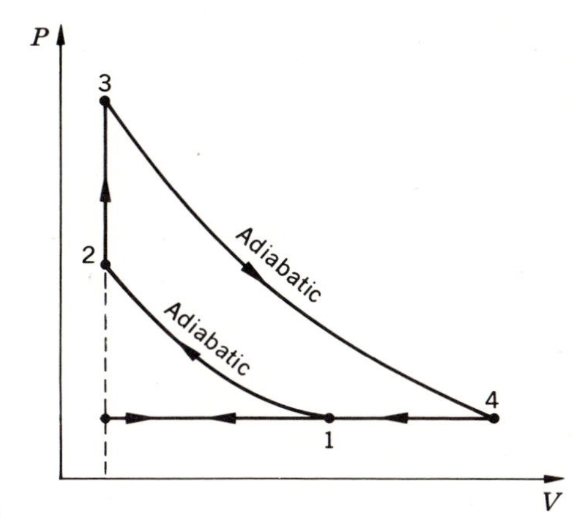

Fig. P6-2
Sargent ideal-gas cycle.

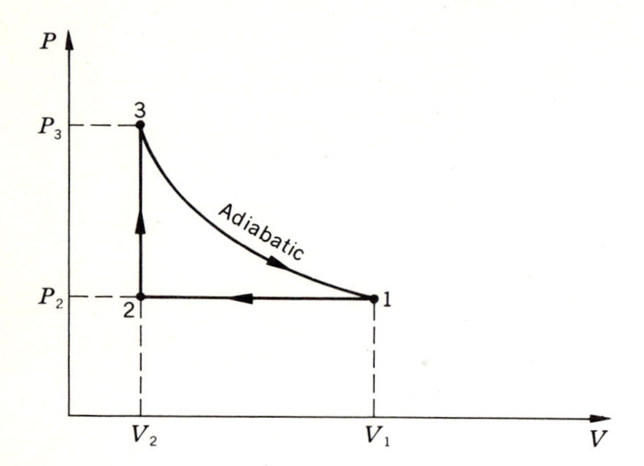

Fig. P6-3

(*c*) Show that the efficiency of this engine is

$$\eta = \frac{\gamma - 1}{\gamma P_2/(P_2 - P_1) + V_1/(V_2 - V_1)} \cdot$$

6-5 A vessel contains $600(\text{cm})^3$ of helium gas at 2(K) and $\frac{1}{36}$(atm). Take the zero of internal energy of helium to be at this point.

(*a*) The temperature is raised at constant volume to 288(K). Assuming helium to behave like an ideal monatomic gas, determine the heat absorbed and the internal energy of the helium. Can this energy be regarded as stored heat or stored work?

(*b*) The helium is now expanded adiabatically to 2(K). How much work is done, and what is the new internal energy? Has heat been converted into work without compensation, thus violating the second law?

(*c*) The helium is now compressed isothermally to its original volume. What are the quantities of heat and work in this process? What is the efficiency of the cycle? Plot on a *PV* diagram.

6-6 Prove that it is impossible for two quasi-static adiabatics to intersect. (*Hint*: Assume that they do intersect, and complete the cycle with an isothermal. Show that the performance of this cycle violates the second law.)

6-7 Show that an engine operating on an ideal-gas Otto cycle has a greater efficiency than an engine operating on an ideal-gas Diesel cycle with the same compression ratio.

7

REVERSIBILITY, CARNOT CYCLE, AND KELVIN TEMPERATURE SCALE

7-1 Reversibility and Irreversibility

In thermodynamics, work is a macroscopic concept. The performance of work may always be described in terms of the raising or lowering of an object or the winding or unwinding of a spring, i.e., by the operation of a device that serves to increase or decrease the potential energy of a mechanical system. Imagine, for the sake of simplicity, a suspended object coupled by means of suitable pulleys to a system so that any work done by or on the system can be described in terms of the raising or lowering of the object. Imagine, further, a series of reservoirs which may be put in contact with the system and in terms of which any flow of heat to or from the system may be described. We shall refer to the suspended object and the series of reservoirs as the *local surroundings* of the system. The local surroundings are, therefore, those parts of the surroundings which interact *directly* with the system. Other mechanical devices and reservoirs which are accessible and which *might* interact with the system constitute the *auxiliary surroundings* of the system—or, for want of a better expression, the *rest of the universe.* The word "universe" is used here in a very restricted technical sense, with no cosmic or celestial implications. The universe merely means a finite portion of the world consisting of the system and those surroundings which may interact with the system.

Now suppose that a process occurs in which (1) the system proceeds from an initial state i to a final state f; (2) the suspended object is lowered to an extent that W units of work are performed; and (3) a transfer of heat Q takes place from the system to the series of reservoirs. If at the conclusion of this process the system may be restored to its initial state i, the object lifted to its former level, and the reservoirs caused to part with the same amount of heat Q, without producing any changes in any other mechanical device or reservoir in the universe, the original process is said to be *reversible.* In other words, *a reversible process is one that is performed in such a way that at the conclusion of the process, both the system and the local surroundings may be restored to their initial states, without producing any changes in the rest of the universe.* A process that does not fulfill these stringent requirements is said to be *irreversible.*

The question immediately arises as to whether natural processes, i.e., the familiar processes of nature, are reversible or not. We shall show that it is a consequence of the second law of thermodynamics that all natural processes are irreversible. By considering representative types of natural processes and examining the features of these processes which are responsible for their irreversibility, we shall then be able to state the conditions under which a process may take place reversibly.

7-2 External Mechanical Irreversibility

There is a large class of processes involving the isothermal transformation of work through a system (which remains unchanged) into internal energy of a reservoir. This type of process is depicted schematically in Fig. 7-1 and is illustrated by the following five examples:

1 Irregular stirring of a viscous liquid in contact with a reservoir.
2 Coming to rest of a rotating or vibrating liquid in contact with a reservoir.
3 Inelastic deformation of a solid in contact with a reservoir.
4 Transfer of electricity through a resistor in contact with a reservoir.
5 Magnetic hysteresis of a material in contact with a reservoir.

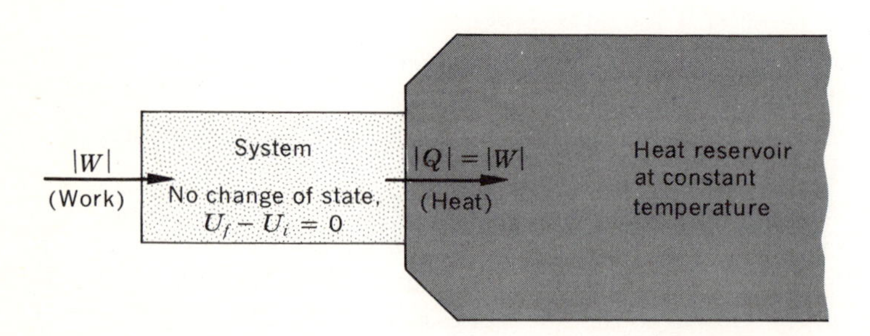

Fig. 7-1
Isothermal transformation of work through a system (which remains unchanged) into internal energy of a reservoir.

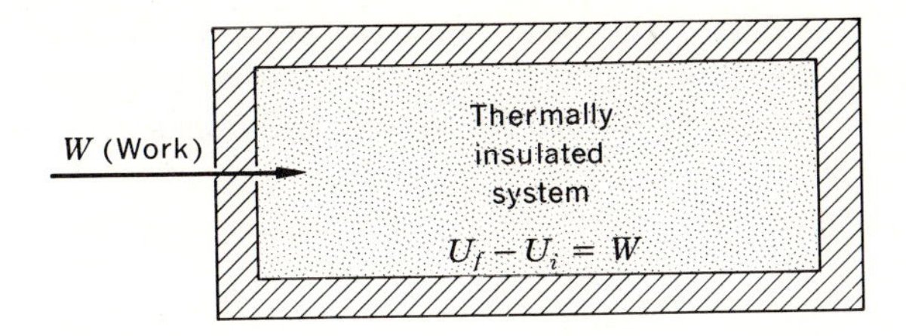

Fig. 7-2
Adiabatic transformation of work into internal energy of a system.

In order to restore the system and its local surroundings to their initial states without producing changes elsewhere, Q units of heat would have to be extracted from the reservoir and converted completely into work. Since this would involve a violation of the second law, all processes of the above type are irreversible.

Another set of processes involves the adiabatic transformation of work into internal energy of a system. This is depicted schematically in Fig. 7-2 and is illustrated by the following examples, similar to the preceding list:

1 Irregular stirring of a viscous thermally insulated liquid.
2 Coming to rest of a rotating or vibrating thermally insulated liquid.
3 Inelastic deformation of a thermally insulated solid.
4 Transfer of electricity through a thermally insulated resistor.
5 Magnetic hysteresis of a thermally insulated material.

A process of this type is accompanied by a rise of temperature of the system from, say, T_i to T_f. In order to restore the system and its local surroundings to their initial states without producing changes elsewhere, the internal energy of the system would have to be decreased by extracting $U_f - U_i$ units of heat, thus lowering the temperature from T_f to T_i, and this heat would have to be completely converted into work. Since this violates the second law, all processes of the above type are irreversible.

The transformation of work into internal energy either of a system or of a reservoir is seen to take place through the agency of such phenomena as viscosity, friction, inelasticity, electric resistance, and magnetic hysteresis. These effects are known as *dissipative effects*, and the work is said to be dissipated. Processes involving the dissipation of work into internal energy are said to exhibit *external mechanical irreversibility*. It is a matter of everyday experience that dissipative effects, particularly friction, are always present in moving devices. Friction, of course, may be reduced considerably by suitable lubrication, but experience has shown that it can never be completely eliminated. If it could, a movable device could be kept in continual operation without violating either of the two laws of thermodynamics. Such a continual motion is known as *perpetual motion of the third kind.*

7-3 Internal Mechanical Irreversibility

The following very important natural processes involve the transformation of internal energy of a system into mechanical energy and then back into internal energy again:

1 Ideal gas rushing into a vacuum (free expansion).
2 Gas seeping through a porous plug (throttling process).
3 Snapping of a stretched wire after it is cut.
4 Collapse of a soap film after it is pricked.

We shall prove the irreversibility of only the first.

During a free expansion, no interactions take place, and hence there are no local surroundings. The only effect produced is a change of state of an ideal gas from a volume V_i and temperature T to a larger volume V_f and the same temperature T. To restore the gas to its initial state, it would have to be compressed isothermally to the volume V_i. If the compression were performed quasi-statically and there were no friction between the piston and cylinder, an amount of work W would have to be done by some outside mechanical device, and an equal amount of heat would have to flow out of the gas into a reservoir at the temperature T. If the mechanical device and the reservoir are to be left unchanged, the heat would have to be extracted from the reservoir and converted completely into work. Since this last step is impossible, the process is irreversible.

At the beginning of a free expansion (see Sec. 5-6) there is a transformation of some of the internal energy into kinetic energy of "mass motion" or "streaming," and then this kinetic energy is dissipated through viscosity into internal energy again. Similarly, when a stretched wire is cut, there is first a transformation of internal energy into kinetic energy of irregular motion and of vibration and then the dissipation of this energy through inelasticity into internal energy again. In all the processes, the first energy transformation takes place as a result of mechanical instability, and the second by virtue of some dissipative effect. A process of this sort is said to exhibit *internal mechanical irreversibility.*

7-4 External and Internal Thermal Irreversibility

Consider the following processes involving a transfer of heat between a system and a reservoir by virtue of a *finite* temperature difference:

1 Conduction or radiation of heat from a system to a cooler reservoir.
2 Conduction or radiation of heat through a system (which remains unchanged) from a hot reservoir to a cooler one.

To restore, at the conclusion of a process of this type, both the system and its local surroundings to their initial states without producing changes elsewhere, heat would have to be transferred by means of a self-acting device from a cooler to a hotter body. Since this violates the second law (Clausius statement), all processes of this type are irreversible. Such processes are said to exhibit *external thermal irreversibility*.

A process involving a transfer of heat between parts of the same system because of nonuniformity of temperature is also obviously irreversible by virtue of the Clausius statement of the second law. Such a process is said to exhibit *internal thermal irreversibility*.

7-5 Chemical Irreversibility

Some of the most interesting processes that go on in nature involve a spontaneous change of internal structure, chemical composition, density, crystal form, etc. Some important examples follow.

Formation of new chemical constituents:
1 All chemical reactions.

Mixing of two different substances:
2 Diffusion of two dissimilar inert ideal gases.
3 Mixing of alcohol and water.

Sudden change of phase:
4 Freezing of supercooled liquid.
5 Condensation of supersaturated vapor.

Transport of matter between phases in contact:
6 Solution of solid in water.
7 Osmosis.

Such processes are by far the most difficult to handle and must, as a rule, be treated by special methods. Such methods constitute what is known as chemical thermodynamics. It can be shown that the diffusion of two dissimilar inert ideal gases is equivalent to two independent free expansions. Since a free expansion is irreversible, it follows that diffusion is irreversible. For the present we will accept the statement that the above processes are irreversible. Processes that involve a spontaneous change of chemical structure, density, phase, etc., are said to exhibit *chemical irreversibility.*

7-6 Conditions for Reversibility

Most processes that occur in nature are included among the general types of processes listed in the preceding articles. Living processes, such as cell division, tissue growth, etc., are no exception. If one takes into account all the interactions that accompany living processes, such processes are irreversible. It is a direct consequence of the second law of thermodynamics that *all natural processes are irreversible.*

A careful inspection of the various types of natural processes shows that all involve one or both of the following features:

1 The conditions for mechanical, thermal, or chemical equilibrium, i.e., thermodynamic equilibrium, are not satisfied.
2 Dissipative effects, such as result from viscosity, friction, inelasticity, electric resistance, and magnetic hysteresis, are present.

For a process to be reversible, it must not possess these features. If a process is performed quasi-statically, the system passes through states of thermodynamic equilibrium, which may be traversed just as well in one direction as in the opposite direction. If there are no dissipative effects, all the work done by the system during the performance of a process in one direction can be returned to the system during the reverse process. We are led, therefore, to the conclusion that a process will be reversible when:

1 It is performed quasi-statically.
2 It is not accompanied by any dissipative effects.

Since it is impossible to satisfy these two conditions perfectly, it is obvious that a reversible process is purely an ideal abstraction, quite devoid of reality. In this sense, the assumption of a reversible process in thermodynamics resembles the assumptions made so often in mechanics, such as those which refer to weightless strings, frictionless pulleys, and point masses.

All the expressions for quasi-static work and heat flow developed previously can be applied to reversible processes. Actually, we have already tacitly used the idealization of a reversible process in Chap. 6. Thus, the quasi-static, frictionless Stirling cycle of Examples 6-2 and 6-3 and the idealized Otto and Diesel cycles of Sec. 6-4 were in fact treated as reversible processes.

A heat reservoir was defined as a body of very large mass capable of absorbing or rejecting an unlimited supply of heat without suffering appreciable changes in its thermodynamic coordinates. The changes that do take place are so very slow and so minute that dissipative actions never develop. *Therefore, when heat enters or leaves a reservoir, the changes that take place* in the reservoir *are the same as those which would take place if the same quantity of heat were transferred reversibly.*

It is possible in the laboratory to approximate the conditions necessary for the performance of reversible processes. For example, if a gas is confined in a cylinder equipped with a well-lubricated piston and is allowed to expand very slowly against an opposing force provided either by an object suspended from a frictionless pulley or by an elastic spring, the gas undergoes an approximately reversible process. Similar considerations apply to a wire and to a surface film.

A reversible transfer of electricity through an electric cell may be imagined as follows: Suppose that a motor whose coils have a negligible resistance is caused to rotate until its back emf is only slightly different from the emf of the cell. Suppose further that the motor is coupled either to an object suspended from a frictionless pulley or to an elastic spring. If neither the cell itself nor the connecting wires to the motor have appreciable resistance, a reversible transfer of electricity takes place.

The reversible process is an abstraction, an idealization, in that it can be approached but never actually realized. It is nevertheless an essential concept and of enormous practical use in engineering thermodynamics, where the assumption that a process is merely quasi-static does not suffice. It is only for the reversible process, for which dissipation effects are presumed absent, that one can calculate work from knowledge of the system properties alone. The choice is between calculations for a reversible process and no calculations at all. Of course, the work calculated for a reversible process will not be correct for the corresponding real process. However, the use of efficiencies based on experience allows reasonable estimates to be made of the work for actual processes. We will frequently invoke the concept of a reversible process throughout the remainder of this book.

7-7 Carnot Cycle

During a part of the cycle performed by the working substance in a heat engine, some heat is absorbed from a hot reservoir; during another part of the cycle, a smaller amount of heat is rejected to a cooler reservoir. The engine is therefore said to operate between these two reservoirs. Since it is a fact of experience that some heat is always rejected to the cooler reservoir, the efficiency of an actual engine is never 100 percent. If we assume that we have at our disposal two reservoirs at given temperatures, it is important to answer the following questions: (1) What is the maximum efficiency that can be achieved by an engine operating between these two reservoirs? (2) What are the characteristics of such an engine? (3) Of what effect is the nature of the working substance?

The importance of these questions was recognized by Nicolas Léonard Sadi Carnot, a brilliant young French engineer who, in the year 1824, before the first law of thermodynamics was firmly established, described in a paper entitled "Sur la puissance motrice du feu" an ideal engine operating in a particularly simple cycle known today as the *Carnot cycle.*

In describing and explaining the behavior of this ideal engine, Carnot made use of three terms: *feu, chaleur,* and *calorique.* By *feu,* he meant fire or flame, and when the word is so translated no misconceptions arise. Carnot gave, however, no definitions for *chaleur* and *calorique,* but in a footnote stated that they had the same meaning. If both these words are translated as heat, then Carnot's reasoning is contrary to the first law of thermodynamics. There is, however, some evidence that in spite of the unfortunate footnote, Carnot did not mean the same thing by *chaleur* and *calorique.* Carnot used *chaleur* when referring to heat in general, but when referring to the motive power of heat that is brought about when heat enters at high temperature and leaves at low temperature, he used the expression *chute de calorique,* never *chute de chaleur.* It is the opinion of a few scientists that Carnot had in the back of his mind the concept of entropy, for which he reserved the term *calorique.* This seems incredible, and yet it is a remarkable circumstance that if the expression *chute de calorique* is translated "fall of entropy," many of the objections to Carnot's work raised by Kelvin, Clausius, and others are no longer valid. (The concept of entropy is treated in the next chapter.) In spite of possible mistranslations, Kelvin recognized the importance of Carnot's ideas and put them in the form in which they appear today.

A Carnot cycle is a set of processes that can be performed by any thermodynamic system whatever, whether chemical, electrical, magnetic, or otherwise. The system or working substance is imagined first to be in thermal equilibrium with a cold reservoir at the temperature T_C. Four processes are then performed in the following order:

1 A *reversible* adiabatic process is performed in such a direction that the temperature rises to that of the hotter reservoir T_H.
2 The working substance is maintained in contact with the reservoir at T_H, and a *reversible* isothermal process is performed in such a direction and to such an extent that heat $|Q_H|$ is absorbed from the reservoir.
3 A *reversible* adiabatic process is performed in a direction opposite to process 1 until the temperature drops to that of the cooler reservoir T_C.
4 The working substance is maintained in contact with the reservoir at T_C, and a *reversible* isothermal process is performed in a direction opposite to process 2 until the working substance is in its initial state. During this process, heat $|Q_C|$ is rejected to the cold reservoir.

An engine operating in a Carnot cycle is called a *Carnot engine*. A Carnot engine operates between two reservoirs in a particularly simple way. All the heat that is absorbed is absorbed at a constant high temperature, namely, that of the hot reservoir. Also, all the heat that is rejected is rejected at a constant lower temperature, that of the cold reservoir. Since all four processes are reversible, the Carnot cycle is a reversible cycle.

7-8 Examples of Carnot Cycles

The simplest example of a Carnot cycle is that for an ideal gas, depicted on a PV diagram in Fig. 7-3. The dashed lines are isotherms at the temperatures T_H and T_C, respectively, T_H being greater than T_C. The gas is originally in the state represented by the point a. The four processes are then:

$a \rightarrow b$ Reversible adiabatic compression until the temperature rises to T_H.
$b \rightarrow c$ Reversible isothermal expansion until any desired point such as c is reached.
$c \rightarrow d$ Reversible adiabatic expansion until the temperature drops to T_C.
$d \rightarrow a$ Reversible isothermal compression until the original state is reached.

During the isothermal expansion $b \rightarrow c$, heat $|Q_H|$ is absorbed from the hot reservoir at T_H. During the isothermal compression $d \rightarrow a$, heat $|Q_C|$ is rejected to the cooler reservoir at T_C.

A mixture of liquid and vapor may also be taken through a Carnot cycle.

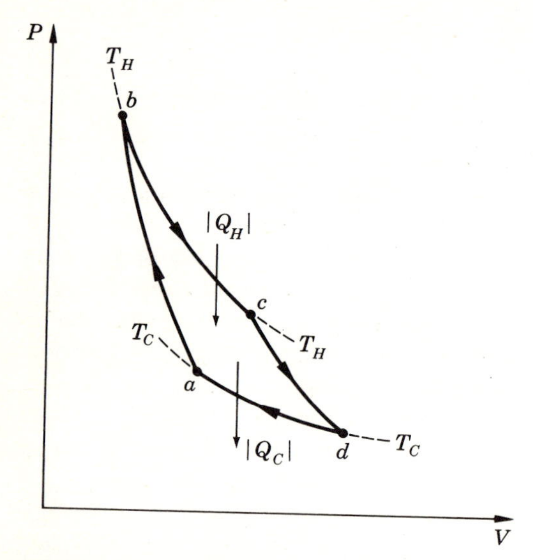

Fig. 7-3
Carnot cycle of an ideal gas.

This is shown on a PV diagram in Fig. 7-4. The dashed lines are the liquid and vapor saturation curves, and the solid horizontal lines are vaporization lines. Starting at point a in the two-phase region, the four processes are as follows:

$a \rightarrow b$ Reversible adiabatic compression until the temperature rises to T_H.
$b \rightarrow c$ Reversible isothermal isobaric vaporization until any arbitrary point such as c is reached.
$c \rightarrow d$ Reversible adiabatic expansion until the temperature drops to T_C.
$d \rightarrow a$ Reversible isothermal isobaric condensation until the initial state is reached.

As a last example of a Carnot cycle, that of a paramagnetic substance is shown on an $\mathcal{H}\mathcal{M}$ diagram in Fig. 7-5. The lines OT_C and OT_H represent isothermals at the temperatures T_C and T_H, respectively. Starting at a, the four processes are:

$a \rightarrow b$ Reversible adiabatic magnetization until the temperature rises to T_H.
$b \rightarrow c$ Reversible isothermal demagnetization until an arbitrary point c is reached.

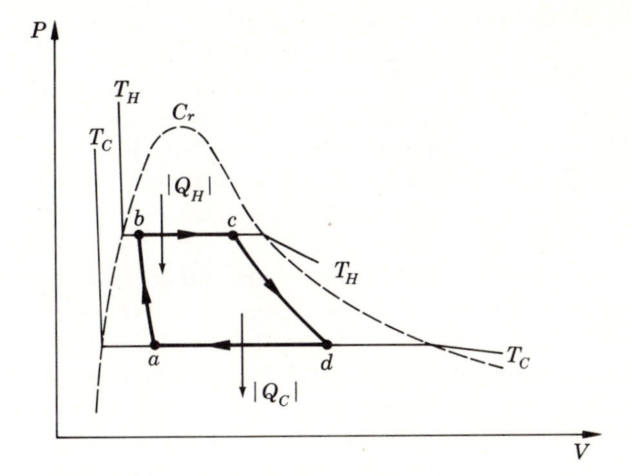

Fig. 7-4
Carnot cycle of a mixture of liquid and vapor.

$c \to d$ Reversible adiabatic demagnetization until the temperature drops to T_C.
$d \to a$ Reversible isothermal magnetization until the initial state is reached.

The net work done in one cycle by a Carnot engine can be adjusted to

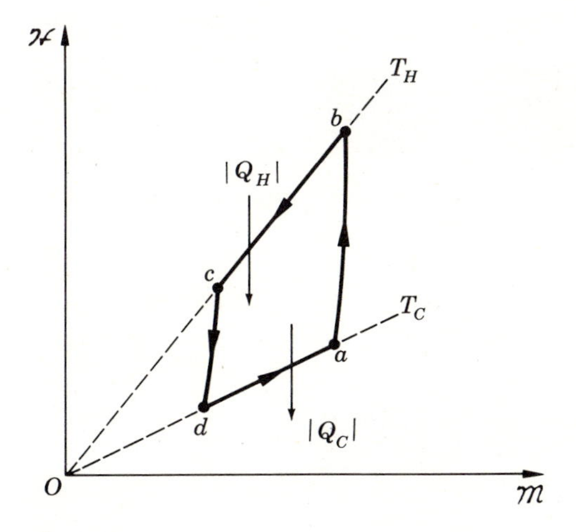

Fig. 7-5
Carnot cycle of a paramagnetic substance.

any arbitrary amount by choosing the position of the point c, that is, by adjusting the extent of the isothermal process $b \rightarrow c$. It is seen that the coordinates used to plot a Carnot cycle and the shape of the cycle depend on the nature of the working substance. It will be shown in the next chapter, however, that it is possible to find two coordinates in terms of which a graph of any Carnot cycle with *any* working substance is a rectangle. Consequently, we shall represent a Carnot engine symbolically with the aid of a rectangle as shown in Fig. 7-6*a*.

If an engine is to operate between only two reservoirs and still operate in a reversible cycle, it must be a Carnot engine. If any other cycle were performed between only two reservoirs, the necessary heat transfer would involve finite temperature differences and therefore could not be reversible. Conversely, if any other cycle were performed reversibly, it would require a series of reservoirs, not merely two. The expression "Carnot engine" therefore means "a reversible engine operating between only two reservoirs." Since a Carnot cycle consists of reversible processes, it may be performed in either direction. When it is performed in a direction opposite to that shown in the examples, it is a refrigeration cycle. A Carnot refrigerator is represented symbolically in Fig. 7-6*b*. *The important feature of reversible refrigeration cycles which distinguishes them from any general reversed engine cycle is that the quantities* $|Q_H|$, $|Q_C|$, *and* $|W|$ *are the same when the cycles are performed in the opposite direction.* For example, exactly the same amount of heat that is absorbed by the Carnot engine from the hot reservoir is rejected to the hot reservoir when the cycle is reversed. This would not be the case if the cycle were not reversible.

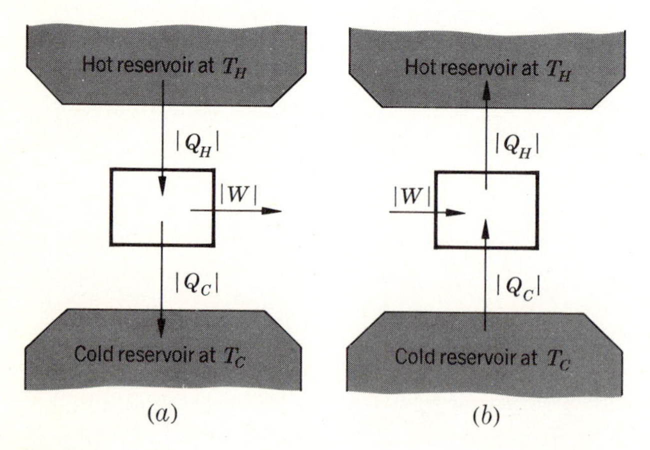

Fig. 7-6
Symbolic representations of (a) *Carnot engine and of* (b) *Carnot refrigerator.*

7-9 Carnot's Theorem and Corollary

We are now in a position to prove Carnot's theorem, which is stated as follows: *No engine operating between two given reservoirs can be more efficient than a Carnot engine operating between the same two reservoirs.*

Imagine a Carnot engine C and any other engine E working between the same two reservoirs and adjusted so that they both deliver the same amount of work, $|W|$. Thus:

	Carnot engine C	Any other engine E
1	Absorbs heat $\lvert Q_H \rvert$ from hot reservoir.	Absorbs heat $\lvert Q'_H \rvert$ from hot reservoir.
2	Performs work $\lvert W \rvert$.	Performs work $\lvert W \rvert$.
3	Rejects heat $\lvert Q_H \rvert - \lvert W \rvert$ to cold reservoir.	Rejects heat $\lvert Q'_H \rvert - \lvert W \rvert$ to cold reservoir.
4	Efficiency $\eta_C = \lvert W/Q_H \rvert$.	Efficiency $\eta_E = \lvert W/Q'_H \rvert$.

Let us assume that the efficiency of the engine E is greater than that of C. Thus

$$\eta_E > \eta_C,$$

$$\frac{|W|}{|Q'_H|} > \frac{|W|}{|Q_H|},$$

and

$$|Q_H| > |Q'_H|.$$

Now let the engine E drive the Carnot engine C backward as a Carnot refrigerator. This is shown symbolically in Fig. 7-7. The engine and the refrigerator coupled together in this way constitute a self-acting device, since all the work needed to operate the refrigerator is supplied by the engine. The net heat extracted from the cold reservoir is

$$|Q_H| - |W| - (|Q'_H| - |W|) = |Q_H| - |Q'_H|.$$

which is positive. The net heat delivered to the hot reservoir is also $|Q_H| - |Q'_H|$. The effect, therefore, of this self-acting device is to transfer $|Q_H| - |Q'_H|$ units of heat from a cold reservoir to a hot reservoir. Since this is a violation of the second law of thermodynamics (Clausius' statement), it cannot be true that $\eta_E > \eta_C$; thus our original assumption is invalid and Carnot's theorem is proved. We may express this result in symbols, thus:

$$\eta_E \leq \eta_C.$$

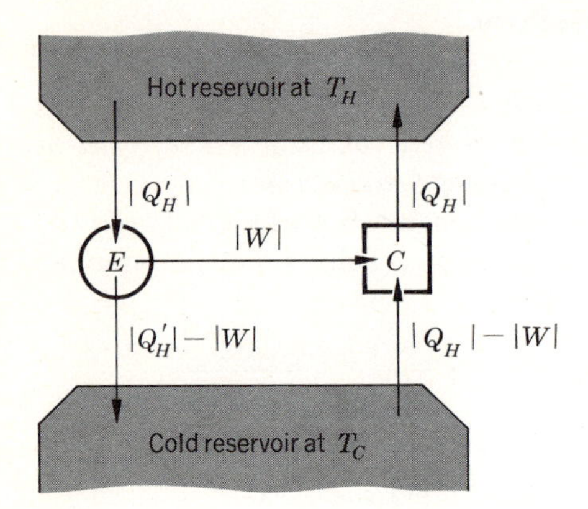

Fig. 7-7
Engine E operating a Carnot refrigerator.

The following corollary to Carnot's theorem may be easily proved: *All Carnot engines operating between the same two reservoirs have the same efficiency.*

Consider two Carnot engines C_1 and C_2, operating between the same two reservoirs. If we imagine C_1 driving C_2 backward, then Carnot's theorem states that

$$\eta_{C1} \leq \eta_{C2}.$$

If C_2 drives C_1 backward, then

$$\eta_{C2} \leq \eta_{C1}.$$

It therefore follows that

$$\eta_{C1} = \eta_{C2}.$$

It is clear from the above result that the nature of the working substance which is undergoing the Carnot cycle has no influence on the efficiency of the Carnot engine.

7-10 The Kelvin Temperature Scale

In demonstrating that the efficiency of a Carnot engine is independent of the nature of the working substance and depends only on the temperatures of the two reservoirs it operates between, we have identified these temperatures with the ideal-gas temperature scale. However, the concept of the Carnot engine provides us with the opportunity to establish a truly *thermodynamic* temperature scale which is quite independent of the properties of any material substance.

Let us denote by θ the temperature measured by some arbitrarily chosen thermometer. Any of the various devices described in Chap. 1 would do. Thus θ represents temperature on some empirical scale. Consider now two Carnot engines, one operating between a hot reservoir at temperature θ_H and a cold reservoir at temperature θ_C, and the second operating between θ_C and a still colder reservoir at temperature θ_0. Let the heat rejected by the first engine $|Q_C|$ be the heat absorbed by the second engine. The two engines working together thus constitute a third Carnot engine, which absorbs heat $|Q_H|$ from a reservoir at θ_H and rejects heat $|Q_0|$ to a reservoir at θ_0.

For the first engine the efficiency is given by

$$\eta = 1 - \frac{|Q_C|}{|Q_H|} = \phi(\theta_H,\theta_C),$$

where ϕ is an unknown function. Rearranging this equation, we get

$$\frac{|Q_H|}{|Q_C|} = \frac{1}{1 - \phi(\theta_H,\theta_C)} = f(\theta_H,\theta_C),$$

where f is also an unknown function.

For the second engine and for the two engines considered together as a third Carnot engine, the same equation must hold; thus

$$\frac{|Q_C|}{|Q_0|} = f(\theta_C,\theta_0),$$

and

$$\frac{|Q_H|}{|Q_0|} = f(\theta_H,\theta_0).$$

Since

$$\frac{|Q_H|}{|Q_C|} = \frac{|Q_H/Q_0|}{|Q_C/Q_0|},$$

we have the result that

$$f(\theta_H,\theta_C) = \frac{f(\theta_H,\theta_0)}{f(\theta_C,\theta_0)}.$$

Now the temperature θ_0 is arbitrarily chosen, and since it does not appear in the left-hand member of the above equation, it must therefore drop out of the ratio on the right. After it has been canceled, the numerator can be written $\psi(\theta_H)$ and the denominator $\psi(\theta_C)$, where ψ is another unknown function. Thus

$$\frac{|Q_H|}{|Q_C|} = \frac{\psi(\theta_H)}{\psi(\theta_C)}.$$

The ratio on the right is defined as the ratio of two *Kelvin* temperatures, and is denoted by T'_H/T'_C. We therefore, have, finally,

$$\frac{|Q_H|}{|Q_C|} = \frac{T'_H}{T'_C}. \tag{7-1}$$

Thus *two temperatures on the Kelvin scale are to each other as the absolute values of the heats absorbed and rejected, respectively, by a Carnot engine operating between reservoirs at these temperatures.* It is seen that the Kelvin temperature scale is independent of the peculiar characteristics of any particular substance. It therefore supplies precisely what is lacking in the ideal-gas scale.

At first thought it might seem that the ratio of two Kelvin temperatures would be impossible to measure, since a Carnot engine is an ideal engine, quite impossible to construct. The situation, however, is not so bad as it seems. The ratio of two Kelvin temperatures is the ratio of two heats that are transferred during two isothermal processes bounded by the same two adiabatics. The two adiabatic boundaries may be located experimentally, and the heats transferred during two isothermal "nearly reversible" processes can be measured with considerable precision. As a matter of fact, this is one of the methods used for measurement of temperatures below 1(K).

To complete the definition of the Kelvin scale we proceed, as in Chap. 1, to assign the arbitrary value of 273.16(K) to the temperature of the triple point of water T'_t. Thus

$$T'_t = 273.16(\mathrm{K}).$$

For a Carnot engine operating between reservoirs at the temperatures T' and T'_t, we have

$$\frac{|Q|}{|Q_t|} = \frac{T'}{T'_t} = \frac{T'}{273.16}$$

or

$$T' = 273.16\frac{|Q|}{|Q_t|}. \tag{7-2}$$

Comparison of Eq. (7-2) with Eq. (1-3) shows that, for the Kelvin scale, $|Q|$ plays the role of a "thermometric property." This does not, however, have the objection attached to a coordinate of an arbitrarily chosen thermometer, inasmuch as the behavior of a Carnot engine is independent of the nature of the working substance.

7-11 Absolute Zero

It follows from Eq. (7-2) that the heat transferred isothermally between two given adiabatics decreases as the temperature decreases. Conversely, the smaller the value of $|Q|$, the lower the corresponding T'. The smallest possible value of $|Q|$ is zero, and the corresponding T' is absolute zero. *Thus, if a system undergoes a reversible isothermal process without transfer of heat, the temperature at which this process takes place is called absolute zero.* In other words, at absolute zero, an isotherm and an adiabatic are identical.

It should be noticed that the definition of absolute zero holds for all substances and is therefore independent of the peculiar properties of any one arbitrarily chosen substance. Furthermore, the definition is in terms of purely macroscopic concepts. No reference is made to molecules or to molecular energy.

A Carnot engine absorbing heat $|Q_H|$ from a hot reservoir at T'_H and rejecting heat $|Q_C|$ to a cooler reservoir at T'_C has an efficiency

$$\eta = 1 - \frac{|Q_C|}{|Q_H|}.$$

Since

$$\frac{|Q_C|}{|Q_H|} = \frac{T'_C}{T'_H},$$

$$\eta = 1 - \frac{T'_C}{T'_H}. \tag{7-3}$$

For a Carnot engine to have an efficiency of 100 percent, it is clear that T'_C must be zero. Only when the lower reservoir is at absolute zero will all the heat be converted into work. Since nature does not provide us with a reservoir at absolute zero, a heat engine with 100 percent efficiency is a practical impossibility.

7-12 Carnot Cycle of an Ideal Gas. Equality of Ideal-Gas Temperature T and Kelvin Temperature T′

A Carnot cycle of an ideal gas is depicted on a PV diagram in Fig. 7-3. The two isothermal processes $b \rightarrow c$ and $d \rightarrow a$ are represented by the ideal-gas equation of state, Eq. (5-12):

$$PV = RT_H \qquad \text{and} \qquad PV = RT_C.$$

For any infinitesimal reversible process of an ideal gas, the first law is given by Eq. (5-19):

$$\delta Q = C_V\,dT + P\,dV. \tag{5-19}$$

Applying this equation to the isothermal process $b \rightarrow c$, we find

$$|\,Q_H\,| = |\,Q_{bc}\,| = \int_{V_b}^{V_c} P\,dV = RT_H \ln \frac{V_c}{V_b}.$$

Similarly, for the isothermal process $d \rightarrow a$, the absolute value of the heat rejected is

$$|\,Q_C\,| = RT_C \ln \frac{V_d}{V_a}.$$

Therefore
$$\frac{|\,Q_H\,|}{|\,Q_C\,|} = \frac{T_H \ln (V_c/V_b)}{T_C \ln (V_d/V_a)}.$$

Equation (5-19) applied to a reversible adiabatic process ($\delta Q = 0$) yields

$$-C_V\,dT = P\,dV,$$

or
$$-C_V\,dT = \frac{RT}{V}\,dV.$$

Integrating for the adiabatic process $a \rightarrow b$, we get

$$\frac{1}{R}\int_{T_C}^{T_H} C_V \frac{dT}{T} = \ln \frac{V_a}{V_b}.$$

Similarly, for the adiabatic process $c \rightarrow d$,

$$\frac{1}{R}\int_{T_C}^{T_H} C_V \frac{dT}{T} = \ln \frac{V_d}{V_c}.$$

But the left-hand sides of these two equations are identical. Therefore

$$\ln \frac{V_a}{V_b} = \ln \frac{V_d}{V_c},$$

or
$$\ln \frac{V_c}{V_b} = \ln \frac{V_d}{V_a},$$

and we get, finally,

$$\frac{|\,Q_H\,|}{|\,Q_C\,|} = \frac{T_H}{T_C}.$$

Comparing this last equation with Eq. (7-1), we find

$$\frac{T_H}{T_C} = \frac{T'_H}{T'_C}.$$

Representing the temperature T_H of any hot reservoir by T' and T, and the temperature T_C of the cold reservoir at the triple point of water by T'_t and T_t, we can write the preceding equation as

$$\frac{T}{T_t} = \frac{T'}{T'_t}.$$

Since $T'_t = T_t = 273.16(\text{K})$, it follows that

$$\boxed{T' = T.}$$

The Kelvin temperature is therefore numerically equal to the ideal-gas temperature. Equation (7-3) for the Carnot efficiency may thus be written

$$\boxed{\eta = 1 - \frac{T_C}{T_H}.} \tag{7-4}$$

Similarly, Eq. (7-1) becomes

$$\frac{|Q_H|}{|Q_C|} = \frac{T_H}{T_C}. \tag{7-5}$$

According to Carnot's theorem (Sec. 7-9), Eq. (7-4) provides an upper limit on the thermal efficiency of a heat engine operating between the two constant temperature levels T_C and T_H.

Example 7-1

In a particular steam power plant, superheated steam enters the turbine at 600(K). The temperature of the water in the condenser is 310(K). If the plant is rated at 7.5×10^5(kW), estimate the minimum rate at which heat must be transferred from the condenser to the surroundings.

From the definition of thermal efficiency, Eq. (6-1), and the first law, Eq. (6-2), we have

$$\eta = \frac{|W|}{|Q_H|} = \frac{|W|}{|W| + |Q_C|},$$

from which

$$|Q_C| = \left(\frac{1-\eta}{\eta}\right)|W|. \tag{A}$$

For a given value of $|W|$, $|Q_C|$ is a minimum when the thermal efficiency of the power cycle is a maximum. By Carnot's theorem, the theoretical maximum efficiency is given by Eq. (7-4). Substitution of this expression into Eq. (A) yields

$$|Q_C|_{\min} = \left(\frac{T_C}{T_H - T_C}\right)|W|. \tag{B}$$

Strictly, Eq. (B) applies to a power cycle for which heat transfer between system and surroundings occurs only at a single high temperature T_H and a single low temperature T_C. We therefore assume hot and cold reservoir temperatures of $T_H = 600(\text{K})$ and $T_C = 310(\text{K})$, the maximum and minimum temperatures of the working fluid in the steam power cycle. Substitution into Eq. (B) gives, with $|W| = 7.5 \times 10^5(\text{kW}) = 7.5 \times 10^8(\text{J})/(\text{s})$,

$$|Q_C|_{\min} = \left(\frac{310}{600 - 310}\right) \times 7.5 \times 10^8(\text{J})/(\text{s}),$$

or

$$|Q_C|_{\min} = 8.0 \times 10^8(\text{J})/(\text{s}).$$

Thus approximately the same amount of energy is discarded as waste heat as appears as work. The theoretical minimum heat supplied by the hot reservoir is found from the first law:

$$|Q_H|_{\min} = |W| + |Q_C|_{\min} = (7.5 + 8.0) \times 10^8(\text{J})/(\text{s}),$$

or

$$|Q_H|_{\min} = 15.5 \times 10^8(\text{J})/(\text{s}).$$

The above computation is based on a thermal efficiency of $\eta = 1 - (310/600) = 0.483$. The actual efficiency of a real plant would be lower, and the corresponding values of $|Q_C|$ and $|Q_H|$ larger. In a real steam power plant, a coal- or oil-fired furnace or a nuclear reactor serves as the hot reservoir; the cold reservoir usually consists of a river or other body of water. The transfer of large quantities of heat to our lakes and rivers leads to substantial increases in their temperatures; this is the best-known form of thermal pollution.

Example 7-2

In Example 6-2 we determined an expression for the efficiency of a reversible ideal-gas Stirling cycle with regeneration. This efficiency is identical to the Carnot efficiency η as given by Eq. (7-4). Repeat the derivation for a reversible ideal-gas Stirling cycle with *no* regenerator, and show that the efficiency η_E for this cycle is less than or equal to η. Assume constant heat capacities.

We refer to Fig. 6-1 and to results obtained in Example 6-2. In the absence of a regenerator, heat is transferred between system and surroundings during the isochoric steps $2 \rightarrow 3$ and $4 \rightarrow 1$ as well as during the isothermal steps $1 \rightarrow 2$ and $3 \rightarrow 4$. Thus the quantity of heat $|Q_H|$ transferred to the system is the sum of *two* terms:

$$|Q_H| = |Q_{23} + Q_{34}| = \int_{T_C}^{T_H} C_V\,dT + RT_H \ln\frac{V_a}{V_b},$$

or

$$|Q_H| = C_V(T_H - T_C) + RT_H \ln\frac{V_a}{V_b}.$$

Similarly, the quantity of heat $|Q_C|$ transferred from the system is given as

$$|Q_C| = |Q_{41} + Q_{12}| = C_V(T_H - T_C) + RT_C \ln\frac{V_a}{V_b}.$$

Substitution into Eq. (6-3) yields

$$\eta_E = 1 - \frac{|Q_C|}{|Q_H|} = \frac{R(T_H - T_C)\ln(V_a/V_b)}{C_V(T_H - T_C) + RT_H\ln(V_a/V_b)}.$$

Dividing the numerator and denominator of this expression by $RT_H \ln(V_a/V_b)$, we find

$$\eta_E = \frac{\eta}{1 + f\eta}, \tag{A}$$

where η is the Carnot efficiency as given by Eq. (7-4), and the function f is defined by

$$f = \frac{C_V}{R\ln(V_a/V_b)},$$

Since $V_a/V_b \geq 1$, the value of f is always positive. Thus we conclude from Eq. (A) that η_E is less than or equal to η, approaching zero in the limit as $V_a/V_b \to 1$ and η in the limit as $V_a/V_b \to \infty$. This conforms with Carnot's theorem, which denies the possibility of an efficiency *greater* than η.

Clearly, the presence of the regenerator results in a more efficient device. The reason for this is that the engine with the regenerator acts as its own source for the heat $|Q_{23}|$, and the only transfer of heat from the surroundings to the engine is from the single hot reservoir at temperature T_H. If the regenerator is not present, reversible heat input to the engine occurs with respect to a series of "hot" reservoirs with temperatures ranging from T_C to T_H, resulting in lower net efficiency. Similarly, the engine with the regenerator acts as its own sink for the heat $|Q_{41}|$, and the only transfer of heat to the surroundings is to the single cold reservoir at temperature T_C. If the regenerator is not present, reversible heat transfer to the surroundings occurs with respect to a series of "cold" reservoirs with temperatures ranging from T_H to T_C, again resulting in lower overall efficiency.

Problems

7-1 Classify and discuss the following natural phenomena from the standpoint of irreversibility: a waterfall, the weathering of rocks, the rusting of iron, a forest fire, the tearing of a piece of cloth, lightning, a compressed spring dissolving in acid, spontaneous combustion of a coal pile, aging of a magnet, shelf aging of an electric cell.

7-2 A gas at pressure P is contained within a cylinder-piston combination. For the following five sets of conditions, tell (a) whether $\delta W = -P\,dV$ or not and (b) whether the process is reversible, quasi-static, or irreversible:

- (i) There is no external pressure on the piston and no friction between the piston and the cylinder wall.
- (ii) There is no external pressure, and friction is small.
- (iii) The piston is jerked out faster than the average molecular speed.
- (iv) The friction is adjusted to allow the gas to expand slowly.
- (v) There is no friction, but the external pressure is adjusted to allow the gas to expand slowly.

7-3 An inventor claims to have developed an engine that takes in 25,000(J)/(s) at a temperature of 400(K), rejects 12,000(J)/(s) at a temperature of 200(K), and delivers 15(kW) of mechanical power. Would you advise investing money to put this engine on the market?

7-4 A Carnot engine absorbs 1,000(kJ) of heat from a reservoir at the temperature of the normal boiling point of water and rejects heat to a reservoir at the temperature of the triple point of water. Find the heat rejected, the work done, and the efficiency.

7-5 Which is the more effective way to increase the efficiency of a Carnot engine: to increase T_H, keeping T_C constant, or to decrease T_C, keeping T_H constant?

7-6 Draw a symbolic diagram of a set of Carnot engines with the following characteristics: Each engine absorbs the heat rejected by the preceding one at the temperature at which it was rejected, and each engine delivers the same amount of work. Show that the temperature intervals between which these engines operate are all equal.

7-7 A body of total heat capacity A is cooled by a Carnot engine, and the heat rejected by the engine is absorbed by another body of total heat capacity B. Starting with the expression

$$\delta W = \frac{T_B - T_A}{T_A} \delta Q_A:$$

(*a*) Derive an expression relating T_A and T_B at any time.
(*b*) Derive an expression for the work obtained as a function of A, B, T_A and the initial temperatures T_{A_0} and T_{B_0}.

7-8 Consider a metal rod, for which the linear expansivity α is known to be positive, to be the working substance of a Carnot engine. Show that adiabatic extension of the rod must cause its temperature to decrease and that adiabatic compression must cause its temperature to increase in order for the laws of thermodynamics to be satisfied. Note that $C_\sigma > C_\epsilon$.

7-9 Calculate the minimum work required to manufacture 5(lb_m) of ice cubes from water initially at 32(°F). Assume that the surroundings are at 80(°F). The latent heat of fusion of water at 32(°F) is 143.4(Btu)/(lb_m).

7-10 (*a*) A reversible heat engine operates between a heat reservoir at T_H = 1200(R) and another heat reservoir at T_C = 500(R). If during one complete cycle of the engine 100(Btu) of heat are transferred to the engine from the reservoir at T_H, what is the work?

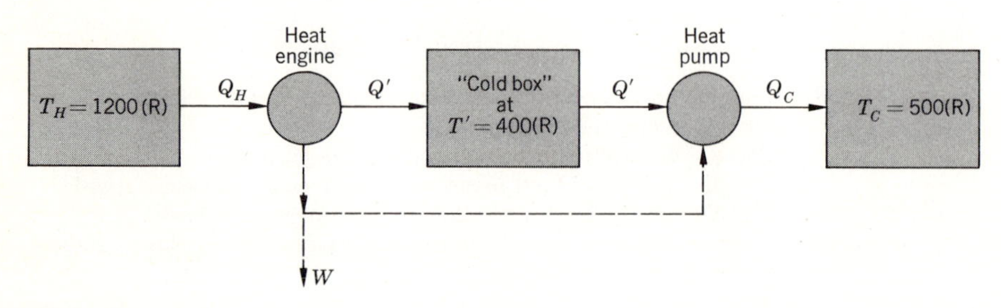

Fig. P7-10

(*b*) It is proposed that the engine will produce more work if the cold reservoir is lowered to 400(R), and it is decided to maintain a finite "cold box" at 400(R) by refrigeration and to use this cold box as the heat sink for the engine. Heat is removed from the cold box by a reversible refrigerator which discharges heat to the original reservoir at $T_C = 500$(R). The scheme is represented in Fig. P7-10.
If $Q_H = 100$(Btu), calculate the *net* work of the process, and compare the result with the answer to (*a*). Assume that both the engine and the refrigerator go through one complete cycle.

8

ENTROPY

8-1 Clausius' Theorem

In Chap. 3 we presented formulas for the calculation of work for various simple thermodynamic systems. The results were summarized in Sec. 3-11 by the statement that W for all such systems involves the product of a generalized force Y (for example, P, σ, $\mathcal{H}$, etc.) and a generalized displacement dX (for example, dV, $d\epsilon$, $d\mathcal{M}$, etc.), and that a generalized work diagram is given by a plot of Y against the corresponding X. In this chapter we use YX diagrams for the development and discussion of some general principles common to all thermodynamic systems.

Consider a reversible process for 1(mol) or a unit mass of material, represented by the smooth curve if on the generalized work diagram shown in Fig. 8-1. The nature of the system is immaterial. The dashed curves through i and f, respectively, represent portions of adiabatic processes. Let us draw a curve ab representing an isothermal process in such a way that the area under the smooth curve if is equal to the area under the zigzag path $iabf$. Then the work done in traversing both paths is the same, or

$$W_{if} = W_{iabf}.$$

Application of the first law separately to the two paths gives

$$Q_{if} = U_f - U_i - W_{if}$$

and

$$Q_{iabf} = U_f - U_i - W_{iabf}.$$

Therefore

$$Q_{if} = Q_{iabf}.$$

But

$$Q_{iabf} = Q_{ia} + Q_{ab} + Q_{bf},$$

and, since ia and bf are adiabatic paths,

$$Q_{ia} = Q_{bf} = 0.$$

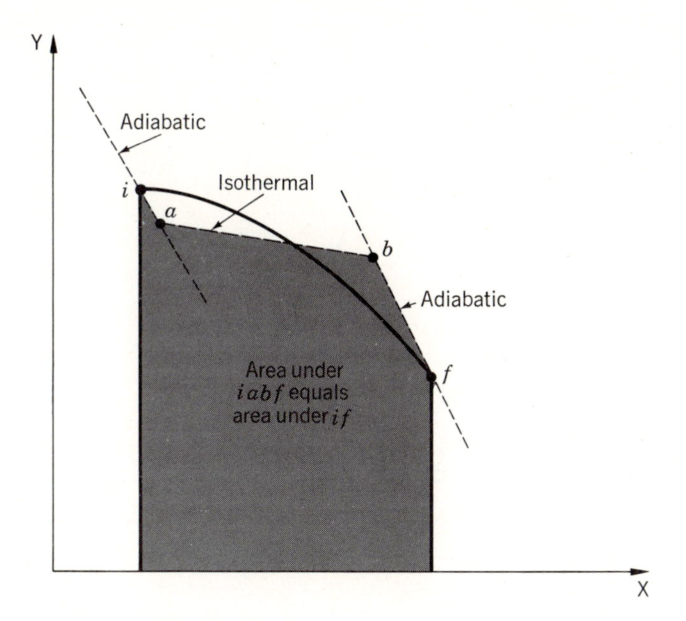

Fig. 8-1
Generalized work diagram. (if = any reversible process; ia = reversible adiabatic process; ab = reversible isothermal process; bf = reversible adiabatic process.)

Combination of the last three equations yields finally

$$Q_{if} = Q_{ab}.$$

Thus for *any* reversible process it is always possible to find a reversible zigzag path between the given initial and final states, consisting of an adiabatic followed by an isothermal followed by an adiabatic, such that the heat transferred during the isothermal portion is the same as that transferred during the original process.

Now consider the smooth closed curve representing a reversible cycle on the work diagram shown in Fig. 8-2. Since no two reversible adiabatic lines can intersect (see Prob. 6-6), a number of adiabatic lines may be drawn, dividing the cycle into a number of adjacent strips. A zigzag closed path may now be drawn, consisting of alternate adiabatic and isothermal portions, such that the heat transferred during all the isothermal portions is equal to the heat transferred in the original cycle. Consider the two isothermal processes *ab* at the temperature T_1, during which heat $|Q_1|$ is absorbed, and *cd* at the temperature T_2, during which heat $|Q_2|$ is rejected. Since the ends of the isothermals *ab* and *cd* are connected by adiabatics,

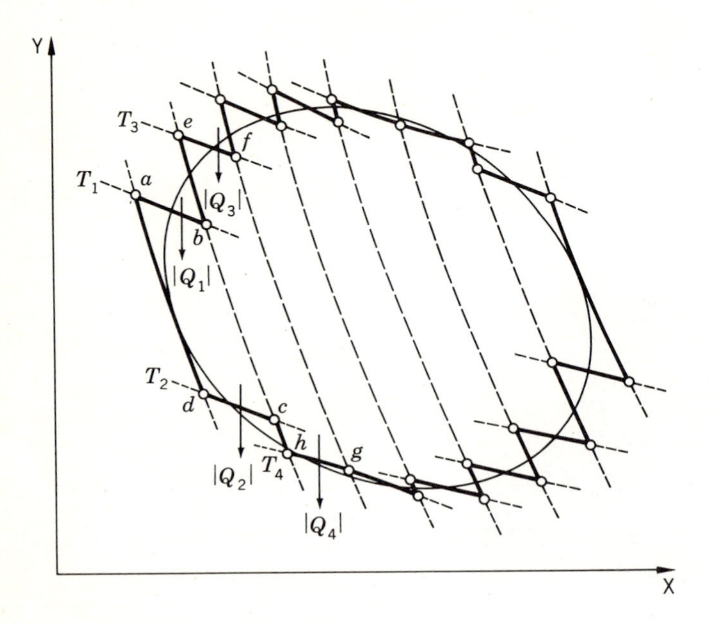

Fig. 8-2

Generalized work diagram. (Smooth closed curve = reversible cycle; zigzag closed path = alternate reversible isothermal and adiabatic processes.)

abcda is a Carnot cycle, and we have by Eq. (7-5)

$$\frac{|Q_1|}{T_1} = \frac{|Q_2|}{T_2}.$$

For the sake of convenience and simplicity, we have been using the absolute values of Q_1 and Q_2. Let us now adhere to the sign convention and regard any Q as an algebraic symbol, positive for heat absorbed and negative for heat rejected. We may then write

$$\frac{Q_1}{T_1} + \frac{Q_2}{T_2} = 0,$$

where Q_1 stands for a positive number, and Q_2 for a negative number. Since the ends of the isothermals *ef* and *gh* are connected by adiabatics, *efghe* is also a Carnot cycle, and

$$\frac{Q_3}{T_3} + \frac{Q_4}{T_4} = 0.$$

If a similar equation is written for each pair of isothermals whose ends are connected by adiabatics and if all the equations are added, the result is obtained that

$$\frac{Q_1}{T_1} + \frac{Q_2}{T_2} + \frac{Q_3}{T_3} + \frac{Q_4}{T_4} + \cdots = 0.$$

Since no heat is transferred during the adiabatic portions of the zigzag cycle, we may write

$$\sum \frac{Q}{T} = 0,$$

where the summation is taken over the *complete* zigzag cycle.

Now imagine the cycle divided into a very large number of strips bounded by adiabatics close together. If we connect these adiabatics with small isothermals in the manner already described, a zigzag path may be traced that may be made to approximate the original cycle as closely as we please. When these isothermal processes become infinitesimal, the ratio $\delta Q/T$ for an infinitesimal isothermal between two adjacent adiabatics is equal to the ratio $\delta Q/T$ for the infinitesimal piece of the original cycle bounded by the same two adiabatics. In the limit, therefore, we have for any reversible cycle,

$$\boxed{\oint_R \frac{\delta Q}{T} = 0.} \tag{8-1}$$

The circle through the integral sign signifies that the integration takes place over the complete cycle, and the letter R emphasizes the fact that the equation is true only for a reversible cycle. This result is known as *Clausius' theorem.*

8-2 Entropy

Let an initial equilibrium state of any thermodynamic system consisting of a unit mass or a mole of material be represented by the point i on any convenient diagram such as the generalized work diagram of Fig. 8-3. Denote a final equilibrium state by the point f. It is possible to take the system from i to f along any number of different reversible paths. Suppose the system is taken from i to f along the reversible path R_1 and then back to i again along another reversible path R_2. The two paths constitute a reversible cycle, and from Clausius' theorem, we may write

$$\oint_{R_1R_2} \frac{\delta Q}{T} = 0.$$

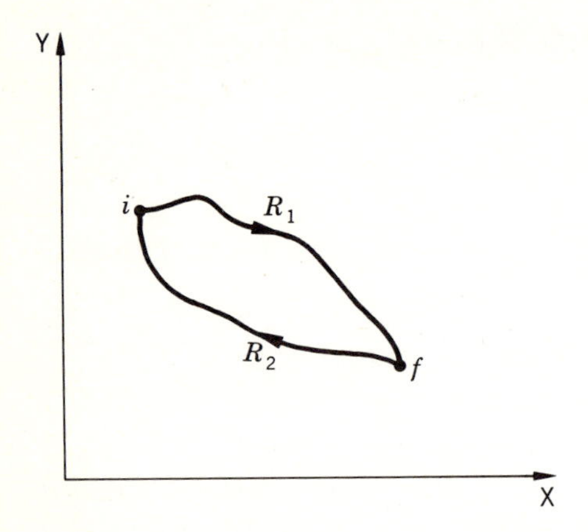

Fig. 8-3
Two reversible paths joining two equilibrium states of a system.

The above integral may be expressed as the sum of two integrals, one for the path R_1 and the other for the path R_2. We then have

$$\int_{R_1\,i}^{f} \frac{\delta Q}{T} + \int_{R_2\,f}^{i} \frac{\delta Q}{T} = 0.$$

But R_2 is a reversible path, and we may write

$$\int_{R_2\,f}^{i} \frac{\delta Q}{T} = -\int_{R_2\,i}^{f} \frac{\delta Q}{T}.$$

Thus we find

$$\int_{R_1\,i}^{f} \frac{\delta Q}{T} = \int_{R_2\,i}^{f} \frac{\delta Q}{T}.$$

Since R_1 and R_2 were chosen at random and represent any two reversible paths, the above equation expresses the important fact that

$$\int_{R\,i}^{f} \frac{\delta Q}{T}$$

is independent of the reversible path connecting i and f. It therefore follows that *there exists a function of the thermodynamic coordinates of a system whose value at the final state minus its value at the initial state equals the integral*

$$\int_{R\,i}^{f} \frac{\delta Q}{T}.$$

This function is therefore a thermodynamic *property*; it is called the *entropy,* and is denoted by S. If S_i is the entropy at the initial state and S_f that at the final state, we have the result that

$$\boxed{\int_{R\,i}^{f} \frac{\delta Q}{T} = S_f - S_i,} \tag{8-2}$$

where the difference $S_f - S_i$ is the *entropy change.*

If the two equilibrium states i and f are infinitesimally near, the integral sign may be eliminated and $S_f - S_i$ becomes dS. The equation then becomes

$$\boxed{\frac{\delta Q_R}{T} = dS,} \tag{8-3}$$

where dS is an exact differential, since it is the differential of an actual function. The subscript R written along with δQ indicates that the preceding equation is true only if the system undergoes a reversible change of state.

It should be noticed that the existence of an entropy function is deduced in the same manner as that of an energy function, i.e., from the observation that a certain quantity is independent of the path. In neither case, however, are we able to calculate an actual value of the function.

The molar entropy has dimensions of (energy)/(mole)(temperature), and the specific entropy has dimensions of (energy)/(mass)(temperature). In this chapter we will deal mainly with systems containing a unit mass or mole of material; for such systems, and for heat reservoirs, we represent the entropy by an unsuperscribed S. When it is necessary to emphasize the extent of a particular *system* [i.e., one containing a mass m or n(mol) of a substance], we will follow the conventions described in Sec. 4-10. For example, Eq. (8-3) could be written for n(mol) as

$$\frac{\delta Q_R}{T} = d(nS) = dS^t.$$

8-3 Entropy of an Ideal Gas

If a system absorbs an infinitesimal amount of heat δQ_R during a reversible process, the differential change in entropy of the system is given by Eq. (8-3). It is significant that although δQ_R is an inexact differential, the ratio $\delta Q_R/T$ is exact. The Kelvin temperature is therefore an *integrating denominator* for δQ_R. If δQ_R is expressed as a sum of differentials involving thermodynamic coordinates, then, upon dividing by T, one may integrate the expression between two states and so obtain the entropy change of the system.

As an example, we consider one of the expressions for δQ_R for 1 (mol) of an ideal gas, Eq. (5-20):

$$\delta Q_R = C_P\, dT - V\, dP \qquad \text{(ideal gas)}. \tag{5-20}$$

Dividing by T, we get

$$\frac{\delta Q_R}{T} = C_P \frac{dT}{T} - \frac{V}{T}\, dP,$$

or

$$dS = C_P \frac{dT}{T} - R \frac{dP}{P} \qquad \text{(ideal gas)}. \tag{8-4}$$

The indefinite integral of Eq. (8-4) is

$$S = \int C_P \frac{dT}{T} - R \ln P + S_0 \qquad \text{(ideal gas)},$$

where S_0 is a constant of integration, which we have no way to determine. However, we always deal with changes of state, for which only *differences* in S are required. The differences may be obtained by application of the last equation separately to two states, followed by subtraction, or by integration of Eq. (8-4) between limits. In either case, the result is the same, viz.,

$$\Delta S = \int_{T_1}^{T_2} C_P \frac{dT}{T} - R \ln \frac{P_2}{P_1} \qquad \text{(ideal gas)}, \tag{8-5}$$

where $\Delta S = S_2 - S_1$.

As an alternative to the above procedure, we could have started with Eq. (5-19) for δQ_R of an ideal gas:

$$\delta Q_R = C_V\, dT + P\, dV \qquad \text{(ideal gas)}. \tag{5-19}$$

Division by T yields

$$\frac{\delta Q_R}{T} = C_V \frac{dT}{T} + \frac{P}{T}\, dV,$$

or

$$dS = C_V \frac{dT}{T} + R \frac{dV}{V} \qquad \text{(ideal gas)}, \tag{8-6}$$

from which we obtain, analogous to Eq. (8-5),

$$\Delta S = \int_{T_1}^{T_2} C_V \frac{dT}{T} + R \ln \frac{V_2}{V_1} \qquad \text{(ideal gas)}. \tag{8-7}$$

Equations (8-4) and (8-6) show that the entropy of an ideal gas is a function of *two* state variables, for example, T and P, or T and V. This is in contrast with the behavior of U and H deduced in Sec. 5-7, where it was shown that these functions depend only upon T for an ideal gas.

Example 8-1

Find an equation which relates P and V for the isentropic expansion or compression of an ideal gas. Assume constant heat capacities.

By definition, an isentropic process is one for which the entropy remains constant. The required relationship between two of the state variables is found by setting $dS = 0$ in either Eq. (8-4) or Eq. (8-6). Starting with Eq. (8-4), we have, with $dS = 0$,

$$C_P \frac{dT}{T} - R \frac{dP}{P} = 0.$$

But we require a relationship between P and V. Since $PV = RT$, we can write

$$\frac{dP}{P} + \frac{dV}{V} = \frac{dT}{T}.$$

Thus the first equation becomes

$$(C_P - R) \frac{dP}{P} + C_P \frac{dV}{V} = 0,$$

or

$$C_V \frac{dP}{P} + C_P \frac{dV}{V} = 0,$$

or

$$d \ln P + \gamma d \ln V = 0,$$

where $\gamma = C_P/C_V$. Integration yields

$$PV^\gamma = k, \tag{A}$$

where k is a constant.

Equation (A) is identical with Eq. (5-29), an expression derived earlier for the quasi-static, adiabatic expansion or compression of an ideal gas with constant heat capacities. The agreement is not fortuitous, for according to Eq. (8-3),

$$dS = \frac{\delta Q_R}{T}.$$

Hence if $\delta Q_R = 0$, then $dS = 0$, which is the definition of an isentropic process. Thus, *during a reversible, adiabatic process the entropy of a system remains constant.* This is why the subscript S was used in Secs. 5-9 and 5-10 for identification of the adiabatic derivative $(\partial P/\partial V)_S$.

8-4 *TS* Diagram

Suppose a system consisting of 1 (mol) or a unit mass of material undergoes a *reversible* change of state from state i to state f. Equation (8-3) then yields an expression for δQ for each infinitesimal portion of the process:

$$\delta Q_R = T\,dS. \tag{8-8}$$

The net heat transfer for the process is found by integration. Thus,

$$Q_R = \int_i^f T\,dS. \tag{8-9}$$

This integral can be interpreted graphically as the area under a curve on a diagram in which T is plotted against S. The nature of the curve on the TS diagram is determined by the kind of reversible process that the system undergoes. Clearly, an isothermal process is represented by a horizontal line, and an isentropic process by a vertical line. The nature of the curves for isobaric and isochoric processes for PVT systems can be deduced as follows.

Division of Eq. (8-8) by dT and restriction to constant P gives

$$\left(\frac{\delta Q}{dT}\right)_P = T\left(\frac{\partial S}{\partial T}\right)_P \qquad \text{(reversible)}.$$

But, by Eq. (4-19), the left-hand side of this equation is equal to the constant-pressure heat capacity C_P. Thus we find

$$C_P = T\left(\frac{\partial S}{\partial T}\right)_P. \tag{8-10}$$

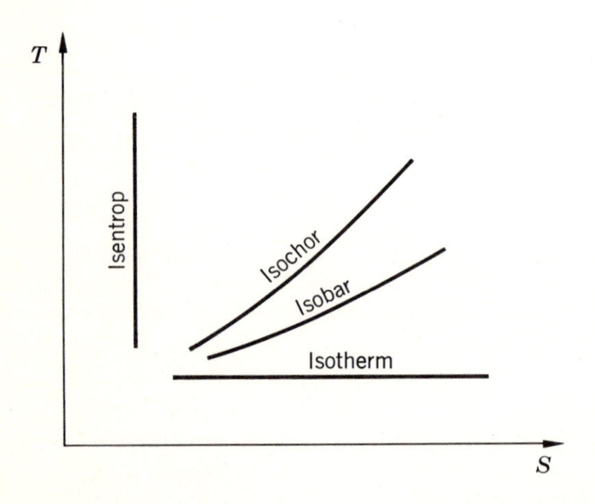

Fig. 8-4
Curves representing reversible processes of a PVT system on a TS diagram.

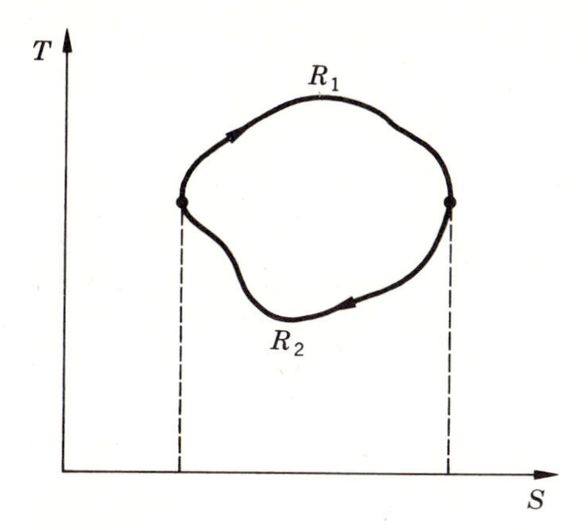

Fig. 8-5
Reversible cycle on a TS diagram.

Since C_P is always positive, a curve representing an isobaric process on a *TS* diagram must be positive in slope.

An expression similar to Eq. (8-10), valid for an isochoric (V = constant) process, is found in an entirely analogous manner. The result is

$$C_V = T\left(\frac{\partial S}{\partial T}\right)_V. \tag{8-11}$$

Again, C_V is always positive, so that a curve representing an isochoric process on a *TS* diagram is also positive in slope. It is known that $C_P \geq C_V$; therefore $(\partial T/\partial S)_V \geq (\partial T/\partial S)_P$. The curves for the four processes considered here are as shown in Fig. 8-4.

The *TS* diagram is convenient for the representation of reversible cycles; the closed curve in Fig. 8-5 represents such a cycle for the working substance of a heat engine. The (positive) area under R_1 is equal to the heat absorbed Q_H, and the (negative) area under R_2, to the heat rejected Q_C. The area inside the closed curve is therefore $Q_C + Q_H$, which is equal to $-W$ by the first law. Since the efficiency η is equal to $1 - (|Q_C|/|Q_H|) = 1 + (Q_C/Q_H)$, η may also be determined directly from areas on the diagram.

Example 8-2

Sketch a Carnot cycle on a *TS* diagram.

A Carnot cycle consists of two reversible, adiabatic steps connected by two reversible, isothermal steps. But a reversible, adiabatic process is one of constant entropy. Therefore *a Carnot cycle for any working substance is represented by a rectangle on a TS diagram.* This is illustrated for a Carnot heat engine in Fig. 8-6. In this figure, the labelled points, *a*, *b*, *c*, and *d*, correspond to those on the work diagrams of Figs. 7-3, 7-4, and 7-5.

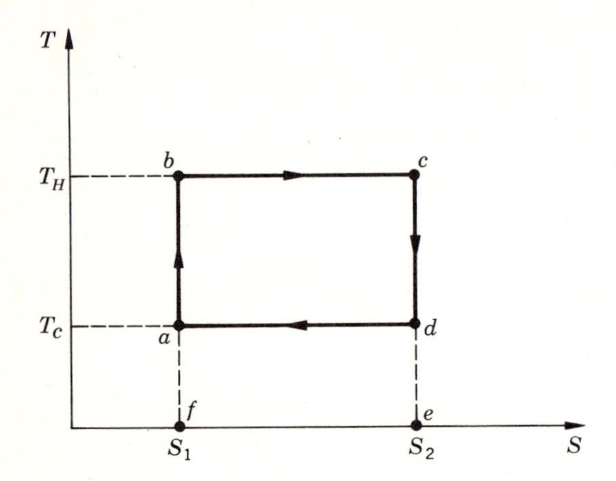

Fig. 8-6
A Carnot cycle of any system is a rectangle when represented on a TS diagram.

The heat and work terms are readily obtained from the diagram. Thus,

$$Q_H = \text{Area } abcdefa = T_H(S_2 - S_1),$$

$$Q_C = \text{Area } defad = -T_C(S_2 - S_1),$$

and

$$W = -\text{Area } abcda = -(T_H - T_C)(S_2 - S_1).$$

The efficiency η is given by

$$\eta = 1 + \frac{Q_C}{Q_H} = 1 + \frac{\text{Area } defad}{\text{Area } abcdefa} = 1 - \frac{T_C}{T_H},$$

in agreement with Eq. (7-4). This method of derivation of Eq. (7-4) verifies the conclusion reached earlier (Sec. 7-9), that the efficiency of a Carnot engine is independent of the working substance.

8-5 Entropy and Reversibility

We will find it necessary to consider *all* the entropy changes that take place when a system undergoes a process. If we calculate the entropy change of the system and add to this the entropy change of the local surroundings, we obtain a quantity that is the sum of all the entropy changes brought about by this particular process. This is the *total* entropy change ΔS_{total} due to the process in question.

When a finite amount of heat is absorbed or rejected by a reservoir, extremely small changes in the coordinates take place in every unit of mass. The entropy change of a unit of mass is therefore very small. Since the total mass of a reservoir is large, however, the total entropy change is finite. Suppose that a reservoir is in contact with a system and that heat Q is absorbed by the reservoir at the temperature T. The reservoir undergoes nondissipative changes determined entirely by the quantity of heat absorbed. Exactly the same changes *in the reservoir* would take place if the same amount of heat Q were transferred reversibly. Hence the entropy change of the reservoir is Q/T.

> *Whenever a reservoir absorbs heat Q at the temperature T from any system during any kind of process, the entropy change of the reservoir is Q/T.*

Consider now the *total* entropy change that is brought about by the performance of any *reversible* process. The process will, in general, be accompanied by a flow of heat between a system and a set of reservoirs ranging in temperature from T_i to T_f. During *any* infinitesimal portion of the process, an amount of heat $|\delta Q_R|$ is transferred between the system and one of the reservoirs at the temperature T. If $|\delta Q_R|$ is absorbed by the system, then

$$dS_{\text{system}} = +\frac{|\delta Q_R|}{T},$$

$$dS_{\text{reservoir}} = -\frac{|\delta Q_R|}{T},$$

and

$$dS_{\text{total}} = 0.$$

If $|\delta Q_R|$ is rejected by the system, then

$$dS_{\text{system}} = -\frac{|\delta Q_R|}{T},$$

$$dS_{\text{reservoir}} = +\frac{|\delta Q_R|}{T},$$

and

$$dS_{\text{total}} = 0.$$

If δQ_R is zero, then neither the system nor the reservoir will have an entropy change, and dS_{total} is still zero. Since this is true for any infinitesimal portion of the reversible process, it is true for all such portions; therefore we may conclude that *when a reversible process is performed, no change in the entropy of the system and its surroundings, taken together, occurs as a result of the process*, that is, $\Delta S_{\text{total}} = 0$.

8-6 Entropy and Irreversibility

When a system undergoes an irreversible process between an initial equilibrium state and a final equilibrium state, the entropy change of the system is given by Eq. (8-2):

$$\Delta S_{\text{system}} = S_f - S_i = \int_{R\,i}^{f} \frac{\delta Q}{T},$$

where R indicates *any reversible process arbitrarily chosen* by which the system may be brought from the given initial state to the given final state. No integration is performed over the original irreversible path. Instead, we imagine the change in state of the system to be brought about by a *reversible* process. This is possible when the initial and the final states of the system are equilibrium states. When either the initial or the final state is a nonequilibrium state, special methods must be used. In this section we shall limit ourselves to irreversible processes all of which involve initial and final states of equilibrium.

Processes exhibiting external mechanical irreversibility (*a*) Processes involving the isothermal dissipation of work through a system (which remains unchanged) into internal energy of a reservoir, such as:

1 Irregular stirring of a viscous liquid in contact with a reservoir.
2 Coming to rest of a rotating or vibrating liquid in contact with a reservoir.
3 Inelastic deformation of a solid in contact with a reservoir.
4 Transfer of electricity through a resistor in contact with a reservoir.
5 Magnetic hysteresis of a material in contact with a reservoir.

In the case of any process involving the isothermal transformation of work W through a system into internal energy of a reservoir, there is no entropy change of the system because the thermodynamic coordinates do not change. There is a flow of heat $|Q|$ into the reservoir where $|Q| = |W|$ (see Fig. 7-1). Since the reservoir absorbs $|Q|$ units of heat at the temperature T, its entropy change is $|Q|/T$ or $|W|/T$. The total entropy change ΔS_{total} is therefore $|W|/T$, which is a positive quantity.

(*b*) Processes involving the adiabatic dissipation of work into internal energy of a system, such as:

1 Irregular stirring of a viscous thermally insulated liquid.
2 Coming to rest of a rotating or vibrating thermally insulated liquid.
3 Inelastic deformation of a thermally insulated solid.

4 Transfer of electricity through a thermally insulated resistor.
5 Magnetic hysteresis of a thermally insulated material.

In the case of any process involving the adiabatic transformation of work W into internal energy of a system whose temperature rises from T_i to T_f at constant pressure, there is no flow of heat to or from the surroundings, and therefore the entropy change of the local surroundings is zero (see Fig. 7-2). Calculation of the entropy change of the system is done for a *reversible* process that is imagined to take the system from the given initial state (temperature T_i, pressure P) to the final state (temperature T_f, pressure P). We consider an isobaric process in which the temperature of the system is raised, not by the irreversible performance of work but by the reversible flow of heat from a series of reservoirs ranging in temperature from T_i to T_f. The entropy change of the system will then be

$$\Delta S_{\text{system}} = {}_R\!\int_{T_i}^{T_f} \frac{\delta Q}{T}.$$

For an isobaric process of 1 (mol) or a unit mass of material,

$$\delta Q_R = C_P \, dT, \tag{4-26a}$$

and thus

$$\Delta S_{\text{system}} = \int_{T_i}^{T_f} C_P \frac{dT}{T}. \tag{8-12}$$

This is the total entropy change, and since $T_f > T_i$, it is a positive quantity.

Processes exhibiting internal mechanical irreversibility Processes involving the transformation of internal energy of a system into mechanical energy and then back into internal energy again, such as:

1 Ideal gas rushing into a vacuum (free expansion).
2 Gas seeping through a porous plug (throttling process).
3 Snapping of a stretched wire after it is cut.
4 Collapse of a soap film after it is pricked.

In the case of a free expansion of an ideal gas there is no temperature change of the gas; in addition the entropy change of the local surroundings is zero. For calculation of the entropy change of the system we consider a reversible isothermal expansion of the gas at temperature T from the actual initial volume V_i to the actual final volume V_f. The entropy change of the system is

$$\Delta S_{\text{system}} = {}_R\!\int_{V_i}^{V_f} \frac{\delta Q}{T}.$$

For the isothermal expansion of 1 (mol) of an ideal gas, we have from Eq. (5-19) that

$$\delta Q_R = P\,dV.$$

Therefore

$$\frac{\delta Q_R}{T} = R\frac{dV}{V};$$

whence

$$\Delta S_{\text{system}} = R \ln \frac{V_f}{V_i}.$$

This expression also follows directly from Eq. (8-7) on restriction to constant T. The total entropy change ΔS_{total} is therefore $R \ln(V_f/V_i)$, a positive number.

Processes exhibiting external thermal irreversibility Processes involving a transfer of heat by virtue of a finite temperature difference, such as:

1 Conduction or radiation of heat from a system to its cooler surroundings.
2 Conduction or radiation of heat through a system (which remains unchanged) from a hot reservoir to a cooler one.

In the case of the conduction of $|Q|$ units of heat through a system (which remains unchanged) from a hot reservoir at T_1 to a cooler reservoir at T_2, the following equations apply:

$$\Delta S_{\text{system}} = 0,$$

$$\Delta S_{\text{hot reservoir}} = -\frac{|Q|}{T_1},$$

$$\Delta S_{\text{cold reservoir}} = +\frac{|Q|}{T_2},$$

and

$$\Delta S_{\text{total}} = \frac{|Q|}{T_2} - \frac{|Q|}{T_1}.$$

Since $T_2 < T_1$, this is a positive number.

Processes exhibiting chemical irreversibility Processes involving a spontaneous change of internal structure, chemical composition, density, etc., such as:

1 A chemical reaction.
2 Diffusion of two dissimilar inert ideal gases.
3 Mixing of alcohol and water.

4 Freezing of supercooled liquid.
5 Condensation of a supersaturated vapor.
6 Solution of a solid in water.
7 Osmosis.

Assuming the diffusion of 1 (mol) each of two dissimilar inert ideal gases to be equivalent to two separate free expansions, for each of which

$$\Delta S = R \ln \frac{V_f}{V_i},$$

and taking $V_f = 2V_i$, we obtain

$$\Delta S_{\text{total}} = 2R \ln 2,$$

Table 8-1 Total Entropy Change Due to Typical Natural Processes

Type of irreversibility	Irreversible process	Entropy change of the system	Entropy change of the local surroundings	Total entropy change
External mechanical irreversibility	Isothermal dissipation of work through a system into internal energy of a reservoir	0	$\frac{\lvert W \rvert}{T}$	$\frac{\lvert W \rvert}{T}$
	Adiabatic dissipation of work into internal energy of a system at constant pressure	$\int_{T_i}^{T_f} C_P \frac{dT}{T}$	0	$\int_{T_i}^{T_f} C_P \frac{dT}{T}$
Internal mechanical irreversibility	Free expansion of an ideal gas	$R \ln \frac{V_f}{V_i}$	0	$R \ln \frac{V_f}{V_i}$
External thermal irreversibility	Transfer of heat through a medium from a hot to a cooler reservoir	0	$\frac{\lvert Q \rvert}{T_C} - \frac{\lvert Q \rvert}{T_H}$	$\frac{\lvert Q \rvert}{T_C} - \frac{\lvert Q \rvert}{T_H}$
Chemical irreversibility	Diffusion of two dissimilar inert ideal gases	$2R \ln 2$	0	$2R \ln 2$

also a positive number. All the results of this section are summarized in Table 8-1.

8-7 Entropy and Nonequilibrium States

The calculation of the entropy changes associated with the irreversible processes discussed in Sec. 8-6 presented no special difficulties because in all cases, the system either did not change at all (in which case only the entropy changes of reservoirs had to be calculated) or the terminal states of a system were equilibrium states that could be connected by a suitable reversible process. Consider, however, the following process involving internal thermal irreversibility. A thermally conducting bar, brought to a nonuniform temperature distribution by contact at one end with a hot reservoir and at the other end with a cold reservoir, is removed from the reservoirs and then thermally insulated and kept at constant pressure. An *internal* flow of heat will finally bring the bar to a uniform temperature, but the transition will be from an initial nonequilibrium state to a final equilibrium state. It is obviously impossible to find one reversible process by which the system may be brought from the same initial to the same final state. What meaning, therefore, may be attached to the entropy change associated with this process?

Let us consider the bar to be composed of an infinite number of infinitesimally thin sections, each of which has a different initial temperature but all of which have the same final temperature. Suppose we imagine all the sections to be insulated from one another and all kept at the same pressure and then each section to be put in contact successively with a series of reservoirs ranging in temperature from the initial temperature of the particular section to the common final temperature. This defines an infinite number of reversible isobaric processes, which may be used to take the system from its initial nonequilibrium state to its final equilibrium state. We shall now define the entropy change as the result of integrating $\delta Q/T$ over all these reversible processes. In other words, in the absence of one reversible process to take the system from i to f, we conceive of an infinite number of reversible processes, one for each volume element.

As an example, consider the uniform bar of length L depicted in Fig. 8-7. A typical volume element at x has a mass

$$dm = \rho A\, dx,$$

where ρ is the mass density, and A the cross-sectional area. The total heat capacity of the section is

$$C_P\, dm = C_P \rho A\, dx.$$

Let us suppose that the initial temperature distribution is linear, so that

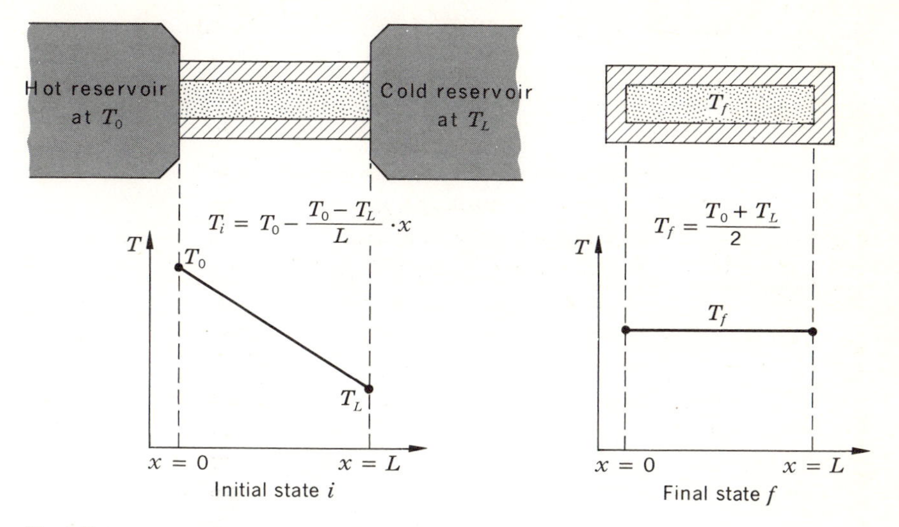

Fig. 8-7
Process exhibiting internal thermal irreversibility.

the section at x has an initial temperature

$$T_i = T_0 - \frac{T_0 - T_L}{L} x.$$

If no heat is lost and if we assume for the sake of simplicity that the thermal conductivity, density and heat capacity of all sections remain constant, the final temperature will be

$$T_f = \frac{T_0 + T_L}{2}.$$

Integrating $\delta Q/T$ over a reversible isobaric transfer of heat between the volume element and a series of reservoirs ranging in temperature from T_i to T_f, we get, for the entropy change of *this one volume element,*

$$C_P \rho A \, dx \int_{T_i}^{T_f} \frac{dT}{T} = C_P \rho A \, dx \ln \frac{T_f}{T_i}$$

$$= C_P \rho A \, dx \ln \frac{T_f}{T_0 - \left[\frac{(T_0 - T_L)}{L}\right] x}$$

$$= -C_P \rho A \, dx \ln \left(\frac{T_0}{T_f} - \frac{T_0 - T_L}{L T_f} x \right).$$

The total entropy change is therefore

$$\Delta S_{\text{total}} = -C_P\rho A \int_0^L \ln\left(\frac{T_0}{T_f} - \frac{T_0 - T_L}{LT_f}x\right) dx,$$

which, after integration† and simplification, becomes

$$\Delta S_{\text{total}} = mC_P\left(1 + \ln T_f + \frac{T_L}{T_0 - T_L}\ln T_L - \frac{T_0}{T_0 - T_L}\ln T_0\right).$$

To show that the entropy change is positive, let us take a convenient numerical case such as $T_0 = 400(\text{K})$, $T_L = 200(\text{K})$; whence $T_f = 300(\text{K})$. Then

$$\begin{aligned}\Delta S_{\text{total}} &= mC_P(1 + 5.704 + 5.298 - 2 \times 5.992)\\ &= 0.019mC_P.\end{aligned}$$

A similar method gives the entropy change of a system for a process leading from an initial nonequilibrium state characterized by a nonuniform pressure distribution to a final equilibrium state where the pressure is uniform.

8-8 Principle of the Increase of Entropy

The total entropy change associated with each of the irreversible processes treated up to now was found to be positive. We are led to believe, therefore, that whenever an irreversible process takes place, the total entropy increases. Since we have already shown that this proposition is true for all processes involving the irreversible transfer of heat, we need now consider only *adiabatic* processes in order to establish its general validity. Consider the special case of an adiabatic irreversible process between two equilibrium states of a system consisting of 1(mol) or a unit mass of material.

1 Let the initial state of the system be represented by the point i on the generalized work diagram of Fig. 8-8, and suppose that the system undergoes an *irreversible adiabatic process* to the state f. Then the entropy change of the system is

$$\Delta S = S_f - S_i.$$

A temperature change may or may not have taken place. Whether or

† $\int \ln(a + bx)\, dx = \frac{1}{b}(a + bx)[\ln(a + bx) - 1].$

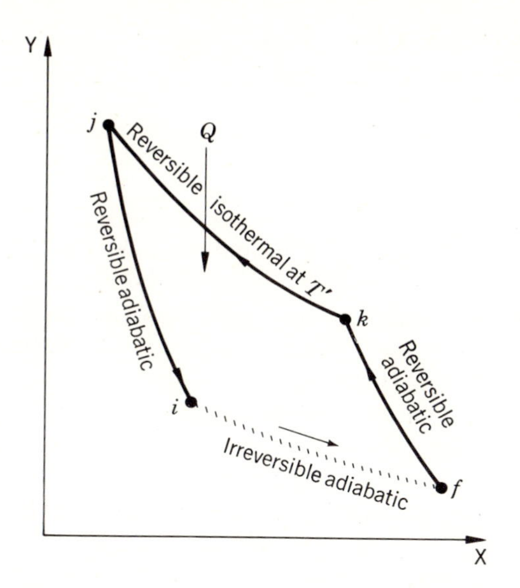

Fig. 8-8
Cycle which contradicts the second law unless $S_f > S_i$.

not, let us cause the system to undergo a *reversible adiabatic process* $f \rightarrow k$ in such a direction as to bring its temperature to that of any arbitrarily chosen reservoir, say, at T'. Then, since $S_f = S_k$,

$$\Delta S = S_k - S_i.$$

Now suppose that the system is brought into contact with the reservoir and caused to undergo a *reversible isothermal process* $k \rightarrow j$ until its entropy is the same as at the beginning. A final *reversible adiabatic process* $j \rightarrow i$ will now bring the system back to its initial state; and since $S_j = S_i$,

$$\Delta S = S_k - S_j.$$

The only heat transfer Q that has taken place in the cycle is during the isothermal process $k \rightarrow j$, where

$$Q = T'(S_j - S_k) = -T'(S_k - S_j).$$

A net amount of work W has been done in the cycle, where

$$-W = Q.$$

It is clear from the Kelvin-Planck statement of the second law that the heat Q cannot have *entered* the system, that is, Q cannot be positive, for then we should have a cyclic process in which no effect has been produced other than the extraction of heat from a reservoir and the performance of an equivalent amount of work. We conclude that $Q \leq 0$, and that

$$-T'(S_k - S_j) \leq 0.$$

Therefore $$S_k - S_j \geq 0.$$

Since the process is adiabatic, $\Delta S_{\text{surroundings}} = 0$, and we have finally $\Delta S_{\text{total}} \geq 0$.

2 If we assume that the original irreversible adiabatic process took place without any change in entropy, it would be possible to bring the system back to i by means of one reversible adiabatic process. Moreover, since the net heat transferred in this cycle is zero, the net work would also be zero. Therefore, under these circumstances, the system and its surroundings would have been restored to their initial states without producing changes elsewhere, which implies that the original process was reversible. Since this is contrary to our original assertion, the entropy of the system cannot remain unchanged. Therefore

$$\Delta S_{\text{total}} > 0.$$

3 Let us now suppose that the system is not homogeneous and not of uniform temperature and pressure and that it undergoes an irreversible adiabatic process in which mixing and chemical reaction may take place. If we assume that the system may be subdivided into parts (each one infinitesimal, if necessary) and that it is possible to ascribe a definite temperature, pressure, composition, etc., to each part, so that each part will have a definite entropy depending on its coordinates, then we may define the entropy of the whole system as the sum of the entropies of its parts. If we now assume that it is possible to take *each part* back to its initial state by means of the reversible processes described in step 1, using the same reservoir for each part, it follows that the entropy change of the whole system is positive and that $\Delta S_{\text{total}} > 0$.

It should be emphasized that we have had to make two assumptions, namely, (1) that the entropy of a system may be defined by subdividing the system into parts and summing the entropies of these parts, and (2) that reversible processes may be found or imagined by which mixtures may be unmixed and reactions may be caused to proceed in the opposite direction. The main justification for these assumptions, and therefore for the entropy principle, lies in the fact that they lead to results in complete agreement with experiment.

The behavior of the entropy as a result of any kind of process may now be represented in the following succinct manner:

$$\boxed{\Delta S_{\text{total}} \geq 0,} \tag{8-13}$$

where the equality sign refers to reversible processes and the inequality sign to irreversible processes. *This equation is the mathematical formulation of the second law.*

Thus the second law provides an answer to the question: In what direction does a process take place? The answer is that a process always takes place in such a direction as to cause an increase in total entropy, i.e., in the entropy of the system plus its surroundings.

Example 8-3

Show that the transfer of heat between a heat reservoir at temperature T_H and one at T_C, where $T_H > T_C$, must be from the hotter to the colder reservoir.

We use the mathematical statement of the second law, Eq. (8-13). Let Q_C be the quantity of heat transferred with respect to the cold reservoir, and Q_H the quantity of heat transferred with respect to the hot reservoir. The two heat terms must be equal in magnitude and opposite in sign; thus,

$$Q_H = -Q_C.$$

The entropy change of the cold reservoir is

$$\Delta S_C = \frac{Q_C}{T_C},$$

and that of the hot reservoir is

$$\Delta S_H = \frac{Q_H}{T_H} = -\frac{Q_C}{T_H}.$$

The *total* entropy change is the sum of these two terms, and must satisfy Eq. (8-13). Thus

$$\Delta S_{\text{total}} = \Delta S_C + \Delta S_H = \frac{Q_C}{T_C} - \frac{Q_C}{T_H} \geq 0,$$

or

$$Q_C\left(\frac{T_H - T_C}{T_C T_H}\right) \geq 0.$$

But $T_H - T_C > 0$ and $Q_C \neq 0$. Therefore by the above inequality we have $Q_C > 0$, which proves that heat of magnitude $|Q_C|$ is transferred *to* the cold reservoir *from* the hot reservoir. This result is entirely in accord with our experience, which tells us that the "spontaneous" flow of heat is always in the direction of decreasing temperature.

Example 8-4

It is desired to lower the temperature of a finite material body from T_1 to T_2 by operation of a refrigerator which discards heat to the surroundings at temperature T_0. Determine an expression for the minimum work required.

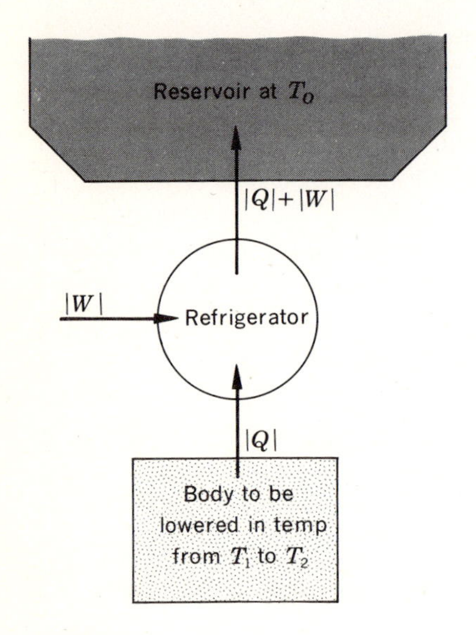

Fig. 8-9
Operation of a refrigerator in lowering the temperature of a finite body from T_1 to T_2.

We use Eq. (8-13), applied to the process represented schematically in Fig. 8-9. After the completion of an integral number of refrigeration cycles, a quantity of heat $|Q|$ has been transferred from the body, a quantity of work $|W|$ has been supplied to the refrigerator, and (by the first law) a quantity of heat $|Q| + |W|$ has been rejected to the surroundings, represented as a heat reservoir. Listing all the entropy changes, we find

$$\Delta S^t \text{ of body} = m(S_2 - S_1) = S_2{}^t - S_1{}^t,$$

$$\Delta S \text{ of refrigerant} = 0,$$

and

$$\Delta S \text{ of reservoir} = \frac{|Q| + |W|}{T_0}.$$

The total entropy change, which is the sum of these terms, must satisfy Eq. (8-13). Thus we have

$$S_2{}^t - S_1{}^t + \frac{|Q| + |W|}{T_0} \geq 0,$$

or

$$|W| \geq T_0(S_1{}^t - S_2{}^t) - |Q|,$$

from which

$$|W|_{\min} = T_0(S_1{}^t - S_2{}^t) - |Q|. \qquad (A)$$

As an example of the application of Eq. (A), we will determine the minimum work required to chill 2(kg) of drinking water to 5(°C) from an initial (ambient) temperature of 25(°C). We assume a constant pressure of 1(atm), and take the specific heat C_P as constant and equal to 4,184(J)/(kg)(K).

Use of Eq. (A) requires expressions (or values) for the entropies and for the heat term $|Q|$. For constant C_P, the isobaric entropy change of the water is found from Eq.

(8-12) to be

$$S_1{}^t - S_2{}^t = mC_P \ln \frac{T_1}{T_2}. \tag{B}$$

For reversible isobaric heat transfer, Q is related to the change in state of the water by Eq. (4-26*b*):

$$Q = m \int_{T_1}^{T_2} C_P \, dT = mC_P(T_2 - T_1),$$

and thus

$$|Q| = mC_P(T_1 - T_2). \tag{C}$$

Combination of Eqs. (A), (B), and (C) gives finally

$$|W|_{\min} = mC_P \left[T_0 \ln \frac{T_1}{T_2} - (T_1 - T_2) \right]. \tag{D}$$

From the given data, we have $T_0 = T_1 = 298(\text{K})$, and $T_2 = 278(\text{K})$. Substitution of numerical values into Eq. (D) then yields

$$|W|_{\min} = 2(\text{kg}) \times 4{,}184(\text{J})/(\text{kg})(\text{K}) \left[298 \ln \frac{298}{278} - (298 - 278) \right](\text{K}),$$

or

$$|W|_{\min} = 5880(\text{J}).$$

Although calculations of this type cannot be expected to give accurate values for work or power requirements, they do provide the basis for more realistic estimates and for the determination of minimum costs associated with real engineering processes.

8-9 Entropy and Unavailable Energy

Consider a system consisting of a heat reservoir at temperature T to exist in surroundings having the uniform and constant temperature T_0, which is lower than T. The surroundings are equivalent to a second heat reservoir. If a Carnot engine is operated between the system and its surroundings, a quantity of heat Q supplied to the engine will be partially converted into work according to the equation

$$|W| = Q\left(1 - \frac{T_0}{T}\right).$$

This represents the maximum amount of work which may be produced from the heat Q supplied to the engine by the system. One may therefore adopt the view that only a part of the heat energy Q is *available* for conversion into work with respect to a surroundings temperature T_0.

Suppose now that the quantity of heat Q is transferred directly from a heat reservoir at temperature T_1 to another heat reservoir at a lower temperature T_2, both existing in surroundings at the still lower temperature

T_0. We wish to show that because of the irreversible transfer of heat from T_1 to T_2, the fraction of the quantity of heat Q available for conversion into work for a given T_0 has decreased. The maximum work which could have been obtained from the heat Q before its irreversible transfer to the temperature level T_2 is given by

$$\text{Maximum work before transfer} = Q\left(1 - \frac{T_0}{T_1}\right).$$

After its irreversible transfer to the reservoir at T_2, the maximum work available is

$$\text{Maximum work after transfer} = Q\left(1 - \frac{T_0}{T_2}\right).$$

The difference between these quantities represents an amount of energy E' that could have been converted to work prior to the irreversible heat-transfer process but which cannot be so converted after the process. Thus E' represents an amount of energy which has become unavailable for conversion into work as the result of the irreversible heat transfer, and is given by

$$E' = Q\left(1 - \frac{T_0}{T_1}\right) - Q\left(1 - \frac{T_0}{T_2}\right)$$

$$= T_0\left(\frac{Q}{T_2} - \frac{Q}{T_1}\right).$$

But the quantity $(Q/T_2 - Q/T_1)$ is the total entropy change of the two heat reservoirs as a result of the irreversible heat-transfer process. Thus

$$E' = T_0\,\Delta S_{\text{total}}.$$

The energy which became unavailable is seen to be directly proportional to the total entropy change resulting from the irreversible process. That the proportionality factor is the absolute temperature of the surroundings T_0 may seem remarkable, because the surroundings were not involved in the irreversible process. But this fact shows clearly that the energy made unavailable for conversion into work depends on, or is determined with respect to, a particular T_0. In practice, this temperature is taken as that naturally available to us, i.e., the temperature of the atmosphere or of cooling water.

The foregoing derivation was specific for the special case of irreversible heat transfer between two heat reservoirs. We shall now present a more abstract derivation to establish the generality of the relation between E' and the total entropy change associated with an irreversible process.

Consider a system which undergoes a change of state by an irreversible process. We assume that the system exists in surroundings which in effect constitute a heat reservoir at temperature T_0. The change in the total internal energy U^t of the system is given by the first law:

$$\Delta U^t = Q + W,$$

where Q is heat exchanged between the system and its surroundings. The entropy change of the surroundings caused by this transfer of heat is

$$\Delta S_0 = \frac{-Q}{T_0}.$$

The minus sign on Q arises because the sign convention for Q is taken with reference to the system, whereas here we are determining the entropy change of the surroundings.

Now, after the irreversible process has occurred, let us consider the restoration of the *system* to its initial state *by a completely reversible process.* For this process the first law may be written

$$\Delta U^t_{\text{rev}} = Q_{\text{rev}} + W_{\text{rev}},$$

where the designation "rev" indicates the reversible restoration process. The entropy change of the surroundings during this process is

$$\Delta S_{0_{\text{rev}}} = \frac{-Q_{\text{rev}}}{T_0}.$$

For the complete cycle—original irreversible process plus reversible restoration—we have

$$\Delta U^t + \Delta U^t_{\text{rev}} = Q + W + (Q_{\text{rev}} + W_{\text{rev}}).$$

But for the cycle there is no net change of the internal energy of the system; therefore

$$0 = Q + W + (Q_{\text{rev}} + W_{\text{rev}}),$$

or

$$Q + Q_{\text{rev}} = -(W + W_{\text{rev}}).$$

Furthermore, there is no net change of the entropy of the *system* for the complete cycle. Thus the total entropy change which occurs as a result of the two processes is the sum of the entropy changes of the *surroundings*:

$$\Delta S_{\text{total}} = \frac{-(Q + Q_{\text{rev}})}{T_0}.$$

Since this is the total entropy change of the cycle, it must also be the total entropy change brought about by the original irreversible process, for the restoration process was reversible and hence caused no change in the total entropy.

Therefore the total entropy change resulting from the original irreversible process is

$$\Delta S_{\text{total}} = \frac{-(Q + Q_{\text{rev}})}{T_0} = \frac{(W + W_{\text{rev}})}{T_0},$$

or $$(W + W_{\text{rev}}) = T_0 \, \Delta S_{\text{total}}.$$

The quantity $W + W_{\text{rev}}$ is the net work of the cycle and as such represents the *extra* work required to restore the system to its initial state, for had the original process been reversible, no *extra* work would have been necessary for restoration of the system by a reversible process. This extra work is therefore the negative of work which would have been obtained had the original process been reversible but which became unavailable because of the irreversibility of the actual process. Thus $(W + W_{\text{rev}})$ is the quantity we have designated E'. Hence we have as a general equation

$$E' = T_0 \, \Delta S_{\text{total}}. \tag{8-14}$$

This equation states that *the energy that becomes unavailable for work during an irreversible process is* T_0 *times the total entropy change in the system and surroundings resulting from an irreversible process.* It must be understood that energy which becomes unavailable for work is not energy which is "lost," for the first law is always valid.

The engineering significance of this result is clear. The greater the irreversibility of a process, the greater the increase in total entropy accompanying it, and the greater the amount of energy which becomes unavailable for work. Since work is a form of energy of high monetary value, every irreversibility in a process carries with it a price.

8-10 Summary

Most of the material in this and preceding chapters has been devoted to the systematic development of the mathematical statements of the first and second laws of thermodynamics. The methodical approach followed here is useful for the beginner, for it illuminates the various concepts necessary to an understanding of the subject (e.g., temperature, work, heat) and also illustrates some of the methods employed in the historical development of the two great generalizations. Classical thermodynamics is an engineering science, and the elucidation of the second law in particular resulted from the efforts of engineers and scientists to discover the limitations imposed by nature on the operation of practical devices. In adopting this same engineering approach, we have extensively employed engine cycles and work diagrams both to rationalize the existence of an entropy function and to develop the principle of the increase of entropy from the Kelvin-Planck statement of the second law.

The next three chapters of the text are devoted to applications, and it is convenient at this point to summarize the content of the first and second laws. We shall do this by the use of concise mathematical assertions, or *axioms*, which not only provide economy of statement but also bring into relief the similarities and differences between the two laws; all of the many observations which led to their development may be regenerated from the axioms by mathematical deduction.[1] The choice of axioms is to some degree arbitrary. We adopt the following four:

Axiom 1 There exists a form of energy, called the internal energy U, which is a property of a system and is functionally related to the measurable coordinates which characterize the system. For a closed system at rest, changes in this property are given by

$$dU = \delta Q + \delta W. \tag{4-9}$$

Axiom 2 **(First law of thermodynamics).** The energy E of any system and its surroundings, considered together, is conserved. Thus,

$$\Delta E_{\text{total}} = 0. \tag{8-15}$$

Axiom 3 There exists a property, called the entropy S, which is functionally related to the measureable coordinates which characterize a system. For a reversible process, changes in this property are given by

$$dS = \frac{\delta Q_R}{T}. \tag{8-3}$$

Axiom 4 **(Second law of thermodynamics).** The entropy change of any system and its surroundings, considered together, is positive and approaches zero for any process which approaches reversibility. Thus,

$$\Delta S_{\text{total}} \geq 0. \tag{8-13}$$

The first and third axioms are similar in form; they assert the *existence*

[1] For conciseness, this is the point of view taken by M. M. Abbott and H. C. Van Ness in "Schaum's Outline of Theory and Problems of Thermodynamics," McGraw-Hill, New York, 1972.

of the two functions U and S, and provide relationships connecting these functions with measurable quantities. These relations do not allow the calculation of absolute values for either U or S; all that is provided in either case is a means for calculation of *changes* in the function.

The second and fourth axioms depend upon the first and third and constitute formal statements of the first and second laws. When a system and its surroundings are considered together (as indicated by the label "total") and when all forms of energy are treated, the first law assumes the form of a *conservation principle*, as given by Eq. (8-15). The second law, however, reduces to a conservation principle only for reversible processes, which are unknown in nature.

Example 8-5

It is stated in Sec. 8-2 and in axiom 3 that the entropy is a thermodynamic property of a system. As such, it is a function of as many thermodynamic coordinates as are necessary to specify the state of a system. Starting from this assertion, we may develop expressions for the differential dS for a PVT system, analogous to those obtained for the internal energy [Eqs. (4-2), (4-3), and (4-23)] and the enthalpy [Eqs. (4-13), (4-14), and (4-24)].

Taking S as a function of T and V, we may write for a differential change of state

$$dS = \left(\frac{\partial S}{\partial T}\right)_V dT + \left(\frac{\partial S}{\partial V}\right)_T dV. \tag{8-16}$$

But, by Eq. (8-11), $(\partial S/\partial T)_V = C_V/T$, and therefore

$$dS = C_V \frac{dT}{T} + \left(\frac{\partial S}{\partial V}\right)_T dV. \tag{8-17}$$

Similarly, considering S a function of T and P, we have

$$dS = \left(\frac{\partial S}{\partial T}\right)_P dT + \left(\frac{\partial S}{\partial P}\right)_T dP, \tag{8-18}$$

which becomes, by Eq. (8-10),

$$dS = C_P \frac{dT}{T} + \left(\frac{\partial S}{\partial P}\right)_T dP. \tag{8-19}$$

We shall show in the next chapter how the derivatives $(\partial S/\partial V)_T$ and $(\partial S/\partial P)_T$ are related to measurable quantities.

For an isochoric process, the integrated form of Eq. (8-17) is

$$\Delta S = \int C_V \frac{dT}{T} \qquad (\text{const } V) \tag{8-20}$$

Similarly, for the important case of an isobaric change of state, Eq. (8-19) yields

$$\Delta S = \int C_P \frac{dT}{T} \qquad (\text{const } P), \tag{8-12}$$

an expression derived in Sec. 8-6 by a different method.

Problems

8-1 Compare the efficiencies of cycles A and B of Fig. P8-1.

8-2 Draw rough TS diagrams for the following ideal-gas cycles: Stirling, Otto, and Diesel; a rectangle on a PV diagram; a "right triangle" on a PV diagram in which the base is an isobar, the altitude an isochor, and the "hypotenuse" an adiabat.

8-3 Show that the molar entropy of an ideal gas with constant heat capacities may be written

$$S = C_V \ln P + C_P \ln V + \text{const.}$$

8-4 An electric current of 10(A) is maintained for 1(s) in a resistor of 25(Ω) while the temperature of the resistor is kept constant at 27(°C).

(*a*) What is the entropy change of the resistor?
(*b*) What is ΔS_{total}?

The same current is maintained for the same time in the same resistor, but now thermally insulated, whose initial temperature is 27(°C). If the resistor has a mass of 10(g) and $C_P = 0.84$(J)/(g)(K),

(*c*) What is the entropy change of the resistor?
(*d*) What is ΔS_{total}?

8-5 (*a*) A kilogram of water at 273(K) is brought into contact with a heat reservoir at 373(K). When the water has reached 373(K), what is the entropy change of the water? Of the heat reservoir? What is ΔS_{total}?
(*b*) If the water had been heated from 273 to 373(K) by first bringing it in contact with a reservoir at 323(K) and then with a reservoir at 373(K), what would have been ΔS_{total}?
(*c*) Explain how the water might be heated from 273 to 373(K) so that $\Delta S_{\text{total}} \approx 0$.

8-6 A body of constant heat capacity $C_P{}^t$ and at temperature T_i is put in contact with a reservoir at a higher temperature T_f. The pressure remains constant while the body comes to equilibrium with the reservoir. Show that

$$\Delta S_{\text{total}} = C_P{}^t[x - \ln(1 + x)],$$

where $x = -(T_f - T_i)/T_f$. Prove that this entropy change is positive.

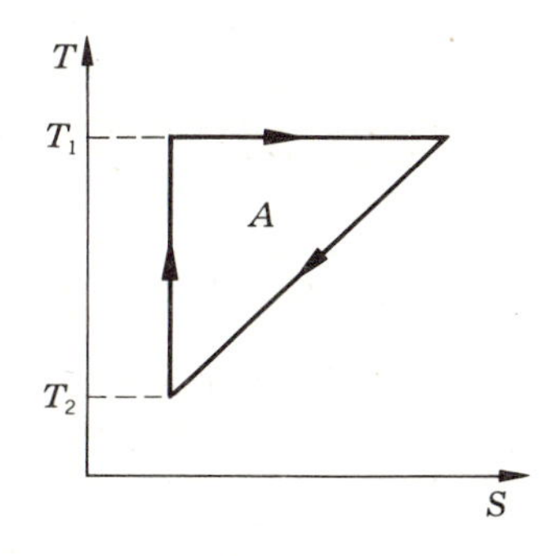

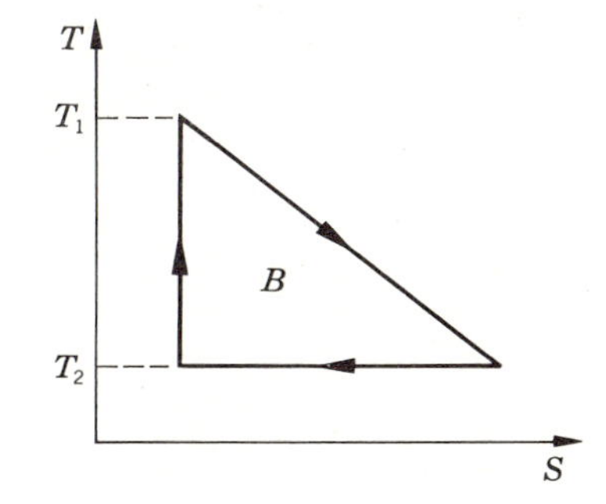

Fig. P8-1

8-7 A thermally insulated cylinder closed at both ends is fitted with a frictionless heat-conducting piston which divides the cylinder into two parts. Initially, the piston is clamped in the center, with 1(ft)^3 of air at 500(R) and 2(atm) pressure on one side and 1(ft)^3 of air at 500(R) and 1(atm) pressure on the other side. The piston is released and the system reaches equilibrium with the piston at a new position. Compute the final pressure and temperature and the increase of entropy. What irreversible process has taken place?

8-8 A piece of hot metal (mass m, specific heat at constant pressure C_P, temperature T_1) is immersed in a cooler liquid (m', C'_P, T'_1) adiabatically and isobarically. Prove that the condition of equilibrium, namely, $T_2 = T'_2$, may be obtained by rendering ΔS_{total} a maximum subject to the condition that the heat lost by the metal equals the heat gained by the liquid.

8-9 A mass m of water at T_1 is isobarically and adiabatically mixed with an equal mass of water at T_2. Show that

$$\Delta S_{\text{total}} = 2mC_P \ln \frac{(T_1 + T_2)/2}{\sqrt{T_1 T_2}}$$

and prove that this is positive by drawing a semicircle of diameter $T_1 + T_2$.

8-10 Solve the problem of Sec. 8-7 when only the hot reservoir is removed, and show that

$$\Delta S_{\text{total}} = mC_P \left(1 + \frac{T_0 - T_L}{2T_L} - \frac{T_0}{T_0 - T_L} \ln \frac{T_0}{T_L}\right).$$

8-11 A body of unit mass is originally at a temperature T_1, which is higher than that of a reservoir at the temperature T_2. Suppose an engine operates in a cycle between the body and the reservoir until it lowers the temperature of the body from T_1 to T_2, thus extracting heat Q from the body. The engine does work $|W|$ and rejects heat to the reservoir at T_2. Show that the maximum work is

$$|W|(\text{max}) = T_2(S_2 - S_1) - Q,$$

where Q, S_1, and S_2 refer to the body.

8-12 One mole of an ideal gas is compressed isothermally but irreversibly at 127(°C) from 1 to 10(atm) by a piston in a cylinder. The heat removed from the gas during compression flows to a heat reservoir at 27(°C). The actual work required is 20 percent greater than the reversible work for the same compression. Calculate the entropy change of the gas, the entropy change of the heat reservoir, and ΔS_{total}.

8-13 A rigid nonconducting tank with a volume of 120(ft)^3 is divided into two equal parts by a thin membrane. Hydrogen gas is contained on one side of the membrane at 50(psia) and 100(°F). The other side is a perfect vacuum. The membrane is suddenly ruptured, and the H_2 gas fills the tank. What is the entropy change of the hydrogen? Consider hydrogen to be an ideal gas for which $C_P = 7$(Btu)/(lb mol)(R).

8-14 Using the data of Table 5-3, calculate ΔS in (J)/(mol)(K) for sulfur dioxide in the ideal-gas state for a temperature change from 25 to 425(°C) at constant pressure.

8-15 Calculate ΔS_{total} as a result of each of the following processes:

(*a*) A copper block of 400(g) and with a total heat capacity at constant pressure of 150(J)/(K) at 100(°C) is placed in a lake at 10(°C).

(*b*) The same block, at 10(°C), is dropped from a height of 100(m) into the lake.

(*c*) Two such blocks, at 100 and 0(°C), are joined together.

8-16 What is ΔS_{total} as a result of each of the following processes:

(*a*) A capacitor of capacitance 1(μF) is connected to a 100(V) reversible battery at 0(°C).

(*b*) The same capacitor, after being charged to 100(V), is discharged through a resistor kept at 0(°C).

8-17 Thirty-six grams of water at a temperature of 20(°C) is converted into steam at 250(°C) at constant atmospheric pressure. Assuming the heat capacity per gram of liquid water to remain practically constant at 4.2(J)/(g)(K) and the heat of vaporization at 100(°C) to be 2260(J)/(g), and using Table 5-3, calculate the entropy change of the system.

8-18 Ten grams of water at 20(°C) is converted into ice at −10(°C) at constant atmospheric pressure. Assuming the heat capacity per gram of liquid water to remain constant at 4.2(J)/(g)(K) and that of ice to be one-half this value, and taking the heat of fusion of ice at 0(°C) to be 335(J)/(g), calculate the entropy change of the system.

8-19 A cylinder closed at both ends, with adiabatic walls, is divided into two parts by a movable frictionless *adiabatic* piston. Originally, the pressure, volume, and temperature are the same on both sides of the piston (P_0, V_0, T_0). The gas is ideal, with C_V independent of T and $\gamma = 1.5$. By means of a heating coil in the gas on the left-hand side, heat is slowly supplied to the gas on the left until the pressure reaches $27P_0/8$. In terms of nR, V_0, and T_0:

(*a*) What is the final right-hand volume?
(*b*) What is the final right-hand temperature?
(*c*) What is the final left-hand temperature?
(*d*) How much heat must be supplied to the *gas* on the left? (Ignore the coil!)
(*e*) How much work is done on the gas on the right?
(*f*) What is the entropy change of gas on the right?
(*g*) What is the entropy change of the gas on the left?
(*h*) What is ΔS_{total}?

8-20 (*a*) If the system in Prob. 3-6 consists of 0.02(lb mol) of an ideal gas with constant heat capacity $C_V = 5$(Btu)/(lb mol)(R), calculate ΔS^t of the gas for steps 12, 23, and 31.

(*b*) What would ΔS^t be for step 12 if this path were instead an isotherm?

8-21 (*a*) The temperature of an ideal gas with constant heat capacity is changed from T_1 to T_2. Show that ΔS of the gas is greater if the change in state occurs at constant pressure than if at constant volume.

(*b*) The pressure of an ideal gas is changed from P_1 to P_2 by an isothermal process and by a constant-volume process. Show that ΔS for the gas is of opposite sign for the two processes.

8-22 Two identical bodies of constant-heat capacity at temperatures T_1 and T_2, respectively, are used as reservoirs for a heat engine. If the bodies remain at constant pressure and undergo no change of phase, show that the work is

$$W = C_P(2T_f - T_1 - T_2),$$

where T_f is the final temperature attained by both bodies. Show that, when $|W|$ is a maximum,

$$T_f = \sqrt{T_1 T_2}.$$

8-23 Two identical bodies of constant heat capacity are at the same initial temperature T_i. A refrigerator operates between these two bodies until one body is cooled to temperature T_2. If the bodies remain at constant pressure and undergo no change of phase, show that the minimum amount of work needed to do this is

$$W(\text{min}) = C_P\left(\frac{T_i^2}{T_2} + T_2 - 2T_i\right).$$

8-24 Consider any system immersed in a medium at the constant temperature T_0 and suppose that *the only reservoir with which the system may exchange heat is this medium.* Let the system undergo a process involving the transfer of heat Q, the performance of work W, and a change of entropy ΔS. Show that:

(*a*) $\Delta S - \dfrac{\Delta U - W}{T_0} \geq 0.$

(*b*) $W(\text{reversible}) = \Delta U - T_0\,\Delta S.$

(*c*) $T_0\,\Delta S_{\text{total}} = W(\text{actual}) - W(\text{reversible}).$

8-25 Three identical finite bodies of constant heat capacity are at temperatures 300, 300 and 100(K). If no work or heat is supplied from the outside, what is the highest temperature to which any one of the bodies can be raised by the operation of heat engines or refrigerators?

8-26 A horizontal cylinder, closed at both ends, is divided in half by a free piston. The cylinder neither absorbs nor conducts heat, and the piston is perfectly lubricated in the cylinder. The left half of the cylinder contains 1(lb mol) of an ideal gas at 2(atm) and 60(°F). The gas has constant heat capacities, $C_V = 5$ and $C_P = 7$(Btu)/(lb mol)(R). The right half of the cylinder is evacuated and the piston is temporarily restrained by latches. The following operations are carried out: (i) The latches are removed, allowing the gas on the left to expand as it pushes the piston to the right-hand end of the cylinder; (ii) a small rod is inserted through the right-hand end of the cylinder and exerts a force on the piston, pushing it slowly back to the initial position.

(*a*) What is the gas temperature after process i once internal equilibrium is established?

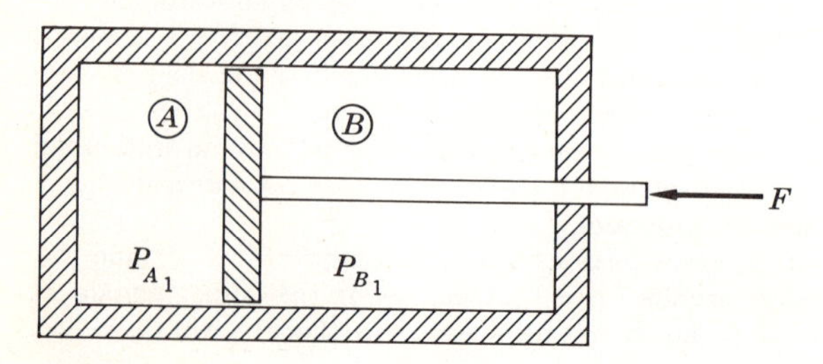

Fig. P8-27

(*b*) What is the gas temperature at the end of process ii, assuming this process to be reversible and adiabatic?
(*c*) What is the gas pressure at the end of process ii?
(*d*) What is W for process ii?
(*e*) What is ΔS for process i and for the combined processes i and ii?

8-27 Figure P8-27 shows a closed cylinder divided into two unequal chambers A and B by a piston that is free to move, except that it is initially kept from moving by a force F applied to a piston rod that extends through the right-hand end of the cylinder. Both the cylinder wall and the piston are perfect heat insulators, so that no heat is exchanged either between the chambers or with the surroundings. The two chambers contain fixed amounts of the same ideal gas, which may be assumed to have the constant heat capacities: $C_V = 5$ and $C_P = 7$(cal)/(mol)(K). The piston has an area of $0.1(\text{m})^2$ and the cylinder has a free length of 1(m). The initial location of the piston is such that chamber A represents $\frac{1}{3}$ of the total volume. The initial conditions are as follows:

$$P_{A_1} = 4(\text{atm}), \qquad P_{B_1} = 2(\text{atm}),$$

$$T_{A_1} = 127(°\text{C}), \qquad T_{B_1} = 27(°\text{C}).$$

The force F is slowly reduced so as to allow the piston to move to the right until the pressures in the two chambers are equal. Assuming the process to be reversible, determine $P_{A_2} = P_{B_2} = P_2$, the temperatures T_{A_2} and T_{B_2}, and the work done by the system against the external force F.

8-28 A closed cylinder is divided into two unequal chambers A and B by a piston that is free to move, except that it is initially kept from moving by a stop (see Fig. P8-28). Both the cylinder wall and the piston are perfect heat insulators, so that no heat is exchanged either between the chambers or with the surroundings. The two chambers contain fixed amounts of the same ideal gas, which may be assumed to have constant heat capacities. Chamber A initially contains n_A(mol) at T_{A_1} and P_{A_1}; chamber B initially contains n_B(mol) at T_{B_1} and P_{B_1}, where $P_{A_1} > P_{B_1}$. If the stop is removed so that the piston is free to move until the pressures in the two chambers equalize, will the change in entropy as a result of the process be less than, equal to, or greater than zero in

(*a*) The entire system, consisting of A and B,
(*b*) Chamber A,
(*c*) Chamber B.

Can the final pressure $P_{A_2} = P_{B_2} = P_2$ be determined by the methods of thermodynamics? Can the final temperatures T_{A_2} and T_{B_2} be determined by the methods

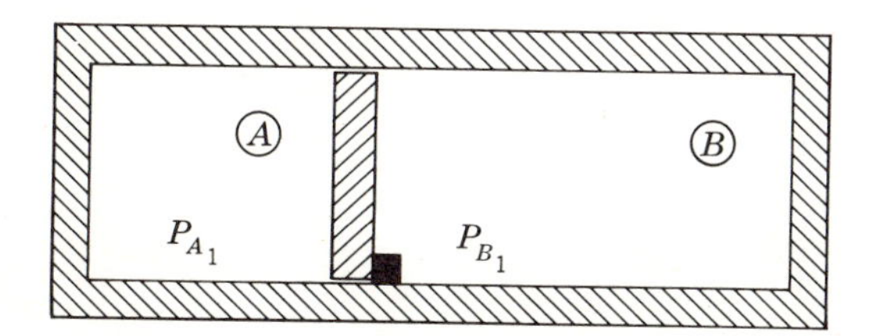

Fig. P8-28

of thermodynamics? How would the problem be changed if the piston were a heat conductor?

8-29 A system consisting of gas contained in a piston-cylinder device undergoes an *irreversible* process between initial and final equilibrium states that causes its internal energy to increase by 30(Btu). During the process the system receives heat in the amount of 100(Btu) from a heat reservoir at 1000(R). The system is then restored to its initial state by a *reversible* process, during which the only heat transfer is between the system and the heat reservoir at 1000(R). The entropy change of the heat reservoir as a result of *both* processes is +0.01 (Btu)/(R). Calculate (*a*) the work done by the system during the first (irreversible) process, (*b*) the heat transfer with respect to the system during the second (reversible) process, (*c*) the work done by the system during the second process.

8-30 One pound mole of an ideal gas with constant heat capacities [C_P = 5(Btu)/(lb mol)(R)] is compressed adiabatically in a piston and cylinder device from 1(atm) and 40(°F) to a final pressure of 5(atm). The process is irreversible and requires 25 percent more work than a reversible, adiabatic compression from the same initial state to the same final pressure. How much work is required, and what is the entropy change of the gas?

9

PURE SUBSTANCES

9-1 Property Relationships. The Helmholtz and Gibbs Functions

The primary thermodynamic functions T, P, V, U, and S have already been introduced in our development and discussions of the zeroth, first, and second laws. In addition, we have defined a number of auxiliary functions, such as the volume expansivity β, the compressibility factor Z, the constant-volume heat capacity C_V, and the enthalpy H. The use of an auxiliary function is usually suggested by the frequent occurrence of the grouping of primary functions which constitutes its definition, and is simply a matter of convenience. In this section we define two more auxiliary functions, the *Helmholtz function* A and the *Gibbs function* G. These functions play an important role in the description of certain types of processes; moreover, their use in conjunction with the first and second laws facilitates the derivation of useful relationships which connect the thermodynamic properties of substances in equilibrium states.

Consider an infinitesimal reversible process undergone by a closed PVT system containing 1(mol) or a unit mass of material. The first law is

$$dU = \delta Q_R + \delta W_R, \tag{4-9}$$

where the subscripts R indicate that the process is reversible. According to Eqs. (8-8) and (3-1), we have

$$\delta Q_R = T\,dS$$

and

$$\delta W_R = -P\,dV.$$

Combination of the three equations gives

$$\boxed{dU = T\,dS - P\,dV.} \tag{9-1}$$

Equation (9-1) is the cornerstone of the classical treatment of the thermodynamic properties of PVT systems. Because it includes only properties, Eq. (9-1) applies to any differential change in the equilibrium state of a closed PVT system. However, only for *reversible* processes can we retain the identifications $\delta Q = T\,dS$ and $\delta W = -P\,dV$ which led to its development.

The enthalpy H was defined in Sec. 4-8:

$$H = U + PV. \tag{4-11}$$

Differentiating, we find

$$dH = dU + P\,dV + V\,dP;$$

combination with Eq. (9-1) yields

$$\boxed{dH = T\,dS + V\,dP.} \tag{9-2}$$

We define the Helmholtz function A by[1]

$$\boxed{A \equiv U - TS.} \tag{9-3}$$

Differention gives

$$dA = dU - T\,dS - S\,dT$$

and combination with Eq. (9-1) yields

$$\boxed{dA = -S\,dT - P\,dV.} \tag{9-4}$$

Finally, the Gibbs function G is defined by[2]

$$\boxed{G \equiv H - TS.} \tag{9-5}$$

Differentiating, we find

$$dG = dH - T\,dS - S\,dT,$$

and combination with Eq. (9-2) gives

$$\boxed{dG = -S\,dT + V\,dP.} \tag{9-6}$$

Since U, H, A, and G are all thermodynamic properties, they may each be considered functions of any other two thermodynamic properties. In particular, each of these functions has associated with it a pair of special independent variables, called *natural variables*. These variables are those whose differentials appear on the right-hand sides of Eqs. (9-1), (9-2),

[1] H. L. F. von Helmholtz (1821–1894), a German physiologist, physicist, and amateur musician. One of the greatest of the nineteenth century generalists, Helmholtz made significant contributions to acoustics, electromagnetic theory, fluid mechanics, optics, physiology, and thermodynamics.

[2] J. W. Gibbs (1839–1903), an American physicist. Also known for his contributions to statistical mechanics and vector analysis, Gibbs is generally recognized as the founder of modern chemical thermodynamics.

(9-4), and (9-6). Thus, the natural variables for U are S and V, those for H are S and P, etc. For any closed PVT system, we therefore have the functional relationships

$$U = U(S,V), \tag{9-7a}$$

$$H = H(S,P), \tag{9-7b}$$

$$A = A(T,V), \tag{9-7c}$$

and

$$G = G(T,P). \tag{9-7d}$$

Since $U = U(S,V)$, we may write for a differential change of state

$$dU = \left(\frac{\partial U}{\partial S}\right)_V dS + \left(\frac{\partial U}{\partial V}\right)_S dV.$$

Comparison of this expression with Eq. (9-1) leads to the following identifications:

$$T = \left(\frac{\partial U}{\partial S}\right)_V \quad \text{and} \quad P = -\left(\frac{\partial U}{\partial V}\right)_S.$$

Similar procedures can be followed for the functions H, A, and G. The results are summarized as follows:

$$T = \left(\frac{\partial U}{\partial S}\right)_V = \left(\frac{\partial H}{\partial S}\right)_P \tag{9-8}$$

$$S = -\left(\frac{\partial A}{\partial T}\right)_V = -\left(\frac{\partial G}{\partial T}\right)_P \tag{9-9}$$

$$P = -\left(\frac{\partial U}{\partial V}\right)_S = -\left(\frac{\partial A}{\partial V}\right)_T \tag{9-10}$$

$$V = \left(\frac{\partial H}{\partial P}\right)_S = \left(\frac{\partial G}{\partial P}\right)_T \tag{9-11}$$

Example 9-1

Two moles of an ideal gas initially at 300(K) and 5(bar) undergo a reversible isothermal expansion to a final pressure of 3(bar). Determine ΔU^t, ΔH^t, ΔA^t, and ΔG^t for the ideal gas.

For an ideal gas, both U and H are functions of T only and, since this is an isothermal process,

$$\Delta U^t = 0$$

and

$$\Delta H^t = 0.$$

For an isothermal process of n(mol) of material, Eq. (9-4) may be integrated to give

$$\Delta A^t = -n \int P\,dV \qquad \text{(const } T\text{)}.$$

But the right-hand side of the equation is just the reversible work W_R for a PVT system, and we have the important result that

$$\Delta A^t = W_R \qquad \text{(const } T\text{)}. \tag{9-12}$$

Thus, the work for an isothermal reversible process is equal to the change in the Helmholtz function of the system. The work for this particular process was determined in Example 3-2, and found to be -2.55(kJ). We therefore have

$$\Delta A^t = -2.55\text{(kJ)}.$$

For an isothermal process of n(mol) of material, Eq. (9-6) is integrated to give

$$\Delta G^t = n \int V\,dP \qquad \text{(const } T\text{)}.$$

For an ideal gas,

$$d(PV) = d(RT),$$

or

$$P\,dV + V\,dP = R\,dT.$$

Thus for an isothermal change in state of an ideal gas,

$$V\,dP = -P\,dV \qquad \text{(const } T\text{)},$$

and we find

$$\Delta G^t = n \int V\,dP = -n \int P\,dV = \Delta A^t.$$

Therefore $\Delta G^t = -2.55\text{(kJ)}.$

Example 9-2

An important use of the Gibbs function is in the characterization of states of phase equilibrium. To illustrate this, we consider the case of the vapor-liquid equilibrium of a single pure substance.

We imagine a closed system containing liquid l in equilibrium with its vapor v, and consider a differential vaporization process during which dn^v moles of vapor is formed. Since the system is closed, the *total* number of moles of substance remains constant, i.e.,

$$n^v + n^l = \text{const},$$

or

$$dn^v = -dn^l. \tag{A}$$

Designating extensive properties of the system (liquid plus vapor) by a superscript t, we have, by Eq. (9-6),

$$dG^t = -S^t\,dT + V^t\,dP.$$

But T and P both remain constant during the vaporization of a pure fluid (see Sec. 5-1), so that $dT = 0$ and $dP = 0$. Thus,

$$dG^t = 0.$$

Now,
$$G^t = n^l G^l + n^v G^v,$$

where G^l and G^v are the molar properties for the liquid and vapor, respectively. Therefore

$$dG^t = n^l\,dG^l + n^v\,dG^v + G^l\,dn^l + G^v\,dn^v = 0.$$

But G^l and G^v remain constant, because both T and P are constant [see Eq. (9-7d)]; hence $dG^l = dG^v = 0$. Additionally, dn^l and dn^v are related by Eq. (A). Thus we obtain finally

$$G^l\,dn^l + G^v\,dn^v = (G^l - G^v)dn^l = 0,$$

or, since dn^l is arbitrary,

$$G^l = G^v. \tag{B}$$

For vapor-liquid equilibrium of a pure substance, the molar or specific Gibbs functions of the two phases are equal. Equation (B) may be considered the equation for the vaporization curve.

The argument above may be extended to any type of two-phase equilibrium of a pure substance for which the equilibrium temperatures and pressures are identical in the two phases. Thus, for solid(s)-vapor equilibrium, we have as the equation for the sublimation curve

$$G^s = G^v.$$

Similarly, the equation of the fusion curve is

$$G^s = G^l.$$

For three-phase (triple-point) equilibrium, we have two simultaneous equations, viz.,

$$G^s = G^l = G^v.$$

Example 9-3

Equations of state for pure substances are usually formulated in terms of the three variables P, V, and T because these are the variables most susceptible to experimental measurement and control. However, a major shortcoming of such a formulation is that it is not *complete*, i.e., given a PVT equation of state, it is not possible to compute values of all thermodynamic functions solely from the equation. For instance, one cannot determine values of the heat capacities C_P or C_V from PVT data alone. This shortcoming does not exist for *fundamental* equations of state—equations which express U, H, A, or G as functions of their natural variables. All thermodynamic properties can be computed from such an equation by differentiation and algebraic manipulation.

As an example, suppose we have a fundamental equation of the form $H = H(S,P)$. The temperature and volume are found by application of Eqs. (9-8) and (9-11):

$$T = \left(\frac{\partial H}{\partial S}\right)_P,$$

$$V = \left(\frac{\partial H}{\partial P}\right)_S.$$

Knowing T and V, we can then determine U, A, and G by use of the definitions, Eqs.

(4-11), (9-3), and (9-5). The results are

$$U = H - P\left(\frac{\partial H}{\partial P}\right)_S,$$

$$A = H - P\left(\frac{\partial H}{\partial P}\right)_S - S\left(\frac{\partial H}{\partial S}\right)_P,$$

and

$$G = H - S\left(\frac{\partial H}{\partial S}\right)_P.$$

Expressions for the heat capacities are also readily found. For example, C_P is given by Eq. (8-10):

$$C_P = T\left(\frac{\partial S}{\partial T}\right)_P.$$

But, by Eq. (9-8),

$$T\left(\frac{\partial S}{\partial T}\right)_P = T\left(\frac{\partial T}{\partial S}\right)_P^{-1} = \left(\frac{\partial H}{\partial S}\right)_P\left[\frac{\partial}{\partial S}\left(\frac{\partial H}{\partial S}\right)_P\right]_P^{-1},$$

and thus

$$C_P = \frac{(\partial H/\partial S)_P}{(\partial^2 H/\partial S^2)_P}.$$

Given the apparent usefulness of fundamental equations of state, one might well ask why they are not used instead of PVT equations. The reason is that not all the natural variables are directly observable. Values for functions such as S, U, H, A, and G must be inferred from other types of measurements, and an analytical formulation is thus indirect and difficult.

9-2 Test for Exactness. Maxwell Equations

Suppose that there exists a functional relationship between the three coordinates x, y, and z:

$$f(x,y,z) = 0.$$

Then z may be considered a function of x and y, and we may write

$$dz = \left(\frac{\partial z}{\partial x}\right)_y dx + \left(\frac{\partial z}{\partial y}\right)_x dy,$$

or

$$dz = M\,dx + N\,dy, \tag{9-13}$$

where

$$M \equiv \left(\frac{\partial z}{\partial x}\right)_y \quad \text{and} \quad N \equiv \left(\frac{\partial z}{\partial y}\right)_x.$$

Differentiating M partially with respect to y and N with respect to x, we get

$$\left(\frac{\partial M}{\partial y}\right)_x = \frac{\partial^2 z}{\partial y\, \partial x}$$

and

$$\left(\frac{\partial N}{\partial x}\right)_y = \frac{\partial^2 z}{\partial x\, \partial y}.$$

But if z and its derivatives are continuous, the two mixed second derivatives are equal. Thus we have

$$\boxed{\left(\frac{\partial M}{\partial y}\right)_x = \left(\frac{\partial N}{\partial x}\right)_y.} \tag{9-14}$$

Equation (9-14) is both a necessary and a sufficient condition for the *exactness* of a differential expression, and it may therefore be used as a test for exactness. When an expression such as $M\,dx + N\,dy$ does not meet this test, *no function z exists* whose differential is equal to the original expression. In thermodynamics, we often develop expressions of the form of Eq. (9-13) in which x, y, and z are *known* to be functions of state, and for which the differential expressions are therefore known to be exact. For such cases, Eq. (9-14) is not used to test exactness but rather to provide additional thermodynamic relationships.

Equations (9-1), (9-2), (9-4), and (9-6) involve thermodynamic properties only, and hence they are exact differential equations. Thus we may apply Eq. (9-14) to each of them and so obtain relationships between derivatives of the differential coefficients which appear on the right-hand side of each equation. For example, application of Eq. (9-14) to Eq. (9-1) provides the relation

$$\left(\frac{\partial T}{\partial V}\right)_S = -\left(\frac{\partial P}{\partial S}\right)_V;$$

this and the analogous equations which follow from Eqs. (9-2), (9-4), and (9-6) are summarized as follows:

$$\left(\frac{\partial T}{\partial V}\right)_S = -\left(\frac{\partial P}{\partial S}\right)_V \tag{9-15}$$

$$\left(\frac{\partial T}{\partial P}\right)_S = \left(\frac{\partial V}{\partial S}\right)_P \tag{9-16}$$

$$\left(\frac{\partial P}{\partial T}\right)_V = \left(\frac{\partial S}{\partial V}\right)_T \tag{9-17}$$

$$\boxed{\left(\frac{\partial V}{\partial T}\right)_P = -\left(\frac{\partial S}{\partial P}\right)_T} \tag{9-18}$$

Equations (9-15) through (9-18) are known as *Maxwell's equations.*[1]

Example 9-4

By invoking the exactness criterion, Eq. (9-14), demonstrate that the reversible heat Q_R is not a thermodynamic property.

We consider the following differential form of the first law for a reversible process of a PVT system:

$$dH = \delta Q_R + V\,dP, \tag{4-16}$$

or

$$\delta Q_R = dH - V\,dP.$$

If H is regarded as a function of T and P, then we have, by Eq. (4-14),

$$dH = \left(\frac{\partial H}{\partial T}\right)_P dT + \left(\frac{\partial H}{\partial P}\right)_T dP.$$

Combination of the last two equations gives

$$\delta Q_R = \left(\frac{\partial H}{\partial T}\right)_P dT + \left[\left(\frac{\partial H}{\partial P}\right)_T - V\right] dP, \tag{A}$$

which is of the form of Eq. (9-13), with $M = (\partial H/\partial T)_P$, and $N = [(\partial H/\partial P)_T - V]$. Differentiating M with respect to P and N with respect to T, we find

$$\left(\frac{\partial M}{\partial P}\right)_T = \frac{\partial^2 H}{\partial P\,\partial T} \quad \text{and} \quad \left(\frac{\partial N}{\partial T}\right)_P = \frac{\partial^2 H}{\partial T\,\partial P} - \left(\frac{\partial V}{\partial T}\right)_P.$$

But H is a function of T and P, so that the mixed second derivatives must be equal. Thus, if the exactness condition Eq. (9-14) is to be satisfied, we must also have

$$\left(\frac{\partial V}{\partial T}\right)_P = 0$$

for all PVT systems at all T and P. But $(\partial V/\partial T)_P = \beta V$, and we know from experiment that $\beta \neq 0$ in general. Thus Eq. (9-14) is not satisfied; Eq. (A) is not an exact differential equation; and Q_R cannot be a thermodynamic property. The use of the δ sign with Q draws express attention to this fact.

9-3 Entropy Equations

The practical value of the Maxwell equations derives from the fact that they relate derivatives of the entropy (a function which cannot be directly

[1] J. C. Maxwell (1831–1879), a Scottish physicist. Maxwell is best known for his statements of the laws of electromagnetism and for his contributions to the kinetic theory of gases.

measured) to measurable or calculable quantities. It was shown in Example 8-5 that

$$dS = C_V \frac{dT}{T} + \left(\frac{\partial S}{\partial V}\right)_T dV. \tag{8-17}$$

Eliminating the entropy derivative by use of the Maxwell relation, Eq. (9-17), we find

$$\boxed{dS = C_V \frac{dT}{T} + \left(\frac{\partial P}{\partial T}\right)_V dV,} \tag{9-19}$$

which is now in a form convenient for use with C_V data and an equation of state. One additional simplification can be made by use of the following identity, developed in Example 2-1:

$$\left(\frac{\partial P}{\partial T}\right)_V = \frac{\beta}{\kappa},$$

where β is the volume expansivity and κ the isothermal compressibility. Equation (9-19) then becomes

$$dS = C_V \frac{dT}{T} + \frac{\beta}{\kappa} dV. \tag{9-20}$$

It was also shown in Example 8-5 that

$$dS = C_P \frac{dT}{T} + \left(\frac{\partial S}{\partial P}\right)_T dP. \tag{8-19}$$

Using the Maxwell relation Eq. (9-18) to eliminate the entropy derivative, we obtain

$$\boxed{dS = C_P \frac{dT}{T} - \left(\frac{\partial V}{\partial T}\right)_P dP.} \tag{9-21}$$

But, from the definition, Eq. (2-2),

$$\left(\frac{\partial V}{\partial T}\right)_P = \beta V.$$

Thus Eq. (9-21) may also be written

$$dS = C_P \frac{dT}{T} - \beta V\, dP. \tag{9-22}$$

A final pair of equivalent expressions can be developed for dS, in which

P and V are the independent variables. They are

$$dS = \frac{C_V}{T}\left(\frac{\partial T}{\partial P}\right)_V dP + \frac{C_P}{T}\left(\frac{\partial T}{\partial V}\right)_P dV \tag{9-23}$$

and

$$dS = \frac{\kappa C_V}{\beta T}\,dP + \frac{C_P}{\beta V T}\,dV. \tag{9-24}$$

For an ideal gas, $PV = RT$, and thus

$$\left(\frac{\partial P}{\partial T}\right)_V = \frac{P}{T} = \frac{R}{V} \quad \text{and} \quad \left(\frac{\partial V}{\partial T}\right)_P = \frac{V}{T} = \frac{R}{P}.$$

Substitution into Eqs. (9-19), (9-21), and (9-23) gives

$$dS = C_V \frac{dT}{T} + R\frac{dV}{V} \qquad \text{(ideal gas)}, \tag{8-6}$$

$$dS = C_P \frac{dT}{T} - R\frac{dP}{P} \qquad \text{(ideal gas)}, \tag{8-4}$$

and

$$dS = C_V \frac{dP}{P} + C_P\frac{dV}{V} \qquad \text{(ideal gas)}. \tag{9-25}$$

Equations (8-6) and (8-4) were developed in Sec. 8-3 by a different but equivalent method.

Example 9-5

Derive expressions for isothermal changes in the entropy for gases described by (*a*) the two-term virial equation, Eq. (5-10), and (*b*) the three-term virial equation, Eq. (5-11).

We have two general expressions for isothermal changes in S, suitable for use with PVT equations of state. The first is obtained by restriction of Eq. (9-19) to constant T and integration over V:

$$\Delta S = \int_{V_1}^{V_2} \left(\frac{\partial P}{\partial T}\right)_V dV \qquad \text{(const } T\text{)}. \tag{9-26}$$

The second follows similarly from Eq. (9-21) and involves an integration over P:

$$\Delta S = -\int_{P_1}^{P_2} \left(\frac{\partial V}{\partial T}\right)_P dP \qquad \text{(const } T\text{)}. \tag{9-27}$$

The choice of which of the equations above to use depends upon the algebraic form of the equation of state. When it can be used, Eq. (9-27) is more convenient, because P and T (rather than V and T) are the usual variables for specification of the thermodynamic state of a PVT system.

(*a*) Equation (5-10) is

$$Z \equiv \frac{PV}{RT} = 1 + \frac{BP}{RT}.$$

Solving for V, we get

$$V = \frac{RT}{P} + B,$$

from which

$$\left(\frac{\partial V}{\partial T}\right)_P = \frac{R}{P} + \frac{dB}{dT}.$$

Substitution into Eq. (9-27) gives

$$\Delta S = -R \int_{P_1}^{P_2} \frac{dP}{P} - \frac{dB}{dT} \int_{P_1}^{P_2} dP,$$

or

$$\Delta S = -R \ln \frac{P_2}{P_1} - \frac{dB}{dT}(P_2 - P_1) \qquad \text{(const } T\text{)}. \tag{9-28}$$

(*b*) Equation (5-11) is

$$Z \equiv \frac{PV}{RT} = 1 + \frac{B}{V} + \frac{C}{V^2}.$$

This equation does not yield a tractable analytical expression for $(\partial V/\partial T)_P$ as a function of just P and T; hence we do not use Eq. (9-27), but we instead employ Eq. (9-26) for computation of ΔS. Solving the equation of state for P, we get

$$P = RT\left(\frac{1}{V} + \frac{B}{V^2} + \frac{C}{V^3}\right),$$

from which

$$\left(\frac{\partial P}{\partial T}\right)_V = \frac{R}{V} + \frac{R}{V^2}\left(B + T\frac{dB}{dT}\right) + \frac{R}{V^3}\left(C + T\frac{dC}{dT}\right).$$

Substitution into Eq. (9-26) gives

$$\Delta S = R \int_{V_1}^{V_2} \frac{dV}{V} + R\left(B + T\frac{dB}{dT}\right) \int_{V_1}^{V_2} \frac{dV}{V^2} + R\left(C + T\frac{dC}{dT}\right) \int_{V_1}^{V_2} \frac{dV}{V^3},$$

or

$$\Delta S = R \ln \frac{V_2}{V_1} - R\left(B + T\frac{dB}{dT}\right)\left(\frac{1}{V_2} - \frac{1}{V_1}\right) - \frac{R}{2}\left(C + T\frac{dC}{dT}\right)\left(\frac{1}{V_2^2} - \frac{1}{V_1^2}\right)$$

$$\text{(const } T\text{)}. \tag{9-29}$$

According to Eqs. (9-28) and (9-29), the calculation of entropy differences from a virial equation requires values not only for the virial coefficients but also for their temperature derivatives. It is also clear from these expressions that the use of the two-term virial equation, Eq. (5-10), affords considerable computational advantages over the three-term equation in inverse powers of V. As already noted, thermodynamic states are usually specified in terms of P and T, and calculation of values of ΔS from Eq. (9-29) for given values of T, P_1, and P_2 requires first the solution of a cubic equation in V for the initial and final volumes V_1 and V_2.

Example 9-6

The pressure on a sample of liquid mercury at 0(°C) is increased reversibly and isothermally from 1(bar) to 1,000(bar). Determine the heat transferred and the work done per (mol) of mercury. For the given range of pressure, β, κ, and V may be assumed essentially constant, with average values

$$\beta = 181 \times 10^{-6}(\text{K})^{-1},$$

$$\kappa = 3.88 \times 10^{-6}(\text{bar})^{-1},$$

$$V = 14.7(\text{cm})^3/(\text{mol}).$$

For a reversible change of state of 1(mol) of material, we have, by Eq. (8-8),

$$\delta Q = T\,dS.$$

Restriction of Eq. (9-22) to constant T gives

$$dS = -\beta V\,dP \qquad (\text{const } T),$$

and combination with the previous equation yields

$$\delta Q = -\beta T V\,dP.$$

Under the assumption of constant β and V, this expression is integrated to give

$$Q = -\beta T V(P_2 - P_1). \tag{A}$$

The corresponding expression for reversible work, valid under the same assumptions as for Eq. (A), was given in Example 3-3:

$$W = \frac{\kappa V}{2}(P_2^2 - P_1^2). \tag{B}$$

Substituting given numerical values into Eq. (A), we find

$$Q = -181 \times 10^{-6}(\text{K})^{-1} \times 273.2(\text{K}) \times 14.7 \times 10^{-6}(\text{m})^3/(\text{mol}) \times (1{,}000 - 1)(\text{bar}),$$

or

$$Q = -7.26 \times 10^{-4}(\text{m})^3(\text{bar})/(\text{mol}) = -72.6(\text{J})/(\text{mol}).$$

Similarly, from Eq. (B), we obtain

$$W = \tfrac{1}{2} \times 3.88 \times 10^{-6}(\text{bar})^{-1} \times 14.7 \times 10^{-6}(\text{m})^3/(\text{mol}) \times [(1{,}000)^2 - (1)^2](\text{bar})^2$$

or

$$W = 2.9(\text{J})/(\text{mol}).$$

During the compression process, 72.6(J) of heat is liberated per (mol) of mercury, while only 2.9(J) of work is done. The source of the extra heat is the internal energy of the mercury, which changes by the amount

$$\Delta U = Q + W = -72.6 + 2.9 = -69.7(\text{J})/(\text{mol}).$$

Whereas κ is always a positive quantity, β may be either positive or negative in sign. In the present example, β is positive and hence heat flows *out* of the system during compression. For a substance with a negative β [e.g., liquid water between 0 and 4(°C)], heat would be absorbed by the system, and its internal energy would increase.

Example 9-7

The pressure on a sample of liquid mercury initially at 0(°C) is increased reversibly and adiabatically from 1(bar) to 1,000(bar). What is the final temperature of the mercury? For the anticipated range of temperature, β, C_P, and V may be assumed constant, with the values

$$\beta = 181 \times 10^{-6}(\text{K})^{-1},$$

$$C_P = 28.0(\text{J})/(\text{mol}),$$

$$V = 14.7(\text{cm})^3/(\text{mol}).$$

A reversible adiabatic process is one of constant entropy. With $dS = 0$, Eq. (9-22) becomes

$$C_P \frac{dT}{T} - \beta V\, dP = 0,$$

or

$$\frac{dT}{T} = \frac{\beta V}{C_P}\, dP.$$

Integration under the assumption of constant β, C_P, and V yields

$$\ln\left(\frac{T_2}{T_1}\right) = \frac{\beta V}{C_P}(P_2 - P_1),$$

$$T_2 = T_1 \exp\left[\frac{\beta V}{C_P}(P_2 - P_1)\right].$$

From the given data, we find

$$T_2 = 273.2(\text{K}) \times \exp\left[\frac{181 \times 10^{-6}(\text{K})^{-1} \times 14.7 \times 10^{-6}(\text{m})^3/(\text{mol}) \times (1{,}000 - 1)(\text{bar})}{28.0(\text{J})/(\text{mol})(\text{K}) \times 10^{-5}(\text{bar})(\text{m})^3/(\text{J})}\right],$$

or

$$T_2 = 275.8(\text{K}); \qquad t_2 = 2.6(°\text{C}).$$

The temperature increases by 2.6(°C); if the system had been a substance with a negative β, the adiabatic increase in pressure would have produced a *decrease* in temperature.

9-4 Internal Energy Equation. Enthalpy Equation

For a differential change of state, dU is given by Eq. (9-1):

$$dU = T\,dS - P\,dV.$$

But, by Eq. (9-19),

$$dS = C_V \frac{dT}{T} + \left(\frac{\partial P}{\partial T}\right)_V dV.$$

Combination of these two expressions yields

$$\boxed{dU = C_V\,dT + \left[T\left(\frac{\partial P}{\partial T}\right)_V - P\right] dV.} \tag{9-30}$$

A similar equation may be developed for dU in which the independent variables are T and P (Prob. 9-36). An alternative form of (9-30), convenient for use with tabulated values of β and κ, is

$$dU = C_V\,dT + \left(\frac{\beta T}{\kappa} - P\right) dV. \tag{9-31}$$

The coefficient of dV in Eqs. (9-30) and (9-31) is just the derivative $(\partial U/\partial V)_T$. Thus,

$$\left(\frac{\partial U}{\partial V}\right)_T = T\left(\frac{\partial P}{\partial T}\right)_V - P = \frac{\beta T}{\kappa} - P. \tag{9-32}$$

Equation (9-32) is used for calculation of isothermal changes in U from PVT data or from a PVT equation of state.

For a differential change of state, dH is given by Eq. (9-2):

$$dH = T\,dS + V\,dP,$$

and, by Eq. (9-21),

$$dS = C_P \frac{dT}{T} - \left(\frac{\partial V}{\partial T}\right)_P dP.$$

Combining the two equations, we find

$$\boxed{dH = C_P\,dT + \left[V - T\left(\frac{\partial V}{\partial T}\right)_P\right] dP,} \tag{9-33}$$

or, equivalently,

$$dH = C_P\,dT + V(1 - \beta T)\,dP, \tag{9-34}$$

from which

$$\left(\frac{\partial H}{\partial P}\right)_T = V - T\left(\frac{\partial V}{\partial T}\right)_P = V(1 - \beta T). \tag{9-35}$$

Equations analogous to Eqs. (9-33) and (9-34), in which the independent variables are T and V, are easily derived (Prob. 9-36).

For an ideal gas,

$$\left(\frac{\partial P}{\partial T}\right)_V = \frac{P}{T} \quad \text{and} \quad \left(\frac{\partial V}{\partial T}\right)_P = \frac{V}{T},$$

and Eqs. (9-32) and (9-35) reduce to expressions presented earlier:

$$\left(\frac{\partial U}{\partial V}\right)_T = 0 \qquad \text{(ideal gas)}, \tag{5-13}$$

$$\left(\frac{\partial H}{\partial P}\right)_T = 0 \qquad \text{(ideal gas)}. \tag{5-16}$$

Example 9-8

Determine an expression for the isothermal change in the enthalpy of a gas described by the two-term virial equation, Eq. (5-10).

For an isothermal process, we have

$$\Delta H = \int_{P_1}^{P_2} \left(\frac{\partial H}{\partial P}\right)_T dP \qquad \text{(const } T\text{)},$$

which becomes, in view of Eq. (9-35),

$$\Delta H = \int_{P_1}^{P_2} \left[V - T\left(\frac{\partial V}{\partial T}\right)_P\right] dP \qquad \text{(const } T\text{)}.$$

For a gas described by Eq. (5-10), we have, from Example 9-5,

$$V = \frac{RT}{P} + B$$

and $$\left(\frac{\partial V}{\partial T}\right)_P = \frac{R}{P} + \frac{dB}{dT}.$$

Hence $$\Delta H = \int_{P_1}^{P_2} \left(\frac{RT}{P} + B - \frac{RT}{P} - T\frac{dB}{dT}\right) dP$$

or $$\Delta H = \left(B - T\frac{dB}{dT}\right)(P_2 - P_1) \qquad \text{(const } T\text{)}. \tag{9-36}$$

In Example 9-11 we shall treat the application of this equation and of the corresponding expression for ΔS, Eq. (9-28).

9-5 Heat-Capacity Equations

Neither C_P nor C_V can be determined from just PVT measurements or from a PVT equation of state. However, we can develop useful relationships for the *difference* between these quantities. Equating the two entropy equations, Eqs. (9-19) and (9-21), we have

$$C_V \frac{dT}{T} + \left(\frac{\partial P}{\partial T}\right)_V dV = C_P \frac{dT}{T} - \left(\frac{\partial V}{\partial T}\right)_P dP,$$

which yields on rearrangement

$$(C_P - C_V)\, dT = T\left(\frac{\partial P}{\partial T}\right)_V dV + T\left(\frac{\partial V}{\partial T}\right)_P dP.$$

Dividing by dT and restricting the result to either constant P or constant V, we find

$$C_P - C_V = T\left(\frac{\partial P}{\partial T}\right)_V \left(\frac{\partial V}{\partial T}\right)_P. \tag{9-37}$$

Alternative equivalent forms of Eq. (9-37) are found by use of the identity

$$\left(\frac{\partial P}{\partial T}\right)_V = -\left(\frac{\partial V}{\partial T}\right)_P \left(\frac{\partial P}{\partial V}\right)_T. \tag{2-11}$$

Thus we have

$$C_P - C_V = -T\left(\frac{\partial P}{\partial T}\right)_V^2 \left(\frac{\partial V}{\partial P}\right)_T \tag{9-38}$$

and

$$\boxed{C_P - C_V = -T\left(\frac{\partial V}{\partial T}\right)_P^2 \left(\frac{\partial P}{\partial V}\right)_T.} \tag{9-39}$$

Equation (9-39) is an important thermodynamic relationship. It shows that:

1 The difference $C_P - C_V$ can never be negative, that is, C_P can never be less than C_V, because $(\partial P/\partial V)_T$ is a negative quantity for all known substances.
2 As the temperature approaches absolute zero, the two heat capacities become equal.

3 When $(\partial V/\partial T)_P = 0$, the two heat capacities are equal. For example, $C_P = C_V$ for liquid water at 4(°C), because at this temperature the density of water exhibits a maximum.

Laboratory measurements of the heat capacities of solids and liquids are usually made for constant-pressure conditions and therefore yield values of C_P. Direct measurements of C_V for a solid or a liquid are very difficult. When such values of C_V are required, they are readily calculated from the more easily measured quantities appearing in Eq. (9-39). An alternative form of this equation follows from the definitions of β and κ:

$$C_P - C_V = \frac{TV\beta^2}{\kappa}. \qquad (9\text{-}40)$$

Example 9-9

The temperature of 1(kg) of liquid mercury is increased from 0 to 5(°C) under conditions of constant volume. How much heat is required? The following data are available for mercury at 0(°C) and 1(bar):

$$\beta = 181 \times 10^{-6}(\text{K})^{-1},$$

$$C_P = 28.0(\text{J})/(\text{mol})(\text{K}),$$

$$\kappa = 3.88 \times 10^{-6}(\text{bar})^{-1},$$

$$V = 14.7(\text{cm})^3/(\text{mol}).$$

We assume constant heat capacities and by Eq. (4-25*b*),

$$Q = n \int C_V \, dT = nC_V \, \Delta T \qquad (\text{const } V). \qquad (A)$$

A value of C_V is computed from the given data by Eq. (9-40). Thus

$$C_V = C_P - \frac{TV\beta^2}{\kappa}$$

$$= 28.0(\text{J})/(\text{mol})(\text{K}) - \frac{273.2(\text{K}) \times 14.7 \times 10^{-6}(\text{m})^3/(\text{mol}) \times (1.81 \times 10^{-4})^2(\text{K})^{-2}}{3.88 \times 10^{-6}(\text{bar})^{-1} \times 10^{-5}(\text{bar})(\text{m})^3/(\text{J})},$$

or

$$C_V = 24.6(\text{J})/(\text{mol})(\text{K}).$$

The molecular weight of mercury is 200.6(g)/(mol) ≡ 0.2006(kg)/(mol). Substitution of numerical values into Eq. (*A*) gives

$$Q = 1(\text{kg}) \frac{1}{0.2006}(\text{mol})/(\text{kg}) \times 24.6(\text{J})/(\text{mol})(\text{K}) \times 5(\text{K}),$$

or $\quad Q = 613(\text{J})$.

Thermodynamics also provides a useful equation for the *ratio* of the heat capacities, $\gamma \equiv C_P/C_V$. We start with Eq. (9-23):

$$dS = \frac{C_V}{T}\left(\frac{\partial T}{\partial P}\right)_V dP + \frac{C_P}{T}\left(\frac{\partial T}{\partial V}\right)_P dV.$$

Division by dV and restriction to constant S yields

$$0 = \frac{C_V}{T}\left(\frac{\partial T}{\partial P}\right)_V \left(\frac{\partial P}{\partial V}\right)_S + \frac{C_P}{T}\left(\frac{\partial T}{\partial V}\right)_P.$$

Solving for γ, we find

$$\gamma = \left[-\frac{(\partial V/\partial T)_P}{(\partial P/\partial T)_V}\right]\left(\frac{\partial P}{\partial V}\right)_S.$$

But the quantity in brackets is just $(\partial V/\partial P)_T$. Thus,

$$\boxed{\gamma = \frac{(\partial P/\partial V)_S}{(\partial P/\partial V)_T}.} \tag{9-41}$$

The derivative $(\partial P/\partial V)_S$ is related to an important physical property, the *adiabatic compressibility* κ_S, defined by

$$\boxed{\kappa_S \equiv -\frac{1}{V}\left(\frac{\partial V}{\partial P}\right)_S.} \tag{9-42}$$

If Eq. (9-41) is rewritten as

$$\gamma = \frac{(\partial P/\partial V)_S}{(\partial P/\partial V)_T} = \frac{-(1/V)(\partial V/\partial P)_T}{-(1/V)(\partial V/\partial P)_S},$$

then by the definitions of Eqs. (2-3) and (9-42) we get

$$\gamma = \frac{\kappa}{\kappa_S}. \tag{9-43}$$

Values for the adiabatic compressibility can be determined from soundspeed and density measurements as discussed in Sec. 9-8.

9-6 Volumetric Properties of Gases. Corresponding States

Use of equations presented in earlier sections of this chapter requires the availability of volumetric and heat-capacity data for real materials. For condensed phases (liquids and solids), these data are usually expressed in the form of values of C_P, V, β, and κ, some or all of which may often be treated as constants over relatively wide ranges of pressure and, to a lesser degree, temperature. The actual behavior of these quantities is treated in Sec. 9-8.

For gases, the situation is more complicated. The molar or specific volume V is a strong function of both T and P. Moreover, both C_P and C_V depend upon pressure, as well as upon temperature, and the direct use of ideal-gas heat capacities in calculations for high-pressure gases can lead to serious errors. Fortunately, all necessary gas-phase property calculations can be done if one has, in addition to ideal-gas heat capacities as a function of T, a complete table of PVT data or, preferably, an accurate PVT equation of state.

Representation of volumetric behavior is greatly facilitated if one has a method for simultaneous correlation of data on many different fluids. The search for such generalized correlations has occupied the attention of scientists and engineers for almost a century and remains an area of lively interest. Most generalized correlations are based on the observation that data for different fluids exhibit a remarkable uniformity when the thermodynamic coordinates are expressed in a suitable dimensionless or *reduced* form. This fact is the experimental basis for the *theorem of corresponding states.*

In one of its forms, the theorem of corresponding states asserts that the compressibility factor Z for all substances is a universal function of the *reduced temperature* T_r and the *reduced pressure* P_r. Hence,

$$Z = Z(T_r, P_r) \qquad \text{(all substances)}, \tag{9-44}$$

where T_r and P_r are, by definition,

$$T_r \equiv \frac{T}{T_c}$$

and

$$P_r \equiv \frac{P}{P_c},$$

and T_c and P_c are the critical temperature and critical pressure of a substance. In other words, the *compressibility factors of all substances are the same when compared at the same values of T_r and P_r.* The approximate validity of this assertion for gases is illustrated in Fig. 9-1.

A more severe test of the theorem of corresponding states is provided by data on the virial coefficients of gases. For sufficiently low pressures, we

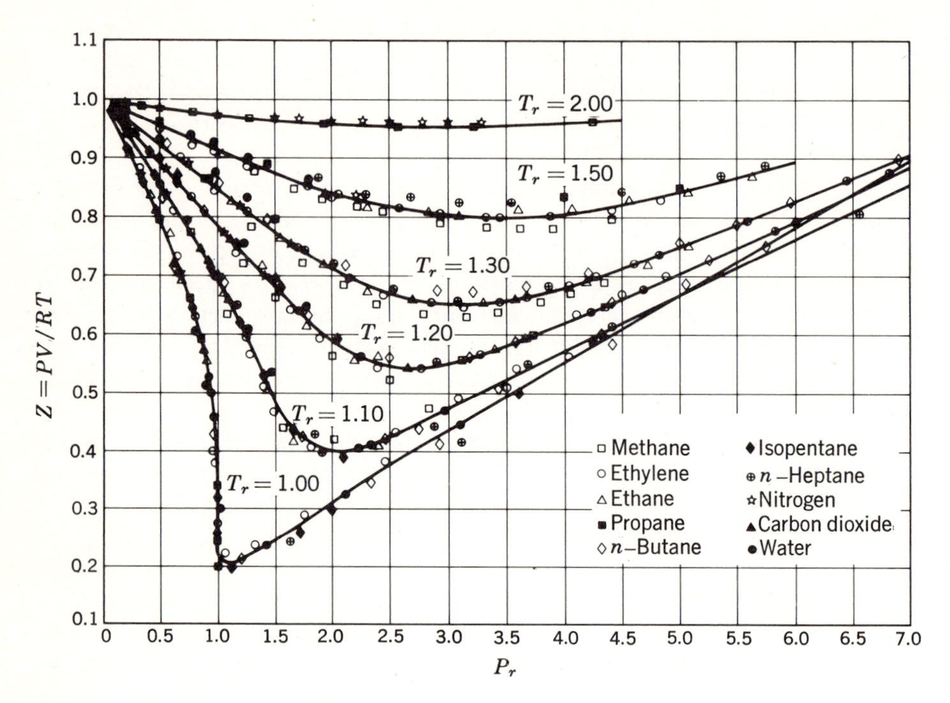

Fig. 9-1

Corresponding-states correlation of the compressibility factor Z [from G. J. Su, Ind. Eng. Chem., **38:** *803 (1946)].*

can assume the validity of the truncated virial equation, Eq. (5-10):

$$Z = 1 + \frac{BP}{RT}. \tag{5-10}$$

A reduced form of this equation is found through the identities $T \equiv T_c T_r$ and $P \equiv P_c P_r$. Thus,

$$Z = 1 + \hat{B}\frac{P_r}{T_r}, \tag{9-45}$$

where $\hat{B}$ is a dimensionless second virial coefficient, defined by

$$\hat{B} \equiv \frac{BP_c}{RT_c}. \tag{9-46}$$

For a pure substance, the virial coefficients are functions of temperature only. If, in the pressure range where Eq. (5-10) is valid, the reduced equation (9-45) is to be compatible with Eq. (9-44), we must have

$$\hat{B} = \hat{B}(T_r) \qquad \text{(all substances)}, \tag{9-47}$$

where $\hat{B}(T_r)$ is a universal function of T_r, valid for all substances.

Figure 9-2, a plot of $\hat{B}$ vs. T_r for several gases, constitutes a test of Eq. (9-47). If the theorem of corresponding states were valid, all the data for $\hat{B}$ would fall on a single curve. Clearly, they do not, although the curves for the various gases tend to converge at higher reduced temperatures. However, the data for the three gases argon, krypton, and xenon *do* define a single curve (the solid line), and reduced correlations of other properties for these substances confirm that they do indeed conform to the theorem of corresponding states. For this and other reasons, Ar, Kr, and Xe are called *simple fluids.*

Corresponding-states correlations based upon Eq. (9-44) are called *two-parameter* correlations, because they require the use of the two "scaling" or reducing parameters T_c and P_c. In view of the results of Fig. 9-2, we

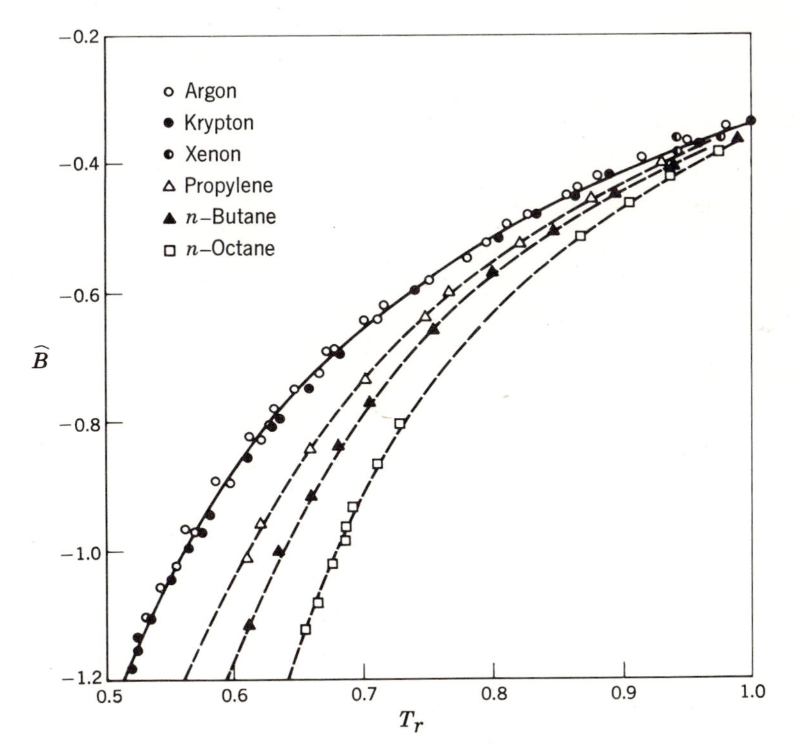

Fig. 9-2
Dimensionless second virial coefficient $\hat{B}(\equiv BP_c/RT_c)$ *for several gases.*

expect two-parameter correlations to be of limited value, particularly at low reduced temperatures. However, simple *three-parameter* extensions of the corresponding-states principle have proved capable of unifying to a considerable degree the PVT behavior of many substances. One of the most useful correlations of this type is due to K. S. Pitzer and his coworkers.[1] It is based on the assumption that

$$Z = Z(T_r, P_r, \omega),$$

where the third parameter ω is a dimensionless number called the *acentric factor*. This factor is a constant for a given substance and is determined from data on vapor-liquid saturation pressures. By definition,

$$\omega \equiv -1 - \log[P_r(\text{sat})]_{T_r=0.7}. \tag{9-48}$$

The argument of the logarithm is the reduced vapor pressure $P(\text{sat})/P_c$, evaluated at a reduced temperature of 0.7. For the three simple fluids, ω is approximately zero; for most other substances it is a positive number less than unity. Values of ω for selected substances are given in Appendix C.

Pitzer's correlation covers a wide range of conditions for gases and liquids and is based on experimental data for many fluids. We consider here only that part of the correlation appropriate to calculations for low-pressure gases via the truncated virial equation, Eq. (5-10). The correlation for second virial coefficients is based on the equation

$$\hat{B} = B^0 + \omega B^1, \tag{9-49}$$

where $\hat{B}$ is the dimensionless second virial coefficient defined by Eq. (9-46), and the terms B^0 and B^1 are functions of T_r only. The B^0 term represents $\hat{B}$ for the simple fluids, Ar, Kr, and Xe, for which $\omega = 0$; it is determined so as to provide a good fit to the solid line in Fig. 9-2. The B^1 term represents the "nonsimple fluid" contribution to $\hat{B}$ for substances for which $\omega \neq 0$. The functions B^0 and B^1 are well represented as functions of T_r by the equations

$$B^0 = 0.083 - \frac{0.422}{T_r^{1.6}} \tag{9-50}$$

and

$$B^1 = 0.139 - \frac{0.172}{T_r^{4.2}}. \tag{9-51}$$

When used in conjunction with Eq. (5-10), Eqs. (9-49) through (9-51) provide good estimates of vapor-phase volumetric properties for reduced

[1] This correlation is presented in Appendix 1 of G. N. Lewis, M. Randall, K. S. Pitzer, and L. Brewer, "Thermodynamics," McGraw-Hill, New York, 1961.

volumes $V_r(\equiv V/V_c)$ greater than about 2.0. For higher densities ($V_r < 2$), Eqs. (9-49) through (9-51) are still valid, but the truncated virial equation, Eq. (5-10), is no longer suitable, and terms containing the third or higher virial coefficients must be considered. Although *generalized correlations* of this sort are intended to be generally applicable, data for strongly polar or associating substances, such as water, ammonia, and the alcohols, are not well represented.

Example 9-10

Determine an estimate for the second virial coefficient of n-heptane vapor at 100(°C).

We have, by Eqs. (9-46) and (9-49),

$$B = \frac{RT_c}{P_c}(B^0 + \omega B^1), \tag{9-52}$$

where B^0 and B^1 are given by Eqs. (9-50) and (9-51). From Appendix C we find, for n-heptane, $T_c = 540.3$(K), $P_c = 27.4$(bar), and $\omega = 0.349$. At 100(°C) or 373.2(K), the reduced temperature is

$$T_r = \frac{373.2}{540.3} = 0.691.$$

From Eqs. (9-50) and (9-51),

$$B^0 = 0.083 - \frac{0.422}{(0.691)^{1.6}} = -0.679,$$

and

$$B^1 = 0.139 - \frac{0.172}{(0.691)^{4.2}} = -0.673.$$

Substitution into Eq. (9-52) gives

$$B = \frac{83.14(\text{cm})^3(\text{bar})/(\text{mol})(\text{K}) \times 540.3(\text{K})}{27.4(\text{bar})} \times [-0.679 + (0.349)(-0.673)],$$

or $B = -1{,}500(\text{cm})^3/(\text{mol})$.

An experimental value is $-1{,}495(\text{cm})^3/(\text{mol})$.

Example 9-11

Propane gas at 100(°C) is compressed isothermally from an initial pressure of 1(bar) to a final pressure of 10(bar). Determine ΔS and ΔH for the gas.

No data are given for the propane gas. We therefore assume the validity of the truncated virial equation, Eq. (5-10), and use the results of Examples 9-5 and 9-8 in conjunction with the generalized correlation for the second virial coefficient. Thus we have

$$\Delta S = -R \ln \frac{P_2}{P_1} - \frac{dB}{dT}(P_2 - P_1) \qquad (\text{const } T), \tag{9-28}$$

and

$$\Delta H = \left(B - T\frac{dB}{dT}\right)(P_2 - P_1) \qquad \text{(const } T\text{)}, \tag{9-36}$$

with B given by

$$B = \frac{RT_c}{P_c}(B^0 + \omega B^1). \tag{9-52}$$

Differentiation of Eq. (9-52) yields the required expression for dB/dT:

$$\frac{dB}{dT} = \frac{RT_c}{P_c}\left(\frac{dB^0}{dT} + \omega\frac{dB^1}{dT}\right),$$

or, since $T = T_cT_r$,

$$\frac{dB}{dT} = \frac{R}{P_c}\left(\frac{dB^0}{dT_r} + \omega\frac{dB^1}{dT_r}\right). \tag{9-53}$$

Expressions for dB^0/dT_r and dB^1/dT_r are obtained by differentiation of Eqs. (9-50) and (9-51). Thus

$$\frac{dB^0}{dT_r} = \frac{0.675}{T_r^{2.6}} \tag{9-54}$$

and

$$\frac{dB^1}{dT_r} = \frac{0.722}{T_r^{5.2}} \tag{9-55}$$

From Appendix C we find for propane that $T_c = 369.9(\text{K})$, $P_c = 42.6(\text{bar})$, and $\omega = 0.152$. At 100(°C), the reduced temperature is

$$T_r = \frac{373.2}{369.9} = 1.01,$$

and by Eqs. (9-50), (9-51), (9-54), and (9-55)

$$B^0 = 0.083 - \frac{0.422}{(1.01)^{1.6}} = -0.332,$$

$$B^1 = 0.139 - \frac{0.172}{(1.01)^{4.2}} = -0.026,$$

$$\frac{dB^0}{dT_r} = \frac{0.675}{(1.01)^{2.6}} = 0.658,$$

and

$$\frac{dB^1}{dT_r} = \frac{0.722}{(1.01)^{5.2}} = 0.686.$$

The values of B and dB/dT are then, by Eqs. (9-52) and (9-53),

$$B = \frac{83.14(\text{cm})^3(\text{bar})/(\text{mol})(\text{K}) \times 369.9(\text{K})}{42.6(\text{bar})}[-0.332 + (0.152)(-0.026)],$$

or $B = -243(\text{cm})^3/(\text{mol})$,

and

$$\frac{dB}{dT} = \frac{83.14(\text{cm})^3(\text{bar})/(\text{mol})(\text{K})}{42.6(\text{bar})}[0.658 + (0.152)(0.686)],$$

or $\frac{dB}{dT} = 1.49(\text{cm})^3/(\text{mol})(\text{K}).$

Substitution of dB/dT into Eq. (9-28) gives finally

$$\begin{aligned}\Delta S &= -8.314(\text{J})/(\text{mol})(\text{K}) \ln \tfrac{10}{1} \\ &\quad -1.49(\text{cm})^3/(\text{mol})(\text{K})[10 - 1](\text{bar}) \times \tfrac{1}{10}(\text{J})/(\text{cm})^3(\text{bar}) \\ &= -19.1 - 1.3,\end{aligned}$$

or $\Delta S = -20.4(\text{J})/(\text{mol})(\text{K}).$

Similarly, substitution of B and dB/dT into Eq. (9-36) gives

$$\Delta H = [-243(\text{cm})^3/(\text{mol}) - 373.2(\text{K}) \times 1.49(\text{cm})^3/(\text{mol})(\text{K})][10 - 1](\text{bar}) \times \tfrac{1}{10}(\text{J})/(\text{cm})^3(\text{bar}),$$

or $\Delta H = -719(\text{J})/(\text{mol}).$

If we had assumed ideal-gas behavior, both B and dB/dT would be zero, and we would have obtained

$$\Delta S = -19.1(\text{J})/(\text{mol})(\text{K})$$

and

$$\Delta H = 0(\text{J})/(\text{mol}).$$

9-7 Residual Properties

In calculating property changes of real gases, we have considered up to now only isothermal changes of state. However, real engineering processes often involve changes of state for which the initial and final temperatures are different. The treatment of such processes requires either the use of *real*-gas heat capacities or of some equivalent procedure which employs ideal-gas heat capacities in conjunction with PVT data or a PVT equation of state. We describe in this section one such procedure, convenient for use with the truncated virial equation, Eq. (5-10).

We wish to compute the change $\Delta M = M_2 - M_1$ of a thermodynamic property M between the two equilibrium states (T_1,P_1) and (T_2,P_2), represented by points 1 and 2 in Fig. 9-3. The actual process which accompanies the change in state is immaterial; we represent it schematically by the solid line on the figure.

Since M is a thermodynamic property, we can use any convenient combination of real or fictitious paths for the purpose of computing ΔM.

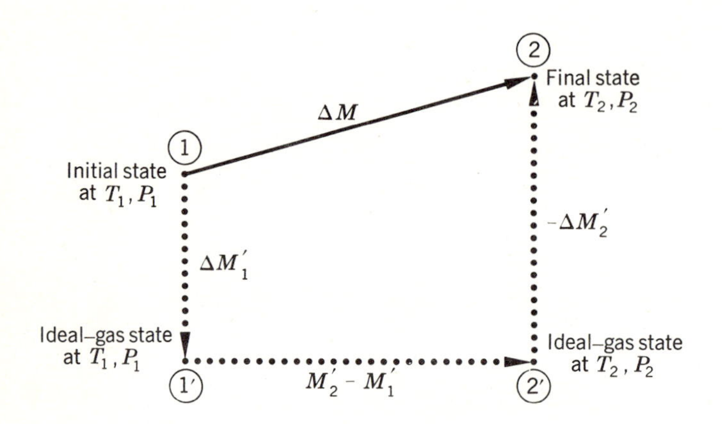

Fig. 9-3
Calculation of the change in a thermodynamic property M. Solid line represents the actual path; dotted lines are the calculational paths.

We choose the three-step path represented by the dotted lines in Fig. 9-3:

$1 \rightarrow 1'$ The *real* gas at T_1 and P_1 is transformed into an *ideal* gas at the same conditions of T_1 and P_1.

$1' \rightarrow 2'$ The *ideal* gas is changed from the state (T_1,P_1) to the state (T_2,P_2).

$2' \rightarrow 2$ The *ideal* gas at T_2 and P_2 is transformed back into a *real* gas at the same conditions of T_2 and P_2.

The total change ΔM is written as the sum of the property changes for the three separate steps:

$$\Delta M = (M_1' - M_1) + (M_2' - M_1') + (M_2 - M_2'),$$

or

$$\Delta M = (M_1' - M_1) + (M_2' - M_1') - (M_2' - M_2),$$

where the primed quantities are ideal-gas properties. The first and third terms in parentheses are of identical form; they represent differences between the value of a thermodynamic property of a substance in the ideal-gas state and the actual value, *both at the same temperature and pressure.* We call these terms *residual properties* and give them the symbol $\Delta M'$. Thus,

$$\Delta M' \equiv M' - M. \tag{9-56}$$

In terms of residual properties, our expression for ΔM can now be written

$$\Delta M = \Delta M_1' + (M_2' - M_1') - \Delta M_2'. \tag{9-57}$$

The utility of Eq. (9-57) is that the residual properties represent isothermal property changes, which can be evaluated from a PVT equation of state; the computation of nonisothermal property changes is done only for the ideal-gas state and involves merely simple formulas already derived in previous chapters. All that remains is to develop the necessary expressions for the residual properties.

We start with the definition of a residual property, Eq. (9-56). Differentiation with respect to P at constant T gives

$$\left(\frac{\partial \Delta M'}{\partial P}\right)_T = \left(\frac{\partial M'}{\partial P}\right)_T - \left(\frac{\partial M}{\partial P}\right)_T.$$

For a constant-temperature change of state, we have

$$d(\Delta M') = \left[\left(\frac{\partial M'}{\partial P}\right)_T - \left(\frac{\partial M}{\partial P}\right)_T\right] dP \qquad (\text{const } T),$$

and integration from $P = 0$ to some finite pressure P yields, on rearrangement,

$$\Delta M' = \lim_{P \to 0} (\Delta M') + \int_0^P \left[\left(\frac{\partial M'}{\partial P}\right)_T - \left(\frac{\partial M}{\partial P}\right)_T\right] dP \qquad (\text{const } T).$$

In order to use this equation, we must first assign a definite value to the zero-pressure limit of $\Delta M'$. In practice, it is *assumed* that the values of certain thermodynamic functions approach their ideal-gas values as P approaches zero, and hence that

$$\lim_{P \to 0} (\Delta M') = 0.$$

This assumption is valid, for example, for $M = H$ and $M = S$, but *not* for $M = V$. If the assumption does hold, then we can write our expression for $\Delta M'$ as

$$\Delta M' = \int_0^P \left[\left(\frac{\partial M'}{\partial P}\right)_T - \left(\frac{\partial M}{\partial P}\right)_T\right] dP \qquad (\text{const } T). \tag{9-58}$$

For most engineering applications, it is sufficient to have expressions for the residual enthalpy $\Delta H'$ and the residual entropy $\Delta S'$. For the enthalpy, we have, by Eqs. (5-16) and (9-35),

$$\left(\frac{\partial H'}{\partial P}\right)_T = 0$$

and

$$\left(\frac{\partial H}{\partial P}\right)_T = V - T\left(\frac{\partial V}{\partial T}\right)_P.$$

Thus Eq. (9-58) gives, with $M = H$,

$$\boxed{\Delta H' = \int_0^P \left[T\left(\frac{\partial V}{\partial T}\right)_P - V\right] dP} \qquad \text{(const } T). \tag{9-59}$$

Similarly, we have for the entropy, by Eqs. (8-4) and (9-18), that

$$\left(\frac{\partial S'}{\partial P}\right)_T = -\frac{R}{P}$$

and

$$\left(\frac{\partial S}{\partial P}\right)_T = -\left(\frac{\partial V}{\partial T}\right)_P.$$

Equation (9-58) then becomes with $M = S$,

$$\boxed{\Delta S' = \int_0^P \left[\left(\frac{\partial V}{\partial T}\right)_P - \frac{R}{P}\right] dP} \qquad \text{(const } T). \tag{9-60}$$

Expressions for the ideal-gas property changes $H_2' - H_1'$ and $S_2' - S_1'$ were given earlier as Eqs. (5-27) and (8-5). They are, in the present notation,

$$H_2' - H_1' = \int_{T_1}^{T_2} C_P' \, dT \tag{5-27}$$

and

$$S_2' - S_1' = \int_{T_1}^{T_2} \frac{C_P'}{T} dT - R \ln \frac{P_2}{P_1}, \tag{8-5}$$

where we now use a prime on C_P to emphasize that it is an ideal-gas heat capacity. Equations (9-57), (9-59), (9-60), (5-27), and (8-5) enable us to compute values of ΔH and ΔS for *any* change in state of a real gas. What is required as input is an equation of state for the gas and an expression for the ideal-gas heat capacity C_P'. We illustrate the method in the following example.

Example 9-12

Hydrogen sulfide gas (H_2S) is compressed from an initial state of 400(K) and 5(bar) to a final state of 600(K) and 25(bar). Determine ΔH and ΔS for the gas.

We are given no data for H_2S, so we will use the truncated virial equation with the generalized correlation for B. The complete expressions for ΔH and ΔS are obtained as special cases of Eq. (9-57):

$$\Delta H = \Delta H_1' + (H_2' - H_1') - \Delta H_2' \tag{9-61}$$

and

$$\Delta S = \Delta S_1' + (S_2' - S_1') - \Delta S_2'. \tag{9-62}$$

For the truncated virial equation, we have from Example 9-5 that

$$V = \frac{RT}{P} + B \qquad \text{and} \qquad \left(\frac{\partial V}{\partial T}\right)_P = \frac{R}{P} + \frac{dB}{dT}.$$

By Eq. (9-59) for $\Delta H'$, then,

$$\Delta H' = \int_0^P \left[T\left(\frac{\partial V}{\partial T}\right)_P - V\right] dP$$

$$= \int_0^P \left[\frac{RT}{P} + T\frac{dB}{dT} - \frac{RT}{P} - B\right] dP,$$

or

$$\Delta H' = P\left(T\frac{dB}{dT} - B\right). \tag{9-63}$$

Similarly, we have by Eq. (9-60) for $\Delta S'$ that

$$\Delta S' = \int_0^P \left[\left(\frac{\partial V}{\partial T}\right)_P - \frac{R}{P}\right] dP$$

$$= \int_0^P \left[\frac{R}{P} + \frac{dB}{dT} - \frac{R}{P}\right] dP,$$

or

$$\Delta S' = P\frac{dB}{dT}. \tag{9-64}$$

The method of calculation of B and dB/dT from the generalized correlation for B was illustrated in Examples 9-10 and 9-11, so we will not repeat the numerical details for the present example. Using the correlation for B, together with the acentric factor and critical properties for H_2S given in Appendix C, we find for the initial state of 400(K) and 5(bar) [$T_{r1} = 1.07$ and $P_{r1} = 0.054$] that

$$\Delta H_1' = 164(\text{J})/(\text{mol})$$

$$\Delta S_1' = 0.283(\text{J})/(\text{mol})(\text{K}).$$

Similarly, for the final state of 600(K) and 25(bar) [$T_{r2} = 1.65$ and $P_{r2} = 0.278$],

$$\Delta H_2' = 369(\text{J})/(\text{mol})$$

$$\Delta S_2' = 0.468(\text{J})/(\text{mol})(\text{K}).$$

It remains to determine $(H_2' - H_1')$ and $(S_2' - S_1')$. The ideal-gas heat capacity C_P' for H_2S is represented by Eq. (5-28) as

$$\frac{C_P'}{R} = a + bT + cT^2, \tag{5-28}$$

with constants a, b, and c as reported in Table 5-3. Substitution of Eq. (5-28) into Eqs. (5-27) and (8-5) gives, on integration,

$$H_2' - H_1' = R\left[a(T_2 - T_1) + \frac{b}{2}(T_2^2 - T_1^2) + \frac{c}{3}(T_2^3 - T_1^3)\right] \tag{9-65}$$

and $$S_2' - S_1' = R\left[a \ln \frac{T_2}{T_1} + b(T_2 - T_1) + \frac{c}{2}(T_2^2 - T_1^2)\right] - R \ln \frac{P_2}{P_1}. \tag{9-66}$$

From Table 5-3, we find that $a = 3.353$, $b = 2.584 \times 10^{-3}(\text{K})^{-1}$, and $c = -0.430 \times 10^{-6}(\text{K})^{-2}$. With the given values for T_1, T_2, P_1, and P_2, we then have from Eqs. (9-65) and (9-66) that

$$H_2' - H_1' = 7{,}530(\text{J})/(\text{mol})$$

$$S_2' - S_1' = 1.861(\text{J})/(\text{mol})(\text{K}).$$

The required values for ΔH and ΔS are found finally from Eqs. (9-61) and (9-62):

$$\Delta H = 164 + 7{,}530 - 369 = 7{,}325(\text{J})/(\text{mol})$$

$$\Delta S = 0.283 + 1.861 - 0.468 = 1.676(\text{J})/(\text{mol})(\text{K}).$$

If we had assumed ideal gas behavior (i.e., $\Delta H' = \Delta S' = 0$), we would have obtained

$$\Delta H = 7{,}530(\text{J})/(\text{mol})$$

and $$\Delta S = 1.861(\text{J})/(\text{mol})(\text{K}).$$

9-8 Properties of Solids and Liquids

For many applications, it is satisfactory to treat the thermodynamic properties of solids and liquids as independent of pressure. Occasionally even their temperature dependence can be neglected. Strictly, however, the properties of all PVT systems in states of single-phase equilibrium depend upon both T and P. In this section we discuss briefly the qualitative behavior of the properties, β, κ, κ_S, C_P, and C_V for the liquid and solid states.

The volume expansivity β of liquids is usually calculated from an empirical equation which relates the density to temperature at constant pressure. Since the molar (or specific) volume V is the reciprocal of the molar (or specific) density ρ, it follows that

$$\beta = -\frac{1}{\rho}\left(\frac{\partial \rho}{\partial T}\right)_P.$$

In modern experiments on the expansion of solids, it is usually the linear expansivity α, rather than the volume expansivity, that is measured.[1] Suppose that the three rectangular dimensions of a unit mass of solid are L_1, L_2, L_3. Then

$$V = L_1 L_2 L_3,$$

$$\left(\frac{\partial V}{\partial T}\right)_P = L_2 L_3 \left(\frac{\partial L_1}{\partial T}\right)_P + L_1 L_3 \left(\frac{\partial L_2}{\partial T}\right)_P + L_1 L_2 \left(\frac{\partial L_3}{\partial T}\right)_P,$$

[1] A brief summary of experimental techniques is given by M. W. Zemansky, "Heat and Thermodynamics," 5th ed., Article 11-9, McGraw-Hill, New York, 1968.

and $$\frac{1}{V}\left(\frac{\partial V}{\partial T}\right)_P = \frac{1}{L_1}\left(\frac{\partial L_1}{\partial T}\right)_P + \frac{1}{L_2}\left(\frac{\partial L_2}{\partial T}\right)_P + \frac{1}{L_3}\left(\frac{\partial L_3}{\partial T}\right)_P.$$

The linear expansivity α was defined in Sec. 2-6:

$$\alpha \equiv \left(\frac{\partial \epsilon}{\partial T}\right)_\sigma = \frac{1}{L}\left(\frac{\partial L}{\partial T}\right)_\sigma,$$

where σ is the stress. But the condition of constant hydrostatic pressure is usually equivalent to a constant-stress condition for a thermal-expansion experiment on an unconstrained solid. Thus we may write

$$\beta = \alpha_1 + \alpha_2 + \alpha_3. \qquad (9\text{-}67)$$

If the solid is *isotropic*, then

$$\alpha_1 = \alpha_2 = \alpha_3 \equiv \alpha,$$

and we obtain as an important special case of Eq. (9-67)

$$\beta = 3\alpha.$$

The pressure dependence of β is quite small for solids and liquids, and β may usually be treated as constant even for pressure changes of the order of several hundred atmospheres. Table 9-1 contains some values of β for liquid mercury at 0(°C) for pressures up to 7,000(atm). The temperature variation of β is more significant. This is illustrated by the data for copper in Table 9-2 and for liquid water in Table 9-3, and by the values for four different metals shown in Fig. 9-4. It is seen that β decreases as the temperature decreases, approaching zero as the temperature approaches absolute zero. Most pure solids behave in a manner similar to that shown in Fig. 9-4, but there are important exceptions. A type of iron known as α iron, for example, behaves normally from 0 to about 1000(K); then its volume expansivity suddenly drops, becoming negative at about 1100(K). Water also shows interesting behavior. The temperature variation of β for both ice and liquid water is plotted in Fig. 9-5. It is a remarkable fact that the volume expansivity is negative at very low temperatures and also in the well-known region from 0 to 4(°C).

The isothermal compressibility κ of liquids is determined from experimental measurements of the volume changes produced by known isothermal changes in pressure. The computed quantity is generally an average compressibility κ(average), defined as

$$\kappa(\text{average}) = -\frac{1}{V_1}\left(\frac{V_2 - V_1}{P_2 - P_1}\right) \qquad (\text{const } T).$$

Table 9-1 Pressure Variation of Properties of Liquid Mercury at 0(°C)

P (atm)	V (cm)^3/(mol)	$\beta \times 10^6$ (K)$^{-1}$	$\kappa \times 10^6$ (bar)$^{-1}$	$\kappa_S \times 10^6$ (bar)$^{-1}$	C_P (J)/(mol) (K)	C_V (J)/(mol) (K)	c (m)/(s)
1	14.72	181	3.88	3.41	28.0	24.6	1,470
1,000	14.67	174	3.79	3.36	28.0	24.8	1,480
2,000	14.62	168	3.69	3.29	28.0	24.9	1,490
3,000	14.57	164	3.60	3.22	27.9	25.0	1,500
4,000	14.51	160	3.48	3.12	27.9	25.0	1,520
5,000	14.45	158	3.38	3.03	27.9	25.1	1,540
6,000	14.42	155	3.25	2.91	27.9	25.1	1,570
7,000	14.38	152	3.12	2.79	27.9	25.1	1,600

Table 9-2 Temperature Variation of Properties of Copper at 1(atm)

T (K)	V (cm)^3/(mol)	$\beta \times 10^6$ (K)$^{-1}$	$\kappa \times 10^6$ (bar)$^{-1}$	$\kappa_S \times 10^6$ (bar)$^{-1}$	C_P (J)/(mol) (K)	C_V (J)/(mol) (K)	c (m)/(s)
50	7.00	11.4	0.713	0.712	6.25	6.24	3,930
100	7.01	31.5	0.721	0.717	16.1	16.0	3,920
150	7.02	40.7	0.734	0.726	20.5	20.3	3,900
200	7.03	45.3	0.749	0.736	22.8	22.4	3,880
250	7.04	48.3	0.763	0.746	24.0	23.5	3,850
300	7.06	50.4	0.778	0.756	24.5	23.8	3,830
500	7.12	54.9	0.839	0.797	25.8	24.5	3,750
800	7.26	60.0	0.922	0.847	27.7	25.4	3,670
1,200	7.45	70.2	1.031	0.885	30.2	25.9	3,640

Table 9-3 Temperature Variation of Properties of Liquid Water at 1(atm)

t (°C)	ρ (g)/(cm)³	$\beta \times 10^6$ (K)⁻¹	$\kappa \times 10^6$ (bar)⁻¹	$\kappa_S \times 10^6$ (bar)⁻¹	C_P (J)/(g)(K)	C_V (J)/(g)(K)	c (m)/(s)
0	0.99984	−68.14	50.89	50.86	4.2174	4.215	1,402
10	0.99970	87.90	47.81	47.76	4.1919	4.188	1,447
20	0.99821	206.6	45.90	45.60	4.1816	4.154	1,482
30	0.99565	303.1	44.77	44.10	4.1782	4.116	1,509
40	0.99222	385.4	44.24	43.12	4.1783	4.073	1,529
50	0.98804	457.8	44.18	42.54	4.1804	4.025	1,542
60	0.98320	523.4	44.50	42.28	4.1841	3.975	1,551
70	0.97777	584.0	45.17	42.31	4.1893	3.924	1,555
80	0.97179	641.3	46.15	42.59	4.1961	3.872	1,555

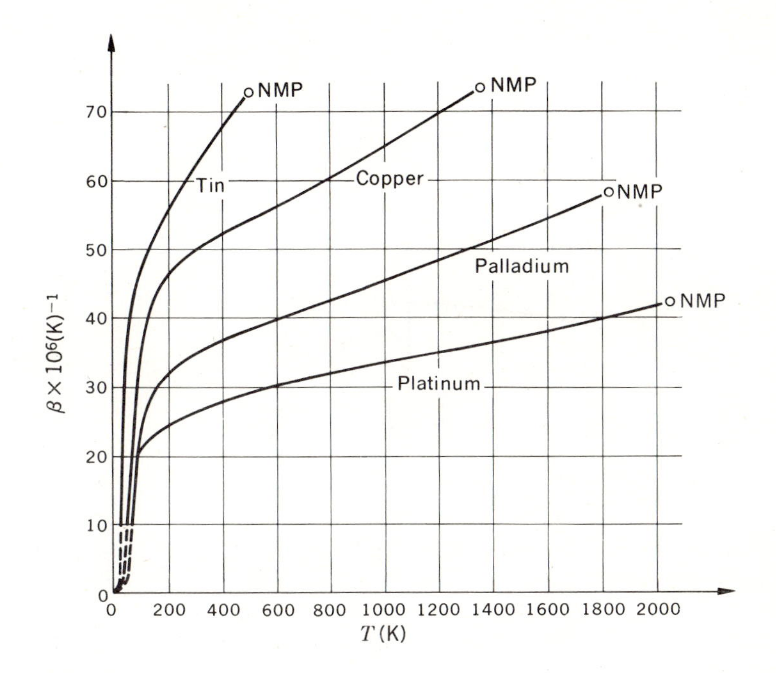

Fig. 9-4
Normal temperature variation of the volume expansivity of metals. (NMP = normal melting point.)

For small changes in pressure, κ(average) is approximately the true isothermal compressibility κ. As with β, values of κ for solids are usually computed from measured values of the corresponding *linear* quantity, in this case the linear compressibility $\delta \equiv -(1/L)(\partial L/\partial P)_T$. Thus we have, analogous to Eq. (9-67), that

$$\kappa = \delta_1 + \delta_2 + \delta_3, \tag{9-68}$$

with the corresponding isotropic special case for which

$$\kappa = 3\delta.$$

The pressure dependence of κ is quite small for solids and liquids; the data for liquid mercury in Table 9-1 indicate that κ changes only by about 20 percent for a 7,000(atm) increase in P. Again, temperature is the more significant variable, as shown by the data for copper in Table 9-2 and for liquid water in Table 9-3 and Fig. 9-6. The observed minimum in κ for water at about 50(°C) represents anomalous behavior, for the isothermal compressibility of liquids usually increases with increasing temperature.

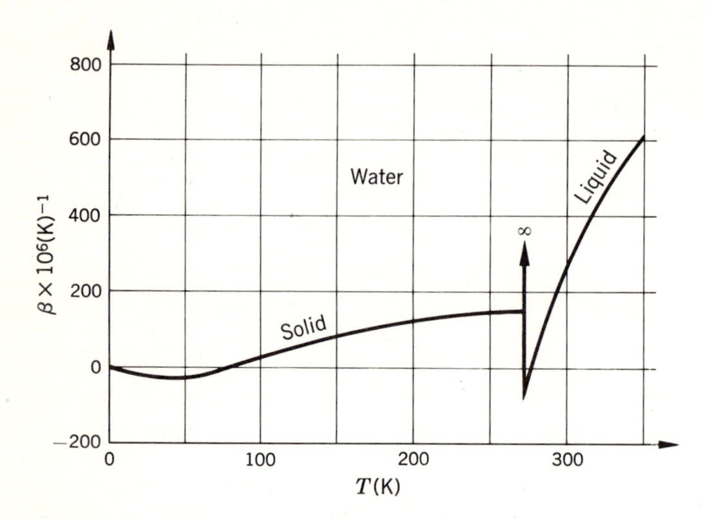

Fig. 9-5
Temperature variation of the volume expansivity of water at 1 (atm) pressure. Break in curve occurs at the normal melting point.

Unlike β, κ approaches a limiting nonzero value as the absolute temperature approaches zero.

The adiabatic compressibility κ_S is obtained most readily from measurements of the soundspeed c. It was shown in Sec. 5-10 that

$$c^2 = -g_c V^2 \left(\frac{\partial P}{\partial V}\right)_S, \qquad (5\text{-}33)$$

where V is the *specific* volume of a substance. Recalling the definition of κ_S, Eq. (9-42), and noting that the mass density ρ is just the reciprocal of V, we can write this equation as

$$\kappa_S = \frac{g_c}{\rho c^2} \qquad (9\text{-}69)$$

Another useful expression may be developed, which relates κ_S to κ. In Sec. 9-5 we found that

$$C_P - C_V = \frac{TV\beta^2}{\kappa}, \qquad (9\text{-}40)$$

and that

$$\frac{C_P}{C_V} = \frac{\kappa}{\kappa_S} \qquad (9\text{-}43)$$

Elimination of C_V between these two expressions gives

$$\kappa - \kappa_S = \frac{TV\beta^2}{C_P} \tag{9-70}$$

This equation bears a remarkable resemblance to Eq. (9-40); it shows that

1 The difference $\kappa - \kappa_S$ can never be negative, that is, κ can never be less than κ_S, because C_P is a positive quantity for all known substances.
2 As the temperature approaches absolute zero, the two compressibilities become equal.
3 When $\beta = 0$, the two compressibilities are equal. For example, $\kappa = \kappa_S$ for liquid water at 4(°C), because at this temperature the density of water exhibits a maximum.

Values of κ_S for mercury, copper, and liquid water are presented in Tables 9-1, 9-2, and 9-3; the κ_S data for water are also plotted in Fig. 9-6.

Experimental measurements of C_P show it to be quite insensitive to pressure for liquids and solids. Thus we see from Table 9-1 that a pressure change of 7,000(atm) results in only a 0.1(J)/(mol)(K) change in C_P for liquid mercury at 0(°C). The temperature dependence of C_P, however, is of considerable importance. In Table 9-2 the values of C_P of copper are given

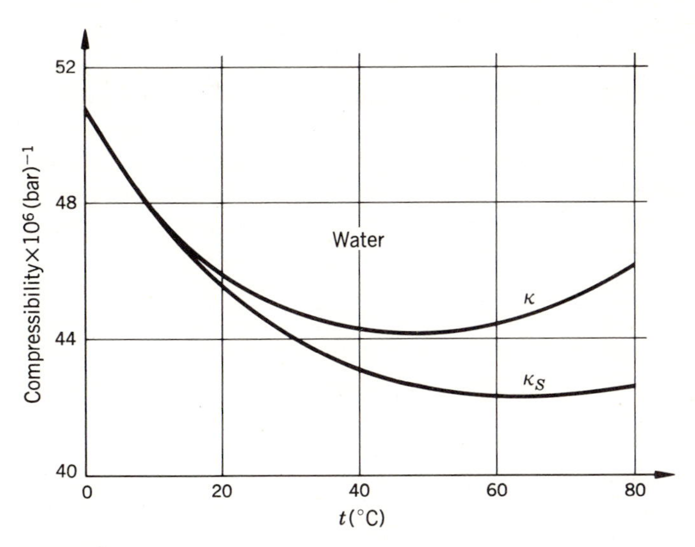

Fig. 9-6

Temperature variation of the isothermal compressibility κ and adiabatic compressibility κ_S of liquid water at 1 (atm) pressure.

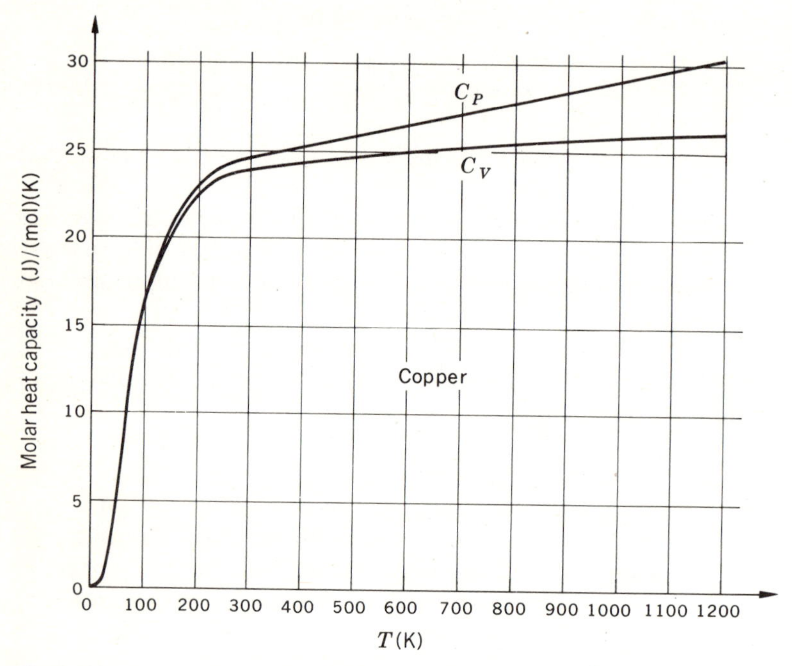

Fig. 9-7
Temperature variation of the molar heat capacities of copper.

over a very wide temperature range. These values are plotted in Fig. 9-7. It is seen that C_P approaches zero as the temperature approaches zero and that as the temperature rises, C_P continuously increases without approaching a constant value. This is the normal behavior of C_P and applies for most pure metals and many compounds. There are, of course, exceptions, among which α iron is interesting. In the temperature range from 1000 to 1100(K), where β for α iron takes a sudden drop, the C_P takes a sudden rise and then drops just as rapidly. Also of interest is the behavior of certain paramagnetic salts in the neighborhood of absolute zero. Below 2(K), C_P rises and then, as the temperature approaches zero, approaches zero again. This behavior is of great importance in connection with the production of very low temperatures by adiabatic demagnetization.

We noted in Sec. 9-5 the difficulty of direct measurement of C_V for solids and liquids. The values of C_V in Tables 9-1 and 9-2 have therefore been *calculated* from the listed values of β, C_P, κ, T, and V by use of Eq. (9-40). Similarly, the C_V values for water in Table 9-3 were computed from listed values of the properties β, c, C_P, T, and ρ.

As with the other properties of solids and liquids, the pressure variation of C_V is slight. The temperature dependence is quite pronounced, however, as illustrated by the C_V values for copper plotted in Fig. 9-7. As shown in this figure, C_V is nearly identical to C_P at low temperatures; moreover, as T approaches absolute zero, both quantities approach a value of zero. The high-temperature behavior of C_V is quite different from that of C_P, however; whereas C_P continues to increase with increasing T, C_V approaches a constant value. This type of limiting high-temperature behavior of C_V is typical for crystalline solids and is of considerable theoretical interest. We cannot, however, pursue the matter here; elementary discussions of the phenomenon are given by Zemansky.[1]

Example 9-13

We wish to illustrate by example the method used in the construction of Tables 9-1, 9-2, and 9-3. The primary data for Tables 9-1 and 9-2 are the listed values of β, C_P, κ, and V. At each pressure or temperature the remaining quantities (κ_S, C_V, and c) are determined by application of Eqs. (9-70), (9-43), and (9-69). Thus, for copper at 300(K), we have $V = 7.06(\text{cm})^3/(\text{mol})$, $\beta = 50.4 \times 10^{-6}(\text{K})^{-1}$, $\kappa = 0.778 \times 10^{-6}(\text{bar})^{-1}$, and $C_P = 24.5(\text{J})/(\text{mol})(\text{K})$; the molecular weight of copper is 63.54. Then

$$\kappa - \kappa_S = \frac{TV\beta^2}{C_P}$$

$$= \frac{300(\text{K}) \times 7.06(\text{cm})^3/(\text{mol}) \times (50.4 \times 10^{-6})^2(\text{K})^{-2} \times 10^{-6}(\text{m})^3/(\text{cm})^3}{24.5(\text{J})/(\text{mol})(\text{K}) \times 10^{-5}(\text{bar})(\text{m})^3/(\text{J})}$$

or

$$\kappa - \kappa_S = 0.022 \times 10^{-6}(\text{bar})^{-1},$$

from which

$$\kappa_S = (0.778 - 0.022) \times 10^{-6} = 0.756 \times 10^{-6}(\text{bar})^{-1}.$$

By Eq. (9-43), we have

$$C_V = \frac{\kappa_S}{\kappa} C_P = \frac{0.756 \times 10^{-6}(\text{bar})^{-1}}{0.778 \times 10^{-6}(\text{bar})^{-1}} 24.5(\text{J})/(\text{mol})(\text{K}),$$

or
$$C_V = 23.8(\text{J})/(\text{mol})(\text{K}).$$

Finally, Eq. (9-69) gives for the soundspeed

$$c = \sqrt{\frac{g_c V}{\kappa_S}} = \sqrt{\frac{1 \times 7.06(\text{cm})^3/(\text{mol}) \times 10^{-6}(\text{m})^3/(\text{cm})^3 \times 10^5(\text{kg})/(\text{m})(\text{s})^2(\text{bar})}{63.54(\text{g})/(\text{mol}) \times 0.756 \times 10^{-6}(\text{bar})^{-1} \times 10^{-3}(\text{kg})/(\text{g})}},$$

or

$$c = 3{,}830(\text{m})/(\text{s}).$$

[1] *Op. cit*, Articles 11-11 through 11-14.

The computations for Table 9-3 were done similarly, except that the primary data were the listed values of β, c, C_P, and ρ. Values of κ_S, κ, and C_V were calculated at each temperature by use of Eqs. (9-69), (9-70), and (9-43).

9-9 Change of Phase. Clapeyron's Equation

In the familiar phase changes of a pure substance—melting, vaporization, and sublimation—as well as in some less familiar transitions, the temperature and the pressure remain constant. Consider the transition of a pure material from a phase α to a second phase β. We designate the molar (or specific) properties U, H, S, and V of the α phase by U^α, H^α, S^α, and V^α, and those of the β phase by U^β, H^β, S^β, and V^β. Each of these properties is a function of T and P, and hence each remains constant during the phase change. However, the molar (or specific) properties of the *two-phase system* will vary according to the relative amounts of the two phases present.

If we let x be the fraction of the total number of moles (or of the total mass) of the system that exists in phase β, then the molar (or specific) properties of the two-phase system are given by

$$U = (1 - x)U^\alpha + xU^\beta, \qquad S = (1 - x)S^\alpha + xS^\beta,$$

$$H = (1 - x)H^\alpha + xH^\beta, \qquad V = (1 - x)V^\alpha + xV^\beta.$$

These equations can be written in the general form

$$M = (1 - x)M^\alpha + xM^\beta, \tag{9-71}$$

where M represents any molar (or specific) property. According to Eq. (9-71), M for the two-phase system is linear in x and can vary from $M = M^\alpha$ (for $x = 0$) to $M = M^\beta$ (for $x = 1$).

Equation (9-71) is often presented in other equivalent forms. Thus, it can be written as

$$M = M^\alpha + x\,\Delta M^{\alpha\beta}, \tag{9-72}$$

where

$$\Delta M^{\alpha\beta} \equiv M^\beta - M^\alpha. \tag{9-73}$$

For a vaporization process, with $\alpha = l$ and $\beta = v$, $\Delta M^{\alpha\beta}$ is denoted by ΔM^{lv}, and x is called the *quality* of the system. Engineers usually employ a subscript notation, by which Eq. (9-72) is written

$$M = M_\alpha + x\,\Delta M_{\alpha\beta} \tag{9-74}$$

where

$$\Delta M_{\alpha\beta} \equiv M_\beta - M_\alpha. \tag{9-75}$$

The subscript notation is used with the *steam tables*, a compilation of thermodynamic data for water, which we will discuss in Sec. 9-11.

In Example 9-2 we showed that for a pure substance all phases in equilibrium have the same value of the molar or specific Gibbs function. Thus, for two-phase equilibrium,

$$G^\alpha = G^\beta \qquad \text{(phase equilibrium).} \tag{9-76}$$

However, the other thermodynamic properties are not the same for equilibrium phases in the familiar transitions; for example $S^\alpha \neq S^\beta$ and $V^\alpha \neq V^\beta$. We have, by Eqs. (9-9) and (9-11),

$$S = -\left(\frac{\partial G}{\partial T}\right)_P \qquad \text{and} \qquad V = \left(\frac{\partial G}{\partial P}\right)_T.$$

Since S and V have different values in the equilibrium phases at a given T and P, it follows that the derivatives of G are different on opposite sides of the point of transition. This is illustrated in Fig. 9-8*a*, a schematic representation of G along a constant-pressure path that includes a liquid-vapor phase change. Also shown in Fig. 9-8 are similar plots for S, C_P, and V.

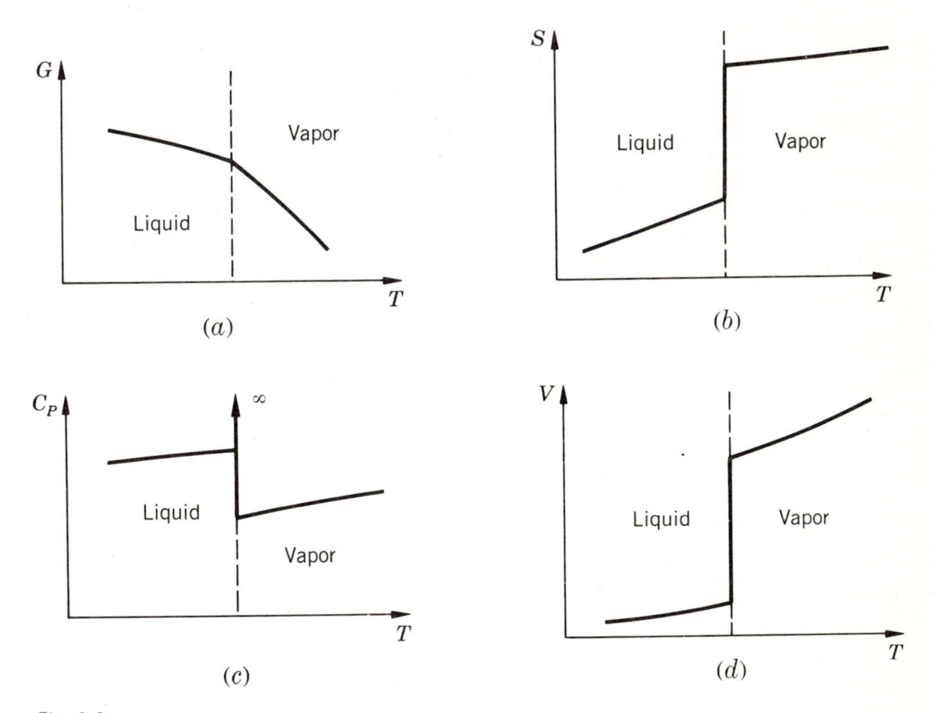

Fig. 9-8

Constant-pressure variation of thermodynamic properties across a phase boundary: (*a*) G, (*b*) $S = -(\partial G/\partial T)_P$, (*c*) $C_P = -T(\partial^2 G/\partial T^2)_P$, (*d*) V.

Consider a system initially in two-phase equilibrium, and imagine its temperature and pressure to be changed by the amounts dT and dP in such a way as to maintain equilibrium between the phases. Then by Eq. (9-76) the differential changes in G are also equal for the two phases, viz.,

$$dG^\alpha = dG^\beta.$$

But, by Eq. (9-6), we have

$$dG = -S\,dT + V\,dP,$$

and therefore

$$-S^\alpha\,dT + V^\alpha\,dP = -S^\beta\,dT + V^\beta\,dP,$$

from which we find

$$\frac{dP}{dT} = \frac{S^\beta - S^\alpha}{V^\beta - V^\alpha}.$$

Since P in this equation is a saturation pressure on the appropriate phase boundary, we will write $P(\text{sat})$ in place of P; additionally, we can simplify the right-hand side of the equation by use of the definitions provided by Eq. (9-73). Thus the last equation becomes

$$\frac{dP(\text{sat})}{dT} = \frac{\Delta S^{\alpha\beta}}{\Delta V^{\alpha\beta}}. \tag{9-77}$$

The entropy term in Eq. (9-77) can be written in a different form. By definition,

$$G = H - TS, \tag{9-5}$$

and application separately to the two equilibrium phases yields

$$G^\alpha = H^\alpha - TS^\alpha \qquad \text{and} \qquad G^\beta = H^\beta - TS^\beta.$$

Subtraction gives

$$G^\beta - G^\alpha = (H^\beta - H^\alpha) - T(S^\beta - S^\alpha),$$

and, since $G^\alpha = G^\beta$ for equilibrium, we find that

$$\Delta H^{\alpha\beta} = T\,\Delta S^{\alpha\beta}. \tag{9-78}$$

Combining Eqs. (9-77) and (9-78), we get *Clapeyron's equation*:

$$\boxed{\frac{dP(\text{sat})}{dT} = \frac{\Delta H^{\alpha\beta}}{T\,\Delta V^{\alpha\beta}}.} \tag{9-79}$$

Clapeyron's equation applies to any phase change of a pure material that takes place at constant T and P.

Example 9-14

The enthalpy change $\Delta H^{\alpha\beta}$ which accompanies a change of phase is usually called a *latent heat.* Why is $\Delta H^{\alpha\beta}$ associated with heat?

A phase transition occurring at constant T and P is usually accompanied by an exchange of heat between the substance (taken as the system) and its surroundings. For a reversible isobaric phase transition of 1 (mol) of a PVT system, we see immediately by Eq. (4-17b) that

$$Q = \Delta H^{\alpha\beta} \qquad \text{(reversible, isobaric)}. \qquad (9\text{-}80)$$

For melting, $\Delta H^{\alpha\beta}$ is called the *latent heat of fusion*; for sublimation, the *latent heat of sublimation*; and for vaporization, the *latent heat of vaporization.* The importance of Eq. (9-80) is that it connects $\Delta H^{\alpha\beta}$ with a measurable quantity, Q. Thus values of the latent heats may be determined by calorimetric measurements. Figure 9-9 is a plot of the heat of vaporization ΔH^{lv} of water. Note that ΔH^{lv} decreases as T increases, becoming zero at the critical point, where the liquid and vapor phases are indistinguishable.

For a reversible isothermal phase transition we have from Eq. (8-3)

$$Q = T\,\Delta S^{\alpha\beta} \qquad \text{(reversible, isothermal)}. \qquad (9\text{-}81)$$

Combination of Eqs. (9-80) and (9-81) gives Eq. (9-78),

$$\Delta H^{\alpha\beta} = T\,\Delta S^{\alpha\beta},$$

which we derived by a different procedure.

The Clapeyron equation provides a vital connection between the properties of different phases. It is usually applied to the calculation of latent heats of vaporization and sublimation from vapor-pressure and volumetric

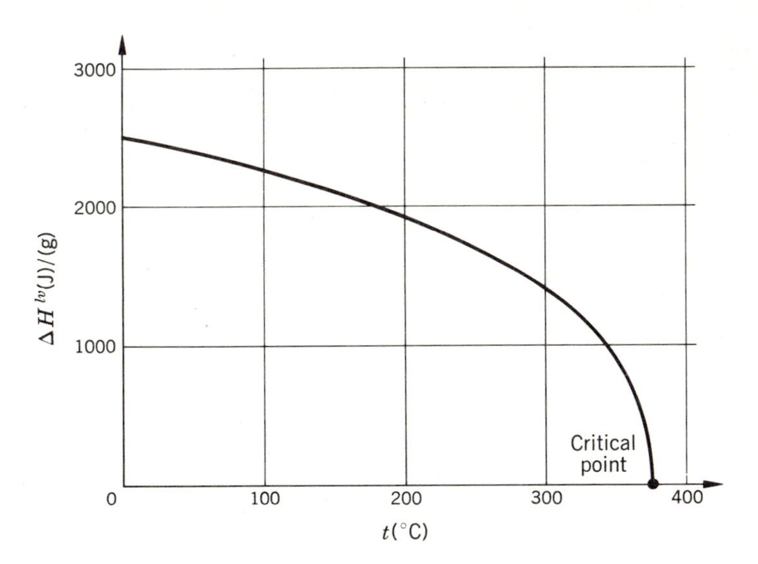

Fig. 9-9
Temperature variation of the latent heat of vaporization for water.

data:

$$\Delta H^{\alpha\beta} = T\,\Delta V^{\alpha\beta}\,\frac{dP(\text{sat})}{dT}.$$

The use of such a procedure requires an accurate representation for vapor pressure as a function of T. The simplest equation for $P(\text{sat})$ capable of providing realistic temperature derivatives is the *Antoine equation*:

$$\ln P(\text{sat}) = A - \frac{B}{T + C}, \tag{9-82}$$

where A, B, and C are constants.[1] For very precise work, more flexible equations are needed. One which is often satisfactory is the five-constant expression

$$\ln P(\text{sat}) = A - \frac{B}{T + C} + DT + E\ln T. \tag{9-83}$$

Generalized correlations are also available for the reduced vapor-liquid saturation pressure $P_r(\text{sat})$ in terms of the reduced temperature T_r and the acentric factor ω.

9-10 Thermodynamic Diagrams

The PV diagram, to which we have already referred a number of times, is the most familiar type of thermodynamic diagram for PVT systems. Actually, a thermodynamic diagram in two dimensions can be constructed with any pair of thermodynamic properties as rectangular coordinates. The phase boundaries on such a plot appear as curves. Superimposed are various other curves which represent the functional relationships connecting the coordinate variables at constant values of another thermodynamic property. For example, on a PV diagram it is common to display curves relating P to V at constant T, or isotherms. One could also include curves of constant enthalpy, or isenthalps, curves of constant entropy, or isentrops, etc. As a matter of fact, such an extensive PV diagram is never used in practice, simply because other choices for the properties representing the rectangular coordinates have proved to be much more convenient. The diagrams in general use are those which plot temperature

[1] Tables of Antoine constants, valid for specified ranges of T, are given in handbooks and data compilations. See for example T. Boublik et al., "The Vapor Pressures of Pure Substances," Elsevier, New York, 1973.

vs. entropy (TS diagram), enthalpy vs. entropy (HS, or Mollier diagram), and the logarithm of pressure vs. enthalpy ($\ln P$ vs. H diagram).

The preparation of an accurate thermodynamic diagram or, alternatively, of a set of thermodynamic tables is an exacting task. One must start with experimental data, such as heat capacities and PVT data and, through the equations of thermodynamics, calculate the required properties, such as enthalpy and entropy. The raw data must themselves be accurate, and the calculations must be carried out with great precision. We outline briefly below the types of procedures followed in constructing engineering diagrams and indicate some of the features of the three major types of diagrams.

We consider first the construction of a TS diagram. For an isobaric change in state, we have from Eq. (9-21) that

$$S_2 - S_1 = \int_{T_1}^{T_2} C_P \frac{dT}{T} \qquad \text{(const } P\text{)}. \tag{9-84}$$

Equation (9-84) yields *changes* in entropy, and if we wish to have numerical values for S itself, we must arbitrarily assign a value to S for some particular state, called the *reference state*. For example, the entropy of water is usually taken as zero for the saturated liquid at 32(°F) [0(°C)]. All other entropies for water are then referred to this state.

A TS diagram for water is shown schematically in Fig. 9-10. A single isobar is represented by the series of lines $ABDEFG$, which corresponds to a series of isobaric reversible processes by which compressed solid water (ice) is transformed into superheated water vapor (steam). Thus:

AB = isobaric heating of solid to its melting point,
BD = isobaric, isothermal melting,
DE = isobaric heating of liquid to its boiling point,
EF = isobaric, isothermal vaporization,
FG = isobaric heating of vapor.

If heat capacities are available, the entropy changes along paths AB, DE, and FG may be computed by Eq. (9-84). However, steps BD and EF are both isobaric *and* isothermal, so that Eq. (9-84) is therefore not applicable. The entropy changes for these phase transitions must be calculated from latent heats by Eq. (9-78); the latent heats in turn must be determined directly by experiment, or by use of the Clapeyron relation, Eq. (9-79). It is clear that to establish each isobar such as $ABDEFG$ in Fig. 9-10 it is necessary to have the required latent heats and heat capacities at the particular pressure of the isobar. Even if such data were available (and they usually are not), it would still be necessary to relate one isobar to

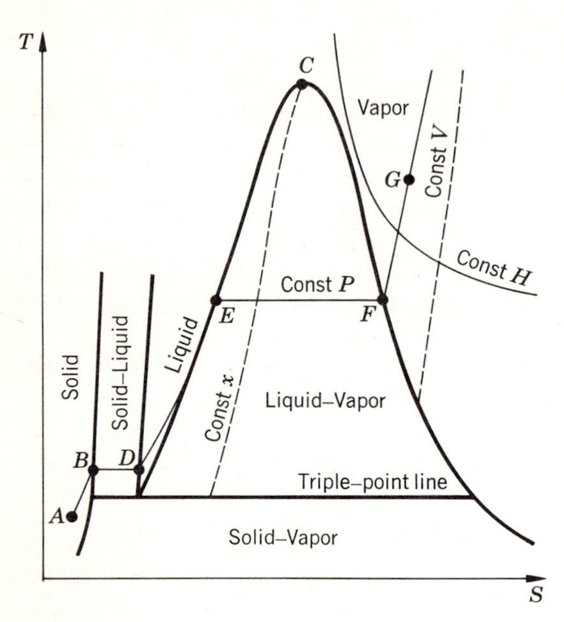

Fig. 9-10
TS diagram.

another so that all values of S could be referred back to the reference state. We must therefore have, at least for one temperature, sufficient data to allow the calculation of the change of entropy with pressure at constant temperature. For a single-phase region such a relation is given by Eq. (9-27):

$$S_2 - S_1 = -\int_{P_1}^{P_2} \left(\frac{\partial V}{\partial T}\right)_P dP \qquad (\text{const } T). \qquad (9\text{-}27)$$

For the evaluation of the integral on the right of Eq. (9-27), one must have volumetric data for the substance as a function of both T and P. Through the use of Eqs. (9-84) and (9-27) one can establish as many isobars on the TS diagram as wanted, within the range of the variables for which data are available.

In addition to isobars, it is common to find isenthalps and isochors included on a TS diagram. We have by Eqs. (9-8) and (9-11)

$$\left(\frac{\partial H}{\partial S}\right)_P = T \qquad \text{and} \qquad \left(\frac{\partial H}{\partial P}\right)_S = V.$$

These expressions apply to two-phase as well as to single-phase regions.

Integrating, we get

$$H_2 - H_1 = \int_{S_1}^{S_2} T\,dS \qquad (\text{const } P), \tag{9-85}$$

and

$$H_2 - H_1 = \int_{P_1}^{P_2} V\,dP \qquad (\text{const } S). \tag{9-86}$$

Once isobars have been established on the TS diagram, enthalpy changes along an isobar are readily calculated from Eq. (9-85). Equation (9-86) relates enthalpies on different isobars so as to refer all values to an arbitrarily assigned value for some reference state. For water, the enthalpy, like the entropy, is usually assigned the value zero for saturated liquid at 32(°F). Calculations of this nature allow one to construct isenthalps on the TS diagram. Isochors may be plotted directly from PVT data. We show on the TS diagram of Fig. 9-10 an isenthalp and an isochor for the vapor phase. Also given is a line of constant quality x.

The actual *shape* of the vapor-liquid dome of the TS curve is of considerable importance for certain engineering applications. In Fig. 9-11 are shown idealized Rankine cycles on TS diagrams for two different types of fluids. The numbered points on the figure correspond to those on the PV diagram of Fig. 6-3*b*. The TS diagram of Fig. 9-11*a* is for a fluid such as water, for which the shape of the vapor-liquid dome is such that the adiabatic expansion step $5 \rightarrow 6$ (here assumed to be reversible) carries the fluid into the two-phase region. Fluids of this type are called *wetting fluids.* Figure 9-11*b* depicts a similar Rankine cycle on a TS diagram for a fluid for

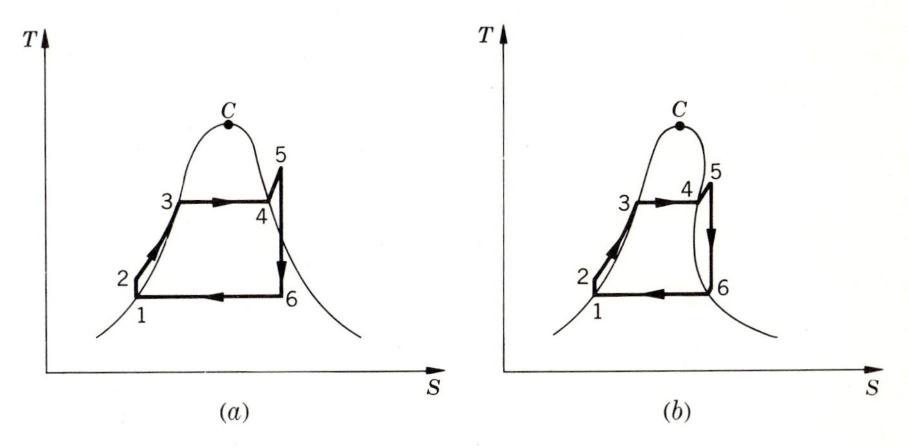

Fig. 9-11
Idealized Rankine cycle on a TS diagram for (a) a wetting fluid, (b) a nonwetting fluid.

which the vapor-liquid dome doubles back on itself. In this case, conditions can conveniently be adjusted so that the adiabatic expansion is entirely in the vapor region; fluids which exhibit a vapor-liquid dome of this shape are therefore called *nonwetting fluids*, or *retrograde fluids*. Examples of nonwetting fluids are *n*-pentane and Freon 113, a refrigerant. The expansion step $5 \rightarrow 6$ is the work-producing part of the Rankine cycle, and it is desirable to minimize the amount of condensate formed in the turbine. Thus nonwetting fluids are attractive candidates for automotive applications of the Rankine power cycle.

Once the thermodynamic properties have been calculated for preparation of one type of diagram, for example, the TS diagram just discussed, the values are available for replotting on any coordinate scale one wishes. The type of diagram perhaps most commonly employed by engineers is the HS, or Mollier, diagram. The general outlines of the Mollier diagram for water are shown schematically in Fig. 9-12. Included for the vapor phase are an isobar, an isotherm, and a line of *constant superheat*. The superheat of a vapor is the number of degrees by which the actual temperature of the vapor exceeds the vapor-liquid saturation temperature at the same pres-

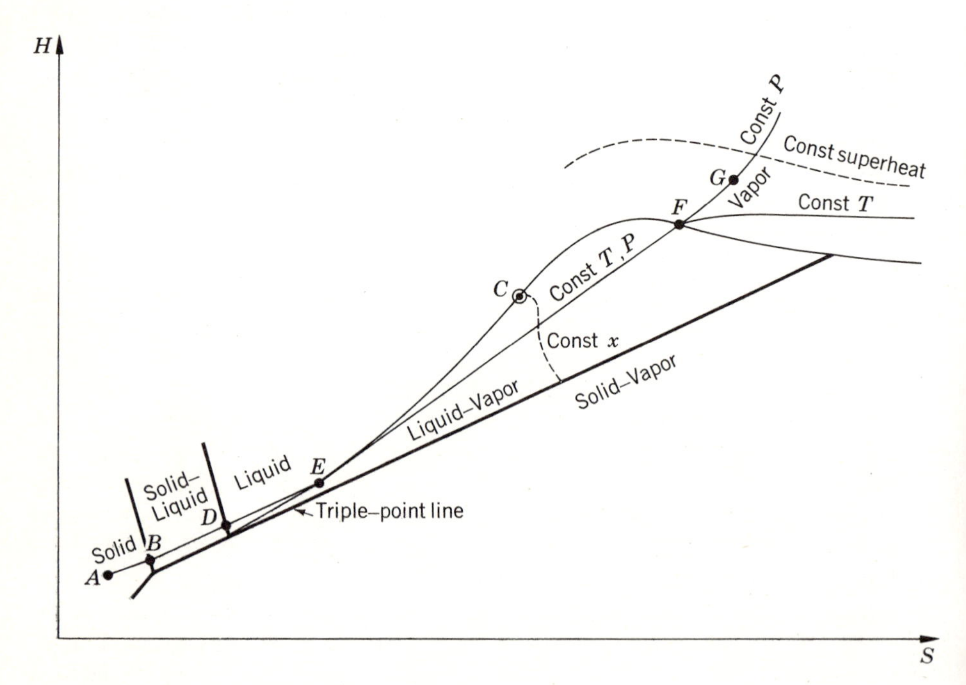

Fig. 9-12
HS diagram (Mollier diagram).

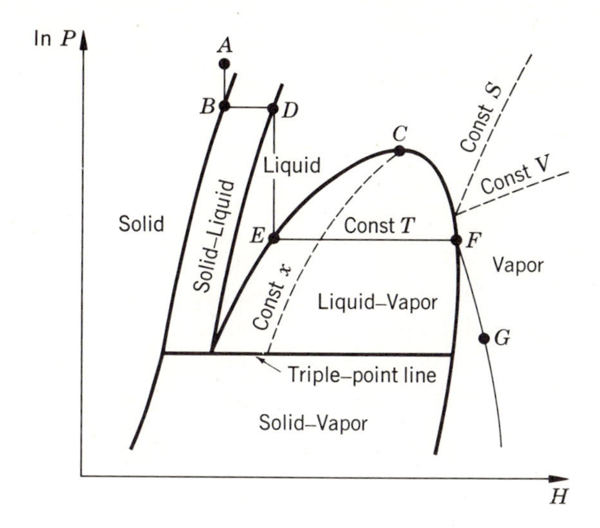

Fig. 9-13
lnP vs. H diagram.

sure. Thus steam at 1(atm) and 125(°C) has 125 − 100 = 25(°C) of superheat.

The path $ABDEFG$ represents a reversible isobaric transition from compressed solid at A to superheated vapor at G. The individual portions of the curve correspond to the same steps shown in the TS diagram of Fig. 9-10. Since the slope of an isobar on a Mollier diagram $= (\partial H/\partial S)_P = T$, all sublimation, fusion, and vaporization lines are straight, and the higher the temperature, the steeper the slope. When a heating process is accompanied by a temperature rise of the fluid, the slope increases throughout the process.

A detailed Mollier diagram for steam is given in the back end-papers of this book. The units of the diagram are those commonly employed by American engineers, viz., temperature in (°F), P in (psia), H in (Btu)/(lb_m), and S in (Btu)/(lb_m)(R). Factors for conversion to SI units are given in Appendix A; a numerical example is given at the end of this section. Note that the relative amounts of vapor and liquid in the vapor-liquid region are represented by lines of constant moisture percent on a mass basis, rather than by lines of constant quality.

The third useful engineering diagram is the ln P vs. H diagram, shown schematically in Fig. 9-13 for the substance water. An isobar on such a diagram is a horizontal line. The path $ABDEFG$ represents a reversible isothermal transition from compressed solid at A to superheated vapor at

G. Thus:

$$
\begin{aligned}
AB &= \text{isothermal expansion of solid to its melting point,}\\
BD &= \text{isothermal, isobaric melting,}\\
DE &= \text{isothermal expansion of liquid to its boiling point,}\\
EF &= \text{isothermal, isobaric vaporization,}\\
FG &= \text{isothermal expansion of vapor.}
\end{aligned}
$$

Since the latent heat of vaporization is the difference between the enthalpy of saturated vapor and the enthalpy of saturated liquid, this quantity is represented by horizontal line segments such as EF. It is obvious from Fig. 9-13 that the heat of vaporization decreases with increasing temperature and becomes zero at the critical point. The heats of sublimation are represented by horizontal lines traversing the solid-vapor region. At the triple point the heat of sublimation is evidently equal to the sum of the heat of fusion and the heat of vaporization.

Example 9-15

Five kilograms of steam are expanded reversibly and adiabatically from initial conditions of 20(bar) and 400(°C) to a final pressure of 1(bar). Determine from the Mollier diagram for steam (*a*) the temperature and the physical state of the water at final conditions, (*b*) the total enthalpy change ΔH^t in (J) of the 5(kg) of water.

Use of the Mollier diagram requires initial conversion of all units into those of the diagram. Thus we have for the given conditions:

$$t_1 = 1.8 \times 400 + 32 = 752(°\text{F}),$$

$$P_1 = 20(\text{bar}) \times 14.5038(\text{psia})/(\text{bar}) = 290(\text{psia}),$$

and

$$P_2 = 1(\text{bar}) \times 14.5038(\text{psia})/(\text{bar}) = 14.5(\text{psia}).$$

The initial state of the water is found from the Mollier diagram to lie above the saturation dome, in the superheated vapor region. The enthalpy and entropy for this state are read from the diagram as

$$H_1 = 1395(\text{Btu})/(\text{lb}_m)$$

and

$$S_1 = 1.70(\text{Btu})/(\text{lb}_m)(\text{R}).$$

The expansion process is reversible and adiabatic, and hence isentropic. The final state of the water is therefore determined by the intersection of the 1.70(Btu)/(lb_m)(R) isentrop (a vertical line) with the 14.5(psia) isobar. We also find from the diagram that in the final state the water is in the two-phase region and consists of 4 percent liquid and 96 percent vapor. Thus its quality x (mass fraction of vapor) is 0.96. The final enthalpy and temperature are read from the diagram as

$$H_2 = 1110(\text{Btu})/(\text{lb}_m)$$

$$t_2 = 210(°\text{F}) = 98.9(°\text{C}),$$

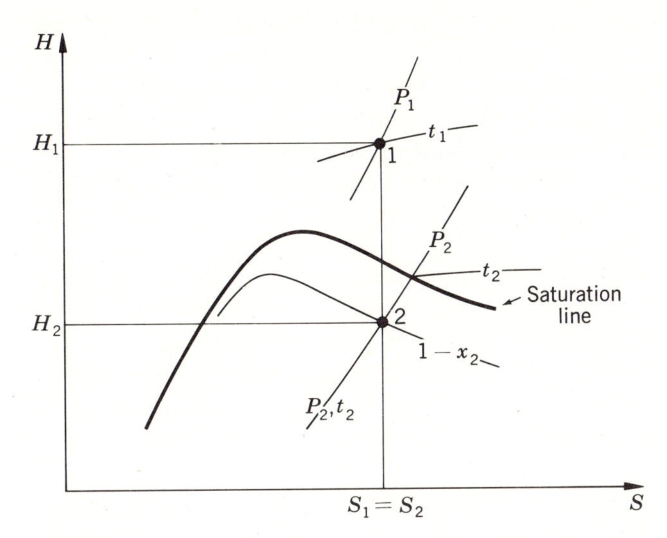

Fig. 9-14
Mollier diagram for steam. Location of initial(1) and final(2) states for Example 9-15.

and ΔH^t is computed to be

$$\Delta H^t = \frac{5\,(\text{kg}) \times 2.205\,(\text{lb}_m)/(\text{kg}) \times [1110 - 1395](\text{Btu})/(\text{lb}_m)}{9.478 \times 10^{-4}(\text{Btu})/(\text{J})},$$

or $\qquad \Delta H^t = -3.31 \times 10^6(\text{J}).$

The methods for location of the thermodynamic states and properties for this example are illustrated schematically in Fig. 9-14.

9-11 Steam Tables

Thermodynamic diagrams can surely be drawn to a scale large enough so that numerical values for the properties can be read from them. Even so, it is convenient to have the same data available in tabular form. This provides greater accuracy than is possible from readings of graphs. Tabulated values can be given at close enough intervals so that linear interpolation is sufficiently accurate for all practical purposes. Both tables and diagrams for the most common substances are found in the engineering literature and in handbooks of data.

Because of the great industrial importance of water, data have been extensively compiled for this substance. The resulting tables, usually called *steam tables*, are in widespread use. Appendix D of this text contains two tables for water. The first (Table D-1) is for states of vapor-liquid equilibrium ("saturation") at temperatures from 32 to 705.34(°F), the critical temperature of water. Listed in this table for each temperature are the saturation pressure and values of V, H, and S for the saturated liquid and vapor phases. Subscript notation is employed (see Sec. 9-9), with the subscript f representing the liquid phase and the subscript g the vapor phase. Also given are values for the three property changes of vaporization: ΔV_{fg}, ΔH_{fg}, and ΔS_{fg}. The units of P, H, and S are the same as those for the Mollier diagram discussed in Sec. 9-10 and the units of V are $(\text{ft})^3/(\text{lb}_m)$. Additionally, saturation pressures are given in units of (in Hg) for temperatures up to the normal boiling point [212(°F)].

The second table (Table D-2) is for superheated steam. In this table values of V, H, and S are displayed as functions of pressure and temperature. In addition, for each value of pressure there are given the corresponding saturation temperature and the values of V, H, and S for saturated liquid and vapor.

The use of the steam tables is illustrated in the following examples.

Example 9-16

One pound *mass* of steam at 350(°F) and 20(psia) is compressed isothermally and reversibly in a piston and cylinder device to a final state of saturated liquid water. Determine the heat transferred and the work of the process.

The work for this nonflow process is given directly by the first law:

$$W = \Delta U - Q.$$

As a result of the definition of the enthalpy, we have

$$\Delta U = \Delta H - \Delta(PV),$$

and therefore

$$W = \Delta H - \Delta(PV) - Q.$$

Since the process is reversible and isothermal, the heat transferred is

$$Q = T\,\Delta S.$$

Since the initial and final states of the steam are fully specified in the problem statement, we can find the following values from the steam tables:

Initial state: Superheated steam at $t = 350(°\text{F})$ and $P_1 = 20(\text{psia})$. $V_1 = 23.91(\text{ft})^3/(\text{lb}_m)$, $H_1 = 1214.8(\text{Btu})/(\text{lb}_m)$, and $S_1 = 1.8101(\text{Btu})/(\text{lb}_m)(\text{R})$.

Final state: Saturated liquid at $t = 350(°\text{F})$. $P_2 = 134.62(\text{psia})$, $V_2 = 0.01799(\text{ft})^3/(\text{lb}_m)$, $H_2 = 321.64(\text{Btu})/(\text{lb}_m)$, and $S_2 = 0.5030(\text{Btu})/(\text{lb}_m)(\text{R})$.

Thus,

$$Q = T(S_2 - S_1) = (350 + 460)(0.5030 - 1.8101) = -1058.8(\text{Btu}),$$

and

$$W = (H_2 - H_1) - (P_2V_2 - P_1V_1) - Q$$

$$= [321.6 - 1214.8](\text{Btu}) - [(134.62)(0.01799) - (20)(23.91)](\text{psia})(\text{ft})^3 \times \frac{144(\text{in})^2/(\text{ft})^2}{778.2(\text{ft})(\text{lb}_f)/(\text{Btu})} + 1058.8(\text{Btu}),$$

or

$$W = 253.7(\text{Btu}).$$

Example 9-17

Five kilograms of steam are expanded reversibly and adiabatically from initial conditions of 20(bar) and 400(°C) to a final pressure of 1(bar). Using data from the steam tables, determine the work for the process.

The process is identical to the one treated in Example 9-15. However, the Mollier diagram used for that example does not contain the volumetric data usually needed for nonflow work calculations. The work for the process is given as in the preceding example by

$$W = \Delta H^t - \Delta(PV^t) - Q,$$

or, since $Q = 0$,

$$W = m[(H_2 - H_1) - (P_2V_2 - P_1V_1)]. \qquad (A)$$

The given conditions were converted to engineering units in Example 9-15:

$$t_1 = 752(°\text{F}),$$

$$P_1 = 290(\text{psia}),$$

$$P_2 = 14.5(\text{psia}).$$

Values of V_1, H_1, and S_1 are found from the superheat tables by linear interpolation with respect to temperature at 290(psia) between listed values for 700 and 800(°F). Thus

$$V_1 = 2.3080 + \left(\frac{752 - 700}{800 - 700}\right)(2.5302 - 2.3080) = 2.4235(\text{ft})^3/(\text{lb}_m).$$

Similarly,

$$H_1 = 1367.9 + (0.52)(1420.1 - 1367.9) = 1395.0(\text{Btu})/(\text{lb}_m),$$

$$S_1 = 1.6784 + (0.52)(1.7215 - 1.6784) = 1.7008(\text{Btu})/(\text{lb}_m)(\text{R}).$$

The values of H_1 and S_1 are the same as those obtained from the Mollier diagram in Example 9-15.

The expansion process is isentropic, and the final state of the system is therefore determined by the thermodynamic coordinates $P_2 = 14.5$(psia) and $S_2 = 1.7008$(Btu)/(lb$_m$)(R). However, the superheat tables show no entropy values this small at the given pressure level. Therefore the final state of the water is one of vapor-liquid equilibrium and the properties are determined from the saturation tables by use of Eq. (9-74), with $\alpha = f$ and $\beta = g$:

$$M = M_f + x\,\Delta M_{fg}. \qquad (B)$$

There are no entries in the saturation tables for $P = 14.5$(psia). We must therefore first determine the values of M_f and ΔM_{fg} at 14.5(psia) by linear interpolation between the given values for $P = 14.123$(psia) [$t = 210$(°F)] and for $P = 14.696$(psia) [$t = 212$(°F)]. Doing this, we find for $P = 14.5$(psia) that

$$V_f = 0.01671, \qquad \Delta V_{fg} = 27.15(\text{ft})^3/(\text{lb}_m),$$

$$H_f = 179.33, \qquad \Delta H_{fg} = 970.8(\text{Btu})/(\text{lb}_m),$$

$$S_f = 0.3110, \qquad \Delta S_{fg} = 1.4467(\text{Btu})/(\text{lb}_m)(\text{R}).$$

The corresponding temperature t_2 is found by interpolation to be 211.3(°F) = 99.6(°C).

We know the entropy at 14.5(psia), so that the quality x_2 in the final state may be computed from Eq. (B) with $M = S$:

$$x_2 = \frac{S_2 - S_f}{\Delta S_{fg}} = \frac{1.7008 - 0.3110}{1.4467} = 0.961.$$

The values of V_2 and H_2 are then determined from Eq. (B) for $x = x_2 = 0.961$ with $M = V$ and $M = H$:

$$V_2 = V_f + x_2 \, \Delta V_{fg} = 0.0167 + (0.961)(27.15) = 26.11(\text{ft})^3/(\text{lb}_m)$$

and

$$H_2 = H_f + x_2 \, \Delta H_{fg} = 179.33 + (0.961)(970.8) = 1112.2(\text{Btu})/(\text{lb}_m).$$

The work for the process is determined by substitution of numerical values into Eq. (A):

$$\begin{aligned} W = {} & 5(\text{kg}) \times 2.205(\text{lb}_m)/(\text{kg})\{[1112.2 - 1395.0](\text{Btu})/(\text{lb}_m) - [(14.5)(26.11) \\ & - (290)(2.4235)](\text{psia})(\text{ft})^3/(\text{lb}_m) \times 144(\text{in})^2/(\text{ft})^2 \\ & \times \frac{1}{778.2}(\text{Btu})/(\text{ft})(\text{lb}_f)\} \\ = {} & -2456.4(\text{Btu}), \end{aligned}$$

or

$$W = -2456.4(\text{Btu}) \frac{1}{0.9478}(\text{kJ})/(\text{Btu}) = -2590(\text{kJ}).$$

Example 9-18

A 3(ft)^3 tank contains superheated steam at 22(psia) and 440(°F). It is cooled until the contents of the tank are just saturated vapor steam. How much heat must be removed from the steam during this process?

The energy equation for the process results immediately from the first law. For m(lb$_m$) of steam in the tank,

$$\Delta U^t = m \, \Delta U = Q + W.$$

For a rigid tank, $W = 0$, and therefore

$$Q = m \, \Delta U.$$

Since

$$U = H - PV,$$

this becomes

$$Q = m\,\Delta H - m\,\Delta(PV)$$

$$= m\,\Delta H - \Delta(PmV).$$

However,

$$mV_1 = mV_2 = V^t = V_{\text{tank}} = \text{constant}.$$

Thus

$$Q = m\,\Delta H - V_{\text{tank}}\,\Delta P.$$

In order to apply this equation we need in addition to the given information the mass m of the steam in the tank, the initial and final enthalpies of the steam, and the final pressure. The mass can be found from the tank volume and the initial specific volume of the steam:

$$m = \frac{V_{\text{tank}}}{V_1}.$$

The initial state of the steam is superheated vapor at 22(psia) and 440(°F). However, entries do not appear in the steam tables at either this pressure or this temperature. Values are given at 20 and 25(psia) and at 400 and 450(°F), and we must do a double interpolation based on the four steam-table entries that lie closest to the initial state. The interpolation scheme for both the specific volume V and the specific enthalpy H is shown by Table 9-4. The four corner entries are the values given in the table for superheated steam. Linear interpolation between pairs of values (as indicated by the arrows) leads to the intermediate values shown in italics. A second interpolation between pairs of intermediate values gives the final values shown in the central box. Actually, two separate interpolation schemes, both of which give the same final results, are indicated in Table 9-4. The solid arrows indicate initial interpolation with respect to temperature and final interpolation with respect to pressure; the dashed arrows, the reverse. The two schemes are equivalent.

Table 9-4 Double Interpolation for Superheated Steam

P(psia)		t(°F) 400	*440*	450
20	V	25.43	*26.65*	26.95
	H	1238.4	*1257.0*	1261.6
22	V	*23.38*	24.50	*24.78*
	H	*1238.2*	1256.8	*1261.4*
25	V	20.30	*21.28*	21.53
	H	1237.9	*1256.5*	1261.1

The values obtained by interpolation are

$$V_1 = 24.50(\text{ft})^3/(\text{lb}_m) \qquad \text{and} \qquad H_1 = 1256.8(\text{Btu})/(\text{lb}_m).$$

The mass of steam in the tank is therefore

$$m = \frac{V_{\text{tank}}}{V_1} = \frac{3(\text{ft})^3}{24.50(\text{ft})^3/(\text{lb}_m)} = 0.1224(\text{lb}_m).$$

Since both m and V_{tank} remain unchanged in the process, the specific volume V_2 of the *saturated* vapor in the tank at the end of the process is equal to V_1. Thus

$$V_2 = V_1 = 24.50(\text{ft})^3/(\text{lb}_m),$$

and we must locate the state of saturated vapor in Table D-1 where V_g has this value. This particular value does not appear, but the values that bracket it are for temperatures of 215 and 220(°F):

t(°F)	P(psia)	$V_g(\text{ft})^3/(\text{lb}_m)$	$H_g(\text{Btu})/(\text{lb}_m)$
215	15.591	25.37	1151.4
217.0	*16.220*	24.50	*1152.6*
220	17.188	23.16	1153.3

The values in italics have been interpolated linearly with respect to V_g, and represent the final state of the steam. Thus

$$P_2 = 16.22(\text{psia}) \qquad \text{and} \qquad H_2 = 1152.6(\text{Btu})/(\text{lb}_m).$$

The heat transferred is therefore given by

$$\begin{aligned} Q &= m(H_2 - H_1) - V_{\text{tank}}(P_2 - P_1) \\ &= 0.1224(1152.6 - 1256.8) - 3(16.22 - 22.00)\frac{(144)}{(778.2)} \\ &= -9.55(\text{Btu}). \end{aligned}$$

Problems

9-1 (*a*) Show that

$$\left(\frac{\partial H}{\partial P}\right)_T = \frac{-RT^2}{P}\left(\frac{\partial Z}{\partial T}\right)_P \qquad \text{and} \qquad \left(\frac{\partial S}{\partial P}\right)_T = -\frac{RZ}{P} - \frac{RT}{P}\left(\frac{\partial Z}{\partial T}\right)_P.$$

(*b*) For a gas which obeys the equation of state

$$Z = \frac{PV}{RT} = 1 + B'P,$$

show that

$$\left(\frac{\partial H}{\partial P}\right)_T = -RT^2\frac{dB'}{dT},$$

and

$$\left(\frac{\partial S}{\partial P}\right)_T = -\frac{R}{P} - R\left(B' + T\frac{dB'}{dT}\right).$$

9-2 For a gas which obeys the equation of state

$$Z = \frac{PV}{RT} = 1 + \frac{B}{V},$$

show that $$\left(\frac{\partial H}{\partial P}\right)_T = \frac{B - T(dB/dT)}{1 + 2B/V},$$

and $$\left(\frac{\partial S}{\partial P}\right)_T = -\frac{V + B + T(dB/dT)}{T(1 + 2B/V)}.$$

9-3 When 1(g) of solid napthalene ($C_{10}H_8$) is burned in oxygen saturated with water vapor at constant volume in a bomb calorimeter, 9620(cal) are evolved at 25(°C). Note that the water formed during the combustion condenses but that the CO_2 formed is a gas. If this combustion process occurred at constant pressure instead of at constant volume at 25(°C), what would be the heat evolved? Oxygen and carbon dioxide may be assumed to be ideal gases.

9-4 The latent heat of vaporization of water at 212(°F) and 1(atm) is 970.3(Btu)/(lb_m). What is ΔS for the vaporization of 1(lb_m) of water at these conditions? If the volume change of vaporization is 26.81(ft)^3/(lb_m), what is ΔU for the vaporization process?

9-5 A gas at high pressure flows through a duct; a thermometer in the duct reads 200(K). A small amount of gas is bled out of this duct through an insulated throttle valve open to the atmosphere. The temperature of the gas leaving the throttle is measured as 180(K). What is the pressure of the gas in the high-pressure duct? The gas obeys the equation of state

$$Z = 1 + \left(b - \frac{a}{T}\right)\frac{P}{RT},$$

where b = 20(cm)^3/(mol),

a = 40,000(K) (cm)^3/(mol).

The molar heat capacity of the gas at low pressure is

$$C_P = 10 + 0.02T,$$

where T is in (K) and C_P is in (cal)/(mol) (K).

9-6 Very pure liquid water can be supercooled at atmospheric pressure to temperatures considerably below 0(°C). Assume that 1(kg) has been cooled as a liquid to −5(°C). A small ice crystal (whose mass may be considered negligible) is now added to "seed" the supercooled liquid. If the subsequent change of state occurs adiabatically and at 1(atm):

(*a*) What is the final state of the system?
(*b*) What fraction of the system freezes?
(*c*) What is the entropy change of the system during this adiabatic process? Is the process reversible?

Data: Latent heat of fusion of water at 0(°C) = 333.4(J)/(g); specific heat of supercooled liquid water = 4.226(J)/(g)(K).

9-7 One mole of an ideal gas originally confined in a vessel at 300(K) and 10(bar) is allowed to expand through a valve into an evacuated vessel whose volume is nine times that of the first. The system is thermally and mechanically insulated from the surroundings. What are the numerical values of W, Q, ΔU, ΔH, ΔS, ΔA, and ΔG in units of (J), (mol), and (K) for this process after sufficient time has elapsed so that the temperatures of the two tanks have equalized?

9-8 (*a*) Derive the equation

$$\left(\frac{\partial C_V}{\partial V}\right)_T = T\left(\frac{\partial^2 P}{\partial T^2}\right)_V.$$

(*b*) Prove that C_V of an ideal gas is a function of T only.
(*c*) In the case of a gas obeying the equation of state

$$Z = 1 + \frac{B}{V},$$

where B is a function of T only, show that

$$\Delta C_V' = \frac{RT}{V}\frac{d^2}{dT^2}BT,$$

where $\Delta C_V'$ is the residual heat capacity at constant volume.

9-9 (*a*) Derive the equation

$$\left(\frac{\partial C_P}{\partial P}\right)_T = -T\left(\frac{\partial^2 V}{\partial T^2}\right)_P.$$

(*b*) Prove that C_P of an ideal gas is a function of T only.
(*c*) In the case of a gas obeying the equation of state

$$Z = 1 + B'P,$$

where B' is a function of T only, show that

$$\Delta C_P' = RTP\frac{d^2}{dT^2}B'T,$$

where $\Delta C_P'$ is the residual heat capacity at constant pressure.
(*d*) Determine an expression for $\Delta C_P'$ of a gas which is described by Eq. (5-10) and by Pitzer's correlation for B.

9-10 The pressure on 500(g) of copper is increased reversibly and isothermally from zero to 5000(atm) at 100(K). (Assume the density, volume expansivity, isothermal compressibility, and heat capacity to remain practically constant. The values are given in Table 9-2.)

(*a*) How much heat is transferred during the compression?
(*b*) How much work is done during the compression?
(*c*) Determine the change of internal energy.
(*d*) What would have been the rise of temperature if the copper had been subjected to a reversible adiabatic compression?

9-11 The pressure on 200(g) of water is increased reversibly and isothermally from 1 to 3,000(atm) at 0(°C). (Property values are given in Table 9-3.)

(*a*) How much heat is transferred?
(*b*) How much work is done?
(*c*) Calculate the change in internal energy.

9-12 The pressure on 1(g) of water is increased from 0 to 1,000(atm) reversibly and adiabatically. Calculate the temperature change when the initial temperature has the three different values given below:

Temperature (°C)	Specific volume V (cm)^3/(g)	$\beta \times 10^6$(K)$^{-1}$	C_P (J)/(g)(K)
0	1.000	−68.1	4.217
10	1.000	+87.9	4.192
50	1.012	+458	4.180

9-13 A gas obeys the equation $P(V - b) = RT$, where b is a constant, and has a constant C_V. Show that

(*a*) U is a function of T only.
(*b*) $\gamma = C_P/C_V$ = constant.
(*c*) A relation that holds during a reversible adiabatic process is

$$P(V - b)^\gamma = \text{const.}$$

9-14 The pressure on 1(ft)^3 of water at 75(°F) [62.2(lb$_m$)] is increased reversibly and isothermally from 1 to 3,000(atm).

(*a*) How much heat is transferred?
(*b*) How much work is done?
(*c*) What is the change in internal energy?
(*d*) What is the change in enthalpy?

Take $\beta = 140 \times 10^{-6}(R)^{-1}$ and $\kappa = 45 \times 10^{-6}(atm)^{-1}$.

9-15 Starting with the first Maxwell equation, derive the remaining three by using only the relations

$$\left(\frac{\partial x}{\partial y}\right)_z \left(\frac{\partial y}{\partial z}\right)_x \left(\frac{\partial z}{\partial x}\right)_y = -1,$$

$$\left(\frac{\partial x}{\partial y}\right)_f \left(\frac{\partial y}{\partial z}\right)_f \left(\frac{\partial z}{\partial x}\right)_f = +1.$$

9-16 The table below gives vapor pressures and specific volumes of saturated liquid and vapor phases for benzene.

$t(°F)$	P(sat) (psia)	V_f $(ft)^3/(lb_m)$	V_g $(ft)^3/(lb_m)$
430	286.3	0.0258	0.3102
440	310.5	0.0262	0.2818
450	336.7	0.0268	0.2565
460	364.2	0.0273	0.2328
470	393.4	0.0279	0.2112

(*a*) Determine the constants A and B to give the best fit to the vapor-pressure data by the equation

$$\ln P(\text{sat}) = A - \frac{B}{T}.$$

Using this equation, determine $dP(\text{sat})/dT$ at 450(°F), and then calculate the latent heat of vaporization of benzene at 450(°F) by Clapeyron's equation.

(*b*) Repeat (*a*) for a vapor-pressure equation of the form

$$\ln P(\text{sat}) = A - \frac{B}{T + C}.$$

(*c*) Determine $dP(\text{sat})/dT$ at 450(°F) from a plot of the vapor-pressure data and calculate the latent heat of vaporization from it.

(*d*) Calculate the latent heat of vaporization of benzene at 450(°F) by means of the approximate equation

$$\Delta H^{lv} = \frac{RT^2}{P}\frac{dP(\text{sat})}{dT}.$$

(*e*) Compare the results of the preceding calculations, and discuss them from the standpoint of accuracy. Can you devise a means of getting a more accurate value? The reported value for the latent heat is 103.1(Btu)/(lb_m).

9-17 Prove that the slope of the sublimation curve [P(sat) vs. T] of a substance at the triple point is greater than that of the vaporization curve at the same point.

9-18 Data for liquid mercury at 0(°C) and 1(atm) are given in Table 9-1. Taking V, β, and κ to be essentially independent of P, calculate

(*a*) The pressure increase above 1(atm) at 0(°C) required to cause a 0.1 percent decrease in V. (For a 0.1 percent change, V may be considered "essentially" constant.)

(*b*) ΔU, ΔH, and ΔS for 1(mol) of mercury for the change described in (*a*).

(*c*) Q and W if the change is carried out reversibly.

9-19 (*a*) Prove that the slope of a curve on a Mollier diagram representing a reversible isothermal process is equal to

$$T - \frac{1}{\beta}.$$

(*b*) Prove that the slope of a curve on a Mollier diagram representing a reversible isochoric process is equal to

$$T + \frac{\gamma - 1}{\beta}.$$

9-20 (*a*) With the aid of a Mollier diagram, show that when a saturated liquid undergoes a throttling process, cooling and partial vaporization result.

(*b*) The pressure of CO_2 at the triple point is 5.1(atm). Show with the aid of a $\ln P$ vs. H diagram that if saturated liquid CO_2 at room temperature undergoes a throttling process to atmospheric pressure, dry ice will be formed.

9-21 Show from the steam tables that at any temperature, the value of the Gibbs function is the same for both saturated liquid and saturated vapor.

9-22 One pound *mass* of steam undergoes the following reversible processes. Determine Q and W in each case:

(*a*) Cooling at constant volume from an initial state of 50(psia) and 500(°F) to a final temperature of 300(°F).

(*b*) The same as (*a*), but at constant pressure.

9-23 Determine the differences between the enthalpy, internal energy, and entropy of steam at 900(°F) and 500(psia) and at 300(°F) and 40(psia). Base your results on 1(lb mol). Compare answers obtained from the steam tables with those calculated, assuming steam to be an ideal gas. See Table 5-3 for C_P of steam.

9-24 Steam at 80(psia) and 400(°F) is expanded at constant enthalpy (as through a valve) to 14.7(psia).

(*a*) What is the temperature of the steam in the final state? If steam under these conditions were an ideal gas, what would be the final temperature?

(*b*) What is its specific volume in the final state?

(*c*) What is the entropy change of the steam?

9-25 A mass of saturated liquid water at a pressure of 1(bar) fills a container. The saturation temperature is 99.63(°C). Heat is added to the water until its temperature reaches 120(°C). If the volume of the container does not change, what is the final pressure? *Data*: The average value of β between 100 and 120(°C) is $80.8 \times 10^{-5}(\text{K})^{-1}$. The value of κ at 1(bar) and 120(°C) is $4.93 \times 10^{-5}(\text{bar})^{-1}$, and may be assumed independent of P. The volume of saturated liquid water at 1(bar) is $1.0432(\text{cm})^3/(\text{g})$.

9-26 Using steam-table data, estimate values for the residual properties $\Delta V'$, $\Delta H'$, and $\Delta S'$ of steam at 400(°F) and 200(psia).

9-27 A closed rigid vessel having a volume of $20(\text{ft})^3$ is filled with steam at 100(psia) and 600(°F). Heat is transferred from the steam until its temperature reaches 350(°F). Determine Q.

9-28 (*a*) One pound *mass* of steam is confined in a cylinder by a piston at 100(psia) and 500(°F). Calculate the work done by the steam if it is allowed to expand isothermally and reversibly to 35(psia). What is the heat absorbed by the steam from the surroundings during the process?
(*b*) Steam at 100(psia) and 500(°F) is allowed to expand adiabatically and reversibly until its pressure is 35(psia). Determine the final temperature of the steam. What is the work done by the steam?
(*c*) Explain the difference between the work values obtained in (*a*) and (*b*).

9-29 At 450(°F) a mixture of saturated steam and liquid water exists in equilibrium. If the specific volume of the mixture is 0.6673(ft)^3/(lb$_m$), calculate the following from data in the steam tables:

(*a*) Percent moisture.
(*b*) Enthalpy of the mixture, (Btu)/(lb$_m$).
(*c*) Entropy of the mixture, (Btu)/(lb$_m$)(R).

9-30 Steam at 50(psia) is known to have an enthalpy of 1038.0(Btu)/(lb$_m$). What is its temperature? What is its entropy? What is its internal energy?

9-31 A piston-and-cylinder engine operating in cycles and using steam as the working fluid executes the following steps:

(*a*) Steam at 75(psia) and 350(°F) is heated at constant volume to a pressure of 110(psia).
(*b*) The steam is then expanded reversibly and adiabatically back to the initial temperature of 350(°F).
(*c*) The steam is finally compressed isothermally and reversibly to the initial pressure of 75(psia). Determine the efficiency of the conversion of heat into work for this engine.

9-32 A tank having a volume of 100(ft)^3 contains 1(ft)^3 of liquid water and 99(ft)^3 of water vapor at standard atmospheric pressure. Heat is transferred to the contents of the tank until the liquid water has just evaporated. How much heat must be added?

9-33 A closed vessel of 10(ft)^3 capacity is filled with saturated steam at 250(psia). If 25 percent of the steam is subsequently condensed, how much heat must be removed and what is the final pressure?

9-34 Describe the characteristics and sketch the paths of the following processes on PV, ln P vs. H, TS, and HS diagrams:

(*a*) Continuous throttling of saturated liquid.
(*b*) Condensing superheated vapor to subcooled liquid.
(*c*) Adiabatic reversible expansion of saturated vapor.
(*d*) Constant-volume heating of saturated liquid.
(*e*) Isothermal expansion, starting at the critical point.

9-35 For a PVT system, show that if V is a function only of the ratio P/T, then U is a function only of T.

9-36 Derive the following equations:

(a) $$dU = \left[C_P - P\left(\frac{\partial V}{\partial T}\right)_P\right]dT - \left[T\left(\frac{\partial V}{\partial T}\right)_P + P\left(\frac{\partial V}{\partial P}\right)_T\right]dP.$$

(b) $$dU = (C_P - \beta PV)dT - (\beta T - \kappa P)\,V\,dP.$$

(c) $$dH = \left[C_V + V\left(\frac{\partial P}{\partial T}\right)_V\right]dT + \left[T\left(\frac{\partial P}{\partial T}\right)_V + V\left(\frac{\partial P}{\partial V}\right)_T\right]dV.$$

(d) $$dH = \left(C_V + \frac{\beta}{\kappa}V\right)dT + \frac{1}{\kappa}(\beta T - 1)\,dV.$$

9-37 A tank has a volume of 2(ft)³ and contains saturated water vapor at 20(psia). Heat transfer at constant volume reduces the temperature to 80(°F). Calculate

(a) The original steam temperature and the final pressure.
(b) The total mass of material in the tank.
(c) The volume and mass of *liquid* water in the tank at the final conditions.

9-38 Explain the principle of operation of a pressure cooker. What is the temperature of the food in a pressure cooker if the steam pressure is 30(psia)?

9-39 Estimate Z, $\Delta H'$, and $\Delta S'$ for steam at 300(°C) and 5(atm) from the following values for the second virial coefficient of water vapor.

t(°C)	B(cm)³/(mol)
290	−125
300	−119
310	−113

9-40 Steam undergoes a change of state from an initial state of 900(°F) and 500(psia) to a final state of 300(°F) and 40(psia). Determine ΔH and ΔS in units of (Btu), (R), and (lb mol) by use of (a) the steam tables, (b) the generalized correlation of Sec. 9-6 and the method illustrated in Example 9-12. The molar heat capacity of steam in the ideal-gas state is given by

$$C'_P = 7.105 + 0.001467T$$

where T is in (R) and C'_P is in (Btu)/(lb mol)(R).

9-41 Verify that an ideal-gas with constant-heat capacities obeys the following fundamental equations of state:

(a) $U = c_1 V^{-R/C_V} e^{S/C_V} + c_2.$
(b) $H = c_3 P^{R/C_P} e^{S/C_P} + c_4.$
(c) $A = (c_5 - R\ln V - C_V \ln T)T + c_6.$
(d) $G = (c_7 + R\ln P - C_P \ln T)T + c_8.$

In these equations, the c_i are constants.

9-42 (a) Show that if β is positive and $(\partial\kappa_S/\partial T)_P$ is negative, then the soundspeed c increases with increasing T at constant P.

(*b*) Show that if $(\partial \kappa_S/\partial P)_T$ is positive, then the soundspeed c decreases with increasing P at constant T.

9-43 Determine the *signs* of the following derivatives for a pure substance:

(*a*) $(\partial G/\partial P)_T$, (*b*) $(\partial^2 G/\partial P^2)_T$, (*c*) $(\partial^2 G/\partial T^2)_P$,
(*d*) $(\partial A/\partial V)_T$, (*e*) $(\partial^2 A/\partial V^2)_T$, (*f*) $(\partial^2 A/\partial T^2)_V$.

9-44 The composition dependence of the volumetric properties of a mixture is introduced into an equation of state by means of *mixing rules* for the equation-of-state parameters. For the virial equation in volume, Eq. (5-8), the mixing rules are known; that for the second virial coefficient B is

$$B = \sum_j \sum_k y_j y_k B_{jk}.$$

Here, the y's are mole fractions, and the sums are taken over all species present in the system. The coefficients B_{ii} with identical repeated subscripts are the second virial coefficients of pure species i; those with mixed subscripts are composition-independent mixture properties called *cross coefficients*. The coefficients B_{jk} have the symmetry property that $B_{jk} = B_{kj}$. Calculate the molar volume V of an equimolar mixture of methane(1), propane(2), and n-pentane(3) at 100(°C) and 1(atm). The following values are available for the B_{jk} [in $(\text{cm})^3/(\text{mol})$] at 100(°C):

$$B_{11} = -20, \qquad B_{22} = -241, \qquad B_{33} = -621,$$

$$B_{12} = -75, \qquad B_{13} = -122, \qquad B_{23} = -399.$$

9-45 The volumetric behavior of real fluids is sometimes approximately represented by the van der Waals equation of state:

$$P = \frac{RT}{V - b} - \frac{a}{V^2}. \tag{A}$$

Use of this equation requires the availability of numerical values for the constants a and b. One way to establish such values is to require that the equation satisfy the two mathematical constraints for the critical state given by Eqs. (5-1*a*) and (5-1*b*):

$$\left(\frac{\partial P}{\partial V}\right)_{T,\text{cr}} = \left(\frac{\partial^2 P}{\partial V^2}\right)_{T,\text{cr}} = 0. \tag{B}$$

(*a*) Show that, if Eq. (*A*) satisfies Eqs. (*B*), then a and b are related to the critical temperature T_c and the critical volume V_c by

$$a = \tfrac{9}{8} R T_c V_c \qquad \text{and} \qquad b = \tfrac{1}{3} V_c. \tag{C}$$

(*b*) Show that Eqs. (*A*) and (*C*) require the critical compressibility factor $Z_c \equiv P_c V_c / R T_c$ to be equal to $3/8$. How does this value compare with the experimental values of Z_c given in Appendix C?

(*c*) Combine Eq. (*A*) with the results of (*a*) and (*b*) and find the following reduced form of the van der Waals equation:

$$P_r = \frac{8T_r}{3V_r - 1} - \frac{3}{V_r^2}.$$

9-46 Calculate Z and V for isopropanol vapor at 200(°C) and 10(atm) by the following methods:

(*a*) Use the truncated virial equation, (5-11), with the experimentally determined virial coefficients
$B = -388(\text{cm})^3/(\text{mol})$ and $C = -26{,}000(\text{cm})^6/(\text{mol})^2$.

(*b*) Use the truncated virial equation, (5-10), employing the generalized correlation of Sec. 9-6 to estimate B.

(*c*) Use the van der Waals equation, estimating a and b from T_c and V_c (see Prob. 9-45).

The following physical constants are available for isopropanol: $T_c = 508.2(\text{K})$, $P_c = 50.0(\text{atm})$, $V_c = 220.4(\text{cm})^3/(\text{mol})$, $\omega = 0.700$.

9-47 One pound *mass* of steam initially at 460(°F) and 290(psia) is expanded in a piston-cylinder device in a reversible adiabatic process to a final pressure of 40(psia). Using the steam tables, determine the final state of the steam and the work of the process. If the steam in its final state is subsequently changed to saturated liquid in a constant-pressure reversible process, how much heat is transferred?

9-48 By assuming that vapor-liquid saturation pressures of normal fluids are described by the empirical equation

$$\log P(\text{sat}) = A + \frac{B}{T},$$

where $P(\text{sat})$ is in (atm) and T is in (K), derive *Edmister's formula* for estimation of the acentric factor:

$$\omega = \frac{3}{7}\left(\frac{\theta}{1-\theta}\right)\log P_c - 1.$$

In this equation, $\theta \equiv T_b/T_c$, where T_b is the normal boiling point [the vapor-liquid saturation temperature for which $P(\text{sat}) = 1(\text{atm})$].

9-49 One method for determination of the second virial coefficient B of pure materials involves the use of the Clapeyron equation with independently measured values of the heat of vaporization ΔH^{lv}, the molar volume $V(\text{sat})$ of the saturated liquid, and the vapor-liquid saturation pressure $P(\text{sat})$. Using the Clapeyron equation, estimate B for methyl ethyl ketone (MEK) at 75(°C). The following data are available for MEK:

(*a*) $\Delta H^{lv} = 7553(\text{cal})/(\text{mol})$ at 75(°C).

(*b*) $V(\text{sat}) = 96.491(\text{cm})^3/(\text{mol})$ for the liquid at 75(°C).

(*c*) $\ln P(\text{sat}) = 43.53916 - (5622.7/T) - 4.70504 \ln T$, where $P(\text{sat})$ is in (atm) and T is in (K).

9-50 Making use of the Pitzer correlation, determine estimates for

(*a*) Z, (*b*) $\Delta H'$, (*c*) $\Delta S'$

of furan vapor at 60(°C) and 1(atm). The following data are available for furan:

$$T_c = 487(\text{K}),$$

$$P_c = 52.5(\text{atm}).$$

Vapor-pressure data:

t(°C)	P(sat) in (mm Hg)
60	1,890
65	2,170
70	2,520

9-51 A 10(ft)^3 tank to contain propane has a bursting pressure of 400(psia). Safety considerations dictate that the tank be charged with no more propane than would exert a pressure half that of the bursting pressure at the temperature of 260(°F). How many (lb$_m$) of propane may be charged to the tank? Note that propane at these conditions cannot be considered an ideal gas.

9-52 The vapor pressure P(sat) of a given liquid is represented by an equation of the form:

$$\ln P(\text{sat}) = A - \frac{B}{T},$$

where A and B are constants, and T is absolute temperature. Show that for this material

$$\Delta S^{lv} = \Delta V^{lv}\left[\frac{BP(\text{sat})}{T^2}\right].$$

9-53 A tank contains exactly 1(lb$_m$) of H_2O consisting of liquid and vapor in equilibrium at 100(psia). If the liquid and the vapor each occupy one-half the volume of the tank, what is the enthalpy of the contents of the tank?

9-54 Ten pounds *mass* of steam initially at a pressure P_1 of 20(psia) and a temperature t_1 of 300(°F) is compressed isothermally and reversibly in a piston-cylinder device to a final pressure such that the steam is just saturated. Calculate Q and W for the process.

10

APPLICATIONS OF THERMODYNAMICS TO ENGINEERING SYSTEMS

10-1 Flow Processes

The development in Chap. 4 of the mathematical expression of the first law in the form

$$\Delta U = U_2 - U_1 = Q + W$$

was based on three related ideas: the existence of an internal-energy function, the principle of conservation of energy, and the thermodynamic definition of heat. Our applications of this equation have been to systems of constant mass which experience no changes in kinetic or gravitational potential energy. Moreover, we have dealt primarily with systems whose properties are uniform throughout at the initial and final conditions of the system.

The engineer must usually deal with much more complex situations. The vast majority of engineering processes involve the flow of fluids, and it is the purpose of this chapter to extend the applications of thermodynamics to such processes. We limit ourselves to systems in which flow is unidirectional and in which surface, electrical, and magnetic effects are negligible.

The general problem of fluid flow is extremely complicated. There may be present any or all of the following conditions:

1 Pressure, temperature, velocity, density, and other property variations in all directions in the fluid.
2 Changes in gravitational potential energy and in kinetic energy of the fluid as it flows from one point to another.
3 Turbulence and other dissipative effects, all encompassed under the general term fluid friction and arising because of the viscous nature of fluids.

We idealize our systems to the extent of imagining that, if we proceed from any given point in a fluid stream in a direction *perpendicular to the flow direction*, there will be no change of pressure, temperature, velocity, density, or other property. Thus the properties assigned at any cross section of a flowing stream are considered uniform over the stream at values representing the average values one would obtain by integration over the actual stream. We also make the basic assumption that the thermodynamic properties of a flowing stream are the same as for the fluid at rest at the same temperature, pressure, and density. These idealizations are necessary to make engineering problems tractable, and experience shows that the errors introduced are negligible for all practical purposes.

10-2 Energy Equations for Closed Systems

A closed system is one which may exchange energy as heat and work with its surroundings but which cannot exchange mass. Since matter does not flow across the boundaries of the system, the mass of the system is necessarily constant. The simple first-law equation

$$\Delta U = Q + W$$

applies to such a system. We wish now to generalize this equation so that it may be applied to systems consisting of several parts between which matter may flow. However, we still impose the restriction of a closed system, so that the total mass in all parts of the system taken together is constant. As we shall see, this restriction does not limit the kinds of processes that may be considered. Rather, it prescribes the proper choice of a system in any given application.

Systems of several parts may interact with the surroundings in such a way that more than one heat term and more than one work term must be included in the energy equation. We therefore replace Q and W by ΣQ and ΣW, where the summation signs merely indicate that all terms of the type considered must be included. These quantities then represent the *total* heat transfer to the system and the *total* work done on the system.

Since the system consists of several parts, we must sum the internal energy over all parts of the system in its original and final states. We then can determine the change in *total* internal energy of the system as the result of carrying out a process.

This internal-energy summation may be done in two ways:

1 Energy may be summed over the identifiable *regions of space* included in

the system. The resulting equation is

$$\Delta \sum_R (mU) = \Sigma Q + \Sigma W, \tag{10-1}$$

or equivalently,

$$\sum_R \Delta(mU) = \Sigma Q + \Sigma W. \tag{10-2}$$

It clearly does not matter whether we take the difference of sums or the sum of the differences. The sign $\sum_R$ indicates a sum over *regions of space.* The quantity (mU) gives the total internal energy of a region, and this may change both in the mass m and in the internal energy per unit mass U from beginning to end of the process.

2 Energy may be summed over the identifiable *mass elements* making up the system. In this case

$$\sum_m (m\,\Delta U) = \Sigma Q + \Sigma W. \tag{10-3}$$

The sign $\sum_m$ indicates a sum over the mass elements which make up the system.

It is very often useful to replace the internal energy U in these equations by the enthalpy H. There are two reasons for this. First, the resulting equations are usually simpler, and second, tables of properties (e.g., steam tables) rarely list values of U but always give complete data for H. By definition,

$$H = U + PV.$$

Therefore

$$\Delta \sum_R (mH) = \Delta \sum_R (mU) + \Delta \sum_R (mPV),$$

or

$$\sum_R \Delta(mH) = \sum_R \Delta(mU) + \sum_R \Delta(mPV),$$

and

$$\sum_m (m\,\Delta H) = \sum_m (m\,\Delta U) + \sum_m (m\,\Delta PV).$$

Combination of each of these expressions with the appropriate form of the energy equation, Eq. (10-1), (10-2), or (10-3), leads immediately to

$$\Delta \sum_R (mH) = \Sigma Q + \Sigma W + \Delta \sum_R (mPV), \tag{10-4}$$

$$\sum_R \Delta(mH) = \Sigma Q + \Sigma W + \sum_R \Delta(mPV), \tag{10-5}$$

and

$$\sum_m (m\,\Delta H) = \Sigma Q + \Sigma W + \sum_m (m\,\Delta PV). \tag{10-6}$$

Example 10-1

A crude method for determining the enthalpy of steam in a steam line is to bleed steam from the line through a hose into a barrel of water. The mass of the barrel with its contents and the temperature of the water are recorded at the beginning and end of the process. The condensation of steam by the water raises the temperature of the water.

Our energy equations are written for a system of constant mass. It is therefore essential to choose the system so that it includes not only the water originally in the barrel, but also the steam in the line that will be condensed by the water during the process. We may imagine a piston in the steam line that separates the steam in the line into two parts: that which will enter the barrel and that which will remain in the line. This view is represented schematically in Fig. 10-1.

In addition to the piston in the steam line, another piston is imagined to separate the water in the barrel from the atmosphere. Both the atmosphere and the steam in the line exert forces (shown as pressures, i.e., forces distributed over an area) which act on the system. These forces move during the process and therefore do work.

The pressure P exerted by the atmosphere on the water in the barrel is taken as constant, but the other properties of the water change during the process. Thus the mass changes from m_1 to m_2; the specific enthalpy, from H_1 to H_2; the specific volume, from V_1 to V_2; and the temperature, from T_1 to T_2. In the steam line, however, the intensive properties T', P', H', and V' remain constant. The mass of steam considered as part of the system goes from an initial value of m' to zero. From a mass balance,

$$m_2 - m_1 = m'.$$

The assumption of mechanical reversibility within the barrel and the steam line allows us to evaluate the work terms as follows (note that $P = P_1 = P_2$):

$$\begin{aligned} W_{\text{barrel}} &= -\int_{m_1V_1}^{m_2V_2} P\,d(mV) = -P\int_{m_1V_1}^{m_2V_2} d(mV) = -P(m_2V_2 - m_1V_1) \\ &= -m_2P_2V_2 + m_1P_1V_1, \end{aligned}$$

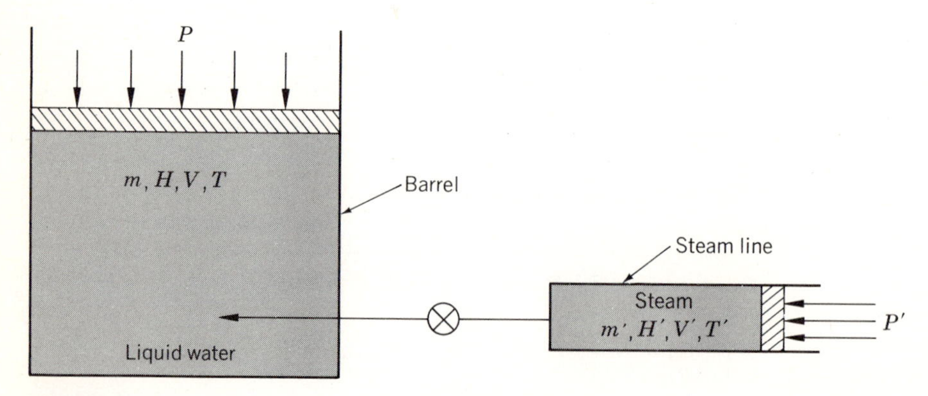

Fig. 10-1
Injection of steam into water.

and

$$W_{\text{line}} = -\int_{m'V'}^{0} P'\,d(m'V') = -P'\int_{m'V'}^{0} d(m'V') = P'm'V'$$
$$= m'P'V'.$$

Therefore $$\Sigma W = -m_2P_2V_2 + m_1P_1V_1 + m'P'V'.$$

If we assume that the steam line and the barrel are well insulated, we may neglect heat transfer and set $\Sigma Q = 0$.

Application of Eq. (10-4) requires the selection of regions of space over which to make the summations $\sum_R$. These are quite obviously the barrel and the steam line. Thus Eq. (10-4) becomes

$$\underbrace{\underbrace{m_2H_2}_{\sum_R \text{final}} - \underbrace{(m_1H_1 + m'H')}_{\sum_R \text{initial}}}_{\Delta \sum_R (mH)} = \underbrace{-m_2P_2V_2 + m_1P_1V_1 + m'P'V'}_{\Sigma W}$$
$$+ \underbrace{\underbrace{m_2P_2V_2}_{\sum_R \text{final}} - \underbrace{(m_1P_1V_1 + m'P'V')}_{\sum_R \text{initial}}}_{\Delta \sum_R (mPV)}.$$

Since the mPV terms all cancel, the final equation is

$$m_2H_2 = m_1H_1 + m'H'.$$

This equation shows that for this process the total enthalpy of the system at the end is the same as the total enthalpy of the system at the start.

Were we to choose to work with Eq. (10-5), we should have

$$\underbrace{\underbrace{(m_2H_2 - m_1H_1)}_{\Delta(mH)_{\text{barrel}}} + \underbrace{(0 - m'H')}_{\Delta(mH)_{\text{line}}}}_{\sum_R \Delta(mH)} = \underbrace{-m_2P_2V_2 + m_1P_1V_1 + m'P'V'}_{\Sigma W}$$
$$+ \underbrace{\underbrace{(m_2P_2V_2 - m_1P_1V_1)}_{\Delta(mPV)_{\text{barrel}}} + \underbrace{(0 - m'P'V')}_{\Delta(mPV)_{\text{line}}}}_{\sum_R \Delta(mPV)}.$$

Again the final equation is

$$m_2H_2 = m_1H_1 + m'H'.$$

Equations (10-4) and (10-5) are two equivalent versions of the same summation process, and there is nothing outside of personal preference to recommend the use of one over the other. Equation (10-6), however, makes use of a different idea, i.e., the summation over mass elements. The obviously identifiable mass elements here are the initial mass of water in the tank m_1 and the initial mass of steam in the line m'. We must now follow these mass elements through the process to see how their properties are altered. At the end of the process the mass m_1 is still liquid water in the barrel and it now has the properties H_2 and V_2. The mass m' has been transferred from the line to the barrel, where it also has the properties H_2 and V_2. With this analysis in mind we may write Eq. (10-6):

$$\underbrace{m_1(H_2 - H_1) + m'(H_2 - H')}_{\sum_m (m\Delta H)} = \underbrace{-m_2P_2V_2 + m_1P_1V_1 + m'P'V'}_{\Sigma W}$$

$$+ \underbrace{m_1(P_2V_2 - P_1V_1) + m'(P_2V_2 - P'V')}_{\sum_m (m\,\Delta PV)}.$$

Upon rearrangement this becomes

$$(m_1 + m')H_2 - m_1H_1 - m'H' = -m_2P_2V_2 + m_1P_1V_1 + m'P'V'$$
$$+ (m_1 + m')P_2V_2 - m_1P_1V_1 - m'P'V'.$$

Substitution of m_2 for $(m_1 + m')$ reduces this equation to the one obtained before:

$$m_2H_2 = m_1H_1 + m'H'.$$

In view of the similarity between the process considered here and the throttling process of Example 4-8, it is not surprising that for both processes the total final enthalpy of the system is unchanged from the total initial enthalpy.

Consider the following experimental results: Steam from a line at 100(psia) is bled into a barrel of water for which $m_1 = 300(\text{lb}_m)$ and $t_1 = 50(°\text{F})$. At the end of the process the measured values are $m_2 = 320(\text{lb}_m)$ and $t_2 = 120(°\text{F})$. Clearly, $m' = 20(\text{lb}_m)$. From the steam tables,

$$H_1 = 18.07(\text{Btu})/(\text{lb}_m),$$
$$H_2 = 87.91(\text{Btu})/(\text{lb}_m),$$

where the enthalpy of liquid water at atmospheric pressure is taken to be the same as *saturated* liquid water at the temperature considered. The error introduced is negligible, because the effect of pressure on liquid properties is slight except near the critical point. Solution for H' gives

$$H' = \frac{m_2H_2 - m_1H_1}{m'} = \frac{(320)(87.91) - (300)(18.07)}{20} = 1135.5(\text{Btu})/(\text{lb}_m).$$

Since saturated steam at 100(psia) has an enthalpy of 1187.3(Btu)/(lb$_m$), the steam in the line is wet. Its quality is easily calculated to be 0.9423, or 94.23 percent vapor.

One observation to be made about this problem is that, in the derivation of the energy equation which applies to it, the mPV terms arising from the substitution of $H - PV$ for U and the mPV terms representing work quantities all canceled. It is readily shown that this is true for any process of this type for which the pressure is constant *in each part* of the system. It is *not* necessary that the pressure be the *same* in all parts of the system. Where P is constant in each part of the system, i.e., in each region of space, the work terms will be of the form

$$W = -P\,\Delta(mV)$$

and

$$\Sigma W = -\sum_R [P\,\Delta(mV)].$$

Equation (10-5) then becomes, since $\Delta(mPV) = P\,\Delta(mV)$,

$$\sum_R \Delta(mH) = \Sigma Q - \sum_R [P\,\Delta(mV)] + \sum_R [P\,\Delta(mV)],$$

or

$$\sum_R \Delta(mH) = \Sigma Q. \tag{10-7}$$

This equation may also be written

$$\Delta \sum_R (mH) = \Sigma Q, \tag{10-8}$$

because taking the difference of the sums is identical with summing the differences. In either case the equation states that the total enthalpy change for the entire system is equal to the total heat transferred. One can equally well sum over the mass elements of the system and write

$$\sum_m (m\,\Delta H) = \Sigma Q. \tag{10-9}$$

This result is a generalization of one of the fundamental properties of the enthalpy function as illustrated in Example 4-5, where it was shown that the change in enthalpy during an isobaric process is equal to the heat that is transferred. It was implicit in that derivation that the pressure be not only constant during the process, but also uniform throughout the system. We now see that uniformity of pressure is not a necessary condition, provided that each part of the system is maintained at *constant* pressure and that the process is mechanically reversible within each part of the system.

For problems of the type considered in this section one may sum over regions of space or over the mass elements of the system. The same answer must always be obtained. The choice may be considered a matter of personal preference. However, in some problems the proper choice may greatly simplify the solution procedure.

Example 10-2

A tank of $50(\text{ft})^3$ capacity contains $1{,}000(\text{lb}_m)$ of liquid water in equilibrium with pure water vapor, which fills the rest of the tank, at a temperature of $212(°\text{F})$ and $1(\text{atm})$

absolute pressure. From a water line at slightly above atmospheric pressure, 1,500(lb_m) of water at 160(°F) is to be bled into the tank. How much heat must be transferred to the contents of the tank during this process if the temperature and pressure in the tank are not to change?

The system, which must be chosen to include the initial contents of the tank and the water to be added during the process, is shown in Fig. 10-2.

Since the tank at all times contains liquid and vapor in equilibrium at 212(°F) and 1(atm), the properties H_f and V_f for the liquid and H_g and V_g for the vapor in the tank are the same at the end as at the start of the process. The properties H' and V' of the liquid water in the line are also constant. As liquid is added to the tank, it displaces some of the vapor that is present initially, and this vapor must condense. Thus the tank at the end of the process contains as liquid all the 1,000(lb_m) of liquid initially present in the tank. This 1,000(lb_m) of liquid is present as saturated liquid at 212(°F) both at the beginning and end of the process, and hence does not change in properties. In addition, the 1,500(lb_m) of liquid added to the tank remains liquid, but it changes from liquid at 160(°F) to saturated liquid at 212(°F). Also, the vapor which condenses adds to the liquid in the tank at the end of the process. Let the amount of vapor which condenses be y(lb_m). This mass of material changes from saturated vapor at 212(°F) to saturated liquid at 212(°F) during the process. The part of the vapor which does not condense remains as saturated vapor at 212(°F) and therefore does not change in properties.

Since the pressure is constant in both the tank and the water line, Eqs. (10-7), (10-8), and (10-9) all apply. However, in the description above of the process, we have identified four masses which constitute the system and have described their changes during the process. This suggests that it may be advantageous to determine the total enthalpy change by summation over the changes which occur in these masses, as indicated by Eq. (10-9).

Thus we first note that two of the masses—the initial 1,000(lb_m) of liquid in the tank and the mass of vapor which does not condense—undergo no change in properties, and may be omitted from consideration. The other two masses are the 1,500(lb_m) of

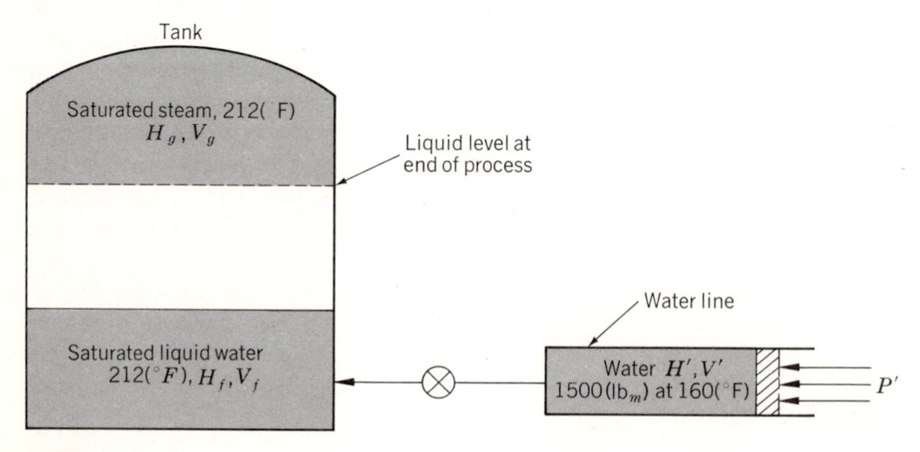

Fig. 10-2
Injection of water into a tank.

water added to the tank from the line and the $y(\mathrm{lb}_m)$ of vapor which condenses. The heat transfer is therefore equal to the enthalpy changes of these two masses:

$$\sum Q = 1{,}500(H_f - H') + y(H_f - H_g).$$

From the steam tables,

$$H_f = 180.07(\mathrm{Btu})/(\mathrm{lb}_m),$$

$$H' = 127.87(\mathrm{Btu})/(\mathrm{lb}_m),$$

$$H_g = 1150.4(\mathrm{Btu})/(\mathrm{lb}_m).$$

Thus

$$\Sigma Q = 1500(180.07 - 127.87) + y(180.07 - 1150.4)$$

$$= 78{,}300 - 970.3y.$$

We must now determine y. This is most easily done by noting that the sum of the volume changes of the mass elements of the system that we have already identified must be equal to the total volume change of the system. The total volume change of the system is just the volume of the $1{,}500(\mathrm{lb}_m)$ of water at $160(^\circ\mathrm{F})$ that is added to the tank. Since the total volume change is negative,

$$\Sigma(m\,\Delta V) = -1{,}500\,V'.$$

The same mass elements that change in enthalpy also change in volume. Thus

$$1{,}500(V_f - V') + y(V_f - V_g) = -1{,}500\,V',$$

or

$$1{,}500\,V_f + y(V_f - V_g) = 0.$$

From the steam tables,

$$V_f = 0.01672(\mathrm{ft}^3)/(\mathrm{lb}_m),$$

$$V_g = 26.83(\mathrm{ft}^3)/(\mathrm{lb}_m).$$

Therefore

$$(1{,}500)(0.01672) + y(0.01672 - 26.83) = 0.$$

Solution for y gives

$$y = 0.9355(\mathrm{lb}_m).$$

For the total heat transferred we have

$$\Sigma Q = 78{,}300 - (970.3)(0.9355)$$

$$= 77{,}390(\mathrm{Btu}).$$

Analysis of this problem through identification of mass elements has led to a very simple solution. If we had summed over regions, we should eventually have reached the same result, but the calculations would have been much more tedious.

Processes of the type considered so far are commonly called *unsteady-state-flow processes,* because of one's natural tendency to focus attention on the barrel or tank. By careful choice of the system for purposes of writing an energy equation, we have been able to treat these processes as though they were "nonflow." Indeed, no flow does occur at the start or end of the process, and therefore we need no kinetic-energy terms in our equations.

Moreover, we have ignored the possibility of changes in gravitational potential energy when material is transferred from one location to another. Terms to account for this are easily included, but are rarely needed.

10-3 Energy Equations for Steady-state-flow Processes

We wish now to consider the kind of process that is referred to as *steady-state flow*. Such processes are of primary importance in engineering because the mass production of materials and energy demands continuous operation of processes. The term steady-state-flow process implies the continuous flow of material through an apparatus. The inflow of mass is at all times exactly matched by the outflow of mass, so that there is no accumulation of material within the apparatus. Moreover, conditions at all points within the apparatus are steady or constant with time. Thus, at any point in the apparatus, the thermodynamic properties are constant; although they may vary from point to point, they do not change with time at a given point.

Consider the schematic diagram of Fig. 10-3, which shows a region of space bounded by the heavy curves and by the line segments 1-1, 2-2, and 3-3. This region is called the *control volume*. The material contained in the control volume is shown by the light shading. In addition, we recognize three other regions of space which communicate with the control volume: one to the left of 1-1, another to the right of 2-2, and a third to the right of 3-3. These we shall call regions 1, 2, and 3, respectively. For the process considered, regions 1 and 2 contain fluid which flows into the control

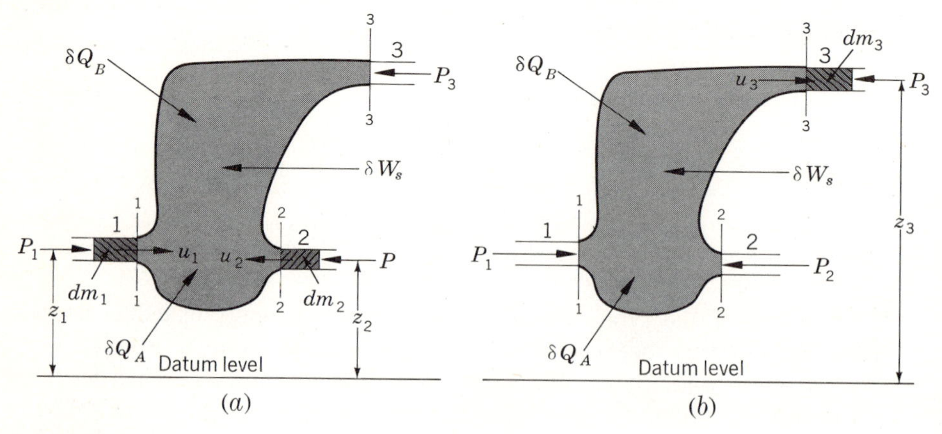

Fig. 10-3
Flow system at (a) time τ and (b) time $\tau + d\tau$.

volume, and region 3 collects fluid which flows from the control volume. There is no limitation on the number of inlets to and outlets from the control volume. We consider here a total of three by way of example.

Our *system* is chosen to include not only the fluid contained initially in the control volume, but in addition all fluid which will enter the control volume during some arbitrarily selected time interval. Thus the initial configuration of the system as shown in Fig. 10-3*a* includes the masses dm_1 and dm_2 which are to flow into the control volume during a differential time interval. The final configuration of the system after time $d\tau$ is shown in Fig. 10-3*b*. Here regions 1 and 2 no longer contain the masses dm_1 and dm_2; however, the mass dm_3 has collected in region 3. Since the process considered is one of steady-state flow, the mass contained in the control volume is constant, and

$$dm_3 = dm_1 + dm_2.$$

The masses in regions 1, 2, and 3 are considered to have the properties of the fluid as measured at 1-1, 2-2, and 3-3, respectively. These properties include a velocity u and an elevation above a datum plane z, as well as the thermodynamic properties. Thus these masses have kinetic and gravitational potential energy as well as internal energy, and an energy equation for steady-flow processes must contain terms for all these forms of energy.

The general form of the first law was developed in Chap. 4 and is given by Eq. (4-6):

$$\Delta U^t + \Delta E_K + \Delta E_P = Q + W,$$

where the superscript t has been added on U to indicate that the internal energy of the entire system must be taken into account. The kinetic energy and potential energy are given by Eqs. (4-7) and (4-8):

$$E_K = \frac{mu^2}{2g_c} \qquad \text{and} \qquad E_P = \frac{mzg}{g_c}.$$

Combining these equations and substituting mU for U^t, we get:

$$\Delta(mU) + \Delta\left(\frac{mu^2}{2g_c}\right) + \Delta\left(\frac{mzg}{g_c}\right) = Q + W.$$

For a system consisting of several parts, summation over the regions of space is required as before; thus

$$\Delta \sum_R (mU) + \Delta \sum_R \left(\frac{mu^2}{2g_c}\right) + \Delta \sum_R \left(\frac{mzg}{g_c}\right) = \sum Q + \sum W. \quad (10\text{-}10)$$

Equation (10-10) is directly applicable to the system shown in Fig. 10-3.

Note that the mass of this system is constant, consisting of the mass contained in the control volume and the mass which is to enter the system during the time interval considered. Equation (10-10) indicates a summation over regions of the system, and the major region is the control volume. However, the mass of fluid in the control volume and the properties of the fluid in the control volume do not change with time. Therefore terms to account for the energy of this region cancel in any energy equation, and may be omitted from the start. The regions that must be considered are those labeled 1, 2, and 3 in Fig. 10-3, and only differential masses leave or enter these regions during the time interval $d\tau$ required for the process that changes the system from the state shown in Fig. 10-3*a* to that of Fig. 10-3*b*. Applied to this process Eq. (10-10) becomes

$$\Delta \sum_R (U\,dm) + \Delta \sum_R \left(\frac{u^2\,dm}{2g_c}\right) + \Delta \sum_R \left(\frac{zg\,dm}{g_c}\right) = \sum \delta Q + \sum \delta W.$$

Written out, the internal-energy term becomes

$$\Delta \sum_R (U\,dm) = U_3\,dm_3 - U_1\,dm_1 - U_2\,dm_2.$$

This result can be expressed more simply as

$$\Delta \sum_R (U\,dm) = \sum (U\,dm)_{\text{out}} - \sum (U\,dm)_{\text{in}} = \Delta(U\,dm)_{\substack{\text{flowing}\\ \text{streams}}}.$$

The kinetic- and potential-energy terms may be expressed similarly. Thus the energy equation may be written

$$\Delta(U\,dm)_{\substack{\text{flowing}\\ \text{streams}}} + \Delta\left(\frac{u^2\,dm}{2g_c}\right)_{\substack{\text{flowing}\\ \text{streams}}} + \Delta\left(\frac{zg\,dm}{g_c}\right)_{\substack{\text{flowing}\\ \text{streams}}} = \sum \delta Q + \sum \delta W,$$

where Δ denotes the difference between streams flowing out and streams flowing in.

The $\Sigma\,\delta Q$ term includes all heat transferred to the system. For the example of Fig. 10-3 it would consist of $\delta Q_A + \delta Q_B$. The $\Sigma\,\delta W$ term includes work quantities of two types. The one shown in Fig. 10-3 and designated δW_s represents *shaft work*, i.e., work transmitted across the boundaries of the system by a rotating or reciprocating shaft. The other work quantities are of the kind already considered, which result from the action of external pressures in regions 1, 2, and 3 on the moving boundaries of the system. Thus, for the process of Fig. 10-3, if we assume mechanical reversibility in regions 1, 2, and 3,

$$\begin{aligned}\sum \delta W &= -P_3V_3\,dm_3 + P_1V_1\,dm_1 + P_2V_2\,dm_2 + \delta W_s \\ &= -\Delta(PV\,dm)_{\substack{\text{flowing}\\ \text{streams}}} + \delta W_s.\end{aligned}$$

Substitution in the energy equation and rearrangement gives

$$\Delta(U\,dm)_{\substack{\text{flowing}\\ \text{streams}}} + \Delta(PV\,dm)_{\substack{\text{flowing}\\ \text{streams}}} + \Delta\left(\frac{u^2\,dm}{2g_c}\right)_{\substack{\text{flowing}\\ \text{streams}}} + \Delta\left(\frac{zg\,dm}{g_c}\right)_{\substack{\text{flowing}\\ \text{streams}}} = \sum \delta Q + \delta W_s.$$

If dm is factored, this equation may be written more simply as

$$\Delta\left[\left(U + PV + \frac{u^2}{2g_c} + \frac{zg}{g_c}\right) dm\right]_{\substack{\text{flowing}\\ \text{streams}}} = \sum \delta Q + \delta W_s.$$

By definition, $H = U + PV$, and we have, finally,

$$\boxed{\Delta\left[\left(H + \frac{u^2}{2g_c} + \frac{zg}{g_c}\right) dm\right]_{\substack{\text{flowing}\\ \text{streams}}} = \sum \delta Q + \delta W_s.} \tag{10-11}$$

Equation (10-11) is the energy equation for a steady-flow process as written for an infinitesimal time interval $d\tau$. If we divide through by $d\tau$ and denote the various rates by

$$\dot{m} \equiv \frac{dm}{d\tau}, \qquad \dot{Q} \equiv \frac{\delta Q}{d\tau}, \qquad \text{and} \qquad \dot{W}_s \equiv \frac{\delta W_s}{d\tau},$$

we have

$$\boxed{\Delta\left[\left(H + \frac{u^2}{2g_c} + \frac{zg}{g_c}\right) \dot{m}\right]_{\substack{\text{flowing}\\ \text{streams}}} = \sum \dot{Q} + \dot{W}_s,} \tag{10-12}$$

which expresses the energy equation in terms of rates, all of which are constant in a steady-state-flow process.

One special case of Eq. (10-12) is very often encountered. If only a single stream enters and a single stream leaves the control volume, then $\dot{m}$ must be the same for both, and Eq. (10-12) becomes

$$\Delta\left(H + \frac{u^2}{2g_c} + \frac{zg}{g_c}\right)\dot{m} = \sum \dot{Q} + \dot{W}_s. \tag{10-13}$$

Division by $\dot{m}$ gives

$$\Delta\left(H + \frac{u^2}{2g_c} + \frac{zg}{g_c}\right) = \frac{\Sigma\dot{Q}}{\dot{m}} - \frac{\dot{W}_s}{\dot{m}} = \sum Q + W_s,$$

or

$$\boxed{\Delta H + \frac{\Delta u^2}{2g_c} + \Delta z\left(\frac{g}{g_c}\right) = \sum Q + W_s.} \tag{10-14}$$

where each term now refers to a *unit mass of fluid* passing through the control volume. All terms in this equation, as in all the energy equations of this chapter, must be expressed in the same energy units.

Example 10-3

Saturated steam at 50(psia) is to be mixed continuously with a stream of water at 60(°F) to produce hot water at 180(°F) at the rate of $500(\text{lb}_m)/(\text{min})$. The inlet and outlet lines to the mixing device all have an internal diameter of 2(in). At what rate must steam be supplied?

Equation (10-12) is appropriate for the solution of this example. Evidently, $\dot{W}_s = 0$. We shall assume that the apparatus is insulated so that, to a good approximation, $\Sigma\dot{Q} = 0$. Presumably, also, the device is small enough so that the inlet and outlet elevations are almost the same and the potential-energy terms may be neglected. The velocity of the outlet hot-water stream is given by $u = \dot{m}V/A$, where $\dot{m}$ is $500(\text{lb}_m)/(\text{min})$, V is the specific volume of water at 180(°F), and $A = (\pi/4)D^2$. From the steam tables $V = 0.01651(\text{ft})^3/(\text{lb}_m)$. Thus

$$u = \frac{500(\text{lb}_m)/(\text{min}) \times 0.01651(\text{ft})^3/(\text{lb}_m)}{\dfrac{\pi}{4} \times \left(\dfrac{2}{12}\right)^2(\text{ft})^2 \times 60(\text{s})/(\text{min})} = 6.3(\text{ft})/(\text{s})$$

From this we can calculate the kinetic energy of the outlet water stream:

$$\frac{\dot{m}u^2}{2g_c} = \frac{500(\text{lb}_m)/(\text{min}) \times 6.3^2(\text{ft})^2/(\text{s})^2}{2 \times 32.174(\text{lb}_m)(\text{ft})/(\text{lb}_f)(\text{s})^2}$$
$$= 308(\text{ft})(\text{lb}_f)/(\text{min}) \qquad \text{or} \qquad 0.396(\text{Btu})/(\text{min}).$$

This is an entirely negligible energy quantity compared with the energy changes being considered. [Note that the process raises the temperature of nearly $500(\text{lb}_m)$ of water per minute from 60 to 180(°F). This requires about 60,000(Btu).] We may therefore neglect the kinetic-energy terms for both water streams. Since we do not yet know the flow rate of the steam, we shall for the present neglect this kinetic-energy term also. We can later see if this is justified. Equation (10-12) therefore reduces to

$$H_3\dot{m}_3 - H_2\dot{m}_2 - H_1\dot{m}_1 = 0,$$

where the subscript 3 refers to the outlet stream; 2, to the inlet water; and 1, to the inlet steam.

From the steam tables:

$$H_1 = 1174.0(\text{Btu})/(\text{lb}_m), \qquad H_2 = 28.07(\text{Btu})/(\text{lb}_m), \qquad H_3 = 147.91(\text{Btu})/(\text{lb}_m).$$

Also $\qquad \dot{m}_3 = 500(\text{lb}_m)/(\text{min}), \qquad \dot{m}_2 = \dot{m}_3 - \dot{m}_1 = 500 - \dot{m}_1.$

Substitution in the energy equation gives

$$(147.91)(500) - (28.07)(500 - \dot{m}_1) - 1174.0\dot{m}_1 = 0 \qquad \text{or} \qquad \dot{m}_1 = 52.29(\text{lb}_m)/(\text{min}).$$

The velocity of this stream is

$$u_1 = \frac{\dot{m}_1 V_1}{(\pi/4)D^2} = \frac{52.29(\text{lb}_m)/(\text{min}) \times 8.522(\text{ft})^3/(\text{lb}_m)}{\dfrac{\pi}{4} \times \left(\dfrac{2}{12}\right)^2(\text{ft})^2 \times 60(\text{s})/(\text{min})} = 340(\text{ft})/(\text{s}).$$

Its kinetic energy is

$$\frac{\dot{m}_1 u_1^2}{2g_c} = \frac{52.29(\text{lb}_m)/(\text{min}) \times 340^2(\text{ft})^2/(\text{s})^2}{2 \times 32.174(\text{lb}_m)(\text{ft})/(\text{lb}_f)(\text{s})^2} = 93{,}940(\text{ft})(\text{lb}_f)/(\text{min}),$$

or 120.8(Btu)/(min). We may now rewrite the energy balance to include the kinetic-energy term for the steam flow:

$$(147.91)(500) - (28.07)(500 - \dot{m}_1) - 1174.0\dot{m}_1 - 120.8 = 0,$$

$$\dot{m}_1 = 52.18(\text{lb}_m)/(\text{min}).$$

The inclusion of the kinetic-energy term results in less than a 0.2 percent change in the answer.

In the preceding example, as in many others, the kinetic-energy term is not significant, even though a stream of rather high velocity is considered. Of course, in cases where kinetic-energy changes are a primary object or result of a process, the kinetic-energy terms are important. For example, the acceleration of an air stream to a high velocity in a wind tunnel requires a considerable work expenditure to produce the kinetic-energy change. For the calculation of this work, one could hardly omit the kinetic-energy terms from the energy equation. The same general comments apply to gravitational potential-energy effects. They are often quite negligible. However, the work generated in a hydroelectric power plant depends directly on the change in elevation of the water flowing through the plant, and the potential-energy terms are of major importance in the energy equation.

10-4 General Energy Equations

In deriving the energy equation for steady-state-flow processes, we found it useful to introduce the concept of a control volume. The resulting energy equation is seen to connect the properties of the streams flowing into and out of the control volume with the heat and work quantities crossing the boundaries of the control volume. We can extend the use of this concept to include unsteady-state-flow processes. The control volume is still a bounded region of space; however, the boundaries may be flexible to allow for expansion or contraction of the control volume. Furthermore, the mass contained in the control volume need no longer be constant. We deal now with transient conditions, where rates and properties vary with time. During an infinitesimal time interval $d\tau$, the mass entering the control volume may be different from the mass leaving the control volume. The difference must clearly be accounted for by the accumulation or depletion of mass within the control volume. Similarly, the difference between the transport of energy out of the control volume and into it by flowing streams need no longer be accounted for solely by the heat and work terms. Energy may be accumulated or depleted within the control volume. To state the situation precisely, the energy crossing the boundaries of the control volume as heat and work, $\Sigma\,\delta Q + \Sigma\,\delta W$, must equal the change in energy of the material contained within the control volume itself plus the net energy transport of the flowing streams. The energy change of the material in the control volume is $d(mU)_{\text{control volume}}$, and the net energy transport of the flowing streams as shown by Eq. (10-11) is

$$\Delta\left[\left(H + \frac{u^2}{2g_c} + \frac{zg}{g_c}\right) dm\right]_{\substack{\text{flowing}\\\text{streams}}}$$

The energy equation therefore becomes

$$\boxed{d(mU)_{\substack{\text{control}\\\text{volume}}} + \Delta\left[\left(H + \frac{u^2}{2g_c} + \frac{zg}{g_c}\right) dm\right]_{\substack{\text{flowing}\\\text{streams}}} = \sum \delta Q + \sum \delta W.} \tag{10-15}$$

The work term $\Sigma\,\delta W$ may include shaft work, but it may also include a term for work resulting from the expansion or contraction of the control volume itself. This equation presumes the control volume to be at rest and requires that z be measured from a datum level through the center of mass of the control volume.

Equation (10-15) is the most general expression for an energy equation that we shall attempt. It has inherent limitations. For example, it assumes no changes in the kinetic and potential energies of the control volume. For the vast majority of applications the control volume may be assumed to change just in internal energy. It is completely impractical to try to write a single energy equation that can be used for all applications. The only suitable guide where complexities arise is strict adherence to the law of conservation of energy, which is the basis for all energy equations. It might be remarked that Eq. (10-15) reduces to (10-11) for steady-state-flow processes. [The first term of (10-15) is zero, and $\Sigma\,\delta W$ becomes δW_s.] For a nonflow process, where the control volume contains the entire system, $dm = 0$, and (10-15) reduces to

$$d(mU) = \sum \delta Q + \sum \delta W.$$

If the system contains several regions, we need to sum the internal-energy term over the regions:

$$\sum_R d(mU) = \sum \delta Q + \sum \delta W,$$

which is the differential form of Eq. (10-2).

Example 10-4

Rework Example 10-2 by application of Eq. (10-15).

We take the tank as our control volume. There is no shaft work, and no expansion work. Therefore, $\Sigma\,\delta W = 0$. There is but one flowing stream, and it flows *into* the control volume. Therefore, the term in (10-15) accounting for the energy of the flowing streams reduces to

$$\Delta\left[\left(H + \frac{u^2}{2g_c} + \frac{zg}{g_c}\right) dm\right]_{\substack{\text{flowing}\\ \text{streams}}} = -\left(H' + \frac{u'^2}{2g_c} + \frac{z'g}{g_c}\right) dm'.$$

However, z' can be taken equal to zero, because the water line and the tank are at essentially the same level. We have no information on u', and can only presume it is small enough so that any contribution from a kinetic-energy term is negligible. This assumption is equivalent to the tacit assumption made in Example 10-2 that the process within the water line is mechanically reversible. This assumption is implicit in the use of (10-9) or any equivalent equation. Equation (10-15) therefore becomes

$$\sum \delta Q = d(mU)_{\text{tank}} - H'\,dm'.$$

We must now integrate this equation over the entire process. Since H' is constant, the result is

$$\sum Q = \Delta(mU)_{\text{tank}} - m'H'.$$

By the definition of enthalpy,

$$\Delta(mU)_{\text{tank}} = \Delta(mH)_{\text{tank}} - \Delta(PmV)_{\text{tank}}.$$

However, both the volume of the tank $(mV)_{\text{tank}}$ and the pressure in the tank are constant. Therefore $\Delta(PmV)_{\text{tank}} = 0$. As a result,

$$\sum Q = \Delta(mH)_{\text{tank}} - m'H'.$$

This equation expresses the fact that for this process the heat transferred is equal to the total enthalpy change caused by the process. This is the same idea upon which we based our earlier solution to this problem.

Clearly, energy equations may be developed from more than one point of view. Different applications yield most readily to different approaches. No formula can substitute for a thorough understanding of the meaning of the law of conservation of energy.

Equation (10-15) may be written in terms of rates by dividing through by $d\tau$:

$$\frac{d(mU)_{\substack{\text{control}\\ \text{volume}}}}{d\tau} + \Delta\left[\left(H + \frac{u^2}{2g_c} + \frac{zg}{g_c}\right)\dot{m}\right]_{\substack{\text{flowing}\\ \text{streams}}} = \sum \dot{Q} + \sum \dot{W}.$$

This equation applies at any instant during processes where conditions and flow rates change continuously. Other forms are, of course, also possible. For example, multiplication by $d\tau$ and integration over the time interval from zero to τ gives

$$\Delta(mU)_{\substack{\text{control}\\\text{volume}}} + \int_0^\tau \Delta\left[\left(H + \frac{u^2}{2g_c} + \frac{zg}{g_c}\right)\dot{m}\right]_{\substack{\text{flowing}\\\text{streams}}} d\tau = \sum Q + \sum W.$$

The integral can be evaluated only if the various quantities of the integrand are known as functions of time. This requires detailed information describing the process as a function of time.

10-5 The Throttling Process (Joule-Kelvin Effect)

The throttling process was discussed briefly in Example 4-8, and in Example 10-1 we made reference to such a process. We shall now review the basic features of a throttling process and then develop the thermodynamic equations which apply to it.

A throttling process is said to occur when fluid flowing steadily in a pipe passes an obstruction, such as a porous plug or a partly closed valve, which causes a discrete pressure drop in the flowing fluid in the absence of heat transfer and without appreciable kinetic-energy change of the fluid. Clearly, no shaft work is accomplished. For such a steady-flow process Eq. (10-14) reduces to

$$\Delta H = H_2 - H_1 = 0.$$

Thus, in a throttling process, the downstream enthalpy is the same as the upstream enthalpy. It should be noted that the downstream station must be far enough removed from the obstruction so as to be located beyond any local high-velocity jet immediately adjacent to the obstruction.

In addition to its engineering applications, the throttling process may be used to provide experimental information on the properties of fluids. This was in fact the object of Joule and Kelvin in their original experiments on throttling processes. In their work a cotton plug was used to provide an obstruction to flow, and gas passed through it parallel to the axis of the pipe. In modern measurements a cup of a strong porous material capable of withstanding a large force allows the gas to seep through in a radial direction. Rigid precautions are taken to provide adequate thermal insulation for the plug and the portion of the pipe near the plug. Suitable manometers and thermometers are used to measure the pressure and temperature of the gas on both sides of the plug.

The experiment is performed in the following way: The pressure and temperature on the high-pressure side of the plug P_i and T_i are chosen arbitrarily. The pressure on the other side of the plug P_f is then set at any value less than P_i, and the temperature of the gas T_f is measured. P_i and

T_i are kept the same, and P_f is changed to another value, and the corresponding T_f is measured. This is done for a number of different values of P_f, the corresponding T_f being measured in each case. P_f is the independent variable of the experiment, and T_f, the dependent variable. The results provide a set of discrete points on a TP diagram, one point being P_iT_i and the others being the various P_f's and T_f's indicated in Fig. 10-4 by numbers (1) to (7). Although the points shown in the figure do not refer to any particular gas, they are typical of most gases. It can be seen that, if a throttling process takes place between the states P_iT_i and P_fT_f (3), there is a rise of temperature. Between P_iT_i and P_fT_f (7), however, there is a drop of temperature. In general, the temperature change of a gas upon seeping through a porous plug depends on the three quantities P_i, T_i, and P_f, and may be an increase or a decrease, or there may be no change whatever.

The eight points plotted in Fig. 10-4 represent equilibrium states of some constant mass of the gas, say, $1(\text{lb}_m)$, at which the gas has the same enthalpy. All equilibrium states of the gas corresponding to this enthalpy must lie on some curve, and it is reasonable to assume that this curve can be obtained by drawing a smooth curve through the discrete points. Such a curve is called an *isenthalpic curve*. It must be understood that *an isenthalpic curve is not the graph of a throttling process.* No such graph can be drawn because in any throttling process the intermediate states traversed by a gas cannot be described by means of thermodynamic coordinates. An isenthalpic curve is the locus of all points representing equilibrium states of the same enthalpy. The porous-plug experiment is performed to provide a few of these points, the rest being obtained by interpolation.

The temperature on the high-pressure side T_i is now changed to another value, P_i being kept the same. P_f is again varied, and the corresponding T_f's measured. Upon plotting the new P_iT_i and the new P_f's and T_f's,

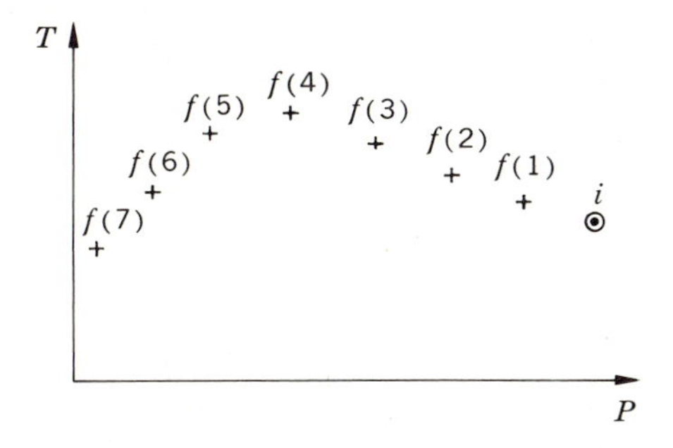

Fig. 10-4
Isenthalpic states of a gas.

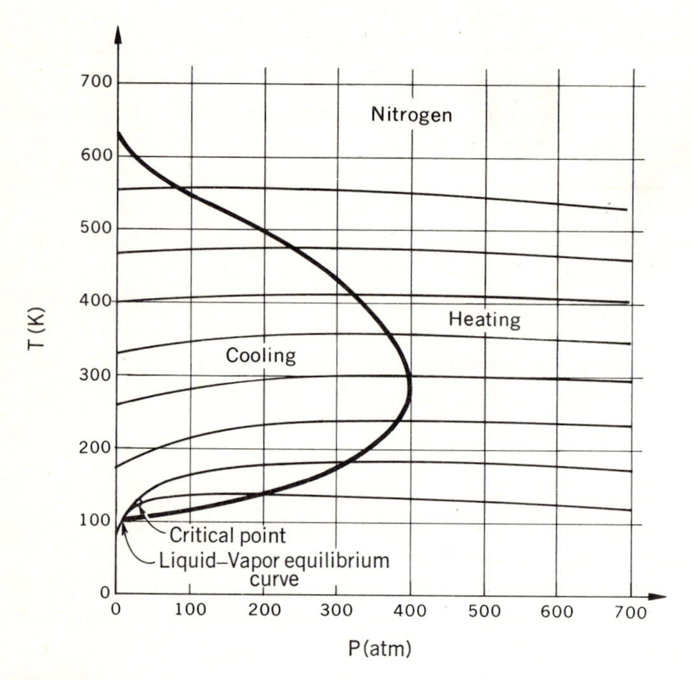

Fig. 10-5

Isenthalps and inversion curve for nitrogen. (From TS diagram prepared by F. Din, 1958.)

another discrete set of points is obtained, which determines another isenthalpic curve corresponding to a different enthalpy. In this way, a series of isenthalpic curves is obtained. Such a series is shown in Fig. 10-5 for nitrogen.

The numerical value of the slope of an isenthalpic curve on a TP diagram at any point is called the *Joule-Kelvin coefficient,* and will be denoted by μ. Thus

$$\mu \equiv \left(\frac{\partial T}{\partial P}\right)_H. \tag{10-16}$$

The locus of all points at which the Joule-Kelvin coefficient is zero, i.e., the locus of the maxima of the isenthalpic curves, is known as the *inversion curve* and is shown in Fig. 10-5 as a heavy curved line. The region inside the inversion curve where μ is positive is called the region of cooling, whereas outside, where μ is negative, is the region of heating.

In general, the difference in enthalpy between two neighboring equilibrium states is given by Eq. (9-33):

$$dH = C_P\,dT - \left[T\left(\frac{\partial V}{\partial T}\right)_P - V\right] dP,$$

or

$$dT = \frac{1}{C_P}\left[T\left(\frac{\partial V}{\partial T}\right)_P - V\right] dP + \frac{1}{C_P}\,dH.$$

Regarding T as a function of P and H,

$$dT = \left(\frac{\partial T}{\partial P}\right)_H dP + \left(\frac{\partial T}{\partial H}\right)_P dH;$$

whence, since $\mu = (\partial T/\partial P)_H$,

$$\boxed{\mu = \frac{1}{C_P}\left[T\left(\frac{\partial V}{\partial T}\right)_P - V\right].} \tag{10-17}$$

This is the thermodynamic equation for the Joule-Kelvin coefficient. It is evident that, for an ideal gas,

$$\mu = \frac{1}{C_P}\left(T\frac{R}{P} - V\right) = 0.$$

For an ideal gas a graph such as Fig. 10-5 would show horizontal isenthalps.

The most important application of the Joule-Kelvin effect is in the liquefaction of gases.

10-6 Liquefaction of Gases by the Joule-Kelvin Effect

An inspection of the isenthalpic curves and the inversion curve of Fig. 10-5 shows that, for the Joule-Kelvin effect to give rise to cooling, the initial temperature of the gas must be below the point where the inversion curve cuts the temperature axis, i.e., below the maximum inversion temperature. For many gases, room temperature is already below the maximum inversion temperature, so that no precooling is necessary. Thus, if air is compressed to a pressure of 200(atm) and a temperature of 52(°C), then, after throttling to a pressure of 1(atm), it will be cooled to 23(°C). On the other hand, if helium, originally at 200(atm) and 52(°C), is throttled to 1(atm), its temperature will rise to 64(°C).

Figure 10-6 shows that, for the Joule-Kelvin effect to produce cooling in hydrogen, the hydrogen must be cooled below 200(K). Liquid nitrogen is used in most laboratories for this purpose. To produce Joule-Kelvin cooling

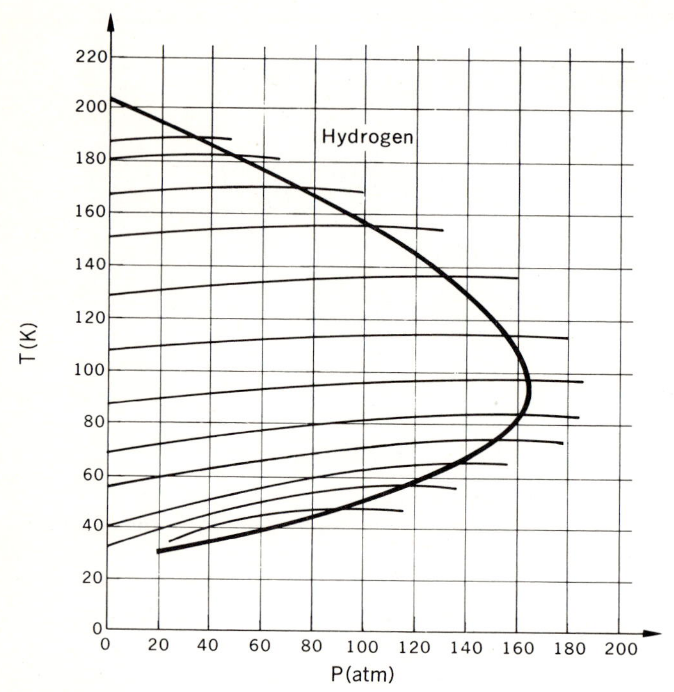

Fig. 10-6
Isenthalps and inversion curve for hydrogen. (Woolley, Scott, and Brickwedde, 1948.)

in helium, the helium is first cooled with the aid of liquid hydrogen. Table 10-1 gives the maximum inversion temperatures of a few gases commonly used in low-temperature work.

It is clear from Fig. 10-5 that, once a gas has been precooled to a temperature lower than the maximum inversion temperature, the optimum pressure from which to start throttling corresponds to a point on the inversion curve. Starting at this pressure and ending at atmospheric pressure, the largest temperature drop is produced. This, however, is not large enough to produce liquefaction. Consequently, the gas that has been cooled by throttling is used to cool the incoming gas, which, after throttling, becomes still cooler. After many repetitions of these successive coolings, the gas is lowered to such a temperature that, after throttling, it becomes partly liquefied. The apparatus used for this purpose is shown schematically in Fig. 10-7. The critical device is a *countercurrent heat exchanger,* shown in more detail in Fig. 10-8.

The gas, after precooling, is sent through the middle tube of a long coil of double-walled pipe. After throttling, it flows back through the outer annular

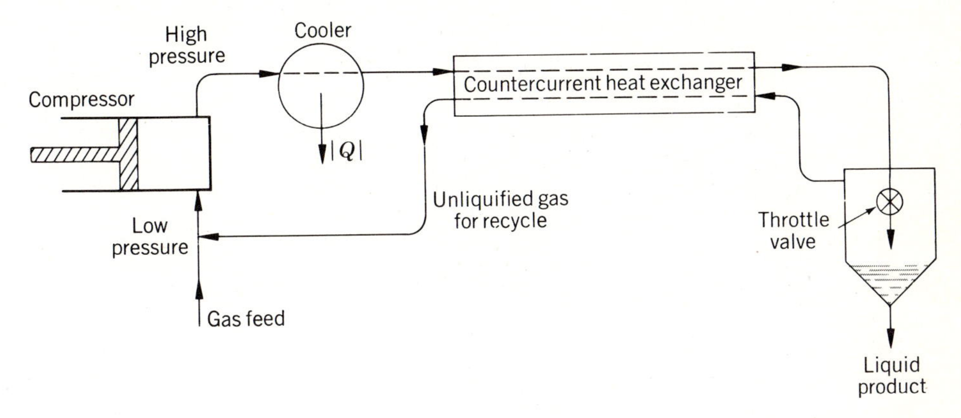

Fig. 10-7
Apparatus for gas liquefaction by means of Joule-Kelvin effect.

space surrounding the middle pipe. For the heat exchanger to be efficient, the temperature of the gas as it leaves must differ only slightly from the temperature at which it entered. To accomplish this, the heat exchanger must be quite long and well insulated.

When steady-state operation is finally reached, liquid is formed at a constant rate. For every pound of gas supplied, a certain fraction y is liquefied; the remaining fraction $1 - y$ is returned to the compressor. Considering the heat exchanger and throttling valve as completely insulated, as shown in Fig. 10-8, we have a flow process in which no shaft work is done and no heat is transferred, the kinetic- and potential-energy terms being negligible. It therefore follows that the enthalpy of the entering

Table 10-1 Maximum Inversion Temperatures

Gas	Maximum inversion temperature (K)
Carbon dioxide	≈1500
Argon	780
Oxygen	764
Air	659
Nitrogen	621
Neon	231
Hydrogen	202
Helium	≈40

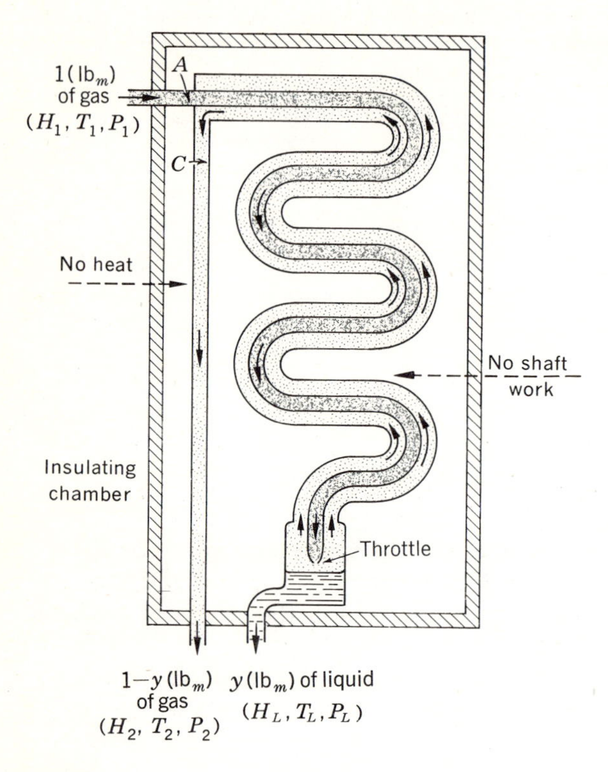

Fig. 10-8
Throttle valve and heat exchanger in steady state.

gas is equal to the enthalpy of $y(\text{lb}_m)$ of emerging liquid plus the enthalpy of $(1 - y)(\text{lb}_m)$ of emerging gas. If

H_1 = enthalpy of entering gas at T_1, P_1,

H_L = enthalpy of emerging liquid at T_L, P_L,

H_2 = enthalpy of emerging gas at T_2, P_2,

then
$$H_1 = yH_L + (1 - y)H_2,$$

or
$$y = \frac{H_2 - H_1}{H_2 - H_L}. \tag{10-18}$$

Now, for steady-state operation, H_L is determined by the pressure on the liquid, which fixes the temperature, and hence is constant. H_2 is determined by the pressure drop in the return tube and the temperature at C, which is only a little below that at A. Hence it remains constant. H_1 refers to a temperature T_1 that is fixed, but at a pressure that may be chosen at will. Therefore the fraction liquefied y may be varied only by varying H_1, which depends on P_1. Since

$$y = \frac{H_2 - H_1}{H_2 - H_L},$$

y will be a maximum when H_1 is a minimum; and since H_1 may be varied only by varying the pressure, the condition that it be a minimum is that

$$\left(\frac{\partial H_1}{\partial P}\right)_{T=T_1} = 0.$$

But
$$\left(\frac{\partial H}{\partial P}\right)_T = -\left(\frac{\partial H}{\partial T}\right)_P \left(\frac{\partial T}{\partial P}\right)_H = -C_P\mu.$$

Hence, for y to be a maximum,

$$\mu = 0 \qquad \text{at} \qquad T = T_1,$$

or the point (T_1,P_1) must lie on the inversion curve.

Example 10-5

The liquefaction of natural gas is carried out on a large scale. Since natural gas is largely methane, we will for simplicity assume it to be pure methane, to be liquefied in an apparatus such as that shown in Fig. 10-7.

The methane enters the apparatus at 1(atm) and 80(°F), is compressed to 1,000(psia), and is cooled again to 80(°F). It then enters the heat exchanger section of the apparatus as shown in Fig. 10-8, where saturated liquid methane at 1(atm) is withdrawn as product. Unliquefied methane returns through the heat exchanger, leaving the exchanger at 1(atm) and 75(°F). What fraction y of the methane is liquefied in the process? Data for methane are as follows:

Superheated methane vapor at 80(°F), 1,000(psia):

$$H_1 = 381.6(\text{Btu})/(\text{lb}_m)$$

Superheated methane vapor at 75(°F), 1(atm):

$$H_2 = 407.7(\text{Btu})/(\text{lb}_m)$$

Saturated liquid methane at -258.6(°F), 1(atm):

$$H_L = 17.7(\text{Btu})/(\text{lb}_m)$$

Eq. (10-18) is directly applicable:

$$y = \frac{H_2 - H_1}{H_2 - H_L} = \frac{407.7 - 381.6}{407.7 - 17.7} = 0.067$$

Thus only 6.7 percent of the methane is liquefied during a single pass through the apparatus. The unliquefied methane is returned to the compressor to be recycled.

10-7 Adiabatic Flow Processes

The flow of real fluids is an inherently irreversible process because of the viscous dissipation effects which inevitably accompany flow; such processes always cause an increase in the total entropy. However, the degree of irreversibility depends on the nature of the process. Consider the *adiabatic* processes represented on the sketch of a Mollier diagram in Fig. 10-9. Assume that point 1 represents the initial state of the fluid for all processes to be considered. From this point the fluid may be compressed, as indicated by the series of dots labeled *a*. This series of dots is not intended to represent the *path* of the process, for it is irreversible, and the fluid passes through nonequilibrium states, which cannot be represented on a thermodynamic diagram. The series of dots is merely a schematic indication of the changes that occur between the initial and final states of the fluid.

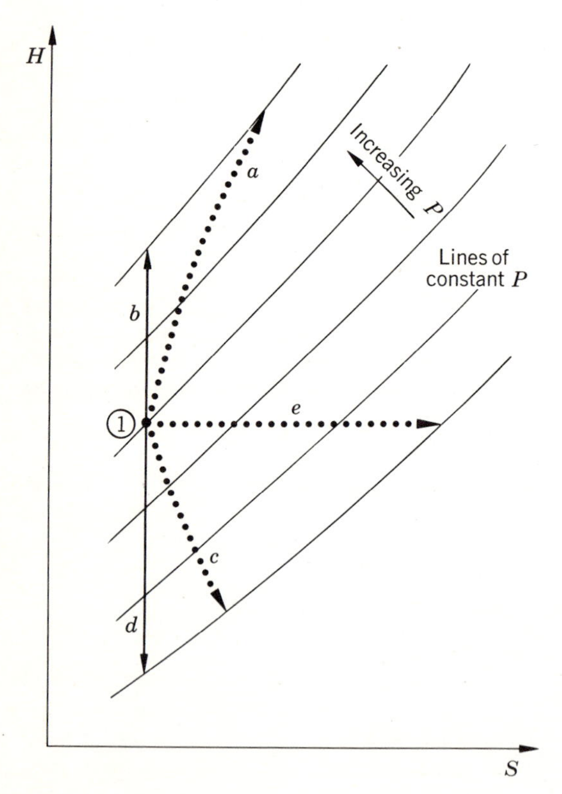

Fig. 10-9
Adiabatic flow processes on a Mollier diagram.

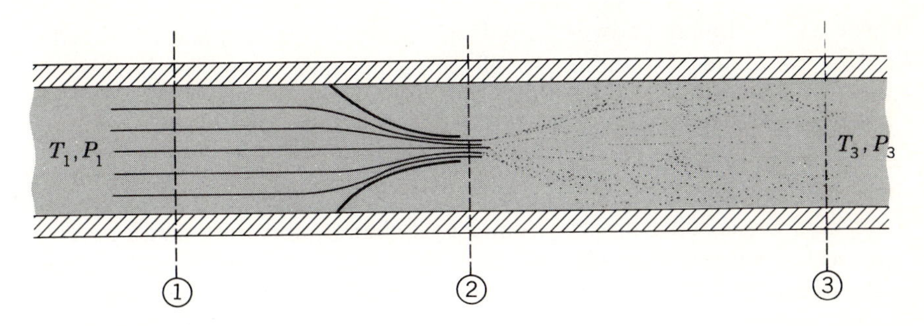

Fig. 10-10
Flow through a nozzle.

The entropy increase caused by the compression process *a* depends on the design of the compressor, i.e., on its efficiency. Although this efficiency cannot in practice approach 100 percent, we can surely *imagine* a compressor whose efficiency approaches this limit and which results in a reversible adiabatic compression, as indicated by the vertical path *b*. No adiabatic compression process can lead to states lying to the left of this line, for this would require a decrease in the entropy of the fluid, contrary to the requirements of the second law.

The fluid at point 1 may also be expanded to lower pressures, either through an expansion engine (e.g., a turbine) that produces work or through a nozzle to produce a high-velocity jet. In either case the process is represented in Fig. 10-9 by *c*. The entropy increase caused by the process depends on the design of the apparatus, on its efficiency. Again, we can *imagine* a device of 100 percent efficiency which causes no entropy change. The path for such a reversible adiabatic expansion is shown by the vertical line *d*. No adiabatic expansion process can lead to states to the left of this line, for this would result in an entropy decrease.

If the expansion device were so designed as to produce no work and no appreciable velocity change, it would merely act as a valve, and a throttling process would result. Such a process occurs with no change in enthalpy, represented in Fig. 10-9 by *e*. It is inevitably irreversible, always causing an entropy increase.

The difference between an expansion process in a nozzle and a throttling process can be explained with the aid of Fig. 10-10, which shows a nozzle with reference points placed upstream from the nozzle and at the nozzle exit. A third reference point is located well downstream from the nozzle. The discharge stream, with its high kinetic energy, can be used for the production of work; however, if no work-producing device is present and if the jet from the nozzle merely discharges into the downstream pipe, then it generates turbulence downstream from the nozzle. In this case the flowing fluid swirls about, expanding to fill the pipe, and dissipates its kinetic energy as a result of the damping effects of viscosity. Thus kinetic energy is reconverted into internal energy and the fluid passes station 3 with a low velocity; and the overall process from 1 to 3 is for all practical purposes a throttling process. Although the change from 1 to 2 can be made nearly reversible, the ensuing process from 2 to 3 cannot, and the overall process must result in an entropy increase.

Equation (10-14) is applicable to all the steady-flow processes considered in this section. Since all are also adiabatic, $Q = 0$. In addition, none of the devices treated (compressors, turbines, nozzles, and throttle valves) causes any appreciable change in elevation of the flowing stream. We may therefore omit the potential-energy term. Thus for the processes considered, Eq. (10-14) reduces to

$$\Delta H + \frac{\Delta u^2}{2g_c} = W_s \qquad \text{(adiabatic flow)}. \tag{10-19}$$

If our device is a nozzle, no shaft work is done, and this equation becomes

$$\Delta H + \frac{\Delta u^2}{2g_c} = 0$$

or

$$\frac{u_2^2 - u_1^2}{2g_c} = -(H_2 - H_1). \tag{10-20}$$

This equation expresses the fact that the velocity change caused by flow through an adiabatic nozzle is directly related to the enthalpy change for the process. Presuming we know the upstream conditions P_1, T_1, and u_1, as well as the exit pressure P_2, we would like to determine the conditions T_2 and u_2. In general, however, we need T_2 to find H_2, and H_2 to calculate u_2. Thus the energy equation alone does not suffice for the calculation we wish to make. If on the other hand we consider the limiting case of reversible flow or expansion of fluid through the nozzle, we can write an additional equation, namely $S_2 = S_1$, because for a reversible adiabatic process there can be no change in total entropy. This condition may be imposed on the energy equation by the further qualification

$$\frac{u_2^2 - u_1^2}{2g_c} = -(H_2 - H_1)_S, \tag{10-21}$$

which shows that the enthalpy change is that which occurs when the expansion process takes place at constant entropy. This specifies a path for the process and allows the enthalpy change $(H_2 - H_1)_S$ to be determined when data for the fluid are available. The exit velocity u_2 is then found from Eq. (10-21). The velocity given by this equation can very nearly be attained in a properly designed nozzle. However, the design of the nozzle depends on other considerations (fluid mechanics) quite outside the scope of thermodynamics, which provides merely the limit of what can be attained without prescribing the means of doing it.

For compressors and turbines, Eq. (10-19) is further simplified because the kinetic-energy term is taken to be negligible. The reason for this is that in properly designed devices of this kind, entrance and exit conduits are made of such a size that fluid velocities are held to relatively low values so as to minimize fluid friction. Equation (10-19) then becomes

$$W_s = \Delta H \qquad \text{(compressors and turbines)}. \tag{10-22}$$

The isentropic expansion of a fluid through a nozzle produces a fluid stream of high kinetic energy. This stream can be made to impinge on a turbine blade so as to provide a force to move the blade. The stream thus does work on the turbine blade at the expense of its kinetic energy. This is the principle on which the operation of a turbine depends. A series of nozzles and blades is arranged to expand the fluid in stages and to convert kinetic energy into shaft work. The overall result of the process is the expansion of a fluid from a high pressure to a low pressure with the production of work rather than the production of a high-velocity stream. The process is represented in Fig. 10-11.

Since Eq. (10-22) is applicable, we see that the shaft work of a turbine is given directly by the enthalpy change of the expanding fluid, and the only problem is the determination of ΔH. In the design of a turbine one usually knows the intake conditions (and therefore the initial properties at 1) and the discharge pressure P_2. We cannot get H_2 without knowing an additional

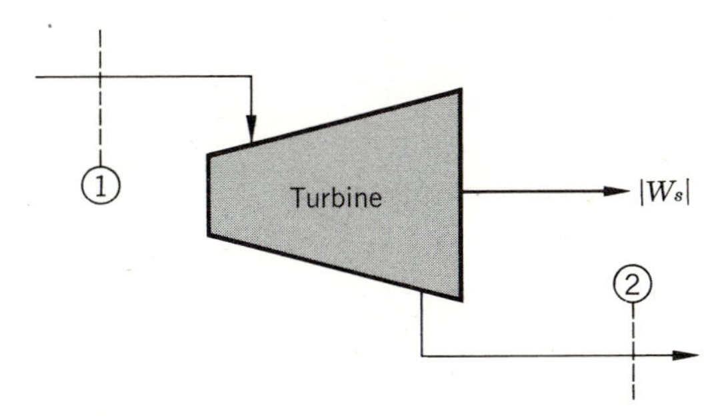

Fig. 10-11
Expansion in a turbine produces shaft work.

condition at the turbine discharge. If the turbine operates reversibly as well as adiabatically, i.e., isentropically, then this additional condition is given by $S_2 = S_1$, and in this case Eq. (10-22) becomes

$$W_s = (\Delta H)_S. \tag{10-23}$$

The work given by Eq. (10-23) is the limiting or *maximum* shaft work that can be produced by adiabatic expansion of a fluid from a given initial state to a given final pressure. Actual machines produce an amount of work equal to 75 or 80 percent of this. Thus we can define an expansion efficiency as

$$\eta \equiv \frac{W_s(\text{actual})}{W_s(\text{isentropic})}.$$

These two work terms are given by (10-22) and (10-23). That is, $W_s(\text{actual}) = \Delta H$ and $W_s(\text{isentropic}) = (\Delta H)_S$. Thus

$$\eta = \frac{\Delta H}{(\Delta H)_S} \qquad \text{(expansion)}, \tag{10-24}$$

where ΔH is the actual enthalpy change of the fluid passing through the turbine. These processes, the actual and the reversible, are represented on a Mollier diagram in Fig. 10-12. The reversible process follows a vertical line of constant entropy from point 1 at the higher pressure P_1 to point 2′ at the discharge pressure P_2. The line representing the actual irreversible process starts again from point 1, but proceeds downward and to the right in the direction of increasing entropy. The process terminates at point 2 on

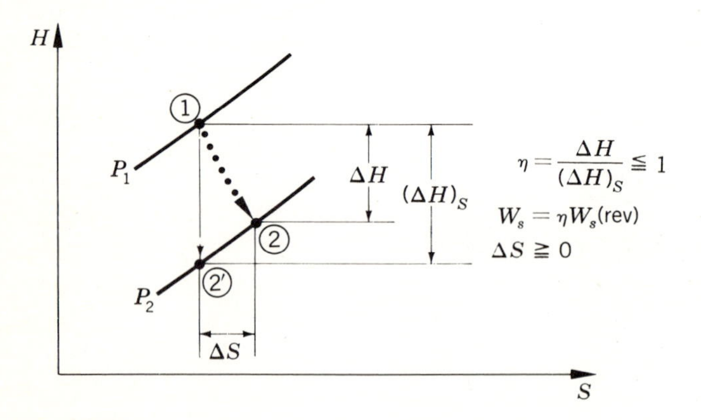

Fig. 10-12
Expansion processes.

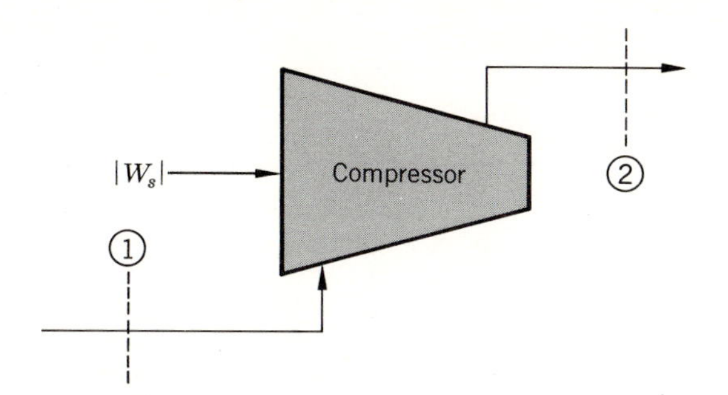

Fig. 10-13
Compression in a flow process as a result of shaft work.

the isobar for P_2. The greater the irreversibility, the further this point will lie to the right on the P_2 isobar, giving lower and lower values for the efficiency η of the process.

An adiabatic compression process, represented schematically in Fig. 10-13, is the opposite of an adiabatic expansion process, because work is done on a fluid so as to raise its pressure. In fact, a reversible, adiabatic expansion process can be made to retrace its path in reverse by a reversible, adiabatic compression process. The energy equation is the same for both processes, because the same assumptions of negligible kinetic-energy and potential-energy changes are made. Thus Eqs. (10-22) and (10-23) are applicable in either case. There is, however, a difference. In the case of adiabatic compression the reversible work is the *minimum* shaft work required for compression of a fluid from its initial state to a given final pressure. An actual irreversible process requires greater work expenditure. Thus the compression efficiency is defined by

$$\eta \equiv \frac{W_s(\text{isentropic})}{W_s(\text{actual})},$$

or by

$$\eta = \frac{(\Delta H)_S}{\Delta H} \qquad \text{(compression)}. \tag{10-25}$$

The adiabatic compression process is represented on a Mollier diagram in Fig. 10-14. Again the vertical line from 1 to 2′ represents the reversible (isentropic) process. Here P_2 is the higher pressure, and a line representing an actual process rises from point 1 and runs toward the right in the direction of increasing entropy. The figure clearly shows that in this case, $(\Delta H)_S$ is the minimum possible enthalpy increase.

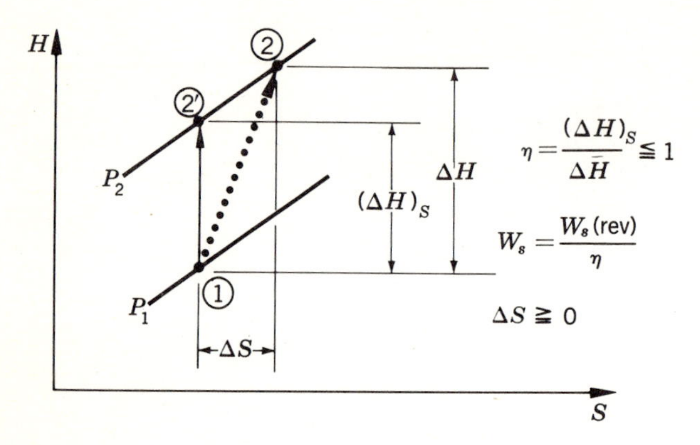

Fig. 10-14
Compression processes.

Example 10-6

A simple Rankine steam power plant was described in Sec. 6-3. Its basic features are again shown schematically in Fig. 10-15. Assume that the steam is generated in the boiler at 500(psia) and superheated to 1000(°F). It is fed to an adiabatic turbine which has an expansion efficiency of 75 percent. The turbine exhausts to a condenser maintained at 5(psia). Condensate is then pumped as liquid back to the boiler. Assume that no subcooling of the condensate occurs in the condenser and that the work of pumping the condensate to the boiler is negligible. For power generation of 200,000(kW) what is the rate of steam circulation? How much heat must be supplied to the steam by the boiler, and how much heat must be given up in the condenser? What is the thermal efficiency of the cycle?

The steam entering the turbine (point 1, Fig. 10-15) is superheated steam at 500(psia)

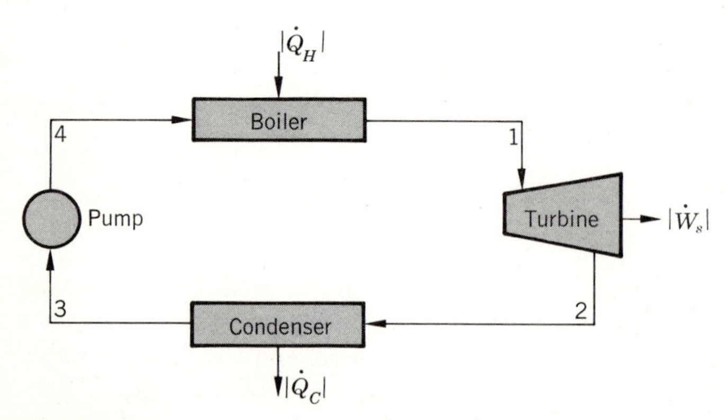

Fig. 10-15
Steam power plant.

and 1000(°F). Values for its enthalpy and entropy are found from the steam tables:

$$H_1 = 1518.8(\text{Btu})/(\text{lb}_m) \qquad \text{and} \qquad S_1 = 1.7356(\text{Btu})/(\text{lb}_m)(\text{R}).$$

Steam exhausts from the turbine (point 2) at 5(psia), and if we assume momentarily that the expansion process in the turbine is isentropic, then (see Fig. 10-12)

$$S_2' = S_1 = 1.7356(\text{Btu})/(\text{lb}_m)(\text{R}).$$

We note from the steam tables that the entropy of *saturated* vapor steam at 5(psia) has a value greater than S_2'. Thus the state at 2′ is wet steam, and we must determine its quality. At 5(psia),

$$S_f = 0.2347 \qquad \text{and} \qquad S_g = 1.8437(\text{Btu})/(\text{lb}_m)(\text{R}).$$

As a result of Eqs. (9-74) and (9-75) we can write

$$S_2' = (1 - x_2')S_f + x_2'S_g$$

or

$$1.7356 = (1 - x_2')(0.2347) + x_2'(1.8437).$$

Solution for the quality gives

$$x_2' = 0.9328.$$

The enthalpy H_2' is now given by

$$H_2' = (1 - x_2')H_f + x_2'H_g.$$

At 5(psia),

$$H_f = 130.13 \qquad \text{and} \qquad H_g = 1130.8(\text{Btu})/(\text{lb}_m).$$

Thus

$$H_2' = (0.0672)(130.13) + (0.9328)(1130.8) = 1063.6(\text{Btu})/(\text{lb}_m),$$

and the enthalpy change as a result of isentropic expansion of the steam is

$$(\Delta H)_S = H_2' - H_1 = 1063.6 - 1518.8 = -455.2(\text{Btu})/(\text{lb}_m).$$

Since the expansion efficiency η is given as 75 percent, we have from Eq. (10-24) that

$$\Delta H = H_2 - H_1 = \eta(\Delta H)_S = (0.75)(-455.2) = -341.4(\text{Btu})/(\text{lb}_m).$$

Thus

$$H_2 = H_1 - 341.4 = 1518.8 - 341.4 = 1177.4(\text{Btu})/(\text{lb}_m),$$

and the steam at point 2 has an enthalpy of 1177.4(Btu)/(lb_m) at a pressure of 5(psia). Location of this state in the steam tables or on the Mollier diagram shows it to be superheated vapor at a temperature of 264(°F).

The work of the turbine follows from Eq. (10-22):

$$W_s = \Delta H = -341.4(\text{Btu})/(\text{lb}_m).$$

This is the work done by each pound *mass* of steam flowing through the turbine. The *power output* $\dot{W}_s$ of the turbine is obtained by multiplication of the absolute value of this work by the mass flow rate $\dot{m}$ of steam through the turbine. For a mass flow rate in (lb_m)/(h) we have

$$\dot{W}_s = |W_s|\,\dot{m} = 341.4(\text{Btu})/(\text{lb}_m) \times \dot{m}(\text{lb}_m)/(\text{h}).$$

However, the power rating of the plant is given in the problem statement as 200,000(kW), or

$$\dot{W}_s = 2 \times 10^5(\text{kW}) \times 56.87(\text{Btu})/(\text{min})(\text{kW}) \times 60(\text{min})/(\text{h})$$
$$= 682.4 \times 10^6(\text{Btu})/(\text{h}).$$

Thus

$$341.4(\text{Btu})/(\text{lb}_m) \times \dot{m}(\text{lb}_m)/(\text{h}) = 682.4 \times 10^6(\text{Btu})/(\text{h}),$$

and

$$\dot{m} = 2.00 \times 10^6(\text{lb}_m)/(\text{h}),$$

a flow rate of 1,000(tons)/(h) or 25.2(kg)/(s).

Since the liquid leaving the condenser at point 3 is not subcooled, it must be saturated at 5(psia). For saturated liquid at this pressure we have from the steam tables that

$$H_3 = 130.13(\text{Btu})/(\text{lb}_m) \qquad \text{at} \qquad t_3 = 162.25(°\text{F}).$$

If the work of pumping this liquid back to the boiler is negligible (see Example 10-9), then $H_4 = H_3$, and the liquid at point 4 is compressed liquid at 500(psia) with the properties

$$H_4 = 130.13(\text{Btu})/(\text{lb}_m) \qquad \text{at} \qquad t_4 = 162.25(°\text{F}).$$

For the condenser and the boiler, the energy equation (10-14) reduces to

$$\Delta H = \Sigma Q,$$

upon assumption that the kinetic and potential-energy terms are negligible. Thus for the condenser, we have

$$H_3 - H_2 = Q_C,$$

or

$$(H_3 - H_2)\dot{m} = \dot{Q}_C.$$

Similarly, for the boiler

$$(H_1 - H_4)\dot{m} = \dot{Q}_H.$$

Thus

$$\dot{Q}_C = (130.1 - 1177.4)(2.00 \times 10^6) = -2.093 \times 10^9(\text{Btu})/(\text{h}),$$

a heat rate out of 613,500(kW), and

$$\dot{Q}_H = (1518.8 - 130.1)(2.00 \times 10^6) = 2.776 \times 10^9(\text{Btu})/(\text{h}),$$

a heat rate in of 813,500(kW).

These quantities are, respectively, the rate of heat transfer from the condenser and the rate of heat transfer to the boiler.

The thermal efficiency η_t of the plant is [see Eq. (6-1)]:

$$\eta_t = \frac{|\dot{W}_s|}{|\dot{Q}_H|} = \frac{200{,}000}{813{,}500} = 0.246.$$

Thus 24.6 percent of the heat supplied to the boiler appears as work.

Example 10-7

Refrigeration at a temperature level of 150(R) is required for a certain process. A cycle using helium gas has been proposed to operate as follows: Helium at 1(atm) is compressed

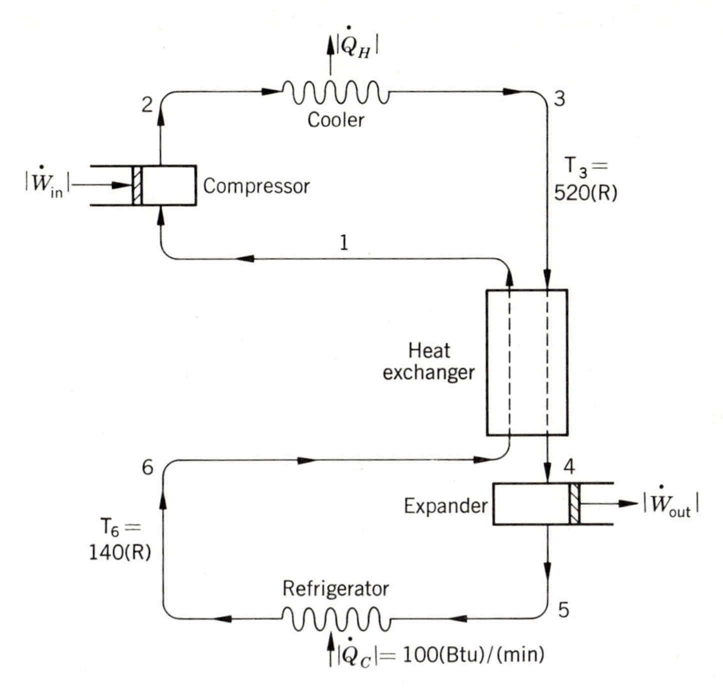

Fig. 10-16
Refrigeration cycle with helium as the working substance.

adiabatically to 5(atm), water-cooled to 60(°F), and sent to a heat exchanger, where it is cooled by returning helium. From there it goes to an adiabatic expander, which delivers work to be used to help drive the compressor. The helium then enters the refrigerator, where it absorbs enough heat to raise its temperature to 140(R). It returns to the compressor by way of the heat exchanger. The apparatus is shown schematically in Fig. 10-16.

Helium may be considered an ideal gas with a constant molar heat capacity at constant pressure of 5(Btu)/(lb mol)(R). If the efficiencies of the compressor and expander are 80 percent and if the minimum temperature difference in the exchanger is 10(°F), at what rate must the helium be circulated to provide refrigeration at a rate of 100 (Btu)/(min)? What is the net power requirement of the process? What is the coefficient of performance of the cycle? How does it compare with the Carnot COP?

We will assume that no heat is exchanged with the surroundings by the compressor, the expander, and the heat exchanger. Only T_3 and T_6 (Fig. 10-16) are given in the problem statement. However, the minimum temperature difference *between the two streams* within the countercurrent heat exchanger is specified to be 10(°F). Since the two streams flowing through the exchanger have the same flow rates and heat capacities, this temperature difference is actually constant over the entire length of the exchanger. Therefore, we have by inspection

$$T_1 = 510(\text{R}) \qquad \text{and} \qquad T_4 = 150(\text{R}).$$

Consider first the compressor. For reversible, adiabatic (isentropic) operation (see Fig. 10-14), we have for an ideal gas

$$\frac{T_2'}{T_1} = \left(\frac{P_2}{P_1}\right)^{(\gamma-1)/\gamma},$$

$$T_2' = 510\left(\frac{5}{1}\right)^{0.4} = 971(\mathrm{R}).$$

Thus

$$(\Delta H)_S = C_P(T_2' - T_1) = 5(971 - 510) = 2305(\mathrm{Btu})/(\mathrm{lb\ mol}).$$

From Eqs. (10-22) and (10-25) we find

$$W_{\mathrm{in}} = \Delta H = \frac{(\Delta H)_S}{\eta} = \frac{2305}{0.8} = 2881(\mathrm{Btu})/(\mathrm{lb\ mol}).$$

Since

$$\Delta H = C_P(T_2 - T_1) = 5(T_2 - 510) = 2881,$$

$$T_2 = 1086(\mathrm{R}).$$

For the cooler

$$\Delta H = Q_H = C_P(T_3 - T_2) = 5(520 - 1086),$$

and

$$Q_H = -2830(\mathrm{Btu})/(\mathrm{lb\ mol}).$$

If the expander operated reversibly and adiabatically (isentropically), then

$$\frac{T_5'}{T_4} = \left(\frac{P_5}{P_4}\right)^{(\gamma-1)/\gamma}$$

and

$$T_5' = 150\left(\frac{1}{5}\right)^{0.4} = 78.8(\mathrm{R}).$$

Therefore

$$(\Delta H)_S = C_P(T_5' - T_4) = 5(78.8 - 150) = -356(\mathrm{Btu})/(\mathrm{lb\ mol}),$$

and from Eqs. (10-22) and (10-24)

$$W_{\mathrm{out}} = \Delta H = \eta(\Delta H)_S = (0.8)(-356) = -285(\mathrm{Btu})/(\mathrm{lb\ mol}).$$

Since

$$\Delta H = C_P(T_5 - T_4) = 5(T_5 - 150) = -285,$$

$$T_5 = 93(\mathrm{R}).$$

Finally, for the refrigerator

$$\Delta H = Q_C = C_P(T_6 - T_5) = 5(140 - 93),$$

and

$$Q_C = 235(\mathrm{Btu})/(\mathrm{lb\ mol}).$$

Since refrigeration is required at the rate

$$\dot{Q}_C = 100(\mathrm{Btu})/(\mathrm{min}),$$

we have

$$\dot{n} = \frac{\dot{Q}_C}{Q_C} = \frac{100(\text{Btu})/(\text{min})}{235(\text{Btu})/(\text{lb mol})} = 0.425(\text{lb mol})/(\text{min}).$$

The net work of the cycle is simply

$$W = W_{\text{in}} + W_{\text{out}} = 2881 - 285 = 2596(\text{Btu})/(\text{lb mol});$$

and the net power requirement is

$$\dot{W} = \dot{n}W = \frac{0.425(\text{lb mol})/(\text{min}) \times 2596(\text{Btu})/(\text{lb mol})}{56.87(\text{Btu})/(\text{min})(\text{kW})}$$

$$= 19.4(\text{kW}).$$

The coefficient of performance of the cycle is defined by

$$\omega = \frac{|Q_C|}{|W|} = \frac{235}{2{,}596} = 0.0905.$$

For a Carnot refrigerator operating between the temperature limits of 150 and 520(R), the coefficient of performance is much higher:

$$\omega = \frac{T_C}{T_H - T_C} = \frac{150}{520 - 150} = 0.405.$$

10-8 Transonic Flow through Nozzles

One of the most interesting applications of the flow equations is to nozzles which operate with gas velocities in the neighborhood of the sonic velocity. The energy equation for a nozzle is given by Eq. (10-20), which in differential form becomes

$$dH = -\frac{u\,du}{g_c}. \tag{10-26}$$

Since $d(\ln u) = du/u$ and $d(\ln V) = dV/V$, this equation may also be written

$$dH = -\frac{u^2}{g_c}\frac{d(\ln u)}{d(\ln V)}\frac{dV}{V}.$$

Solution for u^2 gives

$$u^2 = -g_c V \frac{dH}{dV}\frac{d(\ln V)}{d(\ln u)}.$$

For steady flow the mass flow rate $\dot{m}$ must be constant over the full length of the nozzle. Since $\dot{m}V = uA$, where A is the cross-sectional area of the nozzle, and V is specific volume we have

$$\frac{uA}{V} = \text{constant.}$$

Logarithmic differentiation gives

$$d(\ln u) + d(\ln A) - d(\ln V) = 0,$$

or

$$\frac{d(\ln V)}{d(\ln u)} = \frac{d(\ln A)}{d(\ln u)} + 1.$$

Our equation for u^2 can now be written

$$u^2 = -g_c V \frac{dH}{dV}\left[\frac{d(\ln A)}{d(\ln u)} + 1\right].$$

If we now restrict ourselves to nozzles in which the flow is reversible as well as adiabatic, i.e., isentropic, then dH/dV can be replaced by $(\partial H/\partial V)_S$:

$$u^2 = -g_c V \left(\frac{\partial H}{\partial V}\right)_S \left[\frac{d(\ln A)}{d(\ln u)} + 1\right].$$

Eq. (9-11) shows that

$$\left(\frac{\partial H}{\partial P}\right)_S = V. \tag{9-11}$$

Multiplication by $(\partial P/\partial V)_S$ gives

$$\left(\frac{\partial H}{\partial V}\right)_S = V\left(\frac{\partial P}{\partial V}\right)_S,$$

and our equation for u^2 becomes

$$u^2 = -g_c V^2 \left(\frac{\partial P}{\partial V}\right)_S \left[\frac{d(\ln A)}{d(\ln u)} + 1\right].$$

According to Eq. (5-33) the sonic velocity or soundspeed c is given by

$$c^2 = -g_c V^2 \left(\frac{\partial P}{\partial V}\right)_S;$$

therefore

$$\frac{u^2}{c^2} = \frac{d(\ln A)}{d(\ln u)} + 1.$$

Since the ratio u/c is defined as the Mach number $\mathbf{M}$, we have finally for isentropic flow in nozzles that

$$\mathbf{M}^2 - 1 = \frac{d(\ln A)}{d(\ln u)}. \tag{10-27}$$

A number of qualitative observations can be based on this equation. There are three cases, corresponding to:

$\mathbf{M} < 1$ Subsonic flow,

$\mathbf{M} = 1$ Sonic flow,

$\mathbf{M} > 1$ Supersonic flow.

It is also useful to note that the signs of two key derivatives are invariant, regardless of the case considered. Writing Eq. (10-26) as

$$\frac{dH}{du} = -\frac{u}{g_c},$$

we see that dH/du is always negative. In addition, Eq. (9-11) shows that dH/dP is always positive for isentropic flow. Thus

$$\frac{dH}{du} = \ominus \qquad \text{and} \qquad \frac{dH}{dP} = \oplus.$$

When $\mathbf{M} < 1$, and the flow is subsonic, Eq. (10-27) shows that $d(\ln A)/d(\ln u)$ is negative. There are two cases to be considered: (1) A decreases in the direction of flow while u increases; (2) A increases in the direction of flow while u decreases. If $\mathbf{M} = 1$, the flow is sonic, and by Eq. (10-27) $d(\ln A)/d(\ln u) = 0$. There is but one case to be treated: (3) A is constant. When $\mathbf{M} > 1$, the flow is supersonic, and from Eq. (10-27) we have that $d(\ln A)/d(\ln u)$ is positive. Again there are two cases: (4) A increases as u increases; (5) A decreases as u decreases. The qualitative conclusions which follow from the equations given are displayed for each case in Table 10-2.

Case 1 is the familiar converging nozzle through which a fluid flows with steadily increasing velocity as the pressure drops (see Fig. 10-17*a*). The limiting velocity is the sonic velocity, and for isentropic flow this can be attained only at the exit of the nozzle where the area has become constant.

Case 2 applies to a device that receives a high-velocity (but subsonic) stream and decreases its velocity, converting kinetic energy into internal energy, thus causing an increase in enthalpy and pressure (see Fig. 10-17*b*). Such a device is called a *diffuser*. A common use of the diffuser is for the compression of the intake air for a jet airplane engine.

Table 10-2 Characteristics of Nozzles and Diffusers

$M < 1$		$M = 1$	$M > 1$	
$\frac{d(\ln A)}{d(\ln u)} = -$		$\frac{d(\ln A)}{d(\ln u)} = 0$	$\frac{d(\ln A)}{d(\ln u)} = +$	
Flow is subsonic		Flow is sonic	Flow is supersonic	
(1)	(2)	(3)	(4)	(5)
A decreases (converging) u increases H decreases P decreases	A increases (diverging) u decreases H increases P increases	A is constant u may increase or decrease H and P may decrease or increase	A increases (diverging) u increases H decreases P decreases	A decreases (converging) u decreases H increases P increases
Converging nozzle	Diverging diffuser	Point condition at an end or transition between subsonic and supersonic	Diverging nozzle	Converging diffuser

Case 4 is again a nozzle, but a *diverging* nozzle. Once flow has reached the sonic velocity, further increases in velocity can be attained only if the nozzle area increases. Thus the diverging nozzle is almost always found as part of a converging-diverging nozzle, a combination of case 1 and case 4 (see Fig. 10-17*c*). At the throat where the converging and diverging sections join, the flow is sonic and we have a transition from subsonic to supersonic flow, represented by case 3.

Case 5 is again a diffuser, one that *converges* and operates at supersonic velocities. Just as a diverging cross section is required to increase the velocity of a supersonic flow, so a converging cross section is required to decrease the velocity of a supersonic flow in an isentropic process. Such a diffuser is used on a supersonic jet aircraft for the compression of the intake air. To bring the air to subsonic velocities (relative to the aircraft), case 5 is combined with case 2 to make a converging-diverging diffuser. The throat of such a device is again represented by case 3, where the transition is from supersonic to subsonic flow.

Flow in real nozzles is, of course, irreversible, and is accompanied by an increase in entropy. However, friction in a properly designed nozzle is remarkably small, and the departure from isentropic flow is not serious. Thus, calculations based on the assumption of reversibility represent to a good approximation the results obtained in properly designed nozzles. The

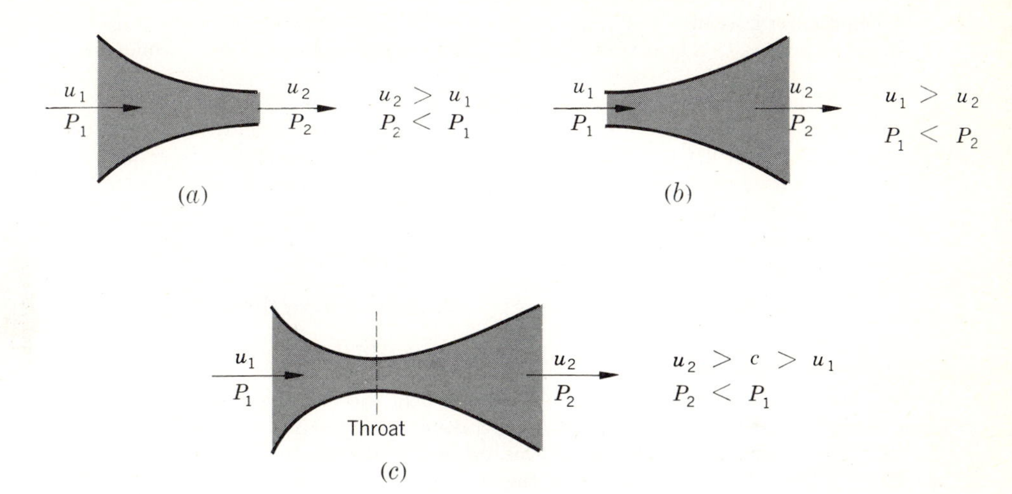

Fig. 10-17
Nozzles and diffusers: (a) converging, (b) diverging, (c) converging-diverging.

proper shape of a nozzle so that it may approach reversible operation depends on design principles that come not from thermodynamics but from fluid mechanics.

10-9 General Restrictions Imposed by the Second Law on Flow Processes

In the two preceding sections we made use of the second law to find the limits of what can be accomplished in adiabatic flow processes. Of course, the second law is universally applicable, and always puts a limit on what can be accomplished by any flow process. Its use requires the assumption of reversibility, for only in this case does it provide an equation. Calculations are based on an appropriate energy equation and on the second law in the form $\Delta S_{\text{total}} = 0$. It is not necessary to know any details about how the process is carried out. One needs merely end conditions and the assumption of reversibility. The results of calculations made on this basis are limiting values, representing the best that is possible. Any actual process is irreversible, accomplishing less than the best.

Example 10-8

A complicated process has been devised to make heat continuously available at a temperature level of 500(°F). The only source of energy is steam, saturated at 250(psia).

Cooling water is available in large supply at 70(°F). How much heat can be transferred from the process to a heat reservoir at 500(°F) for every pound of steam condensed in the process?

We do not need to know any details about how the process is accomplished, but we do need to make a couple of basic assumptions. Therefore we assume that steam flows continuously and that it is condensed and subcooled to the cooling-water temperature by the time it emerges. The properties of the inlet steam and outlet condensate are given by the steam tables as:

$$t_1 = 401(°\text{F}), \qquad t_2 = 70(°\text{F}),$$

$$H_1 = 1201.4(\text{Btu})/(\text{lb}_m), \qquad H_2 = 38.0(\text{Btu})/(\text{lb}_m),$$

$$S_1 = 1.5267(\text{Btu})/(\text{lb}_m)(\text{R}), \qquad S_2 = 0.0745(\text{Btu})/(\text{lb}_m)(\text{R}).$$

The second law denies the possibility that the *only* effect of this process could be the transfer of heat from the steam at temperatures between 70 and 401(°F) to a heat reservoir at 500(°F). However, in this process we have cooling water available at 70(°F). Thus we may consider that the *two* results of the process are transfer of heat from the steam to a heat reservoir at 500(°F) *and* the transfer of heat to another heat reservoir at 70(°F). A representation of the process is given in Fig. 10-18.

If we neglect potential- and kinetic-energy terms, Eq. (10-14) as written for the control volume becomes

$$\Delta H = \sum Q = Q_H + Q_C.$$

Thus $$Q_H + Q_C = 38.0 - 1201.4 = -1163.4(\text{Btu})/(\text{lb}_m).$$

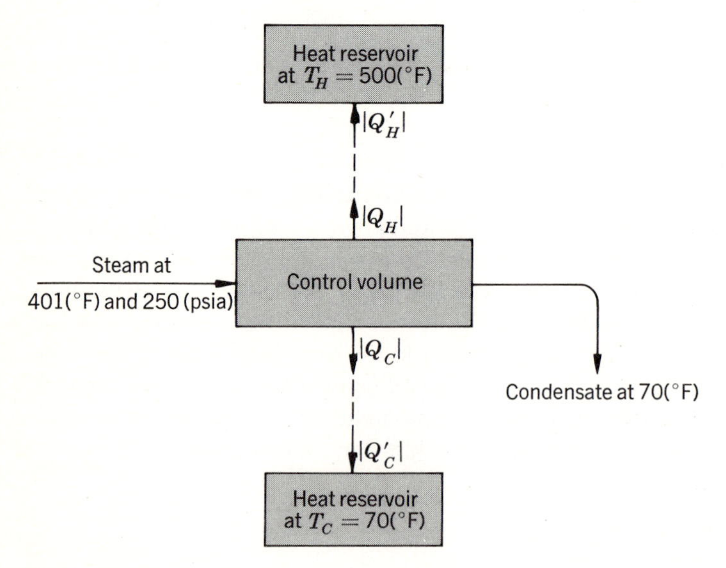

Fig. 10-18
Reversible flow process.

We now apply the second law, $\Delta S_{\text{total}} = 0$. The entropy change of the steam is simply

$$\Delta S = S_2 - S_1 = 0.0745 - 1.5267 = -1.4522(\text{Btu})/(\text{lb}_m)(\text{R}).$$

The entropy change of the heat reservoir at 500(°F) is

$$\frac{Q_H'}{500 + 460}(\text{Btu})/(\text{lb}_m)(\text{R})$$

and the entropy change of the heat reservoir at 70(°F) is

$$\frac{Q_C'}{70 + 460}(\text{Btu})/(\text{lb}_m)(\text{R}).$$

The sum of these three entropy changes is zero. However, Q_H and Q_C, taken with reference to the control volume in the energy equation, are opposite in sign to Q_H' and Q_C', taken with reference to the heat reservoirs in the entropy expressions. Thus, in terms of Q_H and Q_C, the total entropy change is

$$\Delta S_{\text{total}} = -1.4522 - \frac{Q_H}{960} - \frac{Q_C}{530} = 0.$$

The energy equation and this entropy equation constitute two equations in two unknowns, and may be solved for Q_H and Q_C. The results for each pound of steam are

$$Q_H = -879.1(\text{Btu})/(\text{lb}_m), \qquad Q_C = -284.3(\text{Btu})/(\text{lb}_m).$$

The minus signs merely indicate that heat is transferred away from the control volume.

In any actual process more heat would be transferred to the cold reservoir and less to the hot. In order to solve the problem for any case except that of complete reversibility, one would have to know the details of how the process is actually carried out.

10-10 The Mechanical-Energy Balance

Consider the steady-state flow of fluid through a control volume to which there is but one entrance and one exit. The energy equation which applies is Eq. (10-14):

$$\Delta H + \frac{\Delta u^2}{2g_c} + \Delta z\left(\frac{g}{g_c}\right) = \sum Q + W_s.$$

The property relation, Eq. (9-2), gives

$$dH = T\,dS + V\,dP$$

and for reversible processes we may set $T\,dS$ equal to δQ. Then

$$dH = \delta Q + V\,dP.$$

Integration gives

$$H_2 - H_1 = \Delta H = \sum Q + \int_1^2 V\,dP.$$

Substituting for ΔH in the energy equation, we get

$$W_s = \int_1^2 V\,dP + \frac{\Delta u^2}{2g_c} + \Delta z\left(\frac{g}{g_c}\right).$$

The assumption of reversibility was made in order to derive this equation. However, the viscous nature of real fluids leads to frictional effects that make flow processes inherently irreversible. As it stands, this equation is valid only for an imaginary nonviscous fluid; for real fluids it is at best approximate. An additional term may be incorporated in the equation to account for mechanical energy dissipated through fluid friction. The resulting equation is known as the mechanical-energy balance:

$$W_s = \int_1^2 V\,dP + \frac{\Delta u^2}{2g_c} + \Delta z\left(\frac{g}{g_c}\right) + \sum F, \tag{10-28}$$

where ΣF is the friction term. The determination of numerical values for ΣF is a problem in fluid mechanics, not thermodynamics, and will not be considered here.

Bernoulli's famous equation is a special case of the mechanical-energy balance applying to nonviscous, incompressible fluids which do not exchange shaft work with the surroundings. For nonviscous fluids, ΣF is zero, and for incompressible fluids,

$$\int_1^2 V\,dP = V\,\Delta P = \frac{\Delta P}{\rho},$$

where ρ is fluid density. Thus Bernoulli's equation becomes

$$\frac{\Delta P}{\rho} + \frac{\Delta u^2}{2g_c} + \Delta z\left(\frac{g}{g_c}\right) = 0.$$

This may also be written

$$\Delta\left[\frac{P}{\rho} + \frac{u^2}{2g_c} + z\left(\frac{g}{g_c}\right)\right] = 0,$$

or

$$\frac{P}{\rho} + \frac{u^2}{2g_c} + z\left(\frac{g}{g_c}\right) = \text{const.}$$

The severe limitations on Bernoulli's equation should be carefully noted.

Example 10-9

In Example 10-6 the work of pumping liquid water from the condenser at a pressure of 5(psia) to the boiler at a pressure of 500(psia) was neglected. Estimate a value for this work.

If we neglect the kinetic- and potential-energy terms and friction, the mechanical-energy balance, Eq. (10-28), reduces to

$$W_s = \int_1^2 V\,dP.$$

Since the liquid condensate is only slightly compressible, we may take $V = V_1$, and write

$$W_s = V_1(P_2 - P_1),$$

where V_1 is the specific volume of saturated liquid water at 5(psia). The value given in the steam tables is 0.0164(ft)^3/(lb$_m$). Thus

$$W_s = \frac{0.0164(\text{ft})^3/(\text{lb}_m) \times 495(\text{lb}_f)/(\text{in})^2 \times 144(\text{in})^2/(\text{ft})^2}{778(\text{ft})(\text{lb}_f)/(\text{Btu})}$$

$$= 1.5(\text{Btu})/(\text{lb}_m).$$

If one considers the pump as operating with an efficiency of 70 percent, then a better estimate of the work is 1.5/0.7 = 2.1(Btu)/(lb$_m$). Although this value by itself does not appear negligible, it is small enough in comparison with the turbine work of 341.4 (Btu)/(lb$_m$) found in Example 10-6 to be neglected, at least to a first approximation.

Problems

10-1 A tank of 100(ft)^3 capacity contains 3,000(lb$_m$) of liquid water in equilibrium with its vapor, which fills the remainder of the tank. The temperature and pressure are 450(°F) and 422.6(psia), respectively. A quantity of 2,000(lb$_m$) of water at 150(°F) is to be pumped into the tank without removing any steam. How much heat must be added during this process if the pressure and temperature in the tank are to remain at their initial values?

10-2 A tank contains 1(lb$_m$) of steam at a pressure of 300(psia) and a temperature of 700(°F). It is connected through a valve to a vertical cylinder which contains a frictionless piston. The piston is loaded with a weight such that a pressure of 100(psia) is necessary to support it. Initially, the piston is at the bottom of the cylinder. The valve is opened slightly, so that steam flows into the cylinder until the pressure is uniform throughout the system. The final temperature of the steam in the tank is found to be 440(°F). Calculate the temperature of the steam in the cylinder if no heat is transferred from the steam to the surroundings.

10-3 Consider a steady (constant T' and P') supply of gas or vapor connected through a valve to a closed tank containing the same gas or vapor at a lower pressure. The valve is opened so that the gas or vapor flows into the tank and is then shut again.

(a) Develop a general equation relating n_1 and n_2, the moles (or mass) of gas or vapor in the tank at the beginning and end of the process, to the properties U_1 and U_2, the internal energy of the gas or vapor in the tank at the beginning and end of the process, and H', the enthalpy of the gas or vapor in the steady supply, and to Q, the heat transferred to the material in the tank during the process.

(*b*) Reduce the general equation to its simplest form for the special case of an ideal gas with constant heat capacities.
(*c*) Further reduce the equation obtained in (*b*) for the case of $n_1 = 0$.
(*d*) Further reduce the equation obtained in (*c*) for the case where, in addition, $Q = 0$.
(*e*) Apply the appropriate equation to the case where a steady supply of hydrogen at 80(°F) and 2(atm) flows into an evacuated tank of 150(ft)^3 volume, and calculate how many (lb mol) of hydrogen will flow into the tank when the pressures are equalized if:

(*i*) It is assumed that no heat flows from the gas to the tank or through the tank walls.
(*ii*) The tank, mass = 900(lb$_m$), is perfectly insulated, has a specific heat of 0.11(Btu)/(lb$_m$)(°F), has an initial temperature of 80(°F), and is heated by the gas so as always to be at the temperature of the gas in the tank. Assume hydrogen to be an ideal gas with a molar heat capacity at constant pressure of 7(Btu)/(lb mol)(°F).

10-4 Develop equations which may be solved to give the final temperature of the gas remaining in a tank after the tank has been bled from an initial pressure P_1 to a final pressure P_2. Known quantities are the initial temperature, the tank volume, the heat capacity of the gas, the total heat capacity of the containing tank, P_1, and P_2. Assume the tank to be always at the temperature of the gas remaining in the tank, the gas to be ideal with constant heat capacities, and the tank to be perfectly insulated.

10-5 An inventor has developed a complicated process for making heat continuously available at an elevated temperature. Saturated steam at 220(°F) is used as the only source of energy. Assuming that there is plenty of cooling water available at 75(°F), what is the maximum amount of heat which could be made available at a temperature level of 400(°F) for each (Btu) of heat given up by the steam?

10-6 A chemical plant has saturated steam available at 400(psia), but because of a process change, the plant has little use for steam at this pressure. The new process uses steam at 150(psia). In addition, the plant also has exhaust steam available, saturated at 40(psia). It has been suggested that the 40(psia) steam be compressed to 150 (psia). The required work is to be obtained by expansion of the 400(psia) steam to 150(psia). The two streams at 150(psia) could then be mixed. Calculate amounts of each kind of steam required to give enough steam at 150(psia) to supply 1,000,000(Btu)/(h) by condensation only (i.e., no subcooling). Do these calculations for an ideal system. Rework for a practical process, and suggest a method or methods for carrying out the process.

10-7 When nitrogen at 100(atm) and 80(°F) expands adiabatically through a throttle to 1(atm), the temperature drops to 45(°F). A process for liquefying nitrogen is to consist of an exchanger system with no moving parts. The nitrogen will enter at 100(atm) and 80(°F). Unliquefied gas will leave the system at 70(°F) and 1(atm) and liquid nitrogen will be withdrawn saturated at 1(atm). It is estimated that the system can be well enough insulated so that the heat leak from the surroundings will be 25(Btu)/(lb mol) of entering nitrogen. It is claimed that at least 5 percent of the entering nitrogen can be liquefied in one pass through the apparatus. Do you think this is correct? Show the basis for your answer. *Data*: The latent heat of nitrogen at its normal boiling point of

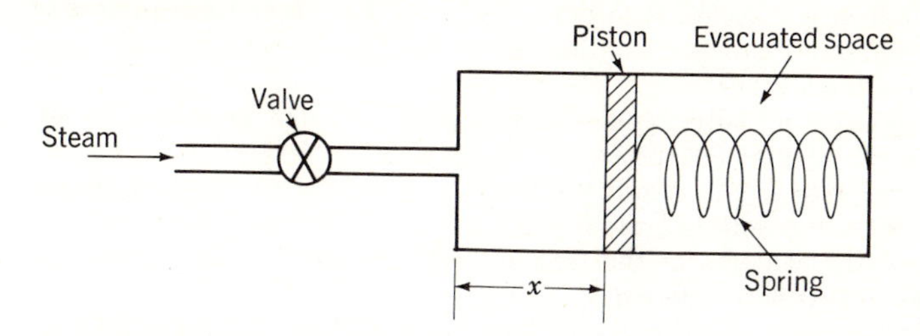

Fig. P10-9

77.4(K) is 1335(cal)/(mol). The specific heat of nitrogen gas may be taken as constant at 0.240(Btu)/(lb$_m$)(°F).

10-8 A lightweight portable power-supply system consists of a 1(ft)^3 bottle of compressed helium connected to a small adiabatic turbine. The bottle is initially charged to 2,000(psia) at 80(°F) and operates the turbine continuously until the pressure declines to 100(psia). The turbine exhausts at 5(psia). Neglecting all heat transfer to the gas, calculate the maximum possible useful work which can be obtained during the process. Assume helium to be an ideal gas with constant heat capacities $C_P = 5$, $C_V = 3$(Btu)/(lb mol)(°F).

10-9 A frictionless piston slides in a cylinder as shown in Fig. P10-9. The piston works to compress a spring, which exerts a force on the piston directly proportional to the distance x. Steam from a line having constant properties (P_1, T_1, U_1, H_1, V_1, etc.) is admitted to the cylinder through the valve and is allowed to flow until the properties of the steam in the cylinder reach the values P_2, T_2, U_2, H_2, V_2, etc. Derive an expression giving $H_2 - H_1 = f(P_2, V_2)$. Neglect heat transfer.

10-10 The vessel shown in Fig. P10-10 is initially evacuated. It is closed at the top by a frictionless piston having a mass of 100(lb$_m$). Initially, it rests on the shoulders of the vessel as shown. The vessel is connected to a steady supply of air at 200(psig) and 70(°F). Air is allowed to flow into the vessel until its volume

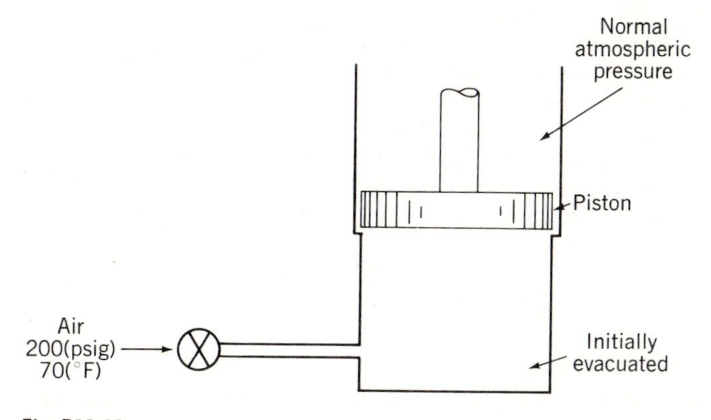

Fig. P10-10

doubles. If no heat is transferred to the air, what is the air temperature in the vessel at the end of the process? Consider air an ideal gas for which $C_P = 7$ and $C_V = 5$(Btu)/(lb mol)(°F).

10-11 A 100(ft)^3 rigid tank initially contains a mixture of saturated vapor steam and saturated liquid water at 500(psia). Of the total mass, 10 percent is vapor. Saturated liquid *only* is bled slowly through a valve from the tank until the final total mass in the tank is $\frac{1}{2}$ the initial total mass. During this process the temperature of the contents of the tank is kept constant by the transfer of heat. How much heat is transferred?

10-12 A stream of air is accelerated to a very high velocity in the following steady-state-flow process: Air is taken from the atmosphere at 1(atm) and 70(°F) and is passed into an adiabatic compressor, where its pressure is raised. It emerges at a fairly high temperature and is then run through a heat exchanger, where some heat is removed. From there it enters a nozzle, in which it expands to its original conditions of 1(atm) and 70(°F), except that now it has a velocity of 1,000(ft)/(s) in an 8(in)-ID pipe. If the power required by the compressor is 40,300(Btu)/(min), how much heat must be removed in the heat exchanger? The specific volume of air as calculated by the ideal-gas law at 1(atm) and 70(°F) is 13.35(ft)^3/(lb$_m$).

10-13 If liquid water at 70(°F) is pumped from a pressure of 1(atm) to a pressure of 50(atm) in a pump that operates reversibly and adiabatically, what is the work required in (Btu)/(lb$_m$)? The specific volume of water may be assumed insensitive to pressure.

10-14 A steam turbine operates adiabatically and produces 4,000(hp). Steam enters the turbine at 300(psia) and 900(°F). Exhaust steam from the turbine is saturated vapor at 5(psia). What is the steam rate in (lb$_m$)/(h) through the turbine? What is the efficiency of the turbine compared with isentropic operation?

10-15 (*a*) An insulated evacuated tank having a volume of 50(ft)^3 is attached to a constant-pressure line containing steam at 50(psia), superheated 10(°F). Steam is allowed to flow into the tank until the pressure has risen to 50(psia). Assuming the tank to be well insulated and to have a negligible heat capacity, how many (lb$_m$) of steam will enter the tank?

(*b*) Prepare graphs showing the mass of steam in the tank and the temperature of the steam in the tank as a function of pressure.

10-16 What is the maximum amount of work that can be obtained in a steady-flow process from steam at 500(psia) and 1000(°F) if in the process H_2O is brought to the conditions of the surroundings at 1(atm) and 70(°F)?

10-17 In a desuperheater, water is sprayed into incoming superheated steam in the proper amount to result in a single stream of saturated vapor leaving the desuperheater. Consider the process to be adiabatic with steady flow through the desuperheater. The following data apply: steam flow rate in of 2000(lb$_m$)/(h); steam entering at 400(psia) and 600(°F); water entering at 420(psia) and 100(°F); steam leaving at 380(psia), saturated.

(*a*) Calculate the mass flow rate of water necessary.
(*b*) What is the difference between the entropy outflow and inflow rates?
(*c*) Is the desuperheating process reversible or irreversible?

10-18 Figure P10-18 shows a vertical tube of cross-sectional area A = 1(ft)^2 containing a frictionless piston of weight w = 763(lb$_f$). Initially, the piston rests

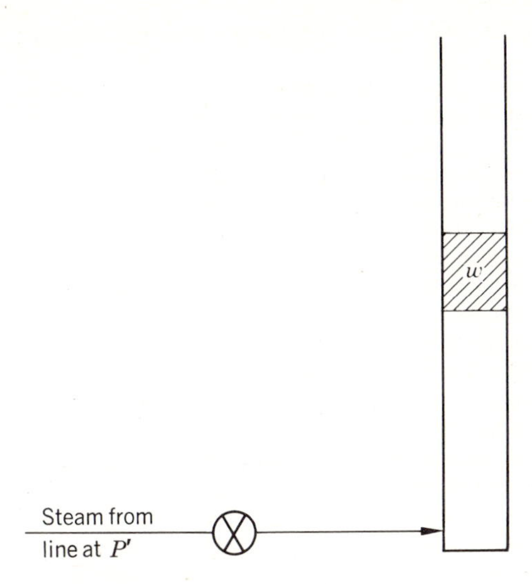

Fig. P10-18

on a mass of 0.1(lb$_m$) of saturated liquid water at the bottom of the tube. The tube is open at the top to the atmosphere, where the pressure is 14.7(psia). Steam is slowly admitted to the bottom of the tube from a steam line containing saturated vapor steam at the constant pressure $P' = 100$(psia). If this non-steady-flow process occurs adiabatically, to what height does the piston rise if just enough steam is admitted to the tube so that after the valve is closed all the steam in the tube is just saturated vapor? Neglect changes in potential and kinetic energies of the H_2O.

10-19 A 100(ft)^3/(min) positive-displacement (piston and cylinder) constant-speed vacuum pump is to be used to evacuate air from a 1,000(ft)^3 tank initially at 1(atm) pressure. If air is essentially an ideal gas and if the entire process is isothermal, what is the absolute minimum time that will be required to reduce the tank pressure to 10(mm Hg)? [*Note*: No specification of the intake volume per stroke has been given; this is part of the problem of determining the absolute minimum time requirement. All that is necessary is for the volume per stroke times the strokes per minute to equal 100(ft)^3/(min). Hence the speed and intake volume of the pump are interdependent.]

10-20 Saturated steam at 25(psia) is to be compressed adiabatically in a centrifugal compressor to 95(psia) at the rate of 200(lb$_m$)/(min). The efficiency of the compressor (compared with isentropic operation) is 70 percent. What is the horsepower requirement of the compressor and what is the final state of the steam, i.e., what are its properties, H, S, T, and V?

10-21 A Hilsch vortex tube is a device with no moving parts which separates a high-pressure gas stream into two low-pressure streams, one at a temperature above and one at a temperature below that of the entering gas stream. It consists of a tube with an inlet port at the center which directs the high-pressure gas stream into the tube perpendicular to the tube axis and tangent to the tube wall. Adjacent to the inlet there is an orifice in the tube. The cooler stream exits from the tube end beyond the orifice, and the warmer stream exits from the other end.

A particular test of a Hilsch tube gave the following results: Air entered at 5(atm) and 19.3(°C). The exit streams left at atmospheric pressure, with temperatures of −21.8 and 26.5(°C), respectively. The ratio of warm air to cool air was 5.39. Are these results consistent with the requirements of the laws of thermodynamics? Take air to be an ideal gas for which $C_P = 7$(cal)/(mol)(K).

10-22 A well-insulated closed tank has a volume of 2,500(ft)^3. Initially, it contains 50,000(lb$_m$) of water distributed between liquid and vapor phases at 80(°F). Saturated steam at 160(psia) is admitted to the tank until the pressure reaches 100(psia). How many (lb$_m$) of steam are added?

10-23 A tank with a volume of 1(m)^3 is initially evacuated. Atmospheric air leaks into the tank through a weld imperfection. The process is slow, and heat transfer with the surroundings keeps the tank and its contents at the ambient temperature of 27(°C). Calculate the amount of heat exchanged with the surroundings during the time it takes for the pressure in the tank to reach 1(atm).

10-24 Consideration is being given to the use of a steady-flow expander (or gas turbine) powered by a stream of hot compressed gases. It is required that the gases should discharge from the turbine at atmospheric pressure and 80(°F). It is also required that the turbine produce 1,250(hp) for a flow rate of 50(lb mol)/(min) of gas. Estimate the initial temperature and pressure required for the gas stream. For the purpose of an estimate, assume that the turbine will operate isentropically and that the gases are ideal with C_P constant at 10.5(Btu)/(lb mol)(°F).

10-25 (*a*) Show that the expression for the Joule-Kelvin coefficient μ, Eq. (10-17), can be written in the equivalent forms

$$\mu = \frac{T^2}{C_P}\left[\frac{\partial(V/T)}{\partial T}\right]_P;$$

$$\mu = \frac{V}{C_P}(\beta T - 1);$$

$$\mu = \frac{RT^2}{C_P P}\left(\frac{\partial Z}{\partial T}\right)_P.$$

(*b*) Show that in the limit as the pressure approaches zero, μ for a real gas approaches the value

$$\mu = \frac{T}{C'_P}\left(\frac{dB}{dT} - \frac{B}{T}\right).$$

Here, C'_P is the heat capacity for the ideal-gas state and B is the second virial coefficient.

10-26 The Joule-Kelvin coefficient μ is a measure of the temperature change during a throttling process. A similar measure of the temperature change produced by an *isentropic* change of pressure is provided by the coefficient μ_S, where

$$\mu_S \equiv \left(\frac{\partial T}{\partial P}\right)_S.$$

Prove that

$$\mu_S - \mu = \frac{V}{C_P}.$$

10-27 (*a*) Show that the inversion curve of a fluid is defined by either of the two expressions

$$\left(\frac{\partial Z}{\partial T}\right)_P = 0$$

or

$$V\left(\frac{\partial Z}{\partial V}\right)_T + T\left(\frac{\partial Z}{\partial T}\right)_V = 0.$$

(*b*) If a gas is described by the truncated virial equation

$$Z = 1 + B'P + C'P^2,$$

show that the equation of the inversion curve is

$$P = -\frac{(dB'/dT)}{(dC'/dT)}.$$

10-28 Saturated steam is compressed continuously from 15 to 70(psia) in a centrifugal compressor which operates adiabatically. For an efficiency of 75 percent compared with isentropic compression, determine the required work and the final state of the steam.

10-29 A power plant employs two adiabatic steam turbines in series. Steam enters the first turbine at 1100(°F) and 950(psia) and discharges from the second turbine at 2(psia). The system was designed so that equal work would be done by the two turbines, and the design was based on an efficiency of 80 percent compared with isentropic operation *in each turbine separately*. If the turbines perform according to these design conditions, what should be the temperature and pressure of the steam between the turbines? What is the overall efficiency of the two turbines considered together, compared with isentropic expansion from the initial state to the final pressure?

10-30 Hot water could in theory be used as a heat source for the generation of work. If hot water is available at 210(°F) and if an infinite heat sink is available at 70(°F), what is the minimum amount of water that would be required for the production of 1(Btu) of work?

10-31 A refrigerator is to be built to cool a brine solution from 70 to 20(°F) continuously. Heat is to be discarded to the atmosphere at a temperature of 80(°F). What is the absolute minimum power requirement of the refrigerator if 100(gal) of brine is to be cooled per minute? How much heat must be discarded to the atmosphere? *Data for the brine*: $C_P = 0.83(\text{Btu})/(\text{lb}_m)(°\text{F})$, $\rho = 71.8(\text{lb}_m)/(\text{ft})^3$.

10-32 Compute the absolute minimum quantity of steam in $(\text{lb}_m)/(\text{h})$ that is required to manufacture ice at the rate of 1(ton)/(h) under the following conditions: water for making ice is supplied at 60(°F); steam is at 200(psia) and 500(°F). (No other energy source is available.) The ice is not subcooled; the temperature of the surroundings is 70(°F); the latent heat of fusion of ice is $143.3(\text{Btu})/(\text{lb}_m)$ at 32(°F).

10-33 An ideal gas with constant heat capacities is caused to flow adiabatically through a nozzle from a region where the pressure and temperature are P_i and T_i, respectively, to a constricted region where the values are P and T. Assuming the initial velocity to be negligible:

(*a*) Show that the final velocity u is

$$u = \sqrt{2g_c C_P T_i \left(1 - \frac{T}{T_i}\right)}.$$

Note that C_P is the *specific* heat capacity of the fluid.

(*b*) Assuming the expansion in the nozzle to be isentropic, show that

$$u = \sqrt{2g_c C_P T_i \left[1 - \left(\frac{P}{P_i}\right)^{(\gamma-1)/\gamma}\right]}$$

$$= \sqrt{2g_c \frac{\gamma}{\gamma - 1} \frac{RT_i}{M} \left[1 - \left(\frac{P}{P_i}\right)^{(\gamma-1)/\gamma}\right]},$$

where M is molecular weight.

10-34 Steam approaches a horizontal nozzle at a pressure of 100(psia) and a temperature of 500(°F) at negligible velocity. It reaches the discharge end of the nozzle at a pressure of 70(psia). Assuming the flow to be isentropic:

(*a*) Find the velocity at the nozzle discharge.

(*b*) Find the area of the nozzle-discharge cross section required to pass a flow of 1(lb_m)/(s).

10-35 Air expands through a nozzle from a negligible initial velocity to a final velocity of 1,100(ft)/(s). Calculate the temperature drop of the air, assuming that air is an ideal gas for which $C_P = 7$(Btu)/(lb mol)(°F). The molecular weight of air is 29.

10-36 Saturated vapor steam at 15(psia) is taken continuously into a compressor at low velocity and is compressed to 45(psia). The steam then enters a horizontal stationary nozzle where it expands adiabatically back to its initial condition of saturated vapor steam at 15(psia). However, the steam now has a velocity of 2,000(ft)/(s). In order to accomplish this steady-flow process it is found necessary to remove heat from the compressor in the amount of 25(Btu)/(lb_m) of steam. What is the work requirement of the compressor?

11

SPECIAL APPLICATIONS OF THERMODYNAMICS

11-1 Simple Systems

In Chaps. 2 to 4 we considered a number of thermodynamic systems describable in terms of just three coordinates. Such *simple systems* include as one class the *PVT* system, which we have discussed in considerable detail. So commonly encountered is the *PVT* system that it is often thought of as *the* type of system to which thermodynamics applies. Hence other types of simple systems, such as a stressed bar, a surface, an electric cell, etc., are thought of as being something special. One of our aims in this chapter is to consider further such "special systems." They are treated in Secs. 11-1 to 11-4 and 11-12 and 11-13.

The differential generalized reversible work for a simple system is

$$\delta W_R = kY\,dX,$$

where Y is the generalized force, dX is the generalized displacement, and k is a constant. For example, for a magnetic solid we have $k = \mu_0$, $Y = \mathcal{H}$, and $X = \mathcal{M}$. Other special cases were treated in Chap. 3, and results are summarized in Table 3-3.

Both temperature and entropy are primitive concepts; thus the differential reversible heat is *always* given by

$$\delta Q_R = T\,dS.$$

Substitution of these equations into the differential first-law expression for a system at rest yields

$$dU = T\,dS + kY\,dX. \tag{11-1}$$

Equation (11-1) is the generalization of Eq. (9-1), and it applies to *any* simple system. It forms the basis for a multitude of derived property relationships, similar to those developed in Chap. 9 for PVT systems. These property relationships need not be derived anew for each simple system, however, for we can proceed by analogy from existing PVT expressions by making the following substitutions:

$$P = -kY \qquad \text{and} \qquad V = X.$$

In subsequent sections of this chapter we shall make use of entropy equations which are analogs of the equations for PVT systems derived in Sec. 9-3. Each of these analogs is a special case of a generalized entropy equation. The two equations we require are extensions of Eqs. (9-19) and (9-21); they are

$$dS = C_X \frac{dT}{T} - k \left(\frac{\partial Y}{\partial T}\right)_X dX \tag{11-2}$$

and

$$dS = C_Y \frac{dT}{T} + k \left(\frac{\partial X}{\partial T}\right)_Y dY. \tag{11-3}$$

The generalized heat capacities C_X and C_Y are

$$C_X \equiv \left(\frac{\partial U}{\partial T}\right)_X = T \left(\frac{\partial S}{\partial T}\right)_X$$

and

$$C_Y \equiv \left(\frac{\partial \tilde{H}}{\partial T}\right)_Y = T \left(\frac{\partial S}{\partial T}\right)_Y,$$

where the generalized enthalpy $\tilde{H}$ is defined by

$$\tilde{H} \equiv U - kYX.$$

Important special cases of the generalized heat capacities are listed in Table 4-3.

11-2 Constant-Volume Stressed Bars

Making the substitutions $k = V = \text{const}$, $Y = \sigma$, and $X = \epsilon$, we find from Eq. (11-3) that

$$dS = C_\sigma \frac{dT}{T} + V \left(\frac{\partial \epsilon}{\partial T}\right)_\sigma d\sigma,$$

or

$$dS = C_\sigma \frac{dT}{T} + V\alpha \, d\sigma, \tag{11-4}$$

Table 11-1

	Tension		Compression	
	Positive α	Negative α	Positive α	Negative α
Change in σ is	+	+	−	−
Change in T is	−	+	+	−

where $\alpha \equiv (\partial\epsilon/\partial T)_\sigma$ is the linear expansivity and σ is the stress. As an application of this expression, we consider the following question: When a bar is stressed reversibly and adiabatically from an initial unstressed state, does its temperature rise or fall?

For a reversible, adiabatic process $dS = 0$, and Eq. (11-4) yields

$$\frac{dT}{T} = -\frac{V\alpha}{C_\sigma}\,d\sigma \qquad (\text{const } S), \tag{11-5}$$

where V, T, and C_σ are all positive quantities, but α may be either positive or negative in sign. Moreover, $d\sigma$ may also be either positive or negative, depending on whether the bar is being stretched or compressed. Thus there are four cases to consider. These are shown in Table 11-1.

From the results in Table 11-1 we see that a bar of steel, with positive α, decreases in temperature when stretched but increases in temperature when compressed. The linear expansivity of rubber is negative when the rubber is in tension, and hence a finite elongation is accompanied by a temperature increase of the rubber.

Equation (11-4) is an exact differential equation; we may therefore apply the exactness criterion, Eq. (9-14), to the coefficients of the differentials dT and $d\sigma$. Doing this, we find an expression for the effect of stress on the constant-stress heat capacity C_σ:

$$\left(\frac{\partial C_\sigma}{\partial \sigma}\right)_T = TV\left(\frac{\partial \alpha}{\partial T}\right)_\sigma.$$

The sign of $(\partial C_\sigma/\partial\sigma)_T$ is thus determined by the sign of $(\partial\alpha/\partial T)_\sigma$. For metals, α is usually a weak function of T; thus C_σ may be considered practically independent of σ, that is, a function of T only. Moreover, for solids it is often a good approximation to take $C_\sigma \approx C_P$.

Example 11-1

An initially unstressed steel wire is stretched reversibly and adiabatically, and its temperature drops from 300 to 299(K). What is the final stress in the wire? For steel at 300(K), take $\alpha = 13 \times 10^{-6}(\mathrm{K})^{-1}$, $C_P = 0.5(\mathrm{J})/(\mathrm{g})(\mathrm{K})$, and $V = 0.128(\mathrm{cm})^3/(\mathrm{g})$.

Integrating Eq. (11-5) under the assumption of constant physical properties, we obtain

$$\sigma_2 = \sigma_1 - \frac{C_\sigma}{V\alpha} \ln \frac{T_2}{T_1}.$$

But $\sigma_1 = 0$, and we can assume that $C_\sigma \approx C_P$. Thus we find

$$\sigma_2 = -\frac{0.5(\mathrm{J})/(\mathrm{g})(\mathrm{K}) \times 10^6(\mathrm{cm})^3/(\mathrm{m})^3 \times 1(\mathrm{N})(\mathrm{m})/(\mathrm{J})}{0.128(\mathrm{cm})^3/(\mathrm{g}) \times 13 \times 10^{-6}(\mathrm{K})^{-1}} \ln \frac{299}{300},$$

or $\sigma_2 = 1{,}000(\mathrm{kN})/(\mathrm{m})^2$.

11-3 Surfaces

Substituting $k = 1$, $Y = \gamma$, and $X = A$ in Eq. (11-2), we obtain on rearrangement

$$T\,dS = C_A\,dT - T\left(\frac{\partial \gamma}{\partial T}\right)_A dA, \tag{11-6}$$

where γ is surface tension and A is the surface area. Imagine a spherical drop of liquid having a surface area A_0, which is the minimum area possible for any given volume. Suppose that the drop is drawn out into a thin film with the aid of a wire framework until the surface area reaches a value A, very much larger than the original value. If this is done reversibly and isothermally and *if the surface tension is a function of temperature only*, then we find from Eq. (11-6), since $\delta Q = T\,dS$, that

$$Q = -T\frac{d\gamma}{dT}(A - A_0).$$

Moreover, the work of the process is, by Eq. (3-8),

$$W = \int_{A_0}^{A} \gamma\,dA = \gamma(A - A_0).$$

From the first law, we then have

$$U - U_0 = \left(\gamma - T\frac{d\gamma}{dT}\right)(A - A_0).$$

Since A is very much larger than A_0, we may equally well write

$$U - U_0 = \left(\gamma - T\frac{d\gamma}{dT}\right)A,$$

where U_0 is the energy of the liquid with practically no surface, and U is the energy of the liquid with the large surface area A. Hence

$$\frac{U - U_0}{A} = \gamma - T\frac{d\gamma}{dT}. \tag{11-7}$$

In the process of expanding the area of the original drop, it is assumed that the *bulk* properties of the liquid enclosed within the surface do not change. Thus $U - U_0$ is interpreted as the energy associated with the surface only, and $(U - U_0)/A$ is the *surface energy per unit area.* This quantity is seen to have the same dimensions as γ. If γ is expressed in (dyn)/(cm), the energy is given in (dyn)(cm)/(cm)^2, or (erg)/(cm)^2. With the aid of the above equation, the values of surface energy per unit area may be calculated once the surface tension has been measured as a function of temperature.

At the critical temperature, the surface tension of all liquids is zero. The surface tension of a pure liquid in equilibrium with its vapor can usually be represented by an empirical formula of the type

$$\gamma = \gamma_0\left(1 - \frac{t}{t_c}\right)^n, \tag{11-8}$$

where γ_0 is the surface tension at 0(°C), t_c is the critical temperature in (°C), and n is a constant having a value between 1 and 2. For liquid water,

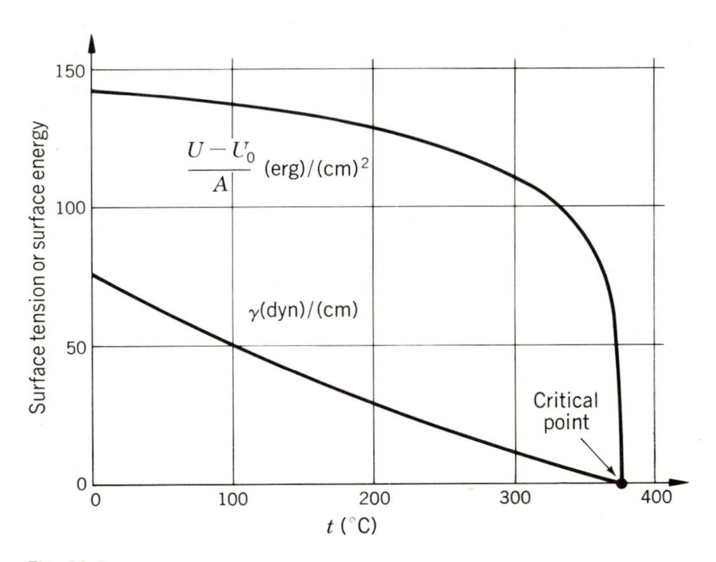

Fig. 11-1
Temperature variation of surface tension and surface energy of water.

$\gamma_0 = 75.5(\text{dyn})/(\text{cm})$, $t_c = 374.1(°\text{C})$, and $n = 1.2$. Since $d\gamma/dT = d\gamma/dt$, all the quantities necessary for calculation of $(U - U_0)/A$ from Eq. (11-7) are at hand; the results for water are shown in Fig. 11-1.

The surface energy per unit area is an important quantity in the kinetic theory of liquids. According to this theory, evaporation occurs when a molecule in the liquid phase possesses sufficient energy to escape through the film at the liquid surface. One would therefore expect a relationship to exist between the heat of vaporization and the surface energy. In fact, the temperature dependence of the two quantities ΔH^{lv} and $(U - U_0)/A$ is quite similar, as is shown by a comparison of Fig. 9-9 with Fig. 11-1.

11-4 Magnetic Material in a Magnetic Field

The second generalized entropy equation, Eq. (11-3), yields, with $k = \mu_0$, $Y = \mathcal{H}$, and $X = \mathcal{M}$,

$$T\,dS = C_{\mathcal{H}}\,dT + \mu_0 T\left(\frac{\partial \mathcal{M}}{\partial T}\right)_{\mathcal{H}} d\mathcal{H}. \tag{11-9}$$

For a reversible, isothermal change of field, this gives

$$\delta Q = \mu_0 T\left(\frac{\partial \mathcal{M}}{\partial T}\right)_{\mathcal{H}} d\mathcal{H}.$$

In the case of a paramagnetic material, such as a crystal containing magnetic ions separated by a large number of nonmagnetic particles, the degree of orientation of the magnetic ions, which determines the magnitude of the magnetization $\mathcal{M}$, is controlled by two factors: the magnetic field strength $\mathcal{H}$, and the temperature T. If $\mathcal{H}$ is kept constant, an increase in temperature is associated with an increase in the energy of vibration, with a consequent reduction in the degree of orientation of the ions. Experiment therefore shows that for paramagnetic materials, $(\partial \mathcal{M}/\partial T)_{\mathcal{H}}$ is *always* negative. Thus the preceding equation shows that such materials reject heat when the field is increased and absorb heat when the field is decreased isothermally.

For a reversible, adiabatic change of field $dS = 0$, and Eq. (11-9) yields

$$\frac{dT}{T} = -\frac{\mu_0}{C_{\mathcal{H}}}\left(\frac{\partial \mathcal{M}}{\partial T}\right)_{\mathcal{H}} d\mathcal{H} \qquad (\text{const } S). \tag{11-10}$$

Thus, since $(\partial \mathcal{M}/\partial T)_{\mathcal{H}}$ is a negative quantity, a paramagnetic material experiences a temperature drop upon adiabatic demagnetization. This phenomenon is called the *magnetocaloric effect* and is used today to produce extremely low temperatures, below 1(K). Experiments of this sort were first performed in 1933 by Giauque in America and were then taken up by Kurti and Simon in England and by De Haas and Wiersma in Holland. In these experiments a paramagnetic salt is cooled to as low a temperature as possible with the aid of liquid helium. A strong magnetic field is then applied, producing a rise of temperature in the substance and a consequent flow of heat to the surrounding helium, some of which is thereby evaporated. After a while, the substance is both strongly magnetized and as cold as possible. At this moment, the space surrounding the substance is evacuated. The magnetic field is now reduced to zero, and the temperature of the paramagnetic salt drops to a very low value.

In the experiments of Kurti and Simon the adiabatic demagnetization was accomplished by switching the magnet off and wheeling it away. In the experiments at Leiden, the whole calorimeter was swung out of the magnetic field.

The final step in an adiabatic demagnetization experiment is to measure the temperature. For temperatures below 1(K), conventional techniques of measurement cannot be used; instead, the paramagnetic salt is in effect used as its own thermometer. The magnetic permeability χ_m of paramagnetic substances is often represented satisfactorily by *Curie's equation*:

$$\chi_m = \frac{C}{T}. \tag{11-11}$$

The constant C is called the *Curie constant*; it has different values for different substances, but its value for any given material can be determined from measurements of χ_m at temperatures greater than 1(K). Curie's equation may therefore be used to define a new temperature scale, suitable for use below 1(K). The new temperature T^*, called the *magnetic temperature*, is defined as

$$T^* \equiv \frac{C}{\chi_m}.$$

Values of T^* for temperatures below 1(K) can be calculated from measurements of the magnetic permeability. In the region where Curie's equation is valid, T^* is identical to the Kelvin temperature; in the region very near absolute zero, T^* can be expected to differ somewhat from the Kelvin temperature.

Example 11-2

The magnetic field strength in a paramagnetic solid at initial temperature T_1 is reversibly and adiabatically changed from $\mathscr{H}_1$ to $\mathscr{H}_2$. Determine an expression for the final temperature T_2. Assume that Curie's equation is valid and that the constant-magnetization heat capacity is given by the expression

$$C_{\mathcal{M}} = \frac{A}{T^2}. \tag{A}$$

We start with Eq. (11-10):

$$\frac{dT}{T} = -\frac{\mu_0}{C_{\mathcal{H}}}\left(\frac{\partial \mathcal{M}}{\partial T}\right)_{\mathcal{H}} d\mathcal{H}. \tag{11-10}$$

Integration of this equation requires expressions for $(\partial \mathcal{M}/\partial T)_{\mathcal{H}}$ and for $C_{\mathcal{H}}$. Combination of Curie's equation with the definition of χ_m, Eq. (3-21), gives

$$\frac{\mathcal{M}}{V} = \chi_m \mathcal{H} = \frac{C}{T}\mathcal{H},$$

or

$$\mathcal{M} = \frac{CV}{T}\mathcal{H}, \tag{B}$$

and differentiation with respect to T yields

$$\left(\frac{\partial \mathcal{M}}{\partial T}\right)_{\mathcal{H}} = -\frac{CV}{T^2}\mathcal{H}. \tag{C}$$

It remains to develop an expression for $C_{\mathcal{H}}$ from the given equation for $C_{\mathcal{M}}$. In Sec. 9-5 we presented general equations for the heat-capacity difference $(C_P - C_V)$; in particular, we showed that

$$C_P - C_V = -T\left(\frac{\partial V}{\partial T}\right)_P^2\left(\frac{\partial P}{\partial V}\right)_T. \tag{9-39}$$

The analogous equation for the present case is obtained from the substitutions $P = -\mu_0 \mathcal{H}$ and $V = \mathcal{M}$. Thus we find that

$$C_{\mathcal{H}} - C_{\mathcal{M}} = \mu_0 T\left(\frac{\partial \mathcal{M}}{\partial T}\right)_{\mathcal{H}}^2\left(\frac{\partial \mathcal{H}}{\partial \mathcal{M}}\right)_T. \tag{D}$$

An expression for $(\partial \mathcal{M}/\partial T)_{\mathcal{H}}$ is given by Eq. (C); the other derivative is similarly found from Eq. (B) to be

$$\left(\frac{\partial \mathcal{H}}{\partial \mathcal{M}}\right)_T = \frac{T}{CV},$$

and Eq. (D) becomes

$$C_{\mathcal{H}} - C_{\mathcal{M}} = \mu_0 CV\frac{\mathcal{H}^2}{T^2}.$$

But $C_{\mathcal{M}}$ is given by Eq. (A); therefore,

$$C_{\mathcal{H}} = \frac{A + \mu_0 CV\mathcal{H}^2}{T^2}. \tag{E}$$

Combination of Eqs. (11-10), (C), and (E) gives finally

$$\frac{dT}{T} = -\mu_0\left(\frac{T^2}{A + \mu_0 CV\mathcal{H}^2}\right)\left(-\frac{CV}{T^2}\mathcal{H}\right)d\mathcal{H},$$

or
$$\frac{dT}{T} = \frac{(\mu_0 C V/A) \mathscr{H} \, d\mathscr{H}}{(\mu_0 C V/A)\mathscr{H}^2 + 1},$$

which yields on integration
$$T_2 = T_1 \sqrt{\frac{(\mu_0 C V/A) \mathscr{H}_2^2 + 1}{(\mu_0 C V/A) \mathscr{H}_1^2 + 1}}.$$

This is the desired expression for the magnetocaloric effect.

The case of greatest interest obtains for $\mathscr{H}_2 = 0$:
$$T_2 = T_1 \left(\frac{\mu_0 C V}{A} \mathscr{H}_1^2 + 1 \right)^{-1/2}. \tag{F}$$

Equation (F) applies for a reversible adiabatic demagnetization process by which the field strength is decreased from an initial value of $\mathscr{H}_1$ to a final value of zero. The goal of such a process is the attainment of extremely low temperatures. As already noted, this requires a low value for the initial temperature T_1, and a large value for the initial field strength $\mathscr{H}_1$. Additionally, the final temperature level is influenced by the properties of the paramagnetic substance; according to Eq. (F), the lowest temperatures will result for substances with large Curie constants and small values of the parameter A, that is, for substances with large permeabilities and small heat capacities.

As a numerical example of the application of Eq. (F), consider the reversible adiabatic demagnetization of a sample of chromium-potassium alum $[Cr_2(SO_4)_3 \cdot K_2SO_4 \cdot 24H_2O]$ from initial conditions of $T_1 = 1.5(\text{K})$ and $\mathscr{H}_1 = 450{,}000(\text{A})/(\text{m})$ to $\mathscr{H}_2 = 0$. We wish to determine the final temperature T_2. For this substance, it is known that $C V = 47.3(\text{K})(\text{cm})^3/(\text{mol})$ and that $A = 0.30(\text{J})(\text{K})/(\text{mol})$. Thus the term $(\mu_0 C V/A)$ has the value

$$\frac{\mu_0 C V}{A} = \frac{1.26 \times 10^{-6}(\text{H})/(\text{m}) \times 1(\text{J})/(\text{A})^2(\text{H}) \times 47.3(\text{K})(\text{cm})^3/(\text{mol}) \times 10^{-6}(\text{m})^3/(\text{cm})^3}{0.30(\text{J})(\text{K})/(\text{mol})},$$

or
$$\frac{\mu_0 C V}{A} = 2.0 \times 10^{-10}(\text{m})^2/(\text{A})^2.$$

Substitution of numerical values into Eq. (F) then gives
$$T_2 = 1.5(\text{K}) \times [2.0 \times 10^{-10}(\text{m})^2/(\text{A})^2 \times (4.5)^2 \times 10^{10}(\text{A})^2/(\text{m})^2 + 1]^{-1/2}$$

or $\quad T_2 = 0.23(\text{K})$.

Substantially lower temperatures can be obtained with stronger initial fields or by use of other paramagnetic salts.

11-5 Mixtures of Ideal Gases

We have not previously considered the effect of composition on the properties of PVT systems. To present a general and comprehensive treatment of the thermodynamics of solutions would carry us far beyond the intended scope of this text. However, it is desirable to introduce the subject and to develop the methods useful for mixtures of ideal gases.

Imagine several vessels, each having the same total volume V^t and each held at the same temperature T. Let the first such vessel contain n_1(mol) of a pure gas at pressure p_1; the second, n_2(mol) of a different pure gas at pressure p_2; etc. In general, each vessel contains n_i(mol) of pure gas i at a pressure p_i. We impose the restriction that in each case the conditions of temperature and pressure are such that the equation of state for an ideal gas is essentially valid. We now combine the contents of all vessels into a single vessel of the same volume V^t and at the same temperature T. Experiment shows that the pressure of the combined gases P is the sum of the initial pressures. That is,

$$P = \sum_i p_i. \tag{11-12}$$

This observation is known as *Dalton's law*. It expresses the fact that the total pressure of a mixture of inert ideal gases is equal to the sum of the pressures exerted by the individual gases as each occupies the total volume alone at the temperature of the mixture.

For each of the pure ideal gases occupying the volume V^t we have

$$p_i = \frac{n_i RT}{V^t}.$$

By Dalton's law,

$$P = \Sigma p_i = (\Sigma n_i)\frac{RT}{V^t}.$$

Thus

$$\frac{p_i}{P} = \frac{n_i}{\Sigma n_i} = y_i,$$

or

$$p_i = y_i P,$$

where y_i is the mole fraction of component i in the mixture, and the product y_iP is called the *partial pressure* of component i in the gas mixture. This equation shows that the partial pressure of a gas in a mixture of ideal gases is the same as the pressure which that gas alone exerts in the total mixture volume at the same temperature.

The significance of the partial pressure is illustrated by the following experiment. If a narrow tube of palladium is closed at one end and the open end is sealed into a glass tube, as shown in Fig. 11-2, the system may be pumped to a very high vacuum. If the palladium remains at room tem-

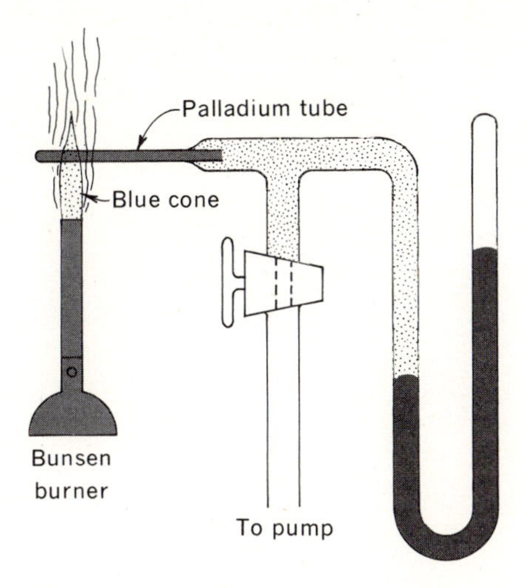

Fig. 11-2
Palladium tube permeable to hydrogen.

perature, the vacuum can be maintained indefinitely. If, however, an ordinary bunsen burner is placed so that the blue cone surrounds part of the tube, with the rest of the flame causing the palladium tube to become red-hot, hydrogen present in the blue cone will pass through the tube, but other gases will not. Red-hot palladium is said to be a *semipermeable membrane,* permeable to hydrogen only.

Experiment shows that the hydrogen continues to flow through the red-hot palladium until the pressure of hydrogen in the vessel reaches a value equal to the partial pressure of the hydrogen in the flame. When the flow stops, *membrane equilibrium* is said to exist. Membrane equilibrium is achieved when the partial pressure of the gas to which the membrane is permeable is the same on both sides of the membrane. We shall suppose that there exists a special membrane permeable to each gas with which we have to deal. Whether this is actually so is not important. We shall make use of the principle of the semipermeable membrane as an ideal device for theoretical purposes.

With the aid of a device equipped with two semipermeable membranes, it is possible to conceive of separating in a reversible manner a mixture of two inert ideal gases. The vessel depicted in Fig. 11-3 is divided into two equal compartments by a rigid wall, which is a membrane permeable only to the gas A_1. Two pistons coupled so that they move together at a constant distance apart are constructed of materials such that one is impermeable to all gases and the other is permeable only to the gas A_2. The initial state is

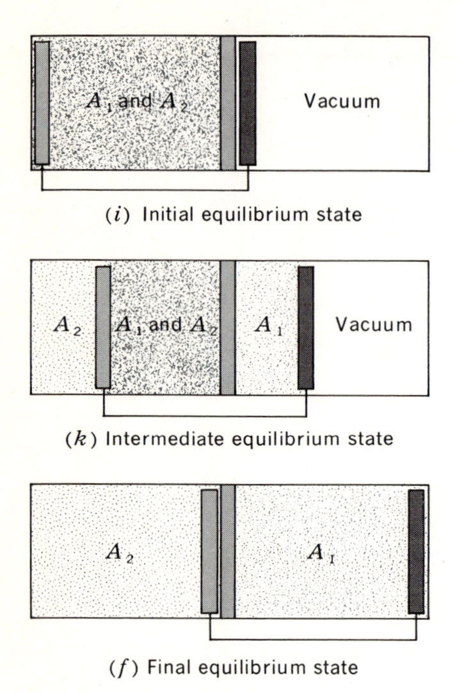

Fig. 11-3
Reversible isothermal separation of two inert ideal gases.

depicted in Fig. 11-3*i*. A mixture of A_1 and A_2 is in the left-hand chamber, and the right-hand chamber is evacuated.

Now imagine pushing the coupled pistons to the right in such a manner that the following conditions are satisfied:

1 The motion is infinitely slow, so that membrane equilibrium exists at all times.
2 There is no friction.
3 The whole system is kept at constant temperature.

These conditions define a *reversible isothermal process*. Consider the system at any intermediate state such as that depicted in Fig. 11-3*k*. If p_1 and p_2 are the partial pressures, respectively, of A_1 and A_2 in the mixture, P_1 is the pressure of A_1 alone, and P_2 is the pressure of A_2 alone, then the forces acting on the coupled pistons are:

$$\text{Force to the left} = (p_1 + p_2) \times \text{area},$$

and $$\text{Sum of the forces to the right} = (P_1 + P_2) \times \text{area}.$$

Since membrane equilibrium exists, $p_1 = P_1$ and $p_2 = P_2$; whence the resultant force acting on the coupled pistons is zero. After the pistons have moved all the way to the right, the gases are completely separated, as shown in Fig. 11-3*f*.

Since the resultant force is infinitesimal in the beginning and zero throughout the remainder of the process, $W = 0$. Also, since the process is isothermal and the internal energy of an ideal gas is a function of T only, the total internal energy of the system must be unchanged during the process. Thus

$$U^t_{\text{final}} = U^t_{\text{initial}}.$$

Thus we conclude that the total internal energy of the initial mixture of ideal gases is just the sum of the internal energies of individual gases in the final state where each occupies alone the initial volume of the mixture at the same temperature.

Since ΔU^t and W for the process are zero, application of the first law shows that Q is also zero. Since the process is reversible and isothermal,

$$Q = T(S^t_{\text{final}} - S^t_{\text{initial}}) = 0,$$

and since T is not zero, we have the result that

$$S^t_{\text{final}} = S^t_{\text{initial}}.$$

Our conclusion with respect to the entropy is therefore analogous to that for the internal energy: The total entropy of the initial mixture of ideal gases at temperature T is the sum of the entropies of the two pure gases at the same temperature when each occupies separately the initial mixture volume. This proposition may be extended to other properties of ideal-gas mixtures. Since

$$H = U + PV = U + RT,$$

it is obviously applicable to the enthalpy. Moreover,

$$G = H - TS,$$

and its extension to the Gibbs function is direct.

One finds that this proposition is valid for all extensive properties of an ideal-gas mixture. Its formal statement is known as Gibbs' theorem: *A total thermodynamic property* (nU, nC_V, nH, nC_P, nS, nA, *or* nG) *of a mixture of ideal gases is the sum of the properties that the individual gases would have if each occupied the total mixture volume alone at the same temperature.*

Suppose that we have a mixture of ideal gases occupying a total volume V^t. Let the numbers of moles of the individual gases be represented by n_1, n_2, n_3, etc. The total number of moles n is given by

$$n = \Sigma n_i.$$

According to Gibbs' theorem the total internal energy of the gas mixture is

$$nU = \Sigma(n_i U_{0_i}),$$

where U_{0_i} is the molar internal energy of pure i when n_i moles of it alone occupy the total volume V^t at the partial pressure $y_i P$.

For an ideal gas, the internal energy is a function of temperature only. Hence U_i is not changed by increasing the pressure from $y_i P$ to P at constant temperature. The above equation may therefore equally well be written

$$nU = \Sigma(n_i U_i),$$

where U_i is the molar internal energy of pure i at the mixture T and P. Division by n yields

$$\boxed{U = \Sigma(y_i U_i)} \quad \text{(ideal gas)}. \tag{11-13}$$

The heat capacities and the enthalpy for ideal gases are also independent of pressure. Therefore completely analogous derivations for these properties show that

$$C_V = \Sigma(y_i C_{V_i}) \quad \text{(ideal gas)}, \tag{11-14}$$

$$H = \Sigma(y_i H_i) \quad \text{(ideal gas)}, \tag{11-15}$$

$$C_P = \Sigma(y_i C_{P_i}) \quad \text{(ideal gas)}. \tag{11-16}$$

The entropy of an ideal gas *is* a function of pressure. It was shown in Sec. 8-3 that for an infinitesimal change of state for an ideal gas,

$$dS = C_P \frac{dT}{T} - R \frac{dP}{P}. \tag{8-4}$$

At constant T, this becomes

$$dS = -R \frac{dP}{P}.$$

For a pressure change from $y_i P$ to P, the entropy changes from S_{0_i} to S_i, and integration gives

$$S_i - S_{0_i} = -R \ln \frac{P}{y_i P} = R \ln y_i,$$

and thus

$$S_{0_i} = S_i - R \ln y_i.$$

By Gibbs' theorem, the total entropy of a mixture of ideal gases is

$$nS = \Sigma(n_i S_{0_i}).$$

Substitution for S_{0_i} yields $nS = \Sigma(n_iS_i) - R\Sigma(n_i \ln y_i)$. We may divide this equation by n to get

$$\boxed{S = \Sigma(y_iS_i) - R\Sigma(y_i \ln y_i)} \qquad \text{(ideal gas)}, \quad (11\text{-}17)$$

or
$$S - \Sigma(y_iS_i) = R\Sigma\left(y_i \ln \frac{1}{y_i}\right).$$

In this equation S is the entropy of 1(mol) of a mixture of ideal gases at temperature T and pressure P, in which the compositions of the constituents are given by the mole fractions y_i. The S_i's are the molar entropies of the pure constituents *at T and P*. Thus the difference $S - \Sigma(y_iS_i)$ represents the entropy change upon mixing ideal gases at constant T and P. According to the last equation, this entropy change of mixing is always positive in sign.

As an example, consider the mixing of $\frac{1}{2}$(mol) of helium and $\frac{1}{2}$(mol) of neon to form 1(mol) of an equimolar mixture at constant T and P. Then

$$S - \Sigma(y_iS_i) = R(\tfrac{1}{2} \ln 2 + \tfrac{1}{2} \ln 2)$$
$$= R \ln 2.$$

On the right-hand side of this expression there are no quantities such as heat capacities that distinguish one gas from another. The result is the same for the mixing of any two inert ideal gases, no matter how similar or dissimilar they are. If, however, the two gases are identical, the concept of mixing has no meaning, and there is no entropy change.

The application of mathematics to the macroscopic processes of nature usually gives rise to continuous results. Our experience suggests that as the two gases become more and more nearly alike, the entropy change due to mixing should get smaller and smaller, approaching zero as the gases become identical. The fact that this is not the case is known as *Gibbs' paradox*. The paradox has been resolved by Bridgman in the following way: To recognize that two gases are dissimilar requires a set of experimental operations. These operations become more and more difficult as the gases become more and more alike, but the operations are possible, at least in principle. In the limit, when the gases become identical, there is a discontinuity in the instrumental operations inasmuch as no instrumental operation exists by which the gases may be distinguished. Hence a discontinuity in a function such as that of an entropy change is to be expected.

The Gibbs function is defined by the equation

$$G = H - TS.$$

If G, H, and S in this equation represent the molar properties of a mixture of ideal gases, we may substitute for H and S by the equations just developed:

$$G = \Sigma(y_i H_i) - T\Sigma(y_i S_i) + RT\Sigma(y_i \ln y_i)$$

$$= \Sigma y_i(H_i - TS_i) + RT\Sigma(y_i \ln y_i)$$

$$= \Sigma(y_i G_i) + RT\Sigma(y_i \ln y_i),$$

or

$$\boxed{G = \Sigma y_i(G_i + RT \ln y_i)} \qquad \text{(ideal gas)}. \qquad (11\text{-}18)$$

11-6 Equilibrium and Nonequilibrium States

Consider a homogeneous mixture of 1(mol) of hydrogen and 1(mol) of oxygen at room temperature and at atmospheric pressure. It is a well-known fact that this mixture will remain indefinitely at the same temperature, pressure, and composition. The most careful measurements over a long period of time will disclose no appreciable spontaneous change of state. One might be inclined to deduce from this that such a mixture represents a system in a state of thermodynamic equilibrium. This, however, is not the case. If a small piece of platinized asbestos is introduced, or if an electric spark is created across two electrodes, an explosion takes place involving a sudden change in the temperature, the pressure, and the composition. If at the end of the explosion the system is brought back to the same temperature and pressure, it will be found that the composition is now $\frac{1}{2}$(mol) of oxygen, no measurable amount of hydrogen, and 1(mol) of water.

The piece of material such as platinized asbestos by whose agency a chemical reaction is promoted is known as a *catalyst*. If chemical combination occurs in a mixture of 1(mol) of hydrogen and 1(mol) of oxygen with different amounts and different kinds of catalysts, and the final composition of the mixture is measured in each case, it is found that (1) the final composition does not depend upon the amount of catalyst used; (2) the final composition does not depend on the kind of catalyst used; (3) the catalyst itself is the same at the end of the reaction as at the beginning. These results lead us to the following conclusions:

1 The initial state of the mixture is a state of mechanical and thermal equilibrium but not of chemical-reaction equilibrium.
2 The final state is a state of thermodynamic equilibrium.
3 The transition from the initial nonequilibrium state to the final equilibrium state is accompanied by a chemical reaction that is too slow to be measured when it takes place spontaneously. Through the agency of the catalyst the reaction is caused to take place more rapidly.

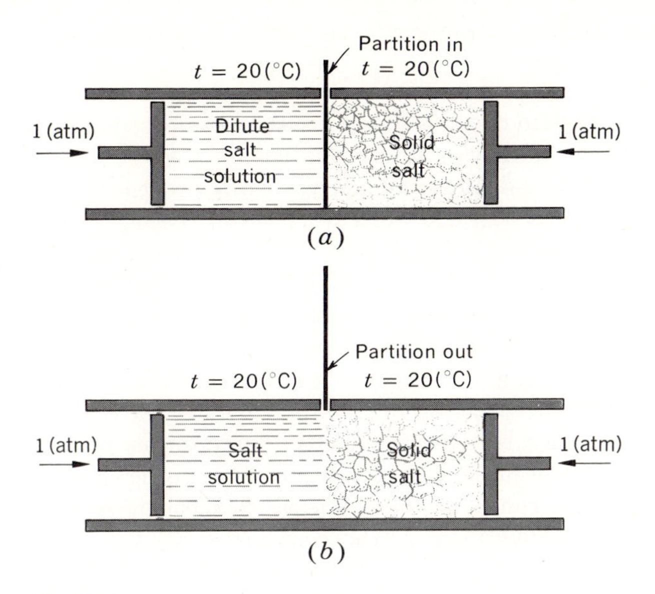

Fig. 11-4
Transport of matter across the boundary between two phases.

Imagine a vessel divided into two compartments by a removable partition as shown in Fig. 11-4*a*. Suppose that one compartment contains a dilute solution of sodium chloride and water which is maintained at a pressure of 1(atm) and at a temperature of 20(°C), the mole fraction of the salt being, say, 0.01. Under these conditions the solution is in thermodynamic equilibrium. Suppose that the other compartment contains solid salt in equilibrium also at a pressure of 1(atm) and a temperature of 20(°C). Now imagine that the partition is removed (Fig. 11-4*b*) and that the pressure and temperature of the whole system are kept constant at the original values. Experiment shows that some solid salt dissolves; i.e., the mole fraction of the salt in the solution increases spontaneously at constant pressure and temperature. After a while, the change ceases and the mole fraction is found to be about 0.1.

Focusing our attention on the solution from the moment it was put in contact with the solid salt, we are led to the following conclusions:

1 The initial state of the solution (at the moment it was put in contact with the solid salt) is one of mechanical and thermal equilibrium but not of phase equilibrium.
2 The final state of the solution is a state of thermodynamic equilibrium.
3 The transition from the initial nonequilibrium state to the final equilib-

rium state is accompanied by a transport of a chemical constituent into the solution.

A phase is defined as a system or a portion of a system composed of any number of chemical species satisfying the requirements that (1) it is homogeneous and (2) it has a definite boundary. The hydrogen-oxygen mixture just described is a gaseous phase of two chemical species and of constant mass. The salt solution is a liquid phase of two chemical species, whose mass, when it is in contact with the solid-salt phase, is variable. Although the initial states of both these phases are nonequilibrium states, it is possible to describe them in terms of thermodynamic coordinates. Since each phase is in mechanical and thermal equilibrium, a definite P and T may be ascribed to each; since each has a definite boundary, each has a definite volume; and since each is homogeneous, the composition of each phase may be described by specifying the number of moles of each constituent. In general, a phase consisting of m species in mechanical and thermal equilibrium may be described by the coordinates P, V, T, n_1, $n_2, \ldots, n_m$.

Under a given set of conditions a phase may undergo a change of state in which some or all of these coordinates change. While this is going on, the phase passes through states, not of thermodynamic equilibrium but of mechanical and thermal equilibrium only. For homogeneous phases these states are connected by an equation of state that is a relation among P, V, T, and the n_i's. Whether a phase is in chemical-reaction or phase equilibrium or not, it has a definite internal energy and enthalpy. Both U and H may be regarded as functions of P, V, T, and the n_i's, and upon elimination of one of the coordinates by means of the equation of state, U and H may be expressed as functions of the n_i's and any two of P, V, and T. We shall assume that the entropy of a phase, and therefore the Helmholtz and Gibbs functions, may be similarly expressed.

During a change of state, the n_i's, which determine the composition of the phase, change either by virtue of a chemical reaction or by virtue of a transport of matter across the boundaries between phases, or both. In general, under given conditions, there is a set of values of the n_i's for which the phase is in *thermodynamic* equilibrium. The functions that express the properties of a phase when it is not in chemical-reaction or phase equilibrium must obviously reduce to those for thermodynamic equilibrium when the equilibrium values of the n_i's are substituted.

Since the properties of a system in mechanical and thermal equilibrium depend only on the n_i's and any two of P, V, and T, it is reasonable to postulate that *these properties are given by functions of the same form as apply when the phase is in thermodynamic equilibrium.* For example, the molar Gibbs function for a mixture of inert ideal gases is

$$G = \Sigma y_i(G_i + RT \ln y_i). \qquad (11\text{-}18)$$

According to the assumption just made, this same equation may be used in connection with an ideal-gas phase in mechanical and thermal equilibrium when the gases are chemically active, when the phase is in contact with other phases, or under both conditions, whether thermodynamic equilibrium exists or not. Under these conditions the y_i's are variables. Whether they are all independent variables or not is a question that cannot be answered until the conditions under which a change of state takes place are specified. It is clear that if the mass of the phase remains constant and the gases are inert, the y_i's are constants. If the mass of the phase remains constant and the gases are chemically active, it will be shown in Sec. 11-9 that each y_i is a function of only one independent variable, called the reaction coordinate.

A system composed of two or more phases is called a *heterogeneous system.* An extensive property, such as V^t, H^t, S^t, or G^t, for any single phase p may be expressed as a function of T, P, and the n_i's for that phase; for example,

$$(G^t)^p = (nG)^p = g(T, P, n_1{}^p, n_2{}^p, \ldots).$$

Since these properties *are* extensive, the total property for the whole of a heterogeneous system is the sum of the properties of the phases. Thus, for example, the total Gibbs function for a two-phase system consisting of liquid (l) and vapor (v) is given by

$$G^t = nG = (nG)^l + (nG)^v.$$

11-7 Conditions for Chemical-Reaction and Phase Equilibrium

Consider any PVT system of constant mass, either homogeneous or heterogeneous, in mechanical and thermal equilibrium but not in phase equilibrium or chemical-reaction equilibrium. Suppose that the system is in contact with a reservoir at temperature T and undergoes an infinitesimal irreversible process involving an exchange of heat δQ with the reservoir. The process may involve a chemical reaction, a transport of matter between phases, or both. Let dS^t denote the total entropy change of the *system,* and dS_0 the total entropy change of the reservoir. The total entropy change of the system plus its surroundings is therefore $dS_0 + dS^t$; and since the performance of an irreversible process is attended by an increase in the total entropy, we have by the second law that

$$dS_{\text{total}} = dS_0 + dS^t > 0.$$

Since $$dS_0 = -\frac{\delta Q}{T},$$

we have $$-\frac{\delta Q}{T} + dS^t > 0,$$

or $$\delta Q - T\,dS^t < 0.$$

During the infinitesimal irreversible process the total internal energy of the system changes by an amount dU^t, and an amount of work $-P\,dV^t$ is performed. The first law for the system is

$$\delta Q = dU^t + P\,dV^t,$$

and the inequality becomes

$$\boxed{dU^t + P\,dV^t - T\,dS^t < 0.} \tag{11-19}$$

This inequality holds during any infinitesimal portion and therefore during all infinitesimal portions of the irreversible process. According to the assumption made in the preceding section, U^t, V^t, and S^t may all be regarded as functions of thermodynamic coordinates.

During the irreversible process for which the above inequality holds, some or all of the coordinates may change. If we restrict the irreversible process by imposing the condition that two of the thermodynamic coordinates remain constant, the inequality can be reduced to a simpler form. Suppose, for example, that the total internal energy and the total volume remain constant. Then the inequality reduces to $dS^t > 0$, which means that the total entropy of a system at constant U^t and V^t increases during an irreversible process, approaching a maximum at the final state of equilibrium. This result, however, is obvious from the entropy principle, since a system at constant U^t and V^t is isolated and therefore, in effect, has no surroundings. The most useful set of conditions is that in which T and P remain constant.

If T and P are constant, the inequality reduces to

$$\left.\begin{aligned} d(U^t + PV^t - TS^t) &< 0, \\ \text{or} \qquad dG^t &< 0, \end{aligned}\right\} \qquad (\text{const } T, P),$$

expressing the result that *the total Gibbs function of a system at constant T and P decreases during an irreversible process.* Every increment of such a process must result in a decrease in the Gibbs function of the system and must bring the system closer to equilibrium. Hence the equilibrium state, if reached along a path of constant T and P, must be the one having the minimum Gibbs function. Thus, at the equilibrium state

$$\boxed{d(nG)_{T,P} = 0,} \tag{11-20}$$

where G is now the *molar* Gibbs function, and n is the total moles of the system. This equation is a general criterion of equilibrium. It is *not* necessary in the application of this criterion that the system considered actually reaches the equilibrium state along a path of constant T and P. Once an equilibrium state is established, no further changes occur, and the system exists in this state at a particular T and P. How this state was *actually* attained is not important, and for purposes of calculation one may as well take the path followed in getting there to have been one of constant T and P.

11-8 Chemical Potential. Phase Equilibrium and Phase Rule

For a closed PVT system containing 1(mol) of material, the total differential of the Gibbs function is given by Eq. (9-6):

$$dG = -S\,dT + V\,dP. \tag{9-6}$$

If the system contains $n(= \Sigma n_i)$ moles of a constant-composition mixture, the corresponding expression is

$$dG^t = -S^t\,dT + V^t\,dP,$$

or

$$d(nG) = -(nS)\,dT + (nV)\,dP.$$

Analogous to Eqs. (9-9) and (9-11), we then have

$$\left[\frac{\partial(nG)}{\partial T}\right]_{P,n} = -nS \qquad \text{and} \qquad \left[\frac{\partial(nG)}{\partial P}\right]_{T,n} = nV.$$

Consider now the more general case where the n_i may vary, either because of chemical reaction or transfer of material across the boundaries of the system. According to the discussion in Sec. 11-6, the total property $G^t = (nG)$ must then be treated as a function of the n_i, as well as of T and P, and we may write for a general differential change of state that

$$d(nG) = \left[\frac{\partial(nG)}{\partial T}\right]_{P,n} dT + \left[\frac{\partial(nG)}{\partial P}\right]_{T,n} dP + \sum \left[\frac{\partial(nG)}{\partial n_i}\right]_{T,P,n_j} dn_i.$$

The subscript n_j on the mole-number derivatives of (nG) indicates that all mole numbers other than n_i are held constant in the differentiation.

The last equation is a purely mathematical statement, and merely represents a formal extension of Eq. (9-6) to an *open* PVT system of variable composition. Replacing the T and P derivatives of (nG) by $-(nS)$ and (nV), we have

$$d(nG) = -(nS)\,dT + (nV)\,dP + \sum \left[\frac{\partial(nG)}{\partial n_i}\right]_{T,P,n_j} dn_i.$$

The mole-number derivatives of (nG) have a special significance to chemical thermodynamics, and therefore they are given a special symbol and name. Thus,

$$\mu_i \equiv \left[\frac{\partial(nG)}{\partial n_i}\right]_{T,P,n_j}, \tag{11-21}$$

where μ_i is the *chemical potential* of chemical species i. Our general equation for $d(nG)$ then becomes

$$\boxed{d(nG) = -(nS)\,dT + (nV)\,dP + \sum \mu_i\,dn_i.} \tag{11-22}$$

Since

$$n_i = nx_i,$$

$$dn_i = x_i\,dn + n\,dx_i.$$

Furthermore,

$$d(nG) = n\,dG + G\,dn,$$

and Eq. (11-22) may be written

$$n\,dG + G\,dn = -nS\,dT + nV\,dP + \Sigma\mu_i n\,dx_i + \Sigma x_i\mu_i\,dn.$$

Rearrangement gives

$$(dG + S\,dT - V\,dP - \Sigma\mu_i\,dx_i)n + (G - \Sigma x_i\mu_i)\,dn = 0.$$

Since n and dn are independent and arbitrary, the quantities in parentheses must each be zero, and we can write

$$dG = -S\,dT + V\,dP + \Sigma\mu_i\,dx_i$$

and

$$\boxed{G = \Sigma x_i\mu_i.} \tag{11-23}$$

The first of these equations is merely a restatement of Eq. (11-22) as written for *one mole of mixture.* The final term is therefore constrained by the condition that $\Sigma\,dx_i = 0$. Equation (11-23) shows that *the molar Gibbs function for a solution is equal to the mole-fraction-weighted sum of the chemical potentials of its constituents.*

Example 11-3

Determine an expression for the chemical potential of component i in an ideal-gas mixture.

An expression for G of an ideal-gas mixture is given by Eq. (11-18):

$$G = \Sigma y_i(G_i + RT \ln y_i).$$

If Eq. (11-23) is rewritten with y_i replacing x_i as the mole fraction, we have

$$G = \Sigma y_i \mu_i.$$

Comparison of these two equations shows that for a component in a mixture of ideal gases the expression for μ_i is simply

$$\mu_i = G_i + RT \ln y_i \qquad \text{(ideal gas)}. \tag{11-24}$$

The value of the introduction of μ_i as a new thermodynamic function is that the general equilibrium criterion given by Eq. (11-20) may be put into more convenient forms by use of chemical potentials. For a single phase at constant T and P, Eq. (11-22) reduces to

$$d(nG)_{T,P} = \Sigma \mu_i \, dn_i,$$

and combination with Eq. (11-20) gives

$$\Sigma \mu_i \, dn_i = 0. \tag{11-25}$$

The application of this equation to problems in single-phase chemical-reaction equilibrium is considered in Sec. 11-10.

If several phases are present in a system, we can write for each phase p that

$$d(nG)^p_{T,P} = \Sigma \mu_i^p \, dn_i^p.$$

For the case of vapor(v)-liquid(l) equilibrium, we thus have

$$d(nG)^l_{T,P} = \Sigma \mu_i^l \, dn_i^l \qquad \text{and} \qquad d(nG)^v_{T,P} = \Sigma \mu_i^v \, dn_i^v.$$

But according to the discussion in Sec. 11-6, nG for the whole system (vapor and liquid) is the sum of the nG's for the two phases. Therefore, by Eq. (11-20),

$$d(nG)_{T,P} = d(nG)^l_{T,P} + d(nG)^v_{T,P} = 0;$$

hence,

$$\Sigma \mu_i^l \, dn_i^l + \Sigma \mu_i^v \, dn_i^v = 0. \tag{11-26}$$

We wish to apply this equilibrium criterion to the special case of vapor-liquid equilibrium in a system containing two chemical species, 1 and 2. We imagine a closed system in which there is a liquid phase consisting of n_1^l(mol) of 1 and n_2^l(mol) of 2, and a vapor phase containing n_1^v(mol) of 1 and n_2^v(mol) of 2.

Whether the system is in equilibrium or not, there is in general a continual exchange of matter between the two phases. The system is closed, and since it is assumed that no chemical reactions occur, the *total* number of moles of each species in the system must remain constant. Thus, for the present case,

$$n_1^l + n_1^v = n_1 = \text{const},$$

and

$$n_2^l + n_2^v = n_2 = \text{const}.$$

For a differential exchange of material between phases, we therefore have

$$dn_1^l + dn_1^v = 0$$

and

$$dn_2^l + dn_2^v = 0,$$

or

$$dn_1^v = -dn_1^l$$

and

$$dn_2^v = -dn_2^l.$$

The last equations are merely equations of *material balance* and have nothing to do with the thermodynamics of the situation. What we must do next is combine them with Eq. (11-26) in order to determine the consequences of the equilibrium criterion.

For the present case, the expanded form of Eq. (11-26) contains four terms, which result from sums taken over two chemical species. Thus,

$$\mu_1^l\, dn_1^l + \mu_1^v\, dn_1^v + \mu_2^l\, dn_2^l + \mu_2^v\, dn_2^v = 0.$$

But the four dn_i^p are not independent; they are related by the two equations of material balance. Combination of the last three equations then gives

$$(\mu_1^l - \mu_1^v)\, dn_1^l + (\mu_2^l - \mu_2^v)\, dn_2^l = 0. \qquad (11\text{-}27)$$

Now the remaining dn_i^p are completely unrestricted and hence may take on arbitrary values. In order for Eq. (11-27) to be satisfied, then, we must have

$$\mu_1^l = \mu_1^v \qquad \text{and} \qquad \mu_2^l = \mu_2^v.$$

The chemical potential of each species must be the same in both phases in order to satisfy the criterion of phase equilibrium.

Although this result has been developed for a two-component, vapor-liquid system, it is readily generalized so as to apply to more complex systems in phase equilibrium. For each additional component, Eq. (11-27) would include an additional term on the left, and we would conclude immediately that

$$\mu_i^\alpha = \mu_i^\beta \qquad (i = 1, 2, \ldots, m),$$

where m is the total number of chemical species in the system, and α and β identify the two equilibrium phases. For additional phases, we would consider the equilibrium requirements for all possible pairs of phases and conclude that for multiple phases at the same T and P, the equilibrium criterion can be satisfied only when the chemical potential of each species in the system is the same in all phases. Mathematically, this is expressed for a π-phase system by the $m(\pi - 1)$ equations

$$\left.\begin{array}{r}\mu_i^{\alpha} = \mu_i^{\pi} \\ \mu_i^{\beta} = \mu_i^{\pi} \\ \cdots\cdots\cdots \\ \mu_i^{\pi-1} = \mu_i^{\pi}\end{array}\right\} \qquad (i = 1, 2, \ldots, m), \tag{11-28}$$

or, more concisely, by

$$\boxed{\mu_i^{\alpha} = \mu_i^{\beta} = \cdots = \mu_i^{\pi}} \qquad (i = 1, 2, \ldots, m). \tag{11-29}$$

Equation (11-28) or (11-29) is the practical or working equation that forms the basis for phase-equilibrium calculations for PVT systems of uniform temperature and pressure. Although developed for the case of just phase equilibrium, both also apply to multiphase systems in which there may occur chemical reactions.

The application of Eq. (11-29) to the solution of specific phase-equilibrium problems requires the availability of *models* of solution behavior, which provide expressions for G or for the μ_i in terms of the measurable coordinates which characterize the system. The simplest such expressions are those presented earlier for ideal-gas mixtures, viz.,

$$G = \Sigma y_i(G_i + RT \ln y_i), \tag{11-18}$$

$$\mu_i = G_i + RT \ln y_i. \tag{11-24}$$

It is convenient to generalize these expressions to include the broader class of *ideal-solution* behavior. Accordingly, we define an ideal solution to be *any* solution (gas, liquid, or solid) for which μ_i is given by

$$\mu_i = G_i + RT \ln x_i \qquad \text{(ideal solution)}, \tag{11-30}$$

for all T and P and for every component. Here x_i is the general symbol for the mole fraction of component i in solution. Although Eq. (11-30) appears formally identical to Eq. (11-24), it does not include in its definition the other special restrictions which were necessary to the development of the ideal-gas-mixture equations. In particular, there are no implied restrictions on the volumetric behavior of pure component i, as there were for the ideal-gas case. Thus an ideal-gas mixture may be considered a special case of an ideal solution.

Example 11-4

Develop an expression which relates the liquid-phase mole fraction x_i to the vapor-phase mole fraction y_i for the case of an ideal liquid solution in phase equilibrium with a vapor phase at a pressure low enough that the vapor is an ideal gas.

The basis for the derivation is Eq. (11-29), which becomes for the present case

$$\mu_i{}^l = \mu_i{}^v \qquad (i = 1, 2, \ldots, m).$$

Substitution of the ideal-solution equation for the μ_i gives

$$G_i{}^l + RT \ln x_i = G_i{}^v + RT \ln y_i,$$

from which we find

$$RT \ln \frac{y_i}{x_i} = G_i{}^l - G_i{}^v \qquad (i = 1, 2, \ldots, m).$$

Differentiating at constant T, we get

$$RT\, d \ln \frac{y_i}{x_i} = dG_i{}^l - dG_i{}^v.$$

Since $dG_i{}^l$ and $dG_i{}^v$ apply to pure i, Eq. (9-6) yields

$$dG_i{}^l = V_i{}^l\, dP \qquad \text{and} \qquad dG_i{}^v = V_i{}^v\, dP.$$

Therefore

$$RT\, d \ln \frac{y_i}{x_i} = (V_i{}^l - V_i{}^v)\, dP.$$

Since $V_i{}^v \gg V_i{}^l$ and since $V_i{}^v = RT/P$, this becomes

$$RT\, d \ln \frac{y_i}{x_i} = -\frac{RT}{P}\, dP = -RT\, d \ln P$$

or

$$d \ln \frac{y_i}{x_i} = -d \ln P.$$

Integration provides

$$\ln \frac{y_i}{x_i} = -\ln P + \ln K,$$

where $\ln K$ represents the constant of integration. Thus

$$y_i P = K x_i.$$

When $y_i = 1$, then $x_i = 1$, and $P = P_i(\text{sat})$, and we find that $K = P_i(\text{sat})$. Therefore

$$\boxed{y_i P = x_i P_i(\text{sat}) \qquad (i = 1, 2, \ldots, m).} \qquad (11\text{-}31)$$

This expression is known as *Raoult's law*. An important feature of Raoult's law is that the composition, pressure, and temperature effects are *separated*; y_i, x_i, and P appear explicitly as separate terms, while T appears implicitly through $P_i(\text{sat})$, which depends upon temperature only. Several assumptions were made in the derivation of Eq. (11-31), the least realistic of these being the assumption of ideal-solution behavior for the liquid phase. Raoult's law works best at low pressures and for systems containing chemically similar species; for other applications, it often provides no more than a first approximation to actual behavior.

For a binary solution of components 1 and 2, Eq. (11-31) is written for each component:

$$y_1 P = x_1 P_1(\text{sat}), \tag{A}$$

$$y_2 P = x_2 P_2(\text{sat}). \tag{B}$$

Since $y_1 + y_2 = 1$, the sum of (A) and (B) gives

$$P = x_1 P_1(\text{sat}) + x_2 P_2(\text{sat}).$$

Substitution of $x_2 = 1 - x_1$ yields

$$P = P_2(\text{sat}) + [P_1(\text{sat}) - P_2(\text{sat})]\, x_1. \tag{C}$$

The pure-component saturation pressures $P_1(\text{sat})$ and $P_2(\text{sat})$ are functions of temperature only, and therefore the relation between P and x_1 at constant T is linear.

Consider as an example the acetone(1)-acetonitrile(2) system at 50(°C). This system is known to conform closely to Raoult's law. At 50(°C) the following vapor

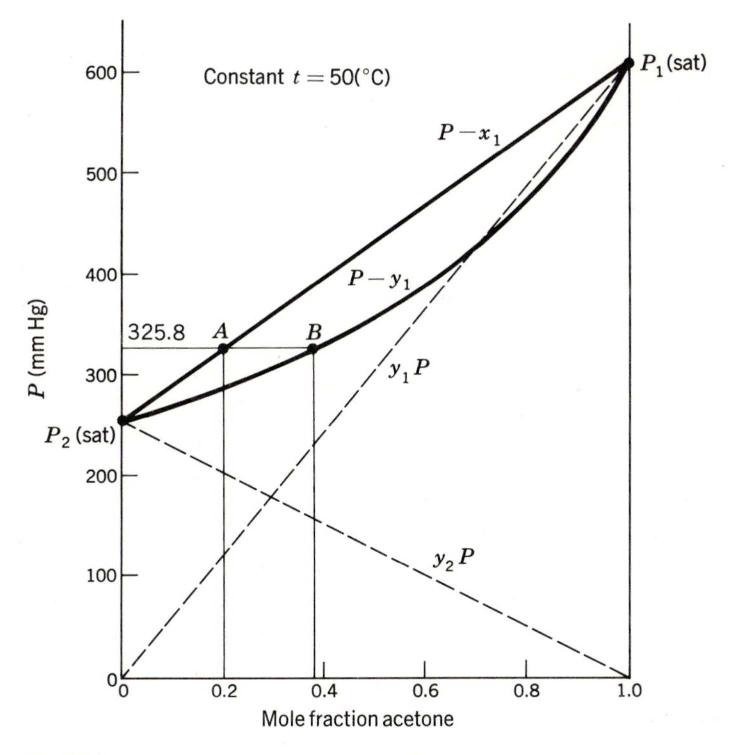

Fig. 11-5

P-x-y diagram for acetone(1)—acetonitrile(2).

pressures are available:

$$P_1(\text{sat}) = 615.0(\text{mm Hg}) \qquad \text{and} \qquad P_2(\text{sat}) = 253.5(\text{mm Hg}).$$

Substitution of these values into Eq. (*C*) yields the expression

$$P = 253.5 + 361.5x_1,$$

from which we may calculate values of P for various values of x_1 for a temperature of 50(°C). The corresponding values of y_1 are then found from Eq. (*A*):

$$y_1 = \frac{x_1 P_1(\text{sat})}{P} = \frac{615.0x_1}{P}.$$

For a value of $x_1 = 0.2$, we have

$$P = 253.5 + (361.5)(0.2) = 325.8(\text{mm Hg}),$$

and

$$y_1 = \frac{(615.0)(0.2)}{325.8} = 0.378.$$

These results mean that at 50(°C) a liquid mixture containing 20 mole percent acetone and 80 mole percent acetonitrile is in equilibrium with a vapor containing 37.8 mole percent acetone at a pressure of 325.8 (mm Hg). Calculations made for a number of values of x_1 at 50(°C) provide the results of the accompanying table. These same results are shown graphically in Fig. 11-5. Points *A* and *B* represent the equilibrium state at $x_1 = 0.2$, $P = 325.8$(mm Hg), and $y_1 = 0.378$ for which calculations were illustrated.

x_1	P(mm Hg)	y_1
0	253.8	0
0.2	325.8	0.378
0.4	398.1	0.618
0.6	470.4	0.785
0.8	542.7	0.907
1.0	615.0	1.000

In addition to its quantitative role in the solution of phase-equilibrium problems, Eq. (11-28) provides the basis for what is known as the *phase rule,* which allows one to calculate the number of independent variables that may be arbitrarily fixed in order to establish the *intensive* state of a *PVT* system. Such a state is established when its temperature, pressure, and the compositions of all phases are fixed. However, for equilibrium states these variables are not all independent, and fixing a limited number of them automatically establishes the others. This number of independent variables is called the *number of degrees of freedom,* or the *variance,* of the system, and is given by the phase rule. It is this number of variables which may be arbitrarily specified and which *must* be specified in order to fix the intensive state of a *PVT* system at equilibrium. This number is merely the difference between the number of independent intensive variables associated with the system and the number of independent equations which may be written connecting these variables.

If m represents the number of chemical species in the system, then there are $(m - 1)$ independent mole fractions for each phase. (The requirement that the sum of the mole fractions must equal unity makes one mole fraction dependent.) For π phases there is a total of $(m - 1)(\pi)$ composition variables. In addition, the temperature and pressure, taken to be uniform throughout the system, are phase-rule variables, which means that the total number of independent variables is $2 + (m - 1)(\pi)$. The masses of the phases are not phase-rule variables, because they have nothing to do with the intensive state of the system.

Equation (11-28) shows that one may write $(\pi - 1)$ independent phase-equilibrium equations for each species and a total of $(\pi - 1)(m)$ such equations for a nonreacting system. Since the μ_i's are functions of temperature, pressure, and the phase compositions, these equations represent relations among the phase-rule variables. Subtraction of the number of independent equations from the number of independent variables gives the number of degrees of freedom F as

$$F = 2 + (m - 1)(\pi) - (\pi - 1)(m)$$

or

$$\boxed{F = 2 - \pi + m,} \tag{11-32}$$

which is the phase rule for a nonreacting PVT system.

The phase rule as given by Eq. (11-32) must be modified for application to PVT systems in which chemical reactions can occur. For each independent chemical reaction, we can write an additional equation which must be satisfied by the system at equilibrium, and the number of degrees of freedom is accordingly decreased. The simplest case is that of a *single* independent reaction, for which the modified form of Eq. (11-32) is

$$F = 1 - \pi + m.$$

If there is only a single phase $(\pi = 1)$, then this expression reduces to

$$F = m.$$

Example 11-5

Discuss with reference to the phase rule the three possible types of phase-equilibrium behavior of a nonreactive pure substance.

For a single pure substance, we have $m = 1$, and the phase rule, Eq. (11-32), gives

$$F = 2 - \pi + m = 2 - \pi + 1,$$

or

$$F = 3 - \pi. \tag{A}$$

The minimum value of π is $\pi = 1$, corresponding to a state of single-phase equilibrium. According to Eq. (A), we have for this case that $F = 2$, which means that *two* variables must be specified in order to fix the intensive state of the system. This is entirely in accord with our experience. For example, determination of the properties of superheated steam from the steam tables requires that we first specify numerical values for any *two* of the displayed thermodynamic coordinates, for example, T and P, T and S, P and H, etc.

The minimum allowable value of F is $F = 0$, corresponding to $\pi = 3$, that is, to a state of *three*-phase ("triple-point") equilibrium. Thus three is the maximum number of different phases of a pure material that can coexist in equilibrium, and, since $F = 0$, the temperature and pressure of such an equilibrium state are *fixed*. For example, there is only a single temperature [0.01(°C)] and a single pressure [4.58(mm Hg)] for which pure ice is in simultaneous equilibrium with pure liquid water and pure water vapor.

The third type of pure-substance phase equilibrium is *two*-phase equilibrium ($\pi = 2$), for which we find by Eq. (A) that there exists but a single degree of freedom. This means, for example, that the intensive states of the separate phases in an equilibrium mixture of liquid water and steam are completely determined by specification of the temperature and that the vapor-liquid saturation pressure P(sat) for any pure substance is a function of temperature only.

11-9 The Reaction Coordinate

If we introduce into a vessel an arbitrary mixture of methane, water vapor, carbon monoxide, and hydrogen, the chemical reaction that becomes possible is indicated by

$$CH_4 + H_2O \rightleftharpoons CO + 3H_2.$$

The chemical species on the left are called *reactants*, and those on the right are known as *products*, regardless of the direction in which the reaction actually proceeds in any particular instance. Symbolic representation of chemical reactions in general may be made as follows:

$$\nu_1' A_1 + \nu_2' A_2 \rightleftharpoons \nu_3' A_3 + \nu_4' A_4,$$

where the A_i's represent chemical species and the quantities ν_i' are called *stoichiometric coefficients*. They are the positive numbers that precede the chemical species and that "balance" the reaction. For the specific reaction considered

$$\nu_{CH_4}' = 1, \qquad \nu_{H_2O}' = 1, \qquad \nu_{CO}' = 1, \qquad \nu_{H_2}' = 3.$$

An alternative algebraic expression for the general chemical reaction is written:

$$\nu_1 A_1 + \nu_2 A_2 + \nu_3 A_3 + \nu_4 A_4 = 0. \qquad (11\text{-}33)$$

The quantities ν_i are called *stoichiometric numbers*, and are numerically equal to the corresponding stoichiometric coefficients. However, a sign convention is adopted that makes the value of ν_i *negative for a reactant* and *positive for a product.* Thus for the reaction

$$CH_4 + H_2O \rightleftharpoons CO + 3H_2,$$

$$\nu_{CH_4} = -1, \qquad \nu_{H_2O} = -1, \qquad \nu_{CO} = 1, \qquad \nu_{H_2} = 3.$$

We have written equations for just four chemical species as a matter of convenience; subsequent equations are of such a character that they are easily applied to reactions among any number of chemical species.

When a reaction as represented by Eq. (11-33) proceeds, the *changes* in the numbers of moles of the species present are directly related to the stoichiometric numbers. Thus for the reaction considered

$$\frac{\Delta n_{H_2O}}{\Delta n_{CH_4}} = \frac{\nu_{H_2O}}{\nu_{CH_4}} = \frac{-1}{-1} = 1,$$

and

$$\Delta n_{H_2O} = \Delta n_{CH_4}.$$

If 0.5(mol) of CH_4 disappears by reaction, then

$$\Delta n_{H_2O} = \Delta n_{CH_4} = -0.5(\text{mol}),$$

and 0.5(mol) of H_2O must also disappear. Similarly

$$\frac{\Delta n_{H_2}}{\Delta n_{CH_4}} = \frac{\nu_{H_2}}{\nu_{CH_4}} = \frac{3}{-1} = -3,$$

and

$$\Delta n_{H_2} = -3\ \Delta n_{CH_4}.$$

For $\Delta n_{CH_4} = -0.5(\text{mol})$, $\Delta n_{H_2} = 1.5(\text{mol})$, and this amount of hydrogen is formed by the reaction.

If we apply this principle for a differential amount of reaction according to Eq. (11-33), we may write

$$\frac{dn_2}{dn_1} = \frac{\nu_2}{\nu_1} \quad \text{or} \quad \frac{dn_2}{\nu_2} = \frac{dn_1}{\nu_1},$$

$$\frac{dn_3}{dn_1} = \frac{\nu_3}{\nu_1} \quad \text{or} \quad \frac{dn_3}{\nu_3} = \frac{dn_1}{\nu_1},$$

and

$$\frac{dn_4}{dn_1} = \frac{\nu_4}{\nu_1} \quad \text{or} \quad \frac{dn_4}{\nu_4} = \frac{dn_1}{\nu_1}.$$

Comparison of these equations shows that

$$\frac{dn_1}{\nu_1} = \frac{dn_2}{\nu_2} = \frac{dn_3}{\nu_3} = \frac{dn_4}{\nu_4}.$$

Each of these terms is related to an amount of reaction as represented by a change in the number of moles of a chemical species. Since all terms are equal, we can identify them collectively with a single expression arbitrarily defined to represent the amount of reaction. Thus we provide a definition of the product $k\,d\epsilon$ by writing

$$\boxed{\frac{dn_1}{\nu_1} = \frac{dn_2}{\nu_2} = \frac{dn_3}{\nu_3} = \frac{dn_4}{\nu_4} = k\,d\epsilon.} \tag{11-34}$$

The general relation between a differential change dn_i in the number of moles of a reacting species and the product $k\,d\epsilon$ is therefore

$$\boxed{dn_i = \nu_i k\,d\epsilon.} \tag{11-35}$$

We have in this way introduced two new quantities, k and ϵ, to which the amount of reaction of all species participating in a given reaction may be related. The quantity ϵ is called the *reaction coordinate,* and it characterizes the extent or degree to which a reaction has taken place.[1] The quantity k is a proportionality constant, which acts as a "scaling factor" and which may be chosen for each particular application so as to restrict values of ϵ to the range from $\epsilon = 0$ to $\epsilon = 1$. Thus when $\epsilon = 0$, the reaction is displaced as far to the left as possible, and when $\epsilon = 1$, it is displaced as far to the right as possible. These limits are reached when the supply of one or more of the reacting species is exhausted; this is determined by the numbers of moles of the various species making up the *initial* state of the system. The procedure by which k is determined is illustrated in the following example.

Example 11-6

For a system in which the reaction

$$CH_4 + H_2O \rightleftharpoons CO + 3H_2$$

occurs, assume that there are present initially 2(mol) of CH_4, 1(mol) of H_2O, 1(mol) of CO, and 4(mol) of H_2. Determine expressions for the n_i's as functions of ϵ.

In the initial state, the reaction coordinate ϵ is not zero. It becomes zero when the reaction shifts to the left to the extent that the original 1(mol) of CO disappears. In this event the system contains at $\epsilon = 0$:

3(mol) of CH_4, 2(mol) of H_2O, no CO, and 1(mol) of H_2.

If the reaction shifts to the right until the original 1(mol) of H_2O is exhausted, then

[1] The reaction coordinate ϵ has been given various other names, such as degree of advancement, degree of reaction, extent of reaction, and progress variable.

Table 11-2

	CH_4 +	$H_2O \rightleftharpoons$	CO +	$3H_2$
Stoichiometric number	−1	−1	1	3
Moles of chemical species initially present	2	1	1	4
Moles present after maximum shift of reaction left, $\epsilon = 0$	3	2	0	1
Moles present after maximum shift of reaction right, $\epsilon = 1$	1	0	2	7

the system contains for $\epsilon = 1$:

1(mol) of CH_4,   no H_2O,   2(mol) of CO,   and   7(mol) of H_2.

These results are summarized in Table 11-2.

For the given reaction, Eq. (11-34) becomes

$$\frac{dn_{CH_4}}{-1} = \frac{dn_{H_2O}}{-1} = \frac{dn_{CO}}{1} = \frac{dn_{H_2}}{3} = k\,d\epsilon,$$

and we may apply Eq. (11-35) to any of the four species, integrating from the state for which $\epsilon = 0$ to that for which $\epsilon = 1$. Thus for methane (CH_4)

$$\int_3^1 dn_{CH_4} = -1k \int_0^1 d\epsilon,$$

and

$$1 - 3 = -k, \qquad \text{or} \qquad k = 2.$$

Similarly for H_2O

$$\int_2^0 dn_{H_2O} = -1k \int_0^1 d\epsilon,$$

and

$$0 - 2 = -k, \qquad \text{or} \qquad k = 2.$$

These integrations produce exactly the same value for k, namely, $k = 2$, regardless of which species is considered.

Thus we have for the particular system of interest the four general relations which follow from Eq. (11-35):

$$dn_{CH_4} = -2d\epsilon,$$

$$dn_{H_2O} = -2d\epsilon,$$

$$dn_{CO} = 2d\epsilon,$$

$$dn_{H_2} = 6d\epsilon.$$

Integration of these expressions from the state where $\epsilon = 0$ (for which the values of n_i have already been determined) to any other state characterized by the value $\epsilon = \epsilon$

yields

$$n_{CH_4} = 3 - 2\epsilon, \qquad y_{CH_4} = \frac{3 - 2\epsilon}{6 + 4\epsilon},$$

$$n_{H_2O} = 2 - 2\epsilon, \qquad y_{H_2O} = \frac{1 - \epsilon}{3 + 2\epsilon},$$

$$n_{CO} = 2\epsilon, \qquad y_{CO} = \frac{\epsilon}{3 + 2\epsilon},$$

$$n_{H_2} = 1 + 6\epsilon, \qquad y_{H_2} = \frac{1 + 6\epsilon}{6 + 4\epsilon},$$

$$\Sigma n_i = 6 + 4\epsilon.$$

The y_i's given on the right are the mole fractions of the chemical species in the mixture for the given value of the reaction coordinate ϵ. Each y_i is given by the defining expression for mole fraction, $y_i = n_i/\Sigma n_i$. The composition of the system is seen to be a function of the single independent variable ϵ.

Example 11-7

Consider a vessel which initially contains just n_0(mol) of water vapor. If dissociation occurs according to the reaction

$$H_2O \rightarrow H_2 + \tfrac{1}{2}O_2,$$

find expressions which relate the number of moles of each chemical species to the reaction coordinate ϵ.

In the initial state $\epsilon = 0$, $n_{H_2O} = n_0$, and $n_{H_2} = n_{O_2} = 0$. If the reaction shifts to the right to the maximum possible extent, dissociation is complete, and we have $\epsilon = 1$, $n_{H_2O} = 0$, $n_{H_2} = n_0$, and $n_{O_2} = \frac{1}{2}n_0$.

Equation (11-34) applied to this reaction becomes

$$\frac{dn_{H_2O}}{-1} = \frac{dn_{H_2}}{1} = \frac{dn_{O_2}}{\frac{1}{2}} = k\,d\epsilon,$$

Integration for H_2O as ϵ goes from 0 to 1 gives

$$\int_{n_0}^{0} dn_{H_2O} = -1k \int_0^1 d\epsilon,$$

or

$$k = n_0.$$

Similar integration for the other species gives the same result.

Application of Eq. (11-35) to each species now provides the relations

$$dn_{H_2O} = -n_0\,d\epsilon,$$

$$dn_{H_2} = n_0\,d\epsilon,$$

$$dn_{O_2} = \tfrac{1}{2}n_0\,d\epsilon.$$

Upon integration from the state for which $\epsilon = 0$ to a state where $\epsilon = \epsilon$, these become:

$$n_{H_2O} = n_0(1 - \epsilon), \qquad y_{H_2O} = \frac{1 - \epsilon}{1 + \epsilon/2},$$

$$n_{H_2} = n_0\epsilon, \qquad y_{H_2} = \frac{\epsilon}{1 + \epsilon/2},$$

$$n_{O_2} = \tfrac{1}{2}n_0\epsilon, \qquad y_{O_2} = \frac{\epsilon/2}{1 + \epsilon/2},$$

$$\Sigma n_i = n_0\left(1 + \frac{\epsilon}{2}\right).$$

In this example, ϵ can be identified with the fractional dissociation of the water vapor initially present.

11-10 Chemical-Reaction Equilibrium in Ideal Gases

We showed in Sec. 11-7 that the total Gibbs function of a closed system at constant T and P must decrease during an irreversible process and that the condition for equilibrium in such a system is

$$d(nG)_{T,P} = 0. \qquad (11\text{-}20)$$

Thus if a mixture of gases is not in chemical equilibrium, any reaction that occurs must be irreversible, and if the system is maintained at constant T and P, the total Gibbs function of the system must decrease. The significance of this is seen from Fig. 11-6, which shows a schematic diagram of nG vs. ϵ, the reaction coordinate. Since ϵ is the single variable that characterizes the extent of reaction, and therefore the composition of the system, the total Gibbs function at constant T and P is determined by ϵ. The arrows in Fig. 11-6 indicate the directions that changes caused by reaction must take and show that there must be a minimum in the curve. It is at this point that the equilibrium criterion of Eq. (11-20) is satisfied, and it is at this point that the reaction coordinate has its equilibrium value ϵ_e. The meaning of Eq. (11-20) is that differential displacements of the chemical reaction can occur at the equilibrium state without causing changes in the total Gibbs function. It is our purpose to use this criterion for the calculation of values of ϵ_e.

According to Eq. (11-22),

$$d(nG) = -(nS)\,dT + (nV)\,dP + \Sigma\mu_i\,dn_i. \qquad (11\text{-}22)$$

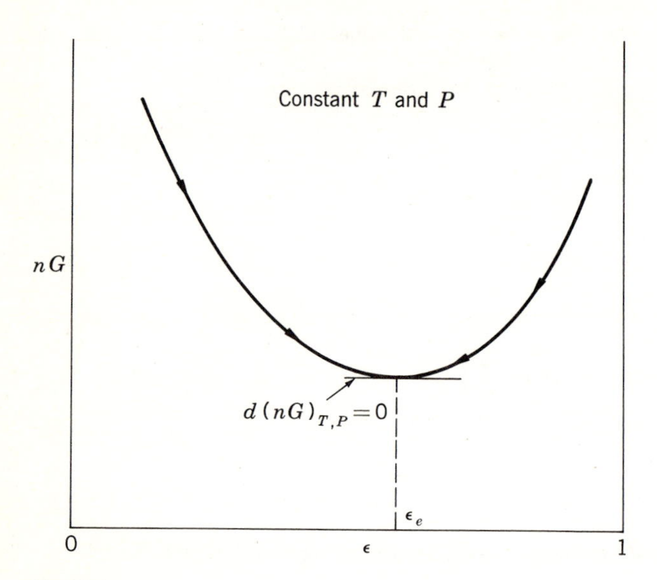

Fig. 11-6
Total Gibbs function in relation to the reaction coordinate.

Therefore at constant temperature and pressure

$$d(nG)_{T,P} = \Sigma \mu_i \, dn_i.$$

The chemical potential for a component in a mixture of ideal gases is

$$\mu_i = G_i + RT \ln y_i. \tag{11-24}$$

Thus for a mixture of ideal gases

$$d(nG)_{T,P} = \Sigma (G_i + RT \ln y_i) \, dn_i,$$

and for equilibrium, Eq. (11-20) requires that this be zero. Therefore at equilibrium a reacting mixture of ideal gases must satisfy the equation

$$\Sigma (G_i + RT \ln y_i) \, dn_i = 0.$$

The dn_i represent changes in the numbers of moles of the chemical species caused by a differential displacement of a chemical reaction around the equilibrium state. Equation (11-35) relates each dn_i to the change in the reaction coordinate:

$$dn_i = k\nu_i \, d\epsilon \qquad (i = 1, 2, 3, \ldots).$$

Thus, for the equilibrium state

$$\Sigma (G_i + RT \ln y_i) k\nu_i \, d\epsilon = 0,$$

which requires that

$$\Sigma(G_i + RT \ln y_i)\nu_i = 0.$$

This is the basic working relation for chemical-reaction equilibrium in a mixture of ideal gases. However, numerical calculations are facilitated if it is recast into quite a different form. First, we break up the left-hand side into two summations:

$$\Sigma(\nu_i G_i) + RT\, \Sigma(\nu_i \ln y_i) = 0.$$

Now the G_i's represent values of the Gibbs function for the pure ideal gases at the temperature and pressure of the mixture. It is convenient to refer these quantities to a *standard state*, which for gases is arbitrarily taken to be the ideal-gas state at 1(atm) and at the temperature of the mixture. The general equation which relates G to T and P is

$$dG = V\, dP - S\, dT. \tag{9-6}$$

For a change in pressure from the standard-state value of $P° = 1$(atm) to the mixture pressure P at constant temperature T for pure ideal gas i,

$$\int_{G_i°}^{G_i} dG_i = \int_{P°}^{P} \frac{RT}{P}\, dP = RT \int_{P°}^{P} \frac{dP}{P},$$

or

$$G_i - G_i° = RT \ln \frac{P}{P°},$$

where the superscript ° indicates a value for the standard state. Since $P° = 1$(atm), we may write

$$G_i = G_i° + RT \ln P,$$

where we must remember that P *represents the pressure in atmospheres and is understood to be divided by* 1(*atm*). *Hence the term* $RT \ln P$ *has the units of* RT *only.*

We may now substitute for G_i in the equilibrium equation

$$\Sigma(\nu_i G_i°) + \Sigma(\nu_i RT \ln P) + RT\, \Sigma(\nu_i \ln y_i) = 0$$

or

$$\frac{\Sigma(\nu_i G_i°)}{RT} + (\Sigma\nu_i) \ln P + \Sigma(\nu_i \ln y_i) = 0.$$

This may equally well be written

$$\frac{\Sigma(\nu_i G_i°)}{RT} + \ln P^{\Sigma\nu_i} + \Sigma(\ln y_i^{\nu_i}) = 0$$

or
$$\frac{\Sigma(\nu_i G_i^\circ)}{RT} + \ln P^{\Sigma\nu_i} + \ln \prod y_i^{\nu_i} = 0,$$

where the sign $\prod$ indicates a continuous or running product of what follows for all values of i. Combination of the second and third terms and rearrangement yields

$$\frac{-\Sigma(\nu_i G_i^\circ)}{RT} = \ln(P^{\Sigma\nu_i} \prod y_i^{\nu_i}).$$

The term on the left depends for a given reaction on the temperature only, and this must therefore also be true of the term on the right. This prompts us to define an *equilibrium constant* K, which is a function of temperature only, as follows:

$$K \equiv P^{\Sigma\nu_i} \prod y_i^{\nu_i}.$$

We may now write

$$\frac{-\Sigma(\nu_i G_i^\circ)}{RT} = \ln K.$$

Finally, we introduce a simplification in nomenclature through the identities

$$\Sigma(\nu_i G_i^\circ) \equiv \Delta G^\circ$$

and
$$\Sigma\nu_i \equiv \nu.$$

As a result we have

$$\frac{-\Delta G^\circ}{RT} = \ln K, \tag{11-36}$$

where
$$K = P^\nu \prod y_i^{\nu_i}. \tag{11-37}$$

As applied to a reacting mixture of four gases according to the reaction

$$\nu_1 A_1 + \nu_2 A_2 + \nu_3 A_3 + \nu_4 A_4 = 0, \tag{11-33}$$

we have

$$\Delta G^\circ = \Sigma(\nu_i G_i^\circ) = \nu_1 G_1^\circ + \nu_2 G_2^\circ + \nu_3 G_3^\circ + \nu_4 G_4^\circ.$$

This quantity is known as the *standard Gibbs-function change of reaction*. Values for many reactions can be found from data in various handbooks and reference works. The value of ν is simply the sum of the stoichiometric numbers:

$$\nu = \Sigma\nu_i = \nu_1 + \nu_2 + \nu_3 + \nu_4.$$

As a specific example consider the reaction

$$CH_4 + H_2O \rightleftharpoons CO + 3H_2.$$

Here

$$\nu = -1 - 1 + 1 + 3 = 2,$$

and

$$\Delta G^\circ = -1G^\circ_{CH_4} - 1G^\circ_{H_2O} + 1G^\circ_{CO} + 3G^\circ_{H_2}.$$

Moreover,

$$K = P^2(y_{CH_4}^{-1} y_{H_2O}^{-1} y_{CO} y_{H_2}^3)$$

$$= P^2 \frac{y_{CO} y_{H_2}^3}{y_{CH_4} y_{H_2O}}.$$

It should be noted that K is dimensionless. In the preceding equation it appears that K has the units of P^2. However, one must recall that P here represents the pressure in (atm) divided by 1(atm), and is therefore dimensionless.

The general procedure used for calculation of the equilibrium composition in a reacting mixture of ideal gases involves several steps. Presuming T and P to be known,

1 Determine a value for ΔG° from a source of data, and find a value for K by means of Eq. (11-36).
2 Relate the various mole fractions to the reaction coordinate as illustrated in Sec. 11-9, and substitute for the y_i's in Eq. (11-37).
3 Knowing K and the ν_i's, solve Eq. (11-37) for ϵ_e, the reaction coordinate at equilibrium.
4 Compute the y_i's from knowledge of ϵ_e.

Example 11-8

The value of ΔG° at 298(K) [25(°C)] for the reaction

$$N_2O_4 \rightleftharpoons 2NO_2$$

is 5,400(J)/(mol). Thus

$$\ln K = \frac{-\Delta G^\circ}{RT} = \frac{-5{,}400}{(8.314)(298)} = -2.180$$

and $K = 0.113$. Find the equilibrium composition at 25(°C) and 2(atm).

Assume that we start with 1(mol) of N_2O_4 and allow the system to reach equilibrium according to the given dissociation reaction at 25(°C) and 2(atm). The equilibrium mole fractions of N_2O_4 and NO_2 are related to the reaction coordinate at equilibrium. In general

$$\frac{dn_{N_2O_4}}{-1} = \frac{dn_{NO_2}}{2} = k\,d\epsilon.$$

When $\epsilon = 0$, $n_{N_2O_4} = 1$, and when $\epsilon = 1$, $n_{N_2O_4} = 0$. Thus

$$\int_1^0 dn_{N_2O_4} = -k \int_0^1 d\epsilon,$$

or

$$k = 1.$$

As a result we may write the general equations

$$dn_{N_2O_4} = -d\epsilon \qquad \text{and} \qquad dn_{NO_2} = 2d\epsilon.$$

Integration from $\epsilon = 0$ to $\epsilon = \epsilon_e$, the reaction coordinate at equilibrium, gives

$$n_{N_2O_4} = 1 - \epsilon_e, \qquad y_{N_2O_4} = \frac{1 - \epsilon_e}{1 + \epsilon_e},$$

$$n_{NO_2} = 2\epsilon_e, \qquad y_{NO_2} = \frac{2\epsilon_e}{1 + \epsilon_e},$$

$$\Sigma n_i = 1 + \epsilon_e.$$

Equation (11-37) for this particular reaction now becomes

$$\frac{y^2_{NO_2}}{y_{N_2O_4}} P^{2-1} = \frac{[2\epsilon_e/(1 + \epsilon_e)]^2}{(1 - \epsilon_e)/(1 + \epsilon_e)} P = \frac{4\epsilon_e^2}{(1 - \epsilon_e)(1 + \epsilon_e)} P = K.$$

Since $P = 2$ and $K = 0.113$, we have

$$\frac{\epsilon_e^2}{1 - \epsilon_e^2} = \frac{0.113}{(2)(4)} = 0.0141,$$

or

$$\epsilon_e^2 = \frac{0.0141}{1.0141} = 0.0139.$$

Thus

$$\epsilon_e = 0.118,$$

$$y_{N_2O_4} = \frac{1 - 0.118}{1 + 0.118} = 0.789,$$

and

$$y_{NO_2} = 0.211.$$

It should be noted that K is a function of temperature, and therefore the equilibrium composition shifts with change in temperature. Although the equilibrium constant is itself independent of pressure, the equilibrium *composition* is seen in the preceding example to depend on pressure, because the term P^ν appears in the equilibrium equation. Reactions of ideal gases for which $\nu = 0$ are not influenced by pressure.

11-11 Standard Heat of Reaction

The equilibrium constant K is related to the standard Gibbs-function change of reaction by Eq. (11-36):

$$\ln K = \frac{-\Delta G^\circ}{RT}.$$

Differentiating with respect to T, we get

$$\frac{d\ln K}{dT} = -\frac{d(\Delta G^\circ/RT)}{dT} = -\sum\left[\nu_i \frac{d(G_i^\circ/RT)}{dT}\right].$$

Now G_i° is the value of the Gibbs function for pure i in the standard state, which for ideal gases is taken to be the state at 1(atm). Thus the pressure of the standard state is constant. The basic differential equation for the Gibbs function,

$$dG = V\,dP - S\,dT, \tag{9-6}$$

applied to the standard state of pure ideal gas i then becomes

$$dG_i^\circ = -S_i^\circ\,dT,$$

or

$$\frac{dG_i^\circ}{dT} = -S_i^\circ,$$

where S_i° is the entropy of i in the standard state.

The enthalpy H_i° of i in the standard state is related to G_i° and S_i° by the definition

$$G_i^\circ = H_i^\circ - TS_i^\circ.$$

Substituting for S_i°, we get

$$G_i^\circ = H_i^\circ + T\frac{dG_i^\circ}{dT},$$

or

$$-H_i^\circ = T\frac{dG_i^\circ}{dT} - G_i^\circ.$$

But

$$T^2\frac{d(G_i^\circ/T)}{dT} = T^2\left[\frac{1}{T}\left(\frac{dG_i^\circ}{dT}\right) - \frac{G_i^\circ}{T^2}\right] = T\frac{dG_i^\circ}{dT} - G_i^\circ.$$

Therefore

$$-H_i^\circ = T^2\frac{d(G_i^\circ/T)}{dT},$$

or

$$\frac{d(G_i^\circ/RT)}{dT} = \frac{-H_i^\circ}{RT^2}.$$

Returning now to our expression for $d \ln K/dT$, we see that it becomes

$$\frac{d \ln K}{dT} = \frac{\Sigma(\nu_i H_i^\circ)}{RT^2},$$

or

$$\boxed{\frac{d \ln K}{dT} = \frac{\Delta H^\circ}{RT^2},} \tag{11-38}$$

where

$$\Delta H^\circ \equiv \Sigma(\nu_i H_i^\circ)$$

is known as the *standard enthalpy change of reaction* or, more commonly, as the *standard heat of reaction*. The latter name is used because ΔH° is simply the total enthalpy change when the reaction proceeds from $\epsilon = 0$ to $\epsilon = 1$ at the standard-state pressure of 1(atm), and it would equal the heat transferred during the process.

Since the change of the equilibrium constant with temperature depends on the heat of reaction, knowledge of this quantity is most important for reacting systems. If the heat of reaction is positive, that is, if the final enthalpy is greater than the initial enthalpy, heat must be added to the system as the reaction proceeds from left to right to maintain constant temperature. Such a reaction is *endothermic*. If the heat of reaction is negative, heat is given off and the reaction is *exothermic*. Combustion reactions are exothermic.

The standard heat of reaction is a function of temperature. Since the standard-state pressure for ideal gases is always 1(atm), changes in temperature of the standard state produce enthalpy changes according to the equation

$$dH_i^\circ = C_{P_i}^\circ \, dT,$$

or

$$H_i^\circ = H_{0_i} + \int C_{P_i}^\circ \, dT,$$

where H_{0_i} is a constant of integration. Thus

$$\Delta H^\circ = \Sigma(\nu_i H_{0_i}) + \int \Sigma(\nu_i C_{P_i}^\circ) \, dT.$$

The first term on the right is a constant for a given reaction and is denoted ΔH_0. Furthermore, the integrand may be written

$$\Delta C_P^\circ = \Sigma(\nu_i C_{P_i}^\circ).$$

So we have, finally,

$$\boxed{\Delta H^\circ = \Delta H_0 + \int (\Delta C_P{}^\circ)\, dT.} \tag{11-39}$$

This may also be written

$$\frac{d(\Delta H^\circ)}{dT} = \Delta C_P{}^\circ,$$

from which we see that the standard heat of reaction increases with temperature when the total heat capacity of the products is greater than the total heat capacity of the reactants. Conversely, when $\Delta C_P{}^\circ$ is negative, ΔH° decreases with an increase in temperature.

$\Delta C_P{}^\circ$ may be expressed as a function of temperature by substituting for the C°_{Pi}'s the empirical equations expressing their temperature dependence. It was pointed out in Sec. 5-8 that such equations often take the form

$$\frac{C_{Pi}{}^\circ}{R} = a_i + b_i T + c_i T^2, \tag{5-28}$$

and values for the constants in this equation are listed for a number of gases in Table 5-3. Combination of these equations gives

$$\frac{\Delta C_P{}^\circ}{R} = \Delta a + (\Delta b)T + (\Delta c)T^2,$$

where Δ has the same significance as it does in ΔG°, ΔH°, et cetera.

The temperature dependence of the equilibrium constant is given by

$$\frac{d \ln K}{dT} = \frac{\Delta H^\circ}{RT^2}. \tag{11-38}$$

Integration yields

$$\boxed{\ln K = \frac{1}{R}\int \frac{\Delta H^\circ}{T^2}\, dT + I,} \tag{11-40}$$

where I is a constant of integration and ΔH° is given as a function of temperature by Eq. (11-39).

Standard heats of reaction for gaseous systems are usually determined calorimetrically, and values of ΔH° at 25(°C) are listed in standard handbooks for the formation of many compounds from their constituent *elements*. These values are denoted by $\Delta H^\circ_{f\,298}$, where the subscript f specifies a *heat of formation* and the 298 indicates the Kelvin temperature. Similarly, standard changes of formation for the Gibbs function at 25(°C) or 298(K),

Table 11-3 Standard Heats of Formation and Standard Gibbs-Function Changes of Formation for Gases at 298(K) in the Ideal-Gas State at 1(atm)

Gas	Formula	$\Delta H^\circ_{f_{298}}$ (J)/(mol)	$\Delta G^\circ_{f_{298}}$ (J)/(mol)
Acetylene	C_2H_2	226,750	209,200
Ammonia	NH_3	−46,190	−16,635
Benzene	C_6H_6	82,930	129,660
Carbon dioxide	CO_2	−393,510	−394,380
Carbon monoxide	CO	−110,525	−137,270
Ethyl alcohol	C_2H_6O	−272,960	−168,620
Hydrogen chloride	HCl	−92,310	−95,265
Methane	CH_4	−74,850	−50,790
Sulfur dioxide	SO_2	−296,900	−300,370
Water	H_2O	−241,825	−228,590

designated by $\Delta G^\circ_{f_{298}}$, are also listed in handbooks and compilations of thermodynamic data. Table 11-3 gives such values for a few common substances.

Example 11-9

We wish to illustrate the use of the preceding material through consideration of the ammonia synthesis reaction

$$\tfrac{1}{2}N_2 + \tfrac{3}{2}H_2 \rightarrow NH_3.$$

This is a formation reaction, and from Table 11-3 we have the following values:

$$\Delta H^\circ_{f_{298}} = -46{,}190\,(\mathrm{J})/(\mathrm{mol})$$

$$\Delta G^\circ_{f_{298}} = -16{,}635\,(\mathrm{J})/(\mathrm{mol}).$$

The heat-capacity data given in Table 5-3 for N_2, H_2, and NH_3 allow the determination of $\Delta C_P{}^\circ$ for this reaction as a function of temperature. Numerical substitution in the defining equation for $\Delta C_P{}^\circ$ yields

$$\Delta C_P{}^\circ = -31.780 + 35.517 \times 10^{-3}T - 9.316 \times 10^{-6}T^2,$$

where T is in (K), and $\Delta C_P{}^\circ$ has the units of (J)/(mol)(K). Since ΔH° is given as a function of temperature by Eq. (11-39),

$$\Delta H^\circ = \Delta H_0 + \int (\Delta C_P{}^\circ)\,dT,$$

we have after integration

$$\Delta H^\circ = \Delta H_0 - 31.780T + 17.758 \times 10^{-3}T^2 - 3.105 \times 10^{-6}T^3.$$

At 298(K), we have $\Delta H^\circ = -46{,}190$. Substitution of these values now provides a value for ΔH_0. The result is a general equation for ΔH°:

$$\Delta H^\circ = -38{,}214 - 31.780T + 17.758 \times 10^{-3}T^2 - 3.105 \times 10^{-6}T^3.$$

A general equation giving $\ln K$ as a function of T is now obtained by substitution for ΔH° in Eq. (11-40),

$$\ln K = \frac{1}{R}\int \frac{\Delta H^\circ}{T^2}\, dT + I. \tag{11-40}$$

Integration gives

$$\ln K = \frac{4{,}596}{T} - 3.822 \ln T + 2.136 \times 10^{-3}T - 0.187 \times 10^{-6}T^2 + I.$$

At $T = 298$(K), we may calculate K from the given data for ΔG°:

$$\ln K = -\frac{\Delta G^\circ}{RT} = \frac{16{,}635}{(8.314)(298)} = 6.715.$$

The integration constant I is now determined by substitution of this value for $\ln K$ at 298(K) into the equation above. The result is an equation for $\ln K$ as a function of T:

$$\ln K = \frac{4{,}596}{T} - 3.822 \ln T + 2.136 \times 10^{-3}T - 0.187 \times 10^{-6}T^2 + 12.431.$$

With the aid of this equation we can calculate the reaction coordinate at equilibrium for the ammonia synthesis reaction as a function of temperature. Assume that $\frac{1}{2}$(mol) of N_2 and $\frac{3}{2}$(mol) of H_2 represent the original constitution of the system, which is to come to equilibrium at pressure P and temperature T. By Eq. (11-34)

$$\frac{dn_{N_2}}{-\frac{1}{2}} = \frac{dn_{H_2}}{-\frac{3}{2}} = \frac{dn_{NH_3}}{1} = k\, d\epsilon.$$

It is easily shown that $k = 1$. Thus, at equilibrium,

$$n_{N_2} = \tfrac{1}{2}(1 - \epsilon_e), \qquad y_{N_2} = \frac{1 - \epsilon_e}{2(2 - \epsilon_e)},$$

$$n_{H_2} = \tfrac{3}{2}(1 - \epsilon_e), \qquad y_{H_2} = \frac{3(1 - \epsilon_e)}{2(2 - \epsilon_e)},$$

$$n_{NH_3} = \epsilon_e, \qquad y_{NH_3} = \frac{\epsilon_e}{2 - \epsilon_e},$$

$$\Sigma n_i = 2 - \epsilon_e.$$

Thus, by Eq. (11-37)

$$\frac{y_{NH_3}}{y_{N_2}^{1/2} y_{H_2}^{3/2}} P^{-1} = \frac{\epsilon_e(2 - \epsilon_e)}{[\frac{1}{2}(1 - \epsilon_e)]^{1/2}[\frac{3}{2}(1 - \epsilon_e)]^{3/2}P} = K,$$

or

$$\frac{\epsilon_e(2 - \epsilon_e)}{(1 - \epsilon_e)^2} = 1.30KP.$$

Rearrangement gives

$$(1 + 1.30KP)\epsilon_e^2 - 2(1 + 1.30KP)\epsilon_e + 1.30KP = 0.$$

Let

$$r = 1 + 1.30KP.$$

Then

$$r\epsilon_e^2 - 2r\epsilon_e + (r - 1) = 0.$$

Substitution of the proper coefficients in the formula for the roots of a quadratic equation gives, after reduction, the simple expression

$$\epsilon_e = 1 \pm \frac{1}{r^{1/2}}.$$

Since ϵ_e must always be less than unity, we need only the smaller root, and

$$\epsilon_e = 1 - \frac{1}{r^{1/2}},$$

or

$$\epsilon_e = 1 - \frac{1}{(1 + 1.3KP)^{1/2}}.$$

We may now evaluate ϵ_e as a function of temperature and pressure. The results of these calculations are shown in Table 11-4. The influence of temperature and pressure on the degree of reaction and on the conversion of reactants to ammonia is clear. It is desirable from the standpoint of the equilibrium conversion to carry out this synthesis at about room temperature. However, at this temperature the reaction rate is negligible, even with the best catalyst yet discovered. At higher temperatures, say 600 or 700(K), rapid reaction can be attained, but the reaction coordinate is very small at atmospheric pressure. Therefore the synthesis is carried out at elevated pressures. The assumption of ideal gases is not really justified at 100(atm). However, the calculated results are certainly qualitatively valid.

For reactions that are not formation reactions, the chemical equation may be expressed as a combination of formation reactions. Values of $\Delta H°$

Table 11-4 Results of Calculations for the Ammonia Synthesis Reaction (Reactants Initially in the Stoichiometric Proportion)

		1(atm)		100(atm)	
T(K)	K	ϵ_e	y_{NH_3}	ϵ_e	y_{NH_3}
300	7.22×10^2	0.968	0.938	0.997	0.994
400	6.37	0.673	0.507	0.964	0.931
500	3.31×10^{-1}	0.164	0.089	0.849	0.738
600	4.31×10^{-2}	0.022	0.014	0.610	0.439
700	9.68×10^{-3}	0.006	0.003	0.336	0.202
800	3.23×10^{-3}	.....	.....	0.155	0.084
900	1.24×10^{-3}	.....	.....	0.072	0.037
1000	5.95×10^{-4}	.....	.....	0.037	0.019

and $\Delta G°$ are then found by a corresponding combination of $\Delta H_f°$ and $\Delta G_f°$ values. For example, the reaction

$$CH_4 + H_2O \rightarrow CO + 3H_2$$

is obtained from the formation reactions

$C + \frac{1}{2}O_2 \rightarrow CO$	$\Delta H_f° = -110{,}525,$	$\Delta G_f° = -137{,}270,$
$-[C + 2H_2 \rightarrow CH_4],$	$-\Delta H_f° = +\ 74{,}850,$	$-\Delta G_f° = +\ 50{,}790,$
$-[H_2 + \frac{1}{2}O_2 \rightarrow H_2O],$	$-\Delta H_f° = +241{,}825,$	$-\Delta G_f° = +228{,}590,$
$CH_4 + H_2O \rightarrow CO + 3H_2,$	$\Delta H° = 206{,}150$(J)/(mol),	$\Delta G° = 142{,}110$(J)/(mol).

11-12 Reversible Cells

Equations for a reversible cell composed of solids and liquids only may be obtained from corresponding equations for a PVT system by replacing V by the charge of the cell q and $-P$ by the emf of the cell e. Thus the generalized entropy equation, Eq. (11-2), can be written

$$T\,dS = C_q\,dT - T\left(\frac{\partial e}{\partial T}\right)_q dq, \tag{11-41}$$

and for a saturated cell whose emf depends on the temperature only, the equation becomes

$$T\,dS = C_q\,dT - T\frac{de}{dT}\,dq.$$

From Eqs. (2-17) and (3-10), we have

$$dq = -j\mathfrak{F}\,dn \qquad \text{and} \qquad \delta W = -ej\mathfrak{F}\,dn,$$

where $j\mathfrak{F}$ is the charge transferred per mole of reacting material, and dn is the moles of material reacted. It will be recalled that by convention dn is taken as positive for discharging the cell and as negative for charging.

For the *reversible isothermal* transfer of charge dq, $T\,dS$ equals δQ, and therefore

$$\delta Q = T\,dS = -T\frac{de}{dT}\,dq$$

$$= j\mathfrak{F}T\frac{de}{dT}\,dn.$$

By the first law

$$dU = \delta Q + \delta W.$$

Substitution for δQ and δW gives

$$dU = j\mathfrak{F}T\frac{de}{dT}\,dn - ej\mathfrak{F}\,dn$$

$$= j\mathfrak{F}\left(T\frac{de}{dT} - e\right)dn.$$

In the equations above, the charge q and the moles of material reacted dn are measures of the extent of the chemical reaction which occurs in the cell, and replace the reaction coordinate. For cells operating at constant pressure and with negligible volume change, the change of internal energy for a differential amount of reaction is equal to the change of enthalpy. This follows from the definition

$$H = U + PV$$

and

$$dH = dU + P\,dV + V\,dP;$$

whence, under the conditions mentioned,

$$dH = dU.$$

Therefore

$$dH = j\mathfrak{F}\left(T\frac{de}{dT} - e\right)dn.$$

For a reversible reaction for which $\Delta n = 1\,(\text{mol})$, this equation integrates to give

$$\Delta H = j\mathfrak{F}\left(T\frac{de}{dT} - e\right).$$

In order to interpret this change of enthalpy, let us take the particular case of the Daniell cell (see Fig. 2-3). The transfer of positive electricity externally from the copper to the zinc electrode is accompanied by the reaction

$$Zn + CuSO_4 \rightarrow Cu + ZnSO_4.$$

When a charge $j\mathfrak{F}$ is transferred, 1 (mol) of each of the reactants disappears and 1 (mol) of each of the products is formed. The change of enthalpy in this case is equal to the enthalpy of 1 (mol) of each of the products minus the enthalpy of 1 (mol) of each of the reactants at the same temperature and pressure, and is the heat of reaction. Since

$$\mathfrak{F} = 96{,}487(\text{C})/(\text{mol}) = 96{,}487(\text{J})/(\text{V})(\text{mol}),$$

we get, finally, for e in (V)

$$\Delta H = -96{,}487\, j \left(e - T\frac{de}{dT}\right) \frac{(\text{J})}{(\text{mol})}.$$

In the case of a saturated reversible cell in which gases are liberated, it can be shown rigorously (see Prob. 11-30) that

$$\boxed{\Delta H = -96{,}487\, j \left[e - T\left(\frac{\partial e}{\partial T}\right)_P\right] \frac{(\text{J})}{(\text{mol})}.} \tag{11-42}$$

The important feature of this equation is that it provides a method of measuring the heat of reaction of a chemical reaction without recourse to calorimetry. If the reaction can be made to proceed in a reversible cell, all that is necessary is measurement of the emf of the cell as a function of the temperature at constant pressure. The heat of reaction is therefore determined with a potentiometer and a thermometer. Both measurements can be made with great accuracy, and hence this method yields highly accurate values of the heat of reaction. It is interesting to compare values of ΔH obtained electrically with those measured calorimetrically. This is shown for a number of cells in Table 11-5.

From Eqs. (11-41) and (2-17), we find, for reversible isothermal operation of a saturated cell,

$$T\, dS = j\mathfrak{F}T \frac{de}{dT}\, dn.$$

Table 11-5 Reversible Cells [Values are based on reaction of 1(mol) of first reactant.]

Reaction	T (K)	j	emf e, (V)	$\frac{de}{dT}$ (mV)/(K)	ΔH, electric method (kJ)/(mol)	ΔH, calorimetric method (kJ)/(mol)
$Zn + CuSO_4 = Cu + ZnSO_4$	273	2	1.0934	−0.453	−235	−232
$Zn + 2AgCl = 2Ag + ZnCl_2$	273	2	1.0171	−0.210	−207	−206
$Cd + 2AgCl = 2Ag + CdCl_2$	298	2	0.6753	−0.650	−168	−165
$Pb + 2AgI = 2Ag + PbI_2$	298	2	0.2135	−0.173	−51.1	−51.1
$Ag + \frac{1}{2}Hg_2Cl_2 = Hg + AgCl$	298	1	0.0455	+0.338	+5.45	+3.77
$Pb + Hg_2Cl_2 = 2Hg + PbCl_2$	298	2	0.5356	+0.145	−96.0	−98.0
$Pb + 2AgCl = 2Ag + PbCl_2$	298	2	0.4900	−0.186	−105	−104

The Gibbs function is defined by

$$G = H - TS,$$

and at constant temperature

$$dG = dH - T\,dS.$$

Since

$$dH = j\mathfrak{F}\left(T\frac{de}{dT} - e\right)dn,$$

we have the result

$$dG = -j\mathfrak{F}e\,dn.$$

For a reversible reaction for which $\Delta n = 1(\text{mol})$, this equation may be integrated, giving

$$\Delta G = -j\mathfrak{F}e,$$

or

$$\boxed{\Delta G = -96{,}487\,je\,\frac{(\text{J})}{(\text{mol})}}, \tag{11-43}$$

where ΔG is the Gibbs-function change for the reaction occurring in the cell.

11-13 Fuel Cells

A fuel cell is a device, similar in some respects to an electrolytic cell, in which the energy released as the result of a combustion reaction appears, at least in part, directly as electric energy. It has the characteristics of a cell in that it consists of a positive and a negative electrode, separated by an electrolyte. However, the reactants are not stored in the cell but are fed to it continuously, and the products of reaction are continuously withdrawn. The fuel cell is thus not given an initial electric charge, and in operation it does not lose an electric charge. It operates as a continuous-flow system, so long as reactants are supplied, and produces a steady electric current.

One electrode is known as the fuel electrode, because fuel, such as hydrogen, methane, propane, etc., is supplied to the cell in such a way as to come into intimate contact with this electrode. The other electrode is known as the oxygen electrode, because oxygen is supplied to it. The cell as a whole acts to effect oxidation of the fuel by means of an electrochemical reaction. The hydrogen-oxygen fuel cell is probably the simplest such device, and will serve to illustrate the principles of operation. A schematic diagram of this cell is shown in Fig. 11-7.

The half-cell reaction occurring at the fuel or hydrogen electrode is

$$H_2 + 2OH^- \rightarrow 2H_2O + 2e^-,$$

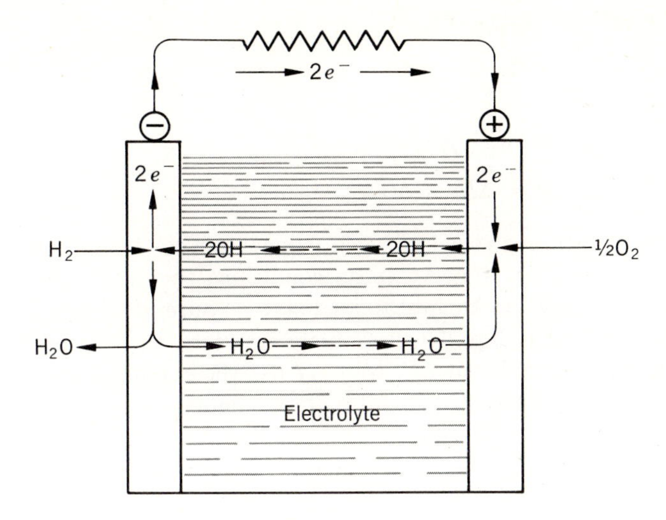

Fig. 11-7
Hydrogen—oxygen fuel cell.

and that at the oxygen electrode is

$$\tfrac{1}{2}O_2 + H_2O + 2e^- \rightarrow 2OH^-.$$

The overall cell reaction is

$$H_2 + \tfrac{1}{2}O_2 \rightarrow H_2O$$

and is merely the "combustion" of the hydrogen.

Since fuel-cell operation is a steady-flow process, the form taken by the first law is

$$\Delta H = Q + W,$$

where potential- and kinetic-energy terms have been omitted as negligible. If the cell operates *reversibly and isothermally*,

$$Q = T\,\Delta S.$$

Thus

$$\Delta H = T\,\Delta S + W,$$

or

$$W = (\Delta H - T\,\Delta S) = \Delta G.$$

Now the work done by a reversible fuel cell is given in terms of its emf by exactly the same expression as for an electrolytic cell:

$$\delta W = -ej\mathfrak{F}\,dn, \qquad (3\text{-}10)$$

or for a reaction for which $\Delta n = 1$(mol),

$$W = -ej\mathfrak{F}.$$

In summary,

$$W = -ej\mathfrak{F} = \Delta G$$

and

$$Q = \Delta H - \Delta G.$$

We may apply these equations to the hydrogen-oxygen fuel cell operating at atmospheric pressure and 25(°C). At these conditions the reactants, hydrogen and oxygen, are gases in their standard states [pure ideal gases at 1(atm)], and the product is liquid water in its standard state [pure liquid at 1(atm)]. Hence, for ΔH and ΔG, we may use the standard heat of reaction ΔH°_{298} and the standard Gibbs-function change of reaction ΔG°_{298} for the reaction

$$H_2 + \tfrac{1}{2}O_2 \rightarrow H_2O\,(\text{liquid}).$$

These are

$$\Delta H^\circ_{298} = -285{,}840(\text{J})/(\text{mol})$$

and

$$\Delta G^\circ_{298} = -237{,}190(\text{J})/(\text{mol}).$$

Thus the reversible work of the hydrogen-oxygen fuel cell under the prescribed conditions for the reaction of 1(mol) of hydrogen and the production of 1(mol) of water is

$$W = \Delta G^\circ_{298} = -237{,}190(\text{J})/(\text{mol}).$$

Since

$$\Delta G^\circ_{298} = -ej\mathfrak{F},$$

the reversible emf of this cell is calculated to be

$$e = \frac{-\Delta G^\circ_{298}}{j\mathfrak{F}} = \frac{237{,}190(\text{J})/(\text{mol})}{2 \times 96{,}487(\text{J})/(\text{V})(\text{mol})}$$

$$= 1.229(\text{V}).$$

The heat transfer between cell and surroundings in order to maintain isothermal operation is

$$Q = \Delta H^\circ_{298} - \Delta G^\circ_{298}$$

$$= -285{,}840 + 237{,}190 = -48{,}650(\text{J})/(\text{mol}).$$

The principles of fuel-cell operation have been understood for many years. However, the intensive research necessary to make them practical engineering devices did not start until after World War II. Considerable effort is still being expended to devise means to make their actual operation approach that which may be predicted on the assumption of reversibility. It is entirely possible that in time fuel cells will become common devices for the generation of electric power.

Problems

11-1 The tension in a steel rod, 3(m) in length and 6.5(cm)^2 in cross-sectional area at 300(K), is increased reversibly and isothermally from 0 to 345,000(kN)/(m)^2.

(a) How much heat is transferred?
(b) How much work is done?
(c) What is the change in internal energy?

Assume the following quantities to remain constant:

$$\alpha = 12 \times 10^{-6}(\text{K})^{-1},$$

$$E = 20 \times 10^{7}(\text{kN})/(\text{m})^2,$$

$$C_\sigma = 0.5(\text{J})/(\text{g})(\text{K}).$$

11-2 (a) For a simple system consisting of a stressed bar, show from Eq. (11-2) that for an isothermal, reversible extension or compression

$$\delta Q = -VT\left(\frac{\partial \sigma}{\partial T}\right)_\epsilon d\epsilon.$$

(b) The equation of state for an ideal elastomer is (see Prob. 3-9)

$$\sigma = \frac{A_0 K T}{A}\left(\lambda - \frac{1}{\lambda^2}\right),$$

where λ is the extension ratio L/L_0, and L_0 is the no-load length and is a function of temperature only. Using this equation together with the result of (a), show for a sample of no-load volume V_0 that the heat transferred in a reversible, isothermal process is

$$\delta Q = -V_0 T K\left(\lambda - \frac{1}{\lambda^2}\right) d\lambda.$$

(c) From this result and that of Prob. 3-9, show that for an ideal elastomer, as for an ideal gas, U is a function of temperature only.

11-3 A mixture of 3(mol) of helium, 4(mol) of neon, and 5(mol) of argon is at a pressure of 1(atm) and a temperature of 300(K). Calculate:

(a) The volume.
(b) The various mole fractions.
(c) The various partial pressures.
(d) The change of entropy due to mixing.
(e) The change in the Gibbs function due to mixing.

11-4 n_1(mol) of an ideal monatomic gas at temperature T_1 and pressure P is in one compartment of an insulated container. In an adjoining compartment, separated by an insulating partition, is n_2(mol) of another ideal monatomic gas at temperature T_2 and pressure P. When the partition is removed:

(a) Show that the final pressure of the mixture is P.
(b) Calculate the entropy change when the gases are identical.
(c) Calculate the entropy change when the gases are different.

11-5 Show, for a magnetic material which obeys Curie's law, that

(a) U and $C_{\mathcal{M}}$ are functions of T only.

(b) $S = \int C_{\mathcal{M}} \frac{dT}{T} - \frac{\mu_0 \mathcal{M}^2}{2CV} + \text{const.}$

(c) $C_{\mathcal{H}} - C_{\mathcal{M}} = \frac{\mu_0 \mathcal{M}^2}{CV}$.

(d) $\left(\frac{\partial C_{\mathcal{H}}}{\partial \mathcal{H}}\right)_T = \frac{2\mu_0 C V \mathcal{H}}{T^2}$.

11-6 (a) Prove that, for a dielectric,

$$\left(\frac{\partial U}{\partial \mathcal{E}}\right)_T = T\left(\frac{\partial \mathcal{P}}{\partial T}\right)_{\mathcal{E}} + \mathcal{E}\left(\frac{\partial \mathcal{P}}{\partial \mathcal{E}}\right)_T.$$

(b) If the equation of state of a dielectric is given by Eq. (3-16), with χ_e a function of T only, show that

$$U = f(T) + \frac{V\epsilon_0 \mathcal{E}^2}{2}\left(\chi_e + T\,\frac{d\chi_e}{dT}\right),$$

where $f(T)$ is an undetermined function of temperature.

(c) What relation must hold between χ_e and T in order that U be a function of T only?

11-7 For the case of a dielectric for which χ_e is given by

$$\chi_e = a + \frac{b}{T},$$

show that

(a) The heat transferred in a reversible isothermal change of field is

$$Q = -\frac{bV\epsilon_0}{2T}(\mathcal{E}_2^2 - \mathcal{E}_1^2),$$

(b) The small temperature change ΔT accompanying a reversible adiabatic change of field is (approximately)

$$\Delta T = \frac{bV\epsilon_0}{2C_{\mathcal{E}}T}(\mathcal{E}_2^2 - \mathcal{E}_1^2).$$

11-8 n_1(mol) of an ideal gas at pressure P_1 and temperature T is in one compartment of an insulated container. In an adjoining compartment, separated by a partition, is n_2(mol) of an ideal gas at pressure P_2 and temperature T. When the partition is removed:

(*a*) Calculate the final pressure of the mixture.
(*b*) Calculate the entropy change when the gases are identical.
(*c*) Calculate the entropy change when the gases are different.
(*d*) Prove that the entropy change in (*c*) is the same as that which would be produced by two independent free expansions.

11-9 What is the minimum amount of work required to separate 1(lb mol) of air at 80(°F) and 1(atm) [21 mol percent O_2 and 79 mol percent N_2] into pure oxygen and pure nitrogen at 80(°F) and 1(atm)?

11-10 For a system made up initially of an equimolar mixture of CO and H_2O and reacting according to the equation

$$CO + H_2O \rightleftharpoons CO_2 + H_2$$

in the gaseous phase, develop general expressions for the mole fractions of the constituents as functions of the reaction coordinate.

11-11 For a system made up initially of 3(mol) of H_2O for every mole of H_2S and undergoing the reaction

$$H_2S + 2H_2O \rightleftharpoons 3H_2 + SO_2$$

in the gaseous phase, develop general expressions for the mole fractions of the constituents as functions of the reaction coordinate.

11-12 Prove that $-\Delta H°/R$ is equal to the slope of a plot of $\ln K$ vs. $1/T$.

11-13 Develop general equations giving $\Delta H°$ and $\ln K$ as functions of T for the reaction

$$\tfrac{1}{2}N_2 + \tfrac{1}{2}O_2 \rightarrow NO.$$

The molar heat capacities as functions of T(K) are given by [298 to 2500(K)]

$$N_2: \quad \frac{C_P^\circ}{R} = 3.35 + 0.513 \times 10^{-3}T,$$

$$O_2: \quad \frac{C_P^\circ}{R} = 3.60 + 0.503 \times 10^{-3}T - 0.201 \times 10^5 T^{-2},$$

$$NO: \quad \frac{C_P^\circ}{R} = 3.54 + 0.463 \times 10^{-3}T - 0.070 \times 10^5 T^{-2}.$$

In addition, for NO

$$\Delta H_{f_{298}}^\circ = 90{,}370(\text{J})/(\text{mol}),$$

$$\Delta G_{f_{298}}^\circ = 86{,}690(\text{J})/(\text{mol}).$$

Prepare a plot of $\ln K$ versus $1/T$ for T from 300 to 2500(K).

11-14 Starting with n_0(mol) of NO, which dissociates according to the equation $NO \rightleftharpoons \tfrac{1}{2}N_2 + \tfrac{1}{2}O_2$, show that at equilibrium

$$K = \frac{1}{2}\frac{\epsilon_e}{1 - \epsilon_e}.$$

11-15 Starting with n_0(mol) of NH_3, which dissociates according to the equation $NH_3 \rightleftharpoons \frac{1}{2}N_2 + \frac{3}{2}H_2$, show that at equilibrium

$$K = \frac{\sqrt{27}}{4} \frac{\epsilon_e^2}{1 - \epsilon_e^2} P.$$

11-16 At 35(°C) and 1(atm), the fractional dissociation of N_2O_4 at equilibrium is 0.27.

(*a*) Calculate K.

(*b*) Calculate ϵ_e at the same temperature when the pressure is 100(mm Hg).

(*c*) The equilibrium constant for the dissociation of N_2O_4 has the values of 0.664 and 0.141 at the temperatures 318 and 298(K), respectively. Calculate the average heat of reaction within this temperature range.

11-17 The equilibrium constant of the reaction $SO_3 \rightleftharpoons SO_2 + \frac{1}{2}O_2$ has the following values:

T(K)	800	900	1000	1100
K	0.0319	0.153	0.540	1.59

Determine the average heat of dissociation graphically.

11-18 Starting with n_0(mol) of CO and $3n_0$(mol) of H_2, which react according to the equation $CO + 3H_2 \rightleftharpoons CH_4 + H_2O$, show that at equilibrium

$$K = \frac{4\epsilon_e^2(2 - \epsilon_e)^2}{27(1 - \epsilon_e)^4 P^2}.$$

11-19 (*a*) A compound X is known to polymerize to the compound X_l in the gas phase according to the reaction $lX(g) \rightarrow X_l(g)$, where l is the number of X units in the polymer $(l > 1)$. For constant l, and assuming ideal-gas behavior, show that the extent of equilibrium polymerization increases with increasing pressure at constant T.

(*b*) The following experimental data were recorded for the equilibrium mole fractions of monomer X in two gas-phase monomer–polymer mixtures:

t(°C)	P(atm)	y_X
100	1.0	0.807
100	1.5	0.750

What is the value of l for the polymerization reaction of part (*a*)?

11-20 In compilations of standard formation data, one also often finds entries for *standard heats of combustion* ΔH_c° for standard combustion reactions. A standard combustion reaction for a given chemical species is the reaction between 1(mol) of that species and oxygen to form specified products, with all the reactant and product species present in their standard states. For compounds containing only carbon, hydrogen, and oxygen, the products are customarily taken as $CO_2(g)$ and $H_2O(l)$. Thus, the standard combustion reaction for gaseous

n-butane $[n\text{-}C_4H_{10}(g)]$ is

$$n\text{-}C_4H_{10}(g) + \tfrac{13}{2}O_2(g) \rightarrow 4CO_2(g) + 5H_2O(l),$$

where all the gaseous species are present as pure ideal gases at 1(atm), and water is present as a pure liquid at 1(atm).

The standard heats of combustion for benzene $[C_6H_6(g)]$, hydrogen, and cyclohexane $[C_6H_{12}(g)]$ at 25(°C) are −789,080(cal)/(mol), −68,317(cal)/(mol), and −944,790(cal)/(mol), respectively. Using these data, calculate ΔH° at 25(°C) for the reaction

$$C_6H_6(g) + 3H_2(g) \rightarrow C_6H_{12}(g).$$

11-21 Tom, Dick, and Harry, members of a thermodynamics class, were given the following chemical-equilibrium problem to solve:

> The ideal-gas ammonia synthesis reaction
>
> $$\tfrac{1}{2}N_2(g) + \tfrac{3}{2}H_2(g) \rightarrow NH_3(g) \qquad (A)$$
>
> is carried out at 298(K) and 1(atm). The equilibrium product gas mixture is found to contain 25 mol percent NH_3. What are the compositions of N_2 and H_2 in the product? At 298(K), $\Delta G_f^\circ = -3976$(cal)/(mol) for $NH_3(g)$.

Inevitably, each of them approached the problem in a different way. Tom based his calculation on the reaction (A) as written. Dick, who likes whole numbers, worked instead with the reaction

$$N_2(g) + 3H_2(g) \rightarrow 2NH_3(g). \qquad (B)$$

Harry, who always does everything backwards, considered the reaction to be

$$NH_3(g) \rightarrow \tfrac{3}{2}H_2(g) + \tfrac{1}{2}N_2(g). \qquad (C)$$

Despite their different approaches to the problem, each arrived at the same (correct) answer. Write the chemical-equilibrium equations for the three ideal-gas reactions (A), (B), and (C), indicate how the three equilibrium constants are related, and show why Tom, Dick, and Harry got the same answer. It is not necessary to determine numerical values for the equilibrium compositions requested in the original statement of the problem.

11-22 The molar Gibbs function G for gas-phase mixtures of normal-pentane (n-C5) and neo-pentane(neo-C5) at 1(atm) and 400(K) is given by the expression

$$G = 9{,}600y_1 + 8{,}990y_2 + 800(y_1 \ln y_1 + y_2 \ln y_2),$$

where 1 ≡ n-C5 and 2 ≡ neo-C5 and G is in (cal)/(mol).

(a) For the isomerization reaction between n-C5 and neo-C5, determine the equilibrium composition at 1(atm) and 400(K).

(b) Sketch a graph of G vs. y_1 for the gaseous n-C5–neo-C5 system at 1(atm) and 400(K). Indicate the location of the equilibrium composition found as the solution to part (a).

11-23 (a) Starting with n_0(mol) of pure A which undergoes the following gas-phase reaction

$$2A \rightleftharpoons A_2,$$

develop an expression for the equilibrium constant K as a function of the equilibrium pressure P and the reaction coordinate ϵ_e. Assume ideal gases.

(*b*) This reaction is carried out in a closed thermostated vessel at constant temperature T and constant total volume. Initially, the vessel is charged with n_0(mol) of pure gas A at a pressure P_0 of 1.5(atm). When the reaction has proceeded to equilibrium, the final (equilibrium) pressure P is found to be 1.0(atm). From this information determine a numerical value for the equilibrium constant K of the reaction at temperature T.

(*c*) Could this same procedure be used if the reaction were

$$A + C \rightleftharpoons D + F?$$

11-24 The water-gas-shift reaction

$$CO(g) + H_2O(g) \rightarrow CO_2(g) + H_2(g)$$

is to be carried out at specified temperature and pressure employing a feed containing only carbon monoxide and steam. Show that the maximum equilibrium composition of hydrogen in the product stream results when the feed contains CO and H_2O in their stoichiometric proportions. Assume ideal-gas behavior.

11-25 Butadiene can be prepared by the gas-phase catalytic dehydrogenation of 1-butene:

$$C_4H_8 \rightarrow C_4H_6 + H_2.$$

In order to suppress side reactions, the butene is diluted with steam before it passes into the reactor. Estimate the temperature at which the reactor must be operated in order to convert 30 percent of the 1-butene to 1,3-butadiene at a reactor pressure of 2(atm) from a feed consisting of 12(mol) of steam per (mol) of 1-butene.

Data for ΔG_f° in (cal)/(mol) as a function of T:

	600(K)	700(K)	800(K)	900(K)
1,3-butadiene	46,780	50,600	54,480	58,400
1-butene	36,070	42,730	49,450	56,250

11-26 (*a*) The partial PT diagram of Fig. P11-26*a* has been pieced together from fragmentary measurements on pure snorkane-dial (SKD). In this figure, S1 and S2 are distinct crystalline forms of SKD, and l and v are liquid and vapor SKD, respectively. Can this partial diagram be correct as shown? Why, or why not?

(*b*) The pure-component PT diagram of Fig. P11-26*b* shows four triple-point states. Can this diagram be correct as shown?

(*c*) A piston-and-cylinder device (Fig. P11-26*c*) is immersed in a constant-temperature bath controlled at temperature T. The device is loaded with n_1(mol) of species 1 and n_2(mol) of 2, and the pressure P is adjusted so that the system is in a state of vapor-liquid equilibrium. The equilibrium mole fractions of species 1 in the liquid and vapor phases are x_1 and y_1,

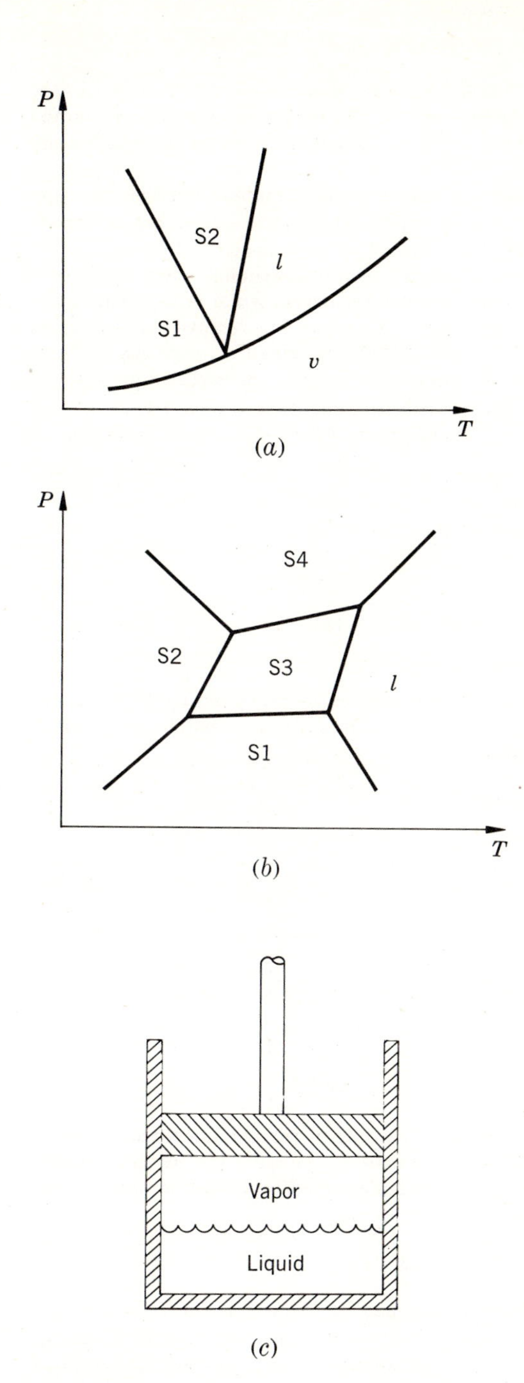

Fig. P11-26

respectively. A number of moles n_1' of component 1 are then added to the system, and the piston is slowly raised until the system pressure is restored to the initial value P. What are the final compositions of the equilibrium liquid and vapor phases?

11-27 Prepare a table similar to that of Table 11-4 for the ammonia synthesis reaction at 100(atm) but for the case where the initial constitution of the system is 2(mol) of N_2 for each (mol) of H_2.

11-28 Hydrogen and oxygen are being produced by the continuous electrolysis of a dilute caustic solution at atmospheric pressure. Liquid water is supplied to the cell at 25(°C), and gaseous hydrogen and oxygen are removed at the same temperature. A current of 3,000(A) at 2.00(V) is passed through the cell at a current efficiency of 90 percent. At what rate must heat be supplied or removed from the cell so that its temperature will be constant at 25(°C)? Current efficiency is defined as the ratio of actual product to that formed according to Faraday's law of electrolysis.

11-29 The emf of the cell

$$\mathrm{Zn} \mid \mathrm{ZnCl_2(sat)} \parallel \mathrm{Hg_2Cl_2(sat)} \mid \mathrm{Hg}$$

is given by the equation

$$e = 1.000 + 0.000094(t - 15),$$

where t is in (°C) and e is in (V).

Write the reaction, and calculate ΔG and the heat of reaction at 100(°C).

11-30 Using the five coordinates P, V, T, e, and q to describe a reversible cell in which gases may be liberated, show that

(*a*) $dG = -S\,dT + V\,dP + e\,dq$, where $G = U + PV - TS$,

(*b*) $$-\left(\frac{\partial S}{\partial q}\right)_{T,P} = \left(\frac{\partial e}{\partial T}\right)_{P,q},$$

(*c*) $dH = T\,dS + V\,dP + e\,dq$,

(*d*) $$\left(\frac{\partial H}{\partial q}\right)_{T,P} = e - T\left(\frac{\partial e}{\partial T}\right)_{P,q},$$

(*e*) For a saturated cell, $\Delta H = -j\mathfrak{F}\left[e - T\left(\frac{\partial e}{\partial T}\right)_P\right]$.

11-31 Equilibrium blackbody radiation in a cavity may be treated as a thermodynamic system characterized by the radiation pressure P, the radiation (cavity) volume V^t, and the radiation temperature T. It is described by the same fundamental exact differential expression as for PVT systems, i.e.,

$$dU^t = T\,dS^t - P\,dV^t, \tag{A}$$

and auxiliary thermodynamic functions may be defined as for PVT systems. Radiation is massless, so that molar or specific properties have no meaning; however, one may deal with properties normalized with respect to V^t. Thus, we can define an energy density $\tilde{U} \equiv U^t/V^t$, an entropy density $\tilde{S} \equiv S^t/V^t$, etc. It is known that

$$\tilde{U} = \tilde{U}(T) \text{ only,} \tag{B}$$

$$P = \tfrac{1}{3}\tilde{U}. \tag{C}$$

Using (B) and (C) in conjunction with (A) (or its consequences), derive the *Stefan-Boltzmann law*:

$$\tilde{U} = aT^4,$$

where a is a constant.

11-32 A small, spherical air bubble is suspended in a pool of water. Because of surface tension γ at the interface between the air bubble and the water, the pressure inside the air bubble is greater than the pressure in the water surrounding the bubble. The relationship between the two pressures is given by

$$P_{\text{air}} = P_{\text{water}} + \frac{2\gamma}{r},$$

where r is the radius of the bubble. If the bubble grows from an initial radius of 10^{-4}(in) to a final radius of 10^{-3}(in), what is the expansion work done by the bubble? The surface tension is constant at $\gamma = 4 \times 10^{-4}(\text{lb}_f)/(\text{in})$, and P_{water} is constant at 18(psia). How much of this expansion work goes to form the additional surface area of the bubble, and how much goes into pushing back the water?

11-33 Vapor-pressure data for acetone(1) and acetonitrile(2) are given in Table P11-33. Assuming the validity of Raoult's law, prepare a *txy* diagram for this system at a constant total pressure P of 400(mm Hg). Compare this diagram with the *Pxy* diagram of Fig. 11-5.

11-34 A very small amount of sugar s is dissolved in liquid water w. The liquid solution is in phase equilibrium with pure water vapor.

(*a*) Assuming ideal-solution behavior for the liquid water, show that the equation of phase equilibrium is

$$G_w{}^v = G_w{}^l + RT\ln(1 - x_s),$$

where $G_w{}^v$ is the molar Gibbs function of the water vapor, $G_w{}^l$ is the molar

Table P11-33

t (°C)	P_1(sat) (mm Hg)	P_2(sat) (mm Hg)
38.45	400	159.4
42	458.3	184.6
46	532.0	216.8
50	615.0	253.5
54	707.9	295.2
58	811.8	342.3
62.33	937.4	400

Gibbs function of pure liquid water, and x_s is the liquid-phase mole fraction of the sugar.

(b) For an infinitesimal change in x_s at constant P, show that

$$-S_w{}^v\,dT = -S_w{}^l\,dT + R\ln(1 - x_s)dT + RT\,d\ln(1 - x_s).$$

(c) Combine the results of (a) and (b) and show that the result of (b) reduces to

$$0 = \frac{H_w{}^v - H_w{}^l}{T}\,dT + RT\,d\ln(1 - x_s).$$

(d) Noting that $x_s \ll 1$ and that $H_w{}^v - H_w{}^l \equiv \Delta H_w{}^{lv}$ (the latent heat of vaporization of water), show that the *boiling-point elevation* of the solution, ΔT, is approximately

$$\Delta T = \frac{RT^2}{\Delta H_w{}^{lv}}\,x_s.$$

11-35 A very small amount of sugar s is dissolved in liquid water w. The liquid solution is in phase equilibrium with pure ice. Show that the *freezing-point depression* of the solution, ΔT, is approximately

$$\Delta T = -\frac{RT^2}{\Delta H_w{}^{sl}}\,x_s,$$

where x_s is the liquid-phase mole fraction of the sugar and $\Delta H_w{}^{sl}$ is the latent heat of fusion of water.

12

THERMODYNAMICS AND STATISTICAL MECHANICS

In the preceding chapters we invariably adopted the macroscopic point of view of matter. Nowhere was it necessary to appeal to any theory of the constitution of matter. Such considerations need never arise in a treatment of classical thermodynamics. However, it is universally accepted that matter consists of atoms of the chemical elements, which interact more or less strongly, depending on conditions. The very strong coupling of atoms through chemical bonds produces larger entities made up of numbers of atoms which are called molecules. Each gram mole of a chemical species consists of about 6×10^{23} molecules.

The atoms and molecules which constitute matter are in continual motion; they translate, rotate, vibrate, and collide. Thus they possess kinetic energy of translation, rotation, and vibration; and furthermore, they exchange energy through collision. In addition, there is potential energy associated with the forces that act between atoms and between molecules. Energy of this form cannot be attributed to single atoms and molecules, but is shared among the interacting particles. Energy is also associated with the electrons and nuclei of atoms. All these forms of energy existing at the microscopic level constitute what in macroscopic thermodynamics is known as internal energy.

The two theories which relate the microscopic behavior of molecules to the macroscopic properties of thermodynamics are *kinetic theory* and *statistical mechanics*. In this chapter we give an elementary treatment of the latter; but first we present a qualitative discussion of certain problems that arise whenever thermodynamics is considered in relation to molecular behavior.

12-1 Maxwell's Demon

Perhaps the most famous problem of this nature was posed by Maxwell in 1871 under the heading "Limitations of the Second Law." He invented (in his mind) a *being* that could deal directly with molecules; this being has since been known as Maxwell's demon. Maxwell suggested that a container filled with gas be divided by a partition in which there was a trapdoor manned by his demon. The demon would observe molecules approaching the trapdoor from both sides and would operate the door so as to allow only fast molecules to pass in one direction and only slow molecules to pass in the other direction. Thus the demon would act to sort molecules according to speed. As a result the gas on one side of the partition would become increasingly warmer and that on the other side cooler. At some point we could start a heat engine to operate between the two temperatures, and it could deliver work continually to the surroundings as long as the demon continued his activities. It would only be necessary to add heat to the system to compensate for the work done, and we would have an engine operating in a cycle that converted all the heat taken in into work, in violation of the second law of thermodynamics. Maxwell imagined that his demon itself did no work; it was a reversible demon that released energy to open a frictionless trapdoor and recovered the same energy when the door closed.

There have been many suggestions as to how to build a device to violate the second law, but not one has ever been demonstrated to do so. However, Maxwell added a new dimension to this endeavor. He postulated a being, intelligent in some sense, that could deal with individual molecules. The easiest way out is to declare that the second law denies that such a being could exist. But life has always been mysterious, and we inevitably suppose that it must have qualities not fully taken into account by the known laws of physics. After all, bacteria are very small beings whose accomplishments are by no means inconsequential. So Maxwell's demon has not been lightly dismissed, and even after almost 100 years it is still a fascinating topic of discussion. We might remark that a demon that sorts molecules according to speed is not the only demon one can imagine. Sorting on molecular species is another example, but the problem with respect to the second law is no different.

Maxwell's demon appears to violate the second law through its ability to deal with individual molecules. Is this the key to "success," or are there other ways to take advantage of the molecular nature of matter in efforts to violate the second law? Let us pose another problem. Suppose again that we have a container divided by a partition. This time there is gas on one side of the partition, but a total vacuum on the other. The partition is removed and the gas expands to fill the total volume. Now the question is whether the gas will ever of its own accord return to its initial location in one part of the container. The overwhelming consensus of informed opinion is that it will, provided one waits long enough! It comes down to a matter of chance. Since the gas molecules are in continual motion, one concludes that there is a finite (but minuscule) probability, a chance, at any instant that the original configuration will be reproduced. One need not even insist on a special initial configuration. It is sufficient to consider the container merely filled with gas and then to ask whether the gas will ever momentarily collect itself in any portion of the container. For if it does, then we can insert a partition and trap the gas in a state of lower entropy than it had initially. Again the overwhelming consensus of informed opinion is that this is possible if one is prepared to wait long enough, say, $10^{10^{10}}$ years. Whether or not you believe this will ever happen is not important. We can imagine it to happen regardless of whether it actually will; and whether real or imaginary, it represents a process seemingly at odds with the second law, and one that does not require dealing with individual molecules. It is important to note, however, that to accomplish the process one must insert a partition into the container at exactly the right instant. Thus the process does require continuous observation of the system by some being or device capable of detecting molecules and taking appropriate action. So we see that the two hypothetical processes just described do have common elements and need not be considered independently.

In particular, they have in common the idea of molecular ordering. In the case of Maxwell's demon, ordering is done on the basis of speed; that is, high-speed molecules are segregated from low-speed molecules. In the second case, ordering is done in the sense that molecules are collected from a larger region of space into a more restricted region. This process, by the way, could also be accomplished by a Maxwell demon, one which allowed molecules to pass only one way through its trapdoor. Another type of ordering process that a Maxwell demon could accomplish results when a gas *mixture* is admitted to the container. The demon could operate his trapdoor so as to allow green molecules to go only one way and red molecules only the other. This would serve to segregate the green from the red molecules. Our use of the word "ordering" gives it perhaps a broader meaning than is found in its everyday use. Our demon is said to bring about ordering whenever he restricts molecules to a given region of space by virtue of some characteristic of the molecule. We have considered molecular speed, molecular species, and even the very characteristic of being a molecule at all. Molecules left to themselves do not become so ordered except, as we have seen, by chance. Ordering at will requires the intervention of some outside agent, of which Maxwell's demon is a very special example.

All of these ordering processes produce a *reduction* in the entropy of the system, and each reduction can easily be calculated by the methods of thermodynamics. We are therefore led to the notion that increasing order corresponds to decreasing entropy, and vice versa; this is the basic idea that underlies statistical mechanics. All that we need in addition is a method of expressing order or disorder in a quantitative way, but this we will leave for later.

There is another aspect of the sorting processes involving Maxwell demons that we have not yet considered. It centers around the fact that the demon must act on the basis of *information*; that is, he cannot act properly until he knows that a molecule in a particular place is a fast one, a slow one, a red one, a green one, directed left, or directed right, etc. Even a demon that sits around waiting for chance to order a system must keep continuously informed of the locations of molecules; otherwise he would never know the moment to insert a partition so as to preserve the long-awaited but otherwise-momentary order. There are two separate ideas which come out of these observations. The first is that there may be some connection between entropy and information. The second is that the information-gathering activities of the demon may be the key to whether or not he can operate so as to cause violations of the second law. The apparent link between information and entropy has been exploited and developed into the subject called *information theory,* which has important applications in the design of communication systems. The fundamental equation of information theory is identical with the equation for entropy in statistical mechanics, and the quantity calculated, having to do with the information content of messages, is even called *entropy.* In statistical mechanics we deduce the properties of matter by applying statistics to large numbers of molecules. In information theory we deduce the information-carrying capacity of communications systems by applying statistics to large numbers of messages.

In our descriptions of the activities and ambitions of Maxwell demons we have implied several questions. Let us state these questions one by one, and provide what are thought to be correct answers:

1 Is it necessary to regard the demon as a *living* being? The answer is that it is not *necessary.* Moreover, it's not even advantageous. The demon is merely an intermediary, a relay mechanism, that responds in a specific way to an information signal. It may therefore be automated or programmed to perform its tasks at least as surely as if it possessed the intelligence of a human being. This is not to deny that an intelligent living being could serve as a sorting demon, but such a demon would be at a disadvantage. In spite of the mysteries of life, every study of life

processes has demonstrated that the laws of physics *do* in fact apply. There may be *additional* laws, but none of those known is violated, not even the second law of thermodynamics. The fantastic ordering of atoms and molecules necessary to produce and maintain a living system is accompanied by a more than compensating disorder created in the surroundings. Thus the ever-increasing order represented by increasing numbers of the human species is more than matched by the trail of disorder left in our surroundings, of which the increasing pollution of our atmosphere, rivers, and oceans is but an example. It is not our problem to explain *how* or *why* the ordering necessary to living systems occurs. The fact is that it does, and it does so without violating the laws of physics. So the attribution of life to Maxwell's demon can only prejudice the case against its ambitions to violate the second law. We can therefore narrow our attention to automated devices.

2 Can an automated device be activated by mechanical interaction with the molecules themselves? The answer to this is no, for the following reason. Any device or portion of a device used to trigger the necessary action to accomplish the sorting of molecules can be no more massive than the molecules themselves; otherwise it would grossly interfere with the motions of the molecules and actually prevent the sorting from being accomplished. On the other hand, any object no more massive than the molecules will be subject to the same thermal motions as the molecules, and as a result cannot be held in its proper location. We are therefore reduced to more massive devices which rely on *information* about the molecules to be sorted, and this can be transmitted only by electromagnetic radiation, of which ordinary light is one example.

3 Can an automated device relying on information sort molecules in violation of the second law? Again, the answer is that it cannot. We may presume our device to be sufficiently massive and to be mechanically reversible, and we concentrate on the problem of how it is to sense the molecules with which it is designed to deal. The device is enclosed within a container, and the only way it can sense its subject molecules without grossly disturbing their motions is by some form of electromagnetic radiation. Thus the device must radiate energy and sense molecules through their reflection of radiation. This energy is absorbed throughout the system and can be shown to cause a greater entropy increase than any decrease caused by the proper working of the device.

Thus we conclude that Maxwell's demon cannot operate in such a way as to violate the second law of thermodynamics. The fact that this problem has been discussed for more than 100 years and is still of interest illustrates the reluctance with which even scientists accept the second law as being inviolate. There is good reason for *wanting* to violate it, for if it could be done, all man's energy requirements could forever be met without any depletion of resources or pollution of his surroundings. Man lives on hope, and does not readily take to restrictions on what he can do. So far, however, he has had to live within the limits defined by the laws of thermodynamics, and all indications are that he will continue to.

12-2 Ensembles and Statistics

The purpose of the preceding discussion has been to illustrate the problems encountered when we merely contemplate dealing with individual molecules. Nevertheless, we would like to be able to use our knowledge of the molecular or microscopic nature of matter to help us understand the macroscopic behavior of matter. The enormous numbers of molecules that make up macroscopic systems and the chaotic motions of these molecules suggest that some sort of statistical treatment might prove useful.

The obvious way to go about developing this subject is to apply statistics to the properties of molecules themselves. However, no such treatment can possibly be general. The reason for this is that molecules interact, and as a result statistical averaging of their *private* properties does not provide any meaningful quantity descriptive of a macroscopic system. For example, molecules do possess their own private kinetic energies, but not their own private potential energies, because potential energy arises through forces acting between molecules and is shared among them. Thus statistical averaging of the private (kinetic) energies of molecules does not in general allow calculation of the internal energy of the system. Only for ideal gases can one say that molecules are independent of one another.

If we are not to deal with individual molecules, what is the alternative? It is to deal with a very large collection of macroscopically identical systems known as an *ensemble.* The word ensemble is most commonly applied to a collection of musicians, but the word has meaning only if all the musicians are playing the same composition. The thing that makes a collection of systems an ensemble is that all members would appear to an outside observer to be identical. If we have a closed (i.e., constant-mass) system containing N identical atoms or molecules and having a fixed volume V and existing in thermal equilibrium with a heat reservoir at temperature T, then we regard the thermodynamic state of the system to be fixed, regardless of what the atoms or molecules inside may be doing.

We imagine this system to be reproduced or replicated a tremendous number of times, and we imagine this collection of macroscopically identical systems to be arranged on a lattice so that the members are in close contact with one another. Then we imagine the entire collection to be *isolated* from its surroundings; that is, we imagine the boundaries of the collection to be impervious to both the passage of matter and of energy. This is our ensemble. But what is its purpose? It is merely an aid to our mental processes, for we now ask how the various members of the ensemble differ from one another at the *microscopic* or molecular level.

We know that molecules move about in a chaotic fashion and that at any instant the particular molecular configuration to be found in any ensemble member is just a matter of chance. If we consider all the members at any one instant, we therefore expect to find a tremendous variety of microscopic configurations. These configurations, seen at any one instant in the ensemble, are presumed to be the same as those we would see in the original real system were we to observe it for a very long time. Furthermore, we assume that the observed macroscopic properties, such as pressure and internal energy, are averages resulting from the various configurations considered, either over a long period of time or over a very large ensemble.

The question yet to be answered is how the various microscopic configurations are to be characterized. The answer is provided by quantum mechanics, and we must here merely accept it. Quantum theory postulates that energy on the microscopic scale is made up of discrete units or *quanta*. Since energy is quantized, the internal energy of a macroscopic system at any instant is the sum of an enormous number of quanta of energy, and because of this, a macroscopic system at any instant is in a particular *quantum state*, characterized by a particular value of its energy, E_q. There is a discrete set of possible energy values, and we will use the notation $\{E_q\}$ to represent the entire set. Any one value E_q of the set $\{E_q\}$ represents the particular energy associated with quantum state q of the system. The set is discrete because one can never get a complete spectrum of values by summing quanta, just as one cannot obtain decimal numbers by adding integers.

Quantum theory also provides the result that *for a closed system of N particles, the set of values $\{E_q\}$ is completely determined by the volume of the system.* Since we have specified the volume V of our system, we have fixed the set $\{E_q\}$, and we can expect the energy of our system to pass through all these possible values over a long period of time and, in fact, to pass through some of them many, many times. Similarly, our ensemble at a given instant is made up of members each in its own quantum state with an energy E_q taken from the set $\{E_q\}$. There may be many members with the same value of E_q. The *total* internal energy of the entire ensemble is just the sum of the energies E_q, each multiplied by the number of members n_q in the particular quantum state q; thus

$$\mathcal{U} = \sum_q n_q E_q = \text{constant}, \tag{12-1}$$

where the summation is over all possible quantum states. The constancy of $\mathcal{U}$ results from the first law of thermodynamics as applied to the ensemble, which is an isolated system. Furthermore, if there are a total of $\mathcal{N}$ members in the ensemble, then

$$\mathcal{N} = \sum_q n_q = \text{constant}. \tag{12-2}$$

The constancy of $\mathcal{N}$ again results because the ensemble is considered isolated. These two equations express the restraints on the ensemble, and we will need to take them into account later.

We now want to consider the makeup of our ensemble, or more precisely the number of *different ways* it can be made up. It is here that statistics enters, and the procedure seems strange indeed. It is probably best explained by taking a specific example that simulates the real situation, but on a very small scale. Assume we have an ensemble that contains 24 members; that is, $\mathcal{N} = 24$, as shown in Fig. 12-1. Consider for a moment the shaded member of this ensemble. We will identify it with the letter a. It is a replica of our original system, which was specified to have the volume V and to be in equilibrium with a heat reservoir at temperature T. Member a also has volume V and is in equilibrium with the rest of the ensemble, which then constitutes a heat reservoir at T for member a. Now each of the 24 ensemble members can be identified by its own letter. It is not the positions

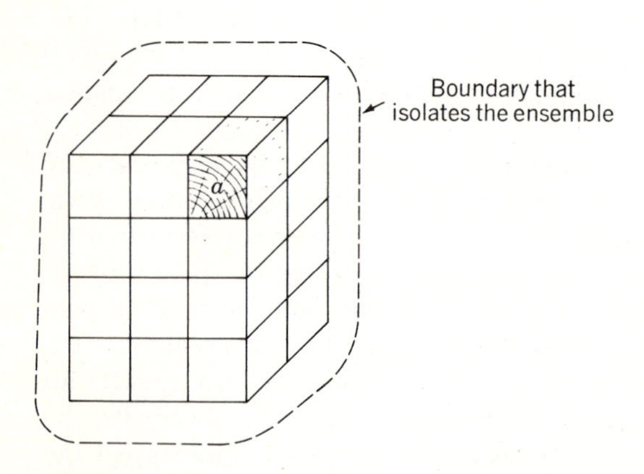

Fig. 12-1
Ensemble.

on the lattice that are being identified by letters, but the ensemble members themselves. Thus Fig. 12-1 shows member a in just one of its 24 possible locations, and once it is put in a particular location there are $\mathcal{N} - 1 = 23$ positions left for placing member b. Moreover, for *each* of the 24 locations for a there are 23 locations for b. Thus there are 24×23 different ways to locate both a and b. Once a and b are placed, there are 22 ways to locate c for each of the 24×23 locations of a and b. Thus there are $24 \times 23 \times 22$ ways to locate a, b, and c. By the time we have located all $\mathcal{N} = 24$ members, we have chosen one arrangement of the ensemble out of $\mathcal{N}! = 24! \approx 6.2 \times 10^{23}$ possible arrangements. Clearly, one does not require a large ensemble in order to generate large numbers.

Each member of the ensemble must at any instant be in a particular quantum state characterized by a particular value of E_q taken from the set of possible values $\{E_q\}$. In our example we will assume that there are only four possible quantum states, identified by setting q equal to 1, 2, 3, or 4. Thus $\{E_q\}$ is made up of the energy values E_1, E_2, E_3, and E_4. Each of our 24 ensemble members must be in one of these four quantum states; so let us assign a quantum state to each ensemble member as indicated in Table 12-1. With this assignment of ensemble members to quantum states, each arrangement of members on the lattice of Fig. 12-1 represents a particular distribution of energy states, which corresponds to a particular sequence in time for the original system to pass through the same states. The complicating factor is that there is more than one ensemble member in each quantum state. If we were to interchange members of the same quantum state, the lattice would look no different. For example, members h and z are both in quantum state 4 and are therefore identical. Interchanging them makes no difference, but our $\mathcal{N}! = 24!$ ways of arranging the lattice counted them as different. Thus for *any* arrangement of the lattice we have a second arrangement that is no different, for we may inter-

Table 12-1

Quantum state q	Energy E_q	Ensemble members in state q	Number of members, n_q
1	E_1	$a, b, f, i, l, m, r, v, y$	$n_1 = 9$
2	E_2	d, e, k, o, p, w, x	$n_2 = 7$
3	E_3	c, g, j, s, t, u	$n_3 = 6$
4	E_4	h, z	$n_4 = 2$
			$\mathcal{N} = 24$

change h and z whatever their locations. Thus to get the number of really different lattice arrangements, we would divide $\mathcal{N}!$ by $2 = 2! = n_4!$ because of the fact that h and z are identical. Similarly, the six members c, g, j, s, t, and u are all in quantum state 3 and are therefore indistinguishable from one another. There are $6! = n_3!$ ways to arrange these six members on their respective lattice sites without making the lattice any different. This is because c could be put on any of six sites, then g on any of the five remaining sites, j on four, etc. Thus, our $\mathcal{N}!$ ways to arrange the lattice is too large by a factor of $6! = n_3!$ and must be divided by this factor to remove the indistinguishable arrangements resulting from the fact that six members are all in quantum state 3. By now it should be clear that we must divide $\mathcal{N}!$ by $n_q!$ for *each* of the q quantum states. Thus the number of ensemble or lattice arrangements that are really different is given by

$$\omega = \frac{\mathcal{N}!}{n_1!n_2!n_3!n_4!} = \frac{24!}{9!7!6!2!} \approx 2.356 \times 10^{11},$$

which is still a big number, though not so large as 24!. It is worth noting at this point that the natural logarithm of ω is not nearly so imposing a number. In fact, $\ln \omega$ is a mere 26.19.

As a result of this example we can now write down a general formula for the number of really different ways an ensemble of $\mathcal{N}$ members can be arranged for q quantum states when n_q is the number of members in a particular quantum state; thus

$$\boxed{\omega = \frac{\mathcal{N}!}{\prod_q n_q!},} \tag{12-3}$$

where the sign $\prod_q$ signifies the running product of all the factorials $n_q!$.

Clearly, in order to evaluate ω we must know not only $\mathcal{N}$ but also the values of all the n_q; that is, we must know the distribution of the ensemble members among the possible quantum states. In our example, used for the purpose of illustration, we assigned values to the n_q's arbitrarily; and had we assigned different values, we would have calculated a different value for ω. For example, we might have set $n_1 = n_2 = n_3 = 0$ and $n_4 = 24$. Then ω would have been 1, and $\ln \omega$ would have been zero. Or we might have set $n_1 = n_2 = n_3 = n_4 = 6$, and ω would have come out to about 2.3×10^{12} and $\ln \omega$ to about 28.5. Evidently, the problem is to find a preferred distribution. The clue to a solution to this problem is found in the second law of thermodynamics.

12-3 The Partition Function and Boltzmann's Distribution

The ensemble of members that we have devised is isolated from its surroundings. Thus at equilibrium its total entropy must be a maximum with respect to all possible internal variations within the ensemble. Thus we conclude that the preferred distribution of ensemble members among the possible quantum states is that distribution which maximizes the entropy of the entire ensemble. The problem now is to find a connection between entropy and some variable which pertains to the ensemble, and here the best we can do is make an educated guess. We have already suggested that there is some connection between entropy and disorder. Moreover, the quantity ω as given by Eq. (12-3) is a measure of the disorder in our original system, for each really different arrangement of our ensemble corresponds to a different sequence of states over a period of time in the original system. The more such possibilities there are, the more chaotic or unpredictable or disordered the original system appears. The limiting case where all ensemble members are in the same quantum state led to a value of $\omega = 1$ or $\ln \omega = 0$. This corresponds to perfect order, for it means that the original system is always in the same quantum state; it never changes its state, and in this sense is in no way chaotic but is completely predictable or ordered. So it makes sense to guess that the total entropy of the ensemble is some function of ω, and all that remains is to find that functional relationship which leads to results that agree with experiment. The relationship which leads to success identifies the ensemble entropy with $\ln \omega$ according to the equation

$$\mathcal{S} = k \ln \omega, \tag{12-4}$$

where $\mathcal{S}$ is the entropy of the entire ensemble and k is a constant, known as *Boltzmann's constant.*

Equation (12-4) is the fundamental postulate of statistical mechanics. There is no way to prove it. We can only guess that it is right and then test the consequences against experiment. The remaining steps are mathematical. First, we use Eq. (12-3) to get an expression for $\ln \omega$, and then we simplify this expression as much as possible and substitute the result into Eq. (12-4) to get an expression for the entropy. Finally, we maximize the entropy to obtain the preferred distribution of quantum states. It is done as follows.

Taking the natural logarithm of both sides of Eq. (12-3), we get

$$\ln \omega = \ln \mathcal{n}\,! - \sum_q \ln n_q!.$$

Since $\mathcal{n}$ is taken to be arbitrarily large for any ensemble of interest, the n_q

are also presumed to be large numbers, and in this case we may use Stirling's formula for the logarithms of factorials; thus

$$\ln X! = X \ln X - X.$$

Our equation for $\ln \omega$ now becomes

$$\ln \omega = n \ln n - n - \sum_q (n_q \ln n_q) + \sum_q n_q.$$

But $n = \sum_q n_q$; therefore

$$\ln \omega = n \ln n - \sum_q (n_q \ln n_q).$$

By a little manipulation we can put this equation into an even simpler form. First, we factor n; thus

$$\ln \omega = n \left[\ln n - \frac{1}{n} \sum_q (n_q \ln n_q)\right].$$

Multiplying the first term in brackets by $\sum_q n_q / n = 1$, we obtain

$$\ln \omega = n \left[\frac{\sum_q n_q}{n} \ln n - \frac{1}{n} \sum_q (n_q \ln n_q)\right].$$

Now n and $\ln n$ are the same for all terms in the summations, and we may therefore write

$$\ln \omega = n \left[\sum_q \left(\frac{n_q}{n} \ln n\right) - \sum_q \left(\frac{n_q}{n} \ln n_q\right)\right],$$

or

$$\ln \omega = -n \sum_q \left[\frac{n_q}{n} (\ln n_q - \ln n)\right],$$

or

$$\ln \omega = -n \sum_q \left(\frac{n_q}{n} \ln \frac{n_q}{n}\right).$$

We now define the *probability* of quantum state q by

$$P_q = \frac{n_q}{n}.$$

So that we have finally

$$\ln \omega = -n \sum_q (P_q \ln P_q). \tag{12-5}$$

Substitution of Eq. (12-5) into Eq. (12-4) gives the following expression for the entropy of the ensemble:

$$\mathcal{S} = -k\mathcal{n} \sum_q (P_q \ln P_q). \tag{12-6}$$

We have come now to the problem of finding the set of probabilities $\{P_q\}$ which maximizes $\mathcal{S}$. Unfortunately, this is not the simple maximum problem that it may seem. The reason is that there are two restraints on the system imposed by Eqs. (12-1) and (12-2). However, the problem of finding a maximum subject to restraints has a standard solution through Lagrange's method of undetermined multipliers. Equation (12-1) may be written

$$\sum_q (n_q E_q) = \mathcal{n} \sum_q \left(\frac{n_q}{\mathcal{n}} E_q\right) = \mathcal{n} \sum_q (P_q E_q) = \mathcal{U},$$

or

$$\mathcal{n} \sum_q (P_q E_q) - \mathcal{U} = 0.$$

We now multiply this equation by an undetermined constant, say λ; thus

$$\lambda[\mathcal{n} \sum_q (P_q E_q) - \mathcal{U}] = 0.$$

Since the left side of this equation is zero, it may be added to the right side of Eq. (12-6) without changing anything to get

$$\mathcal{S} = -k\mathcal{n} \sum_q (P_q \ln P_q) + \lambda[\mathcal{n} \sum_q (P_q E_q) - \mathcal{U}].$$

This equation now incorporates one of the restraints on the system. To maximize $\mathcal{S}$ subject to this restraint, we differentiate and set $d\mathcal{S} = 0$ (note that k, $\mathcal{n}$, λ, $\mathcal{U}$, and the E_q's are all constant) as follows:

$$d\mathcal{S} = -k\mathcal{n}[\sum_q (P_q \, d \ln P_q) + \sum_q (\ln P_q \, dP_q)] + \lambda \mathcal{n} \sum_q (E_q \, dP_q) = 0.$$

Thus since $d \ln P_q = dP_q/P_q$, we have

$$[\sum_q dP_q + \sum_q (\ln P_q \, dP_q)] - \frac{\lambda}{k} \sum_q (E_q \, dP_q) = 0.$$

For simplicity, set $\lambda/k = -\beta$ and collect like terms; thus

$$\sum_q (1 + \ln P_q + \beta E_q) \, dP_q = 0. \tag{12-7}$$

Now we impose the second restraint on our system. Dividing Eq. (12-2) by $\mathscr{N}$, we get

$$1 = \sum_q \frac{n_q}{\mathscr{N}} = \sum_q P_q.$$

Differentiating, we have

$$\sum_q dP_q = 0. \tag{12-8}$$

In order to satisfy both Eqs. (12-7) and (12-8), we must have

$$1 + \ln P_q + \beta E_q = \text{constant},$$

or

$$\ln P_q + \beta E_q = \text{constant} - 1 = A,$$

or

$$\ln P_q = A - \beta E_q.$$

In exponential form this equation becomes

$$P_q = e^A e^{-\beta E_q}.$$

If we now sum the P_q's over all q, we get

$$\sum_q P_q = e^A \sum_q e^{-\beta E_q} = 1.$$

Thus

$$e^A = \frac{1}{\sum_q e^{-\beta E_q}}$$

and

$$\boxed{P_q = \frac{e^{-\beta E_q}}{\sum_q e^{-\beta E_q}} = \frac{e^{-\beta E_q}}{\mathscr{Z}},} \tag{12-9}$$

where $\mathscr{Z} \equiv \sum_q e^{-\beta E_q}$ is called the *system partition function.*

The distribution of probabilities for the possible quantum states of our ensemble, as given by Eq. (12-9), is known as the *Boltzmann distribution.* It is by no means the only distribution of probabilities that will lead to the required ensemble energy $\sum_q (n_q E_q)$; it is the particular distribution that maximizes the ensemble entropy and which therefore conforms to the laws of thermodynamics. It is clear from Eq. (12-9) that the only variable on which P_q depends is E_q, because β is constant by definition and $\mathscr{Z}$ is a summation that is the same for all the P_q's. Thus the probability of a quantum state for a system of fixed volume in equilibrium with a heat reservoir depends only on the energy E_q of the quantum state, and all quantum states with the same energy have the same probability. We could have used this as an alternative basic postulate of statistical mechanics, and we would have reached the same distribution of probabilities.

12-4 Thermodynamic Properties

Having found the distribution of probabilities, our only remaining question is what to do with it. Since its main use is in the calculation of thermodynamic properties, we should look for equations which give these properties in terms of the variables of statistical mechanics. We start with the two basic expressions, Eq. (12-1) for internal energy and Eq. (12-6) for entropy. If we divide Eq. (12-1) by $\mathcal{N}$, we get the average internal energy of an ensemble member or the time-averaged or macroscopic internal energy of the original system U; thus

$$U = \frac{\mathcal{U}}{\mathcal{N}} = \sum_q \left(\frac{n_q}{\mathcal{N}} E_q \right)$$

or

$$\boxed{U = \sum_q (P_q E_q).} \tag{12-10}$$

Similarly, if we divide Eq. (12-6) by $\mathcal{N}$, we get the average entropy of an ensemble member or the entropy of the original system S; thus

$$\boxed{S = -k \sum_q (P_q \ln P_q).} \tag{12-11}$$

By Eq. (12-9) we eliminate the P_q's from Eq. (12-10); this gives

$$U = \sum_q \frac{e^{-\beta E_q} E_q}{\mathcal{Z}}.$$

We also have

$$\left(\frac{\partial \ln \mathcal{Z}}{\partial \beta} \right)_V = \frac{1}{\mathcal{Z}} \left(\frac{\partial \mathcal{Z}}{\partial \beta} \right)_V = \frac{1}{\mathcal{Z}} \left(\frac{\partial \Sigma e^{-\beta E_q}}{\partial \beta} \right)_V = \sum_q \frac{e^{-\beta E_q}(-E_q)}{\mathcal{Z}}.$$

Comparison of the last two equations shows that

$$\left(\frac{\partial \ln \mathcal{Z}}{\partial \beta} \right)_V = -U. \tag{12-12}$$

Since $\mathcal{Z} = \sum_q e^{-\beta E_q}$, we see that $\mathcal{Z}$ is in general a function of β and the E_q's. But the E_q's are functions of volume. Thus $\mathcal{Z} = \mathcal{Z}(\beta, V)$. Therefore

$$d \ln \mathcal{Z} = \left(\frac{\partial \ln \mathcal{Z}}{\partial \beta} \right)_V d\beta + \left(\frac{\partial \ln \mathcal{Z}}{\partial V} \right)_\beta dV,$$

or

$$d \ln \mathcal{Z} = -U \, d\beta + \left(\frac{\partial \ln \mathcal{Z}}{\partial V} \right)_\beta dV.$$

But $$d(U\beta) = U\,d\beta + \beta\,dU,$$

or $$-U\,d\beta = \beta\,dU - d(U\beta).$$

Thus $$d \ln \mathfrak{z} = \beta\,dU - d(U\beta) + \left(\frac{\partial \ln \mathfrak{z}}{\partial V}\right)_\beta dV,$$

and $$\beta\,dU = d(\ln \mathfrak{z} + U\beta) - \left(\frac{\partial \ln \mathfrak{z}}{\partial V}\right)_\beta dV.$$

We also have the thermodynamic equation

$$dU = T\,dS - P\,dV. \tag{9-1}$$

Thus

$$\beta T\,dS = \beta\,dU + \beta P\,dV$$

and

$$\beta T\,dS = d(\ln \mathfrak{z} + U\beta) + \left[\beta P - \left(\frac{\partial \ln \mathfrak{z}}{\partial V}\right)_\beta\right] dV,$$

or finally

$$dS = \frac{1}{\beta T}\,d(\ln \mathfrak{z} + U\beta) + \frac{1}{\beta T}\left[\beta P - \left(\frac{\partial \ln \mathfrak{z}}{\partial V}\right)_\beta\right] dV. \tag{12-13}$$

The next step is to develop another general equation for dS. We start by substituting $e^{-\beta E_q}/\mathfrak{z}$ for P_q in the logarithm of Eq. (12-11) as follows:

$$\begin{aligned} S &= -k \sum_q P_q \ln \frac{e^{-\beta E_q}}{\mathfrak{z}} \\ &= k \sum_q (P_q \ln \mathfrak{z}) + k \sum_q (P_q \beta E_q) \\ &= k \ln \mathfrak{z} \sum_q P_q + k\beta \sum_q (P_q E_q). \end{aligned}$$

However, $\sum_q P_q = 1$, and by Eq. (12-10), $\sum_q (P_q E_q) = U$. Thus

$$S = k \ln \mathfrak{z} + k\beta U = k(\ln \mathfrak{z} + \beta U), \tag{12-14}$$

and $$dS = kd(\ln \mathfrak{z} + \beta U). \tag{12-15}$$

Comparison of Eqs. (12-13) and (12-15) shows that

$$k = \frac{1}{\beta T} \qquad \text{or} \qquad \boxed{\beta = \frac{1}{kT}} \tag{12-16}$$

and that

$$\beta P - \left(\frac{\partial \ln \mathcal{Z}}{\partial V}\right)_\beta = 0 \qquad \text{or} \qquad P = \frac{1}{\beta}\left(\frac{\partial \ln \mathcal{Z}}{\partial V}\right)_\beta. \tag{12-17}$$

From Eq. (12-12)

$$U = -\frac{(\partial \ln \mathcal{Z}/\partial T)_V}{d\beta/dT} = -\frac{(\partial \ln \mathcal{Z}/\partial T)_V}{-1/kT^2}.$$

Thus

$$\boxed{U = kT^2\left(\frac{\partial \ln \mathcal{Z}}{\partial T}\right)_V} \tag{12-18}$$

and by combining Eq. (12-16) with Eqs. (12-14) and (12-17), we get

$$S = k \ln \mathcal{Z} + \frac{U}{T}, \tag{12-19}$$

$$P = kT\left(\frac{\partial \ln \mathcal{Z}}{\partial V}\right)_T. \tag{12-20}$$

Equations (12-18) to (12-20) show that the internal energy, the entropy, and the pressure may be calculated once the partition function for a system is known as a function of T and V. Knowing U, S, P, T, and V, we may readily calculate any other thermodynamic property from its definition. Thus statistical mechanics provides a formalism for the calculation of thermodynamic properties from the partition function. Unfortunately, it does not provide the means for the determination of partition functions. This is a problem in quantum mechanics, one that has been solved only for special cases. The relative simplicity of ideal gases allows them to be treated rather completely, and statistical mechanics is widely used for the calculation of the thermodynamic properties of ideal gases from spectroscopic data. Its value here is not that it makes unnecessary the taking of data but that it allows use of a different sort of data (data that are more readily taken) than would be required by classical thermodynamics. For nonideal gases less progress has been made, but statistical mechanics does show that the correct form for an equation of state is the virial form. However, for almost all cases the coefficients in this equation must still be determined from measurements of macroscopic properties.

For liquids, relatively little progress has been made, because one encounters great difficulty in the evaluation of the partition function. For this reason most work on liquids has been directed toward development of approximate methods, none of which is yet regarded as generally satisfactory. Crystalline solids, however, because of their highly ordered state, have been dealt with more successfully. Statistical mechanics has also been applied to the electron gas to provide useful results with respect to the electrical properties of solids. Yet another application is to the photon gas, and this yields important results with respect to radiation. The properties of plasmas, because of their high temperatures, could hardly be determined except by statistical mechanics.

Although statistical mechanics is based on the presumed reality of atoms and molecules, it does *not* provide, any more than does thermodynamics, a detailed description of atomic and molecular behavior and of atomic and molecular interactions. However, it does provide, as thermodynamics does not, the means by which thermodynamic properties may be calculated whenever detailed descriptions of atomic and molecular behavior are provided from other studies, either theoretical or experimental. Thus statistical mechanics adds something very useful to thermodynamics, but it neither explains thermodynamics nor replaces it.

12-5 Energy Levels

We have so far expressed our equations in terms of the energies and probabilities of the individual quantum states. However, there can be more than one quantum state with the same energy. In fact, the number of different quantum states possible for a particular energy level is called the *degeneracy* g of the level. Thus we may readily develop an alternative set of equations in terms of the probabilities of the energy levels instead of the probabilities of the quantum states. These two sets of probabilities are not the same because many quantum states may have the same energy. If the list of energy levels E_0, E_1, E_2, E_3, etc., is set up so that each energy value appears but once, the subscripts on E then denote energy levels rather than quantum states. Thus $\{E_i\}$ represents the set of energy levels as distinct from $\{E_q\}$, which represents an enumeration of energies of all quantum states. The same numbers appear in each set, but in the list of energy levels each number appears but once, whereas in a listing of the energies of the quantum states a number may appear over and over again.

The probability of an energy level P_i is simply related to the probability P_q of a quantum state having an energy $E_q = E_i$. We have already shown that all quantum states having the same energy must have the same probability. If there are g_i quantum states with the same energy E_i, the probability of each of these quantum states is P_q and the probability of the energy level is

$$P_i = g_i P_q.$$

Therefore Eq. (12-9) gives

$$P_i = g_i P_q = \frac{g_i e^{-\beta E_q}}{\mathfrak{z}},$$

or since $E_q = E_i$ for all the g_i quantum states,

$$P_i = \frac{g_i e^{-\beta E_i}}{\mathfrak{z}}, \tag{12-21}$$

where

$$\mathfrak{z} = \sum_i g_i e^{-\beta E_i}. \tag{12-22}$$

Equation (12-22) can be obtained from Eq. (12-21) by summing over all energy levels i and noting that $\Sigma P_i = 1$. The quantity g_i, the number of different quantum states having energy E_i, has already been designated as the degeneracy of the energy level. Furthermore, in terms of energy levels Eq. (12-10) becomes

$$U = \sum_i P_i E_i. \tag{12-23}$$

The connection between statistical mechanics and thermodynamics can be given greater physical meaning through the following numerical example, which applies to a small-scale imaginary system. Quite obviously, we cannot deal with a real macroscopic system because of the enormous number of energy levels and quantum states associated with a system of some 10^{23} molecules. However, the system considered does display the same general behavior as a real system.

Example 12-1

Consider a system whose energy levels are given by the expression

$$E_i = j + k + l,$$

where E_i is the energy level of i. The quantities j, k, and l may independently have any integral value 1, 2, 3, 4, etc. Determine values for the thermodynamic properties U and S for a value of $\beta = \frac{1}{3}$.

Since $E_i = j + k + l$, the lowest energy level represents an energy $E_0 = 3$, and there is but one way it may be attained: j, k, and l must each be unity. Thus the degeneracy of the level $g_0 = 1$. The next energy level represents an energy $E_1 = 4$, and it may be attained in three ways, so that the degeneracy of the level $g_1 = 3$. The next energy level represents an energy $E_2 = 5$, attainable in six ways, so that $g_2 = 6$. Table 12-2 illustrates the counting scheme, where each row represents a quantum state. The procedure shown may be continued without limit, but one soon realizes that g_i increases in a systematic way. A general formula for g_i for the particular system considered is

Table 12-2

Level i	Energy E_i	j	k	l	Degeneracy g_i
0	3	1	1	1	1
1	4	1	1	2	3
		1	2	1	
		2	1	1	
2	5	1	2	2	6
		2	1	2	
		2	2	1	
		3	1	1	
		1	3	1	
		1	1	3	
3	6	2	2	2	10
		1	2	3	
		1	3	2	
		2	1	3	
		3	1	2	
		2	3	1	
		3	2	1	
		1	1	4	
		4	1	1	
		1	4	1	

found to be

$$g_i = \frac{(i + 1)(i + 2)}{2}.$$

Equation (12-21) may be written

$$P_i \mathfrak{Z} = g_i e^{-\beta E_i}.$$

Since g_i is now known for each E_i, values of $P_i \mathfrak{Z}$ are readily calculated for any given value of β. A set of such computations has been carried out for a value of $\beta = \frac{1}{3}$. The results are shown in column 4 of Table 12-3. Column 1 identifies the value of the index i; column 2 gives the energy values E_i; and column 3 shows the calculated values of the degeneracy g_i. The values of $P_i \mathfrak{Z}$ may be summed to give

$$\sum_i P_i \mathfrak{Z} = \mathfrak{Z}\, \Sigma P_i = \mathfrak{Z},$$

the partition function. The value of $\mathfrak{Z}$ so determined is shown at the bottom of column 4.

The number of energy levels which might be considered is infinite, and the summation $\sum_i P_i \mathfrak{Z}$ contains an infinite number of terms; i.e., it is an infinite series. However, the series converges, and its sum has a limiting value. Table 12-3 must obviously terminate after the listing of a limited number of values. It has been carried only far enough to show that succeeding values of $P_i \mathfrak{z}$ are decreasing rapidly. In determining $\mathfrak{z}$ we have included an adequate number of terms, so that the sum of those omitted is entirely negligible.

Having determined $\mathfrak{z}$, we can now divide the values in column 4 by $\mathfrak{z}$ to obtain the probabilities P_i of the various energy levels. Thus

$$P_i = \frac{P_i \mathfrak{z}}{\mathfrak{z}}.$$

Values of the P_i are listed in column 5.

Table 12-3

$\beta = \frac{1}{3}$

(1) i	(2) E_i	(3) g_i	(4) $P_i \mathfrak{z} = g_i e^{-\frac{1}{3}E_i}$	(5) $P_i = (P_i \mathfrak{z})/\mathfrak{z}$	(6) $P_q = P_i/g_i$	(7) P_iE_i	(8) $P_iE_i^2$
0	3	1	0.368	0.0227	0.0227	0.0681	0.20
1	4	3	0.792	0.0489	0.0163	0.196	0.78
2	5	6	1.134	0.0701	0.01167	0.350	1.75
3	6	10	1.350	0.0834	0.00834	0.500	3.00
4	7	15	1.460	0.0901	0.00602	0.630	4.42
5	8	21	1.453	0.0897	0.00427	0.718	5.75
6	9	28	1.393	0.0860	0.00307	0.775	6.96
7	10	36	1.290	0.0796	0.00221	0.796	7.96
8	11	45	1.148	0.0710	0.001573	0.782	8.59
9	12	55	1.007	0.0622	0.001130	0.748	8.96
10	13	66	0.869	0.0537	0.000815	0.699	9.09
11	14	78	0.733	0.0452	0.000580	0.634	8.86
12	15	91	0.614	0.0379	0.000416	0.569	8.57
13	16	105	0.507	0.0313	0.000298	0.501	8.02
14	17	120	0.414	0.0256	0.000213	0.435	7.40
15	18	136	0.338	0.0208	0.000153	0.376	6.75
⋮	⋮	⋮	⋮	⋮	⋮	⋮	⋮
18	21	190	0.173	0.0107	0.0000563	0.224	4.72
⋮	⋮	⋮	⋮	⋮	⋮	⋮	⋮
21	24	253	0.0848	0.00523	0.0000207	0.126	3.01
⋮	⋮	⋮	⋮	⋮	⋮	⋮	⋮
24	27	325	0.0400	0.00247	0.00000760	0.0667	1.80
⋮	⋮	⋮	⋮	⋮	⋮	⋮	⋮
30	33	496	0.0083	0.000512	0.00000103	0.0169	0.56
⋮	⋮	⋮	⋮	⋮	⋮	⋮	⋮
			$\sum_i P_i \mathfrak{z} = \mathfrak{z} = 16.2$			$\sum_i P_iE_i = U = 10.55$	$\sum_i P_iE_i^2 = 138.3$

The probability of an individual quantum state of energy E_i is

$$P_q = \frac{P_i}{g_i}.$$

Values of the P_q for each energy level are listed in column 6.

A comparison of columns 5 and 6 shows clearly that although the probability of a given quantum state P_q is largest for the lowest energy level and decreases rapidly at higher energy levels, the probability of a given energy level goes through a maximum at an energy value $E_i = 7$. This results from the fact that P_i is a product:

$$P_i = P_q g_i,$$

and as P_q decreases monotonically, g_i increases monotonically. This causes P_i to pass through a maximum. The change of P_q and P_i with the index i is shown graphically in Fig. 12-2.

From the values of P_i and E_i, we may calculate the product P_iE_i. Results are shown in column 7 of Table 12-3. Since by Eq. (12-23)

$$U = \sum_i P_iE_i,$$

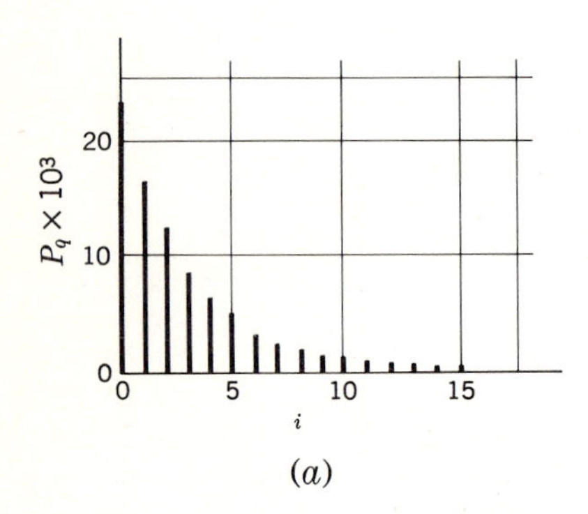

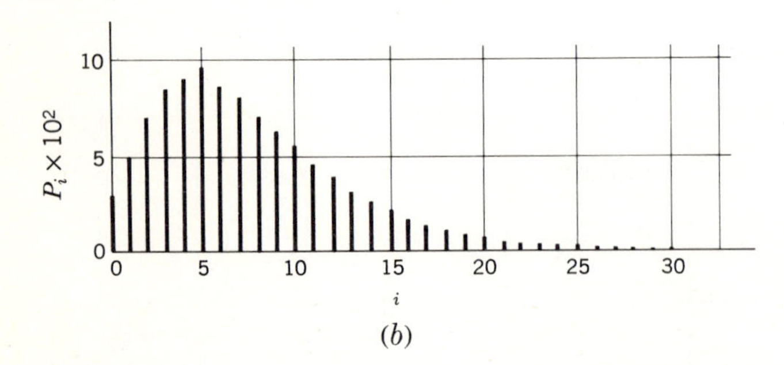

Fig. 12-2
Variation of (a) P_q and (b) P_i with energy level.

we may determine U for the system by summing the values of P_iE_i over all energy levels. Again the series converges to a limiting value, which is shown at the bottom of column 7 to be 10.55. The variation of the term P_iE_i with the index i is shown graphically in Fig. 12-3*a*.

It is interesting to consider the deviation of the values E_i from the mean value $\sum_i P_iE_i = U$. Let the deviation be

$$\sigma_i = E_i - U.$$

Then the mean deviation is

$$\sum_i P_i\sigma_i = \sum_i P_iE_i - U \sum_i P_i = U - U = 0.$$

Thus the mean deviation is identically zero, because the + and − deviations cancel. This difficulty can be avoided if we use the root-mean-square deviation, i.e., the square root of the mean value of the squares of the deviations. Let the deviation be

$$\sigma_i = E_i - U.$$

Then $$\sigma_i^2 = (E_i - U)^2$$

and
$$\begin{aligned}\sigma^2 &= \sum_i P_i\sigma_i^2 = \sum_i P_i(E_i - U)^2 \\ &= \sum_i P_i(E_i^2 - 2E_iU + U^2) \\ &= \sum_i P_iE_i^2 - 2U \sum_i P_iE_i + U^2 \sum_i P_i \\ &= \sum_i P_iE_i^2 - 2U^2 + U^2 \\ &= \sum_i P_iE_i^2 - U^2.\end{aligned}$$

Thus $$\sigma = \sqrt{\sum_i P_iE_i^2 - U^2}. \tag{12-24}$$

The values necessary for the calculation of σ are listed in Table 12-3. Column 8 lists the values of $P_iE_i^2$ and shows the limiting value of the sum $\sum P_iE_i^2$ to be 138.3. The value of U is 10.55. Thus

$$\sigma = \sqrt{138.3 - (10.55)^2} = 5.2.$$

The values of $P_iE_i^2$ are also shown graphically by Fig. 12-3*b*.

Sufficient information is available in Table 12-3 to allow calculation of the entropy by Eq. (12-19):

$$S = k \ln \mathcal{Z} + \frac{U}{T}$$

or
$$\begin{aligned}\frac{S}{k} &= \ln \mathcal{Z} + \frac{U}{kT} \\ &= \ln \mathcal{Z} + \beta U.\end{aligned}$$

Thus $$\frac{S}{k} = \ln 16.2 + (\tfrac{1}{3})(10.55) = 6.30.$$

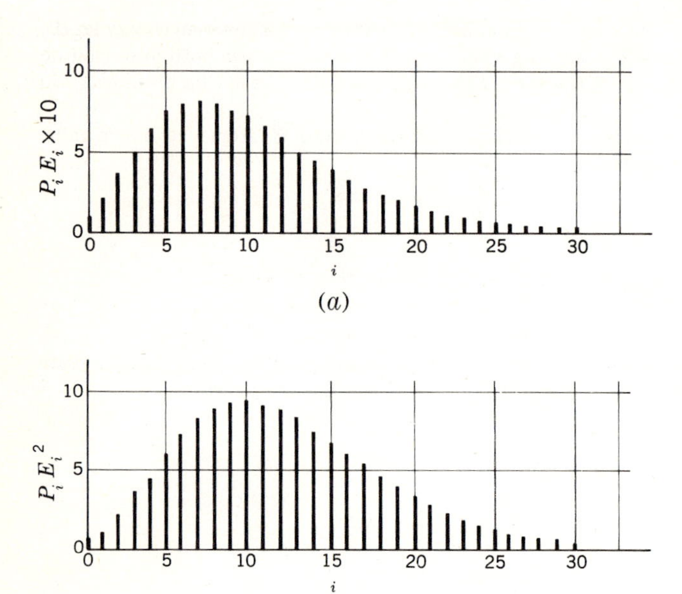

Fig. 12-3
Variations of (a) P_iE_i and (b) $P_iE_i^2$ with energy level.

All the calculations just described, made with $\beta = \frac{1}{3}$, have been repeated for $\beta = \frac{1}{4}$. Since $\beta = 1/kT$, this corresponds to a higher temperature. A summary of results of all calculations, both for $\beta = \frac{1}{3}$ and $\beta = \frac{1}{4}$, is shown in Table 12-4. The increase in temperature (decrease in β from $\frac{1}{3}$ to $\frac{1}{4}$) is seen to result in an increase in the partition function $\mathcal{Z}$, in the internal energy U, in the entropy S, and in the root-mean-square deviation σ of the E_i's from U.

Table 12-4

	$\beta = \frac{1}{3}$	$\beta = \frac{1}{4}$
$\mathcal{Z}$	16.2	43.6
$U = \sum_i P_iE_i$	10.55	13.5
$\sum_i P_iE_i^2$	138.3	232
U^2	111.4	182
$\sigma^2 = \sum_i P_iE_i^2 - U^2$	26.9	50
σ	5.2	7.1
$S/k = \ln \mathcal{Z} + \beta U$	6.30	7.14

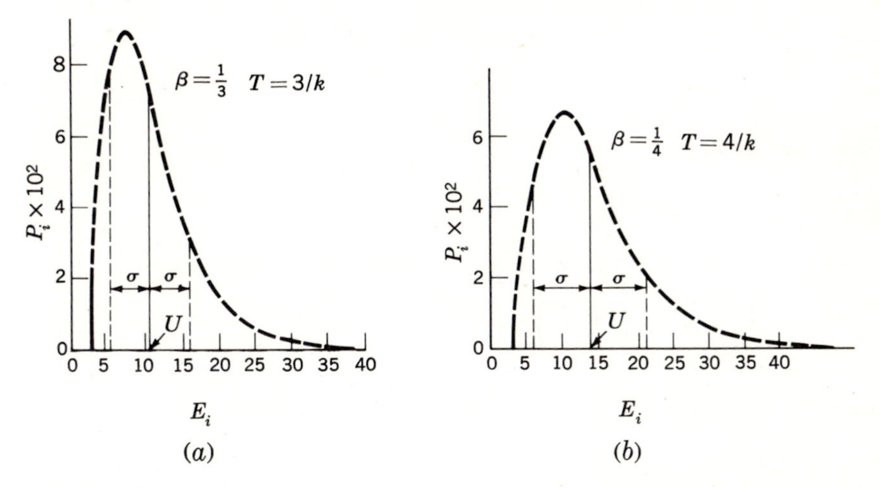

Fig. 12-4

Probabilities P_i as a function of E_i: (a) for $T = 3/k$ and (b) for $T = 4/k$.

In Fig. 12-4 the probabilities P_i of the energy levels are plotted against the energy values E_i for both $\beta = \frac{1}{3}$ and $\beta = \frac{1}{4}$. It is seen that one effect of increasing temperature (decreasing β) is to broaden the peak and to increase the root-mean-square deviation σ. Clearly, σ is a measure of the broadness of the peak, with larger values of σ corresponding to broader peaks.[1] It is also seen from Fig. 12-4 that the effect of a temperature increase is to raise the probabilities of finding the system in higher energy levels. If we think of the ensemble of systems described in Sec. 12-2, we say that an increase in temperature of the ensemble changes the "populations" of systems in the allowed energy levels, shifting them so that more systems are in higher energy states and fewer are in the lower energy states.

Example 12-2

The partition function for 1 (mol) of an ideal monatomic gas, as provided by quantum mechanics, is given by the expression

$$\mathcal{Z} = \frac{V^{N_0}}{N_0!}\left(\frac{2\pi mkT}{h^2}\right)^{\frac{3}{2}(N_0)},$$

where N_0 = Avogadro's number, i.e., the number of atoms contained in the molar volume V,

m = atomic mass,

h = Planck's constant.

[1] For real PVT systems it can be shown that σ is extremely small. This means that the peak is very sharp and that it must occur at a value of E_i exceedingly close to the value of U. The significance of this is that the probability of an appreciable fluctuation of the energy of a real PVT system from its average value U is very minute.

From this expression determine expressions for P, U, H, C_V, and C_P.

The pressure exerted by the system is given by Eq. (12-20):

$$P = kT\left(\frac{\partial \ln \mathfrak{Z}}{\partial V}\right)_T.$$

Taking the logarithm of $\mathfrak{Z}$, we have

$$\ln \mathfrak{Z} = N_0 \ln V - \ln N_0! + \tfrac{3}{2}N_0 \ln \frac{2\pi mkT}{h^2}.$$

At constant T and N_0, the last two terms on the right are constants. Thus

$$\left(\frac{\partial \ln \mathfrak{Z}}{\partial V}\right)_T = N_0\left(\frac{\partial \ln V}{\partial V}\right)_T = \frac{N_0}{V}$$

and

$$P = \frac{N_0 kT}{V}.$$

However, for 1(mol) of an ideal gas, we know that

$$P = \frac{RT}{V}.$$

Comparison of these two equations shows that

$$\boxed{k = \frac{R}{N_0}.}$$

Thus k, known as Boltzmann's constant, is the gas constant divided by Avogadro's number. It has the value 1.381×10^{-23}(J)/(K).

Equation (12-18) allows us to calculate the internal energy of our system:

$$U = kT^2\left(\frac{\partial \ln \mathfrak{Z}}{\partial T}\right)_V.$$

Differentiation of $\ln \mathfrak{Z}$ at constant V and N_0 gives

$$\left(\frac{\partial \ln \mathfrak{Z}}{\partial T}\right)_V = \frac{3}{2}\frac{N_0}{T}.$$

Thus

$$U = \tfrac{3}{2}N_0 kT$$
$$= \tfrac{3}{2}RT.$$

By definition of the enthalpy, for 1(mol) of an ideal gas,

$$H = U + PV = U + RT.$$

Therefore

$$H = \tfrac{5}{2}RT.$$

By Eqs. (4-21) and (4-22) we have

$$C_V = \left(\frac{\partial U}{\partial T}\right)_V \quad \text{and} \quad C_P = \left(\frac{\partial H}{\partial T}\right)_P.$$

Substitution for U and H yields

$$C_V = \tfrac{3}{2}R \qquad \text{and} \qquad C_P = \tfrac{5}{2}R$$

for the heat capacities of ideal monatomic gases, in agreement with the observations described in Sec. 5-8.

12-6 Statistical Interpretation of the First and Second Laws

We have from Eq. (12-23) the basic relation between internal energy and the statistical mechanical coordinates P_i and E_i:

$$U = \sum_i P_i E_i,$$

where

$$\sum_i P_i = 1.$$

We now wish to consider the effects of a differential change in the *equilibrium* state of a system as brought about by changes in its volume and temperature. The thermodynamic state of a closed system in equilibrium with a heat reservoir is established by its volume and by the temperature of the heat reservoir. If either changes, the state of the system will change, and the change of internal energy is given by the total differential of U:

$$dU = \sum_i E_i \, dP_i + \sum_i P_i \, dE_i. \tag{12-25}$$

From the viewpoint of statistical mechanics, the properties of the system are seen to depend on the P_i and the E_i.

Again we make use of the result of quantum mechanics that for a closed system, the E_i's depend only on the volume of the system. Thus changes in volume produce changes in the energy levels E_i and influence U through the term $\sum_i P_i \, dE_i$ as shown by Eq. (12-25). Since volume changes of the system involve work, we may associate the term $\sum_i P_i \, dE_i$ with δW. Thus, from the point of view of statistical mechanics, work is done on a system when the surroundings interact with the system in such a way as to change its set of allowed energy levels to higher values, that is, to increase the values of the E_i.

Consider now a differential increase in temperature of the heat reservoir. This change causes heat to flow to the system until it once again comes to thermal equilibrium with the heat reservoir. If this process occurs at constant volume, we have, from Eq. (12-25) and the first law of thermodynamics,

$$\delta Q = dU = \sum_i E_i \, dP_i \qquad (\text{const } V).$$

Thus changes in temperature influence U through the term $\sum_i E_i\, dP_i$, and this term is clearly to be associated with δQ. We have already seen that an increase in temperature of a system results in a redistribution of the probabilities associated with the energy levels, increasing the P_i's for the higher levels and decreasing those for the lower levels. Thus heat is interpreted as an interaction between the system and surroundings which alters the probabilities associated with the various energy levels.

We turn now to the statistical interpretation of entropy and the second law of thermodynamics. By Eq. (12-11) we have

$$S = -k \sum_q (P_q \ln P_q).$$

Consider now the application of this equation to an *isolated* system, i.e., a system which can exchange neither mass nor energy with its surroundings, and is therefore a system of constant mass and energy. Such a system can exist at but a single energy level, an energy level equal to the internal energy of the system. The number of quantum states accessible to the system must equal the degeneracy of the energy level at which the system exists. Let the number of such states be g. Since each has the same energy, all have the same probability, $P_q = 1/g$. Setting $P_q = 1/g$ and summing over the g quantum states gives

$$S = -kg\left(\frac{1}{g}\ln\frac{1}{g}\right),$$

or

$$\boxed{S = k \ln g.} \tag{12-26}$$

According to the second law of thermodynamics, an irreversible change in an isolated system must result in an increase in the entropy of the system. Equation (12-26) indicates that this increase in entropy results from an increase in the number of quantum states accessible to the system. This is entirely reasonable, for a system which undergoes a change of state at constant mass and energy has available to it all the quantum states accessible to it in its initial state plus additional states that were not accessible in the initial state but which become accessible in the final state. Consider, for example, a system consisting of an insulated vessel partitioned into two equal parts and containing 1 (mol) of helium on one side of the partition and 1 (mol) of argon on the other side. Removal of the partition permits the mixing of the helium and argon to form quantum states not accessible to the partitioned system, but does not disallow those states accessible to the partitioned system.

Since any system plus its surroundings is an isolated system, the most general form of the second law, which states that the total entropy change in a system plus its surroundings must be positive when irreversible changes occur, has as its counterpart in statistical mechanics the statement that the number of accessible quantum states of the system plus its surroundings must increase.

Problems

12-1 Combination of Eqs. (12-21) and (12-23) gives

$$U\mathfrak{Z} = \sum_i g_i e^{-\beta E_i} E_i.$$

Differentiate this equation partially with respect to β at constant volume (and hence constant g_i and E_i). Using this result together with Eqs. (12-21), (12-12), and (12-24), show that for a real material

$$\sigma = T\sqrt{kC_V}.$$

For a monatomic gas at 300(K), what is the value of σ? What is the value of σ/U?

12-2 Prepare a table similar to Table 12-3 but for $\beta = \frac{1}{4}$, and confirm the results of the last column of Table 12-4.

APPENDIX A

APPENDIX A

CONVERSION FACTORS

For conciseness, the conversion factors given below for each quantity are referred to a single basic or derived SI or cgs unit. Conversions between other pairs of units for a given quantity are made by employing the usual rules for manipulation of units.

Example

Find the factor for converting (Btu) to (ft)(lb_f). From the energy entries we find

$$1(\mathrm{J}) = 9.47831 \times 10^{-4}(\mathrm{Btu}) = 0.737562(\mathrm{ft})(\mathrm{lb}_f),$$

from which

$$1(\mathrm{Btu}) = \frac{0.737562}{9.47831 \times 10^{-4}} = 778.158(\mathrm{ft})(\mathrm{lb}_f).$$

Quantity		Conversion
Length	1 (m)	= 100 (cm)
		= 3.28084 (ft)
		= 39.3701 (in)
Mass	1 (kg)	= 10^3 (g)
		= 2.20462 (lb_m)
Force	1 (N)	= 1 (kg) (m) / $(s)^2$
		= 10^5 (dyn)
		= 0.224809 (lb_f)
Pressure	1 (bar)	= 10^5 (kg) / (m) $(s)^2$
		= 10^5 (N) / $(m)^2$
		= 10^6 (dyn) / $(cm)^2$
		= 0.986923 (atm)
		= 14.5038 (psia)
		= 750.061 (mm Hg)
Volume	1 $(m)^3$	= 10^6 $(cm)^3$
		= 10^3 (l)
		= 35.3147 $(ft)^3$
		= 264.172 (gal)
Density	1 (g) / $(cm)^3$	= 10^3 (kg) / $(m)^3$
		= 10^3 (g) / (l)
		= 62.4278 (lb_m) / $(ft)^3$
		= 8.34540 (lb_m) / (gal)
Energy	1 (J)	= 1 (kg) $(m)^2$/ $(s)^2$
		= 1 (N) (m)
		= 1 (W) (s)
		= 10^7 (dyn) (cm)
		= 10^7 (erg)
		= 10 $(cm)^3$ (bar)
		= 10^{-2} (l) (bar)
		= 10^{-5} $(m)^3$ (bar)
		= 0.239006 (cal)
		= 9.86923 $(cm)^3$ (atm)
		= 5.12197 × 10^{-3} $(ft)^3$ (psia)
		= 0.737562 (ft) (lb_f)
		= 9.47831 × 10^{-4} (Btu)
Power	1 (kW)	= 10^3 (kg) $(m)^2$/ $(s)^3$
		= 10^3 (W)
		= 10^3 (J) / (s)
		= 10^3 (V) (A)
		= 239.006 (cal) / (s)
		= 737.562 (ft) (lb_f) / (s)
		= 56.8699 (Btu) / (min)
		= 1.34102 (hp)

NOTE: atm ≡ standard atmosphere; cal ≡ thermochemical calorie; Btu ≡ International Steam Table Btu.

APPENDIX B

APPENDIX **B**

PHYSICAL CONSTANTS

Universal gas constant:

$$
\begin{aligned}
R &= 8.314(\mathrm{J})/(\mathrm{mol})(\mathrm{K}) = 83.14(\mathrm{cm})^3(\mathrm{bar})/(\mathrm{mol})(\mathrm{K}) \\
&= 0.08314(\mathrm{l})(\mathrm{bar})/(\mathrm{mol})(\mathrm{K}) \\
&= 82.06(\mathrm{cm})^3(\mathrm{atm})/(\mathrm{mol})(\mathrm{K}) \\
&= 0.08206(\mathrm{l})(\mathrm{atm})/(\mathrm{mol})(\mathrm{K}) \\
&= 1.987(\mathrm{cal})/(\mathrm{mol})(\mathrm{K}) = 1.986(\mathrm{Btu})/(\mathrm{lb\ mol})(\mathrm{R}) \\
&= 0.7302(\mathrm{ft})^3(\mathrm{atm})/(\mathrm{lb\ mol})(\mathrm{R}) \\
&= 10.73(\mathrm{ft})^3(\mathrm{psia})/(\mathrm{lb\ mol})(\mathrm{R}) \\
&= 1545(\mathrm{ft})(\mathrm{lb}_f)/(\mathrm{lb\ mol})(\mathrm{R})
\end{aligned}
$$

Avogadro constant: $N_0 = 6.0225 \times 10^{23}(\mathrm{mol})^{-1}$
Boltzmann constant: $k = R/N_0 = 1.381 \times 10^{-23}(\mathrm{J})/(\mathrm{K})$
Charge of electron: $e = 1.60210 \times 10^{-19}(\mathrm{C})$
Faraday constant: $\mathfrak{F} = N_0 e = 96{,}487(\mathrm{C})/(\mathrm{mol})$
Speed of light in vacuum: $c_0 = 2.99793 \times 10^8(\mathrm{m})/(\mathrm{s})$
Magnetic permeability of vacuum:

$$\mu_0 = 4\pi \times 10^{-7}(\mathrm{H})/(\mathrm{m})$$

Electric permittivity of vacuum:

$$\epsilon_0 = 1/\mu_0 c_0^2 = 8.85419 \times 10^{-12}(\mathrm{F})/(\mathrm{m})$$

Planck constant: $h = 6.6256 \times 10^{-34}(\mathrm{J})(\mathrm{s})$
Standard acceleration due to gravity:

$$
\begin{aligned}
g &= 9.80665(\mathrm{m})/(\mathrm{s})^2 \\
&= 32.1740(\mathrm{ft})/(\mathrm{s})^2
\end{aligned}
$$

APPENDIX C

APPENDIX C

CRITICAL CONSTANTS AND ACENTRIC FACTOR

Compound	T_c(K)	P_c(bar)	V_c(cm)^3/(mol)	Z_c	ω
Argon	151.0	48.6	75.2	0.290	−0.002
Krypton	209.4	55.0	92.2	0.291	0.000
Xenon	289.8	58.8	118.8	0.290	0.002
Methane	190.7	46.4	99.4	0.290	0.013
Oxygen	154.8	50.8	74.4	0.293	0.021
Nitrogen	126.2	33.9	90.1	0.291	0.040
Carbon monoxide	133.0	35.0	93.1	0.294	0.049
Ethylene	283.1	51.2	124	0.270	0.085
Hydrogen sulfide	373.6	90.1	98.0	0.284	0.100
Ethane	305.4	48.8	148	0.285	0.105
Carbon disulfide	552.0	79.0	170	0.276	0.123
Propylene	365.1	46.0	181	0.274	0.139
Propane	369.9	42.6	200	0.277	0.152
Isobutane	408.1	36.5	263	0.283	0.187
Acetylene	309.5	62.4	113	0.274	0.190
1-Butene	419.6	40.2	240	0.277	0.190
neo-Pentane	433.8	32.0	303	0.269	0.195
n-Butane	425.2	38.0	255	0.274	0.200
Cyclohexane	553.2	40.5	308	0.271	0.209
Benzene	562.1	49.2	260	0.274	0.211
Isopentane	460.4	33.3	308	0.268	0.215
Carbon dioxide	304.2	73.9	94.0	0.274	0.225
Ammonia	405.6	114	72.5	0.245	0.250
n-Pentane	469.5	33.7	311	0.269	0.252
Sulfur dioxide	430.7	78.7	122	0.268	0.273
n-Hexane	507.3	30.3	368	0.264	0.298
Acetone	509.1	47.6	211	0.237	0.318
Water	647.3	220.9	56.8	0.233	0.344
n-Heptane	540.3	27.4	426	0.259	0.349
n-Octane	568.6	24.9	486	0.256	0.398
Methanol	513.2	79.5	118	0.220	0.556
Ethanol	516.3	63.8	167	0.248	0.635

APPENDIX D

APPENDIX D

STEAM TABLES

Table D-1 Saturated Steam†

V = specific volume $(ft)^3/(lb_m)$
H = specific enthalpy $(Btu)/(lb_m)$
S = specific entropy $(Btu)/(lb_m)(R)$
1(in Hg) at 32(°F) = 0.4912(psia)

Temp. t, (°F)	Absolute pressure P		Specific volume			Enthalpy			Entropy		
	Psia	In Hg (32°F)	Sat. liquid V_f	Evap. ΔV_{fg}	Sat. vapor V_g	Sat. liquid H_f	Evap. ΔH_{fg}	Sat. vapor H_g	Sat. liquid S_f	Evap. ΔS_{fg}	Sat. vapor S_g
32	0.0886	0.1806	0.01602	3305.7	3305.7	0	1075.1	1075.1	0	2.1865	2.1865
34	0.0961	0.1957	0.01602	3060.4	3060.4	2.01	1074.0	1076.0	0.0041	2.1755	2.1796
36	0.1041	0.2120	0.01602	2836.6	2836.6	4.03	1072.9	1076.9	0.0082	2.1645	2.1727
38	0.1126	0.2292	0.01602	2632.2	2632.2	6.04	1071.7	1077.7	0.0122	2.1533	2.1655
40	0.1217	0.2478	0.01602	2445.1	2445.1	8.05	1070.5	1078.6	0.0162	2.1423	2.1585

42	0.1315	0.2677	0.01602	2271.8	2271.8	10.06	1069.3	1079.4	0.0203	2.1314	2.1517
44	0.1420	0.2891	0.01602	2112.2	2112.2	12.06	1068.2	1080.3	0.0242	2.1207	2.1449
46	0.1532	0.3119	0.01602	1965.5	1965.5	14.07	1067.1	1081.2	0.0282	2.1102	2.1384
48	0.1652	0.3364	0.01602	1829.9	1829.9	16.07	1065.9	1082.0	0.0322	2.0995	2.1317
50	0.1780	0.3624	0.01602	1704.9	1704.9	18.07	1064.8	1082.9	0.0361	2.0891	2.1252
52	0.1918	0.3905	0.01603	1588.4	1588.4	20.07	1063.6	1083.7	0.0400	2.0786	2.1186
54	0.2063	0.4200	0.01603	1482.4	1482.4	22.07	1062.5	1084.6	0.0439	2.0684	2.1123
56	0.2219	0.4518	0.01603	1383.5	1383.5	24.07	1061.4	1085.5	0.0478	2.0582	2.1060
58	0.2384	0.4854	0.01603	1292.7	1292.7	26.07	1060.2	1086.3	0.0517	2.0479	2.0996
60	0.2561	0.5214	0.01603	1208.1	1208.1	28.07	1059.1	1087.2	0.0555	2.0379	2.0934
62	0.2749	0.5597	0.01604	1129.7	1129.7	30.06	1057.9	1088.0	0.0594	2.0278	2.0872
64	0.2949	0.6004	0.01604	1057.1	1057.1	32.06	1056.8	1088.9	0.0632	2.0180	2.0812
66	0.3162	0.6438	0.01604	989.6	989.6	34.06	1055.7	1089.8	0.0670	2.0082	2.0752
68	0.3388	0.6898	0.01605	927.0	927.0	36.05	1054.5	1090.6	0.0708	1.9983	2.0691
70	0.3628	0.7387	0.01605	868.9	868.9	38.05	1053.4	1091.5	0.0745	1.9887	2.0632
72	0.3883	0.7906	0.01606	814.9	814.9	40.04	1052.3	1092.3	0.0783	1.9792	2.0575
74	0.4153	0.8456	0.01606	764.7	764.7	42.04	1051.2	1093.2	0.0820	1.9697	2.0517
76	0.4440	0.9040	0.01607	718.0	718.0	44.03	1050.1	1094.1	0.0858	1.9603	2.0461
78	0.4744	0.9659	0.01607	674.4	674.4	46.03	1048.9	1094.9	0.0895	1.9508	2.0403
80	0.5067	1.032	0.01607	633.7	633.7	48.02	1047.8	1095.8	0.0932	1.9415	2.0347
82	0.5409	1.101	0.01608	595.8	595.8	50.02	1046.6	1096.6	0.0969	1.9321	2.0290
84	0.5772	1.175	0.01608	560.4	560.4	52.01	1045.5	1097.5	0.1006	1.9230	2.0236
86	0.6153	1.253	0.01609	527.6	527.6	54.01	1044.4	1098.4	0.1042	1.9139	2.0181
88	0.6555	1.335	0.01609	497.0	497.0	56.00	1043.2	1099.2	0.1079	1.9047	2.0126
90	0.6980	1.421	0.01610	468.4	468.4	58.00	1042.1	1100.1	0.1115	1.8958	2.0073

† Reproduced by permission from "Steam Tables: Properties of Saturated and Superheated Steam," copyright 1940, Combustion Engineering, Inc.

Table D-1 Saturated Steam (Continued)

Temp. t, (°F)	Absolute pressure P		Specific volume			Enthalpy			Entropy		
	Psia	In Hg (32°F)	Sat. liquid V_f	Evap. ΔV_{fg}	Sat. vapor V_g	Sat. liquid H_f	Evap. ΔH_{fg}	Sat. vapor H_g	Sat. liquid S_f	Evap. ΔS_{fg}	Sat. vapor S_g
92	0.7429	1.513	0.01611	441.7	441.7	59.99	1040.9	1100.9	0.1151	1.8867	2.0018
94	0.7902	1.609	0.01611	416.7	416.7	61.98	1039.8	1101.8	0.1187	1.8779	1.9966
96	0.8403	1.711	0.01612	393.2	393.2	63.98	1038.7	1102.7	0.1223	1.8692	1.9915
98	0.8930	1.818	0.01613	371.3	371.3	65.98	1037.5	1103.5	0.1259	1.8604	1.9863
100	0.9487	1.932	0.01613	350.8	350.8	67.97	1036.4	1104.4	0.1295	1.8517	1.9812
102	1.0072	2.051	0.01614	331.5	331.5	69.96	1035.2	1105.2	0.1330	1.8430	1.9760
104	1.0689	2.176	0.01614	313.5	313.5	71.96	1034.1	1106.1	0.1366	1.8345	1.9711
106	1.1338	2.308	0.01615	296.5	296.5	73.95	1033.0	1107.0	0.1401	1.8261	1.9662
108	1.2020	2.447	0.01616	280 7	280.7	75.94	1032.0	1107.9	0.1436	1.8179	1.9615
110	1.274	2.594	0.01617	265.7	265.7	77.94	1030.9	1108.8	0.1471	1.8096	1.9567
112	1.350	2.749	0.01617	251.6	251.6	79.93	1029.7	1109.6	0.1506	1.8012	1.9518
114	1.429	2.909	0.01618	238.5	238.5	81.93	1028.6	1110.5	0.1541	1.7930	1.9471
116	1.512	3.078	0.01619	226.2	226.2	83.92	1027.5	1111.4	0.1576	1.7848	1.9424
118	1.600	3.258	0.01620	214.5	214.5	85.92	1026.4	1112.3	0.1610	1.7767	1.9377
120	1.692	3.445	0.01620	203.45	203.47	87.91	1025.3	1113.2	0.1645	1.7687	1.9332
122	1.788	3.640	0.01621	193.16	193.18	89.91	1024.1	1114.0	0.1679	1.7606	1.9285
124	1.889	3.846	0.01622	183.44	183.46	91.90	1023.0	1114.9	0.1714	1.7526	1.9240
126	1.995	4.062	0.01623	174.26	174.28	93.90	1021.8	1115.7	0.1748	1.7446	1.9194
128	2.105	4.286	0.01624	165.70	165.72	95.90	1020.7	1116.6	0.1782	1.7368	1.9150
130	2.221	4.522	0.01625	157.55	157.57	97.89	1019.5	1117.4	0.1816	1.7289	1.9105

132	2.343	4.770	0.01626	149.83	149.85	99.89	1018.3	1118.2	0.1849	1.7210	1.9059
134	2.470	5.029	0.01626	142.59	142.61	101.89	1017.2	1119.1	0.1883	1.7134	1.9017
136	2.603	5.300	0.01627	135.73	135.75	103.88	1016.0	1119.9	0.1917	1.7056	1.8973
138	2.742	5.583	0.01628	129.26	129.28	105.88	1014.9	1120.8	0.1950	1.6980	1.8930
140	2.887	5.878	0.01629	123.16	123.18	107.88	1013.7	1121.6	0.1984	1.6904	1.8888
142	3.039	6.187	0.01630	117.37	117.39	109.88	1012.5	1122.4	0.2017	1.6828	1.8845
144	3.198	6.511	0.01631	111.88	111.90	111.88	1011.3	1123.2	0.2050	1.6752	1.8802
146	3.363	6.847	0.01632	106.72	106.74	113.88	1010.2	1124.1	0.2083	1.6678	1.8761
148	3.536	7.199	0.01633	101.82	101.84	115.87	1009.0	1124.9	0.2116	1.6604	1.8720
150	3.716	7.566	0.01634	97.18	97.20	117.87	1007.8	1125.7	0.2149	1.6530	1.8679
152	3.904	7.948	0.01635	92.79	92.81	119.87	1006.7	1126.6	0.2181	1.6458	1.8639
154	4.100	8.348	0.01636	88.62	88.64	121.87	1005.5	1127.4	0.2214	1.6384	1.8598
156	4.305	8.765	0.01637	84.66	84.68	123.87	1004.4	1128.3	0.2247	1.6313	1.8560
158	4.518	9.199	0.01638	80.90	80.92	125.87	1003.2	1129.1	0.2279	1.6241	1.8520
160	4.739	9.649	0.01639	77.37	77.39	127.87	1002.0	1129.9	0.2311	1.6169	1.8480
162	4.970	10.12	0.01640	74.00	74.02	129.88	1000.8	1130.7	0.2343	1.6098	1.8441
164	5.210	10.61	0.01642	70.79	70.81	131.88	999.7	1131.6	0.2376	1.6029	1.8405
166	5.460	11.12	0.01643	67.76	67.78	133.88	998.5	1132.4	0.2408	1.5958	1.8366
168	5.720	11.65	0.01644	64.87	64.89	135.88	997.3	1133.2	0.2439	1.5888	1.8327
170	5.990	12.20	0.01645	62.12	62.14	137.89	996.1	1134.0	0.2471	1.5819	1.8290
172	6.272	12.77	0.01646	59.50	59.52	139.89	995.0	1134.9	0.2503	1.5751	1.8254
174	6.565	13.37	0.01647	57.01	57.03	141.89	993.8	1135.7	0.2535	1.5683	1.8218
176	6.869	13.99	0.01648	54.64	54.66	143.90	992.6	1136.5	0.2566	1.5615	1.8181
178	7.184	14.63	0.01650	52.39	52.41	145.90	991.4	1137.3	0.2598	1.5547	1.8145
180	7.510	15.29	0.01651	50.26	50.28	147.91	990.2	1138.1	0.2629	1.5479	1.8108

Table D-1 Saturated Steam (Continued)

Temp. t, (°F)	Absolute pressure P		Specific volume			Enthalpy			Entropy		
	Psia	In Hg (32°F)	Sat. liquid V_f	Evap. ΔV_{fg}	Sat. vapor V_g	Sat. liquid H_f	Evap. ΔH_{fg}	Sat. vapor H_g	Sat. liquid S_f	Evap. ΔS_{fg}	Sat. vapor S_g
182	7.849	15.98	0.01652	48.22	48.24	149.92	989.0	1138.9	0.2661	1.5412	1.8073
184	8.201	16.70	0.01653	46.28	46.30	151.92	987.8	1139.7	0.2692	1.5346	1.8038
186	8.566	17.44	0.01654	44.43	44.45	153.93	986.6	1140.5	0.2723	1.5280	1.8003
188	8.944	18.21	0.01656	42.67	42.69	155.94	985.3	1141.3	0.2754	1.5213	1.7967
190	9.336	19.01	0.01657	40.99	41.01	157.95	984.1	1142.1	0.2785	1.5147	1.7932
192	9.744	19.84	0.01658	39.38	39.40	159.95	982.8	1142.8	0.2816	1.5081	1.7897
194	10.168	20.70	0.01659	37.84	37.86	161.96	981.5	1143.5	0.2847	1.5015	1.7862
196	10.605	21.59	0.01661	36.38	36.40	163.97	980.3	1144.3	0.2877	1.4951	1.7828
198	11.057	22.51	0.01662	34.98	35.00	165.98	979.0	1145.0	0.2908	1.4885	1.7793
200	11.525	23.46	0.01663	33.65	33.67	167.99	977.8	1145.8	0.2938	1.4822	1.7760
202	12.010	24.45	0.01665	32.37	32.39	170.01	976.6	1146.6	0.2969	1.4759	1.7728
204	12.512	25.47	0.01666	31.15	31.17	172.02	975.3	1147.3	0.2999	1.4695	1.7694
206	13.031	26.53	0.01667	29.99	30.01	174.03	974.1	1148.1	0.3029	1.4633	1.7662
208	13.568	27.62	0.01669	28.88	28.90	176.04	972.8	1148.8	0.3059	1.4570	1.7629
210	14.123	28.75	0.01670	27.81	27.83	178.06	971.5	1149.6	0.3090	1.4507	1.7597
212	14.696	29.92	0.01672	26.81	26.83	180.07	970.3	1150.4	0.3120	1.4446	1.7566
215	15.591	....	0.01674	25.35	25.37	183.10	968.3	1151.4	0.3165	1.4352	1.7517
220	17.188	....	0.01677	23.14	23.16	188.14	965.2	1153.3	0.3239	1.4201	1.7440
225	18.915	....	0.01681	21.15	21.17	193.18	961.9	1155.1	0.3313	1.4049	1.7362

230	20.78	....	0.01684	19.371	19.388	198.22	958.7	1156.9	0.3386	1.3900	1.7286
235	22.80	....	0.01688	17.761	17.778	203.28	955.3	1158.6	0.3459	1.3751	1.7210
240	24.97	....	0.01692	16.307	16.324	208.34	952.1	1160.4	0.3531	1.3607	1.7138
245	27.31	....	0.01696	15.010	15.027	213.41	948.7	1162.1	0.3604	1.3462	1.7066
250	29.82	....	0.01700	13.824	13.841	218.48	945.3	1163.8	0.3675	1.3320	1.6995
255	32.53	....	0.01704	12.735	12.752	223.56	942.0	1165.6	0.3747	1.3181	1.6928
260	35.43	....	0.01708	11.754	11.771	228.65	938.6	1167.3	0.3817	1.3042	1.6859
265	38.54	....	0.01713	10.861	10.878	233.74	935.3	1169.0	0.3888	1.2906	1.6794
270	41.85	....	0.01717	10.053	10.070	238.84	931.8	1170.6	0.3958	1.2770	1.6728
275	45.40	....	0.01721	9.313	9.330	243.94	928.2	1172.1	0.4027	1.2634	1.6661
280	49.20	....	0.01726	8.634	8.651	249.06	924.6	1173.7	0.4096	1.2500	1.6596
285	53.25	....	0.01731	8.015	8.032	254.18	921.0	1175.2	0.4165	1.2368	1.6533
290	57.55	....	0.01735	7.448	7.465	259.31	917.4	1176.7	0.4234	1.2237	1.6471
295	62.13	....	0.01740	6.931	6.948	264.45	913.7	1178.2	0.4302	1.2107	1.6409
300	67.01	....	0.01745	6.454	6.471	269.60	910.1	1179.7	0.4370	1.1980	1.6350
305	72.18	....	0.01750	6.014	6.032	274.76	906.3	1181.1	0.4437	1.1852	1.6289
310	77.68	....	0.01755	5.610	5.628	279.92	902.6	1182.5	0.4505	1.1727	1.6232
315	83.50	....	0.01760	5.239	5.257	285.10	898.8	1183.9	0.4571	1.1587	1.3158
320	89.65	....	0.01765	4.897	4.915	290.29	895.0	1185.3	0.4637	1.1479	1.6116
325	96.16	....	0.01771	4.583	4.601	295.49	891.1	1186.6	0.4703	1.1356	1.6059
330	103.03	....	0.01776	4.292	4.310	300.69	887.1	1187.8	0.4769	1.1234	1.6003
335	110.31	....	0.01782	4.021	4.039	305.91	883.2	1189.1	0.4835	1.1114	1.5949
340	117.99	....	0.01788	3.771	3.789	311.14	879.2	1190.3	0.4900	1.0994	1.5894
345	126.10	....	0.01793	3.539	3.557	316.38	875.1	1191.5	0.4966	1.0875	1.5841
350	134.62	....	0.01799	3.324	3.342	321.64	871.0	1192.6	0.5030	1.0757	1.5787

Table D-1 Saturated Steam (Continued)

Temp. t, (°F)	Absolute pressure P		Specific volume			Enthalpy			Entropy		
	Psia	In Hg (32°F)	Sat. liquid V_f	Evap. ΔV_{fg}	Sat. vapor V_g	Sat. liquid H_f	Evap. ΔH_{fg}	Sat. vapor H_g	Sat. liquid S_f	Evap. ΔS_{fg}	Sat. vapor S_g
355	143.58	....	0.01805	3.126	3.144	326.91	866.8	1193.7	0.5094	1.0640	1.5734
360	153.01	....	0.01811	2.940	2.958	332.19	862.5	1194.7	0.5159	1.0522	1.5681
365	162.93	....	0.01817	2.768	2.786	337.48	858.2	1195.7	0.5223	1.0406	1.5629
370	173.33	....	0.01823	2.607	2.625	342.79	853.8	1196.6	0.5286	1.0291	1.5577
375	184.23	....	0.01830	2.458	2.476	348.11	849.4	1197.5	0.5350	1.0176	1.5526
380	195.70	....	0.01836	2.318	2.336	353.45	844.9	1198.4	0.5413	1.0062	1.5475
385	207.71	....	0.01843	2.189	2.207	358.80	840.4	1199.2	0.5476	0.9949	1.5425
390	220.29	....	0.01850	2.064	2.083	364.17	835.7	1199.9	0.5540	0.9835	1.5375
395	233.47	....	0.01857	1.9512	1.9698	369.56	831.0	1200.6	0.5602	0.9723	1.5325
400	247.25	....	0.01864	1.8446	1.8632	374.97	826.2	1201.2	0.5664	0.9610	1.5274
405	261.67	....	0.01871	1.7445	1.7632	380.40	821.4	1201.8	0.5727	0.9499	1.5226
410	276.72	....	0.01878	1.6508	1.6696	385.83	816.6	1202.4	0.5789	0.9390	1.5179
415	292.44	....	0.01886	1.5630	1.5819	391.30	811.7	1203.0	0.5851	0.9280	1.5131
420	308.82	....	0.01894	1.4806	1.4995	396.78	806.7	1203.5	0.5912	0.9170	1.5082
425	325.91	....	0.01902	1.4031	1.4221	402.28	801.6	1203.9	0.5974	0.9061	1.5035
430	343.71	....	0.01910	1.3303	1.3494	407.80	796.5	1204.3	0.6036	0.8953	1.4989
435	362.27	....	0.01918	1.2617	1.2809	413.35	791.2	1204.6	0.6097	0.8843	1.4940
440	381.59	....	0.01926	1.1973	1.2166	418.91	785.9	1204.8	0.6159	0.8735	1.4894
445	401.70	....	0.01934	1.1367	1.1560	424.49	780.4	1204.9	0.6220	0.8626	1.4846
450	422.61	....	0.01943	1.0796	1.0990	430.11	774.9	1205.0	0.6281	0.8518	1.4799

455	444.35	....	0.0195	1.0256	1.0451	435.74	769.3	1205.0	0.6342	0.8410	1.4752
460	466.97	....	0.0196	0.9745	0.9941	441.42	763.6	1205.0	0.6403	0.8303	1.4706
465	490.43	....	0.0197	0.9262	0.9459	447.10	757.8	1204.9	0.6463	0.8195	1.4658
470	514.70	....	0.0198	0.8808	0.9006	452.84	751.9	1204.7	0.6524	0.8088	1.4612
475	539.90	....	0.0199	0.8379	0.8578	458.59	745.9	1204.5	0.6585	0.7980	1.4565
480	566.12	....	0.0200	0.7972	0.8172	464.37	739.8	1204.2	0.6646	0.7873	1.4519
485	593.28	....	0.0201	0.7585	0.7786	470.18	733.6	1203.8	0.6706	0.7766	1.4472
490	621.44	....	0.0202	0.7219	0.7421	476.01	727.3	1203.3	0.6767	0.7658	1.4425
495	650.59	....	0.0203	0.6872	0.7075	481.90	720.8	1202.7	0.6827	0.7550	1.4377
500	680.80	....	0.0204	0.6544	0.6748	487.80	714.2	1202.0	0.6888	0.7442	1.4330
505	712.19	....	0.0206	0.6230	0.6436	493.8	707.5	1201.3	0.6949	0.7334	1.4283
510	744.55	....	0.0207	0.5932	0.6139	499.8	700.6	1200.4	0.7009	0.7225	1.4234
515	777.96	....	0.0208	0.5651	0.5859	505.8	693.6	1199.4	0.7070	0.7116	1.4186
520	812.68	....	0.0209	0.5382	0.5591	511.9	686.5	1198.4	0.7132	0.7007	1.4139
525	848.37	....	0.0210	0.5128	0.5338	518.0	679.2	1197.2	0.7192	0.6898	1.4090
530	885.20	....	0.0212	0.4885	0.5097	524.2	671.9	1196.1	0.7253	0.6789	1.4042
535	923.45	....	0.0213	0.4654	0.4867	530.4	664.4	1194.8	0.7314	0.6679	1.3993
540	962.80	....	0.0214	0.4433	0.4647	536.6	656.7	1193.3	0.7375	0.6569	1.3944
545	1003.6	....	0.0216	0.4222	0.4438	542.0	648.9	1191.8	0.7436	0.6459	1.3895
550	1045.6	....	0.0218	0.4021	0.4239	549.3	640.9	1190.2	0.7498	0.6347	1.3845
555	1088.8	....	0.0219	0.3830	0.4049	555.7	632.6	1188.3	0.7559	0.6234	1.3793
560	1133.4	....	0.0221	0.3648	0.3869	562.2	624.1	1186.3	0.7622	0.6120	1.3742
565	1179.3	....	0.0222	0.3472	0.3694	568.8	615.4	1184.2	0.7684	0.6006	1.3690
570	1226.7	....	0.0224	0.3304	0.3528	575.4	606.5	1181.9	0.7737	0.5890	1.3627
575	1275.7	....	0.0226	0.3143	0.3369	582.1	597.4	1179.5	0.7810	0.5774	1.3584

Table D-1 Saturated Steam (Continued)

Temp. t, (°F)	Absolute pressure P		Specific volume			Enthalpy			Entropy		
	Psia	In Hg (32°F)	Sat. liquid V_f	Evap. ΔV_{fg}	Sat. vapor V_g	Sat. liquid H_f	Evap. ΔH_{fg}	Sat. vapor H_g	Sat. liquid S_f	Evap. ΔS_{fg}	Sat. vapor S_g
580	1326.1	....	0.0228	0.2989	0.3217	588.9	588.1	1177.0	0.7872	0.5656	1.3528
585	1378.1	....	0.0230	0.2840	0.3070	595.7	578.6	1174.3	0.7936	0.5538	1.3474
590	1431.5	....	0.0232	0.2699	0.2931	602.6	568.8	1171.4	0.8000	0.5419	1.3419
595	1486.5	....	0.0234	0.2563	0.2797	609.7	558.7	1168.4	0.8065	0.5297	1.3362
600	1543.2	....	0.0236	0.2432	0.2668	616.8	548.4	1165.2	0.8130	0.5175	1.3305
605	1601.5	....	0.0239	0.2306	0.2545	624.1	537.7	1161.8	0.8196	0.5050	1.3246
610	1661.6	....	0.0241	0.2185	0.2426	631.5	526.6	1158.1	0.8263	0.4923	1.3186
615	1723.4	....	0.0244	0.2068	0.2312	638.9	515.3	1154.2	0.8330	0.4795	1.3125
620	1787.0	....	0.0247	0.1955	0.2202	646.5	503.7	1150.2	0.8398	0.4665	1.3063
625	1852.4	....	0.0250	0.1845	0.2095	654.3	491.5	1145.8	0.8467	0.4531	1.2998
630	1919.8	....	0.0253	0.1740	0.1993	662.2	478.8	1141.0	0.8537	0.4394	1.2931
635	1989.0	....	0.0256	0.1638	0.1894	670.4	465.5	1135.9	0.8609	0.4252	1.2861
640	2060.3	....	0.0260	0.1539	0.1799	678.7	452.0	1130.7	0.8681	0.4110	1.2791
645	2133.5	....	0.0264	0.1441	0.1705	687.3	437.6	1124.9	0.8756	0.3961	1.2717
650	2208.8	....	0.0268	0.1348	0.1616	696.0	422.7	1118.7	0.8832	0.3809	1.2641
655	2286.4	....	0.0273	0.1256	0.1529	705.2	407.0	1112.2	0.8910	0.3651	1.2561
660	2366.2	....	0.0278	0.1167	0.1445	714.4	390.5	1104.9	0.8991	0.3488	1.2479
665	2448.0	....	0.0283	0.1079	0.1362	724.5	372.1	1096.6	0.9074	0.3308	1.2382
670	2532.4	....	0.0290	0.0991	0.1281	734.6	353.3	1087.9	0.9161	0.3127	1.2288
675	2619.2	....	0.0297	0.0904	0.1201	745.5	332.8	1078.3	0.9253	0.2933	1.2186

680	2708.4	....	0.0305	0.0810	0.1115	757.2	310.0	1067.2	0.9352	0.2720	1.2072
685	2800.4	....	0.0316	0.0716	0.1032	770.1	284.5	1054.6	0.9459	0.2485	1.1944
690	2895.0	....	0.0328	0.0617	0.0945	784.2	254.9	1039.1	0.9579	0.2217	1.1796
695	2992.7	....	0.0345	0.0511	0.0856	801.3	219.1	1020.4	0.9720	0.1897	1.1617
700	3094.1	....	0.0369	0.0389	0.0758	823.9	171.7	995.6	0.9904	0.1481	1.1385
705	3199.1	....	0.0440	0.0157	0.0597	870.2	77.6	947.8	1.0305	0.0661	1.0966
705.34‡	3206.2	....	0.0541	0	0.0541	910.3	0	910.3	1.0645	0	1.0645

‡ Critical temperature.

Table D-2 Superheated Steam†

Abs press., psia (sat. temp.)	V H S	Sat. water	Sat. steam	Temperature, °F 200	250	300	350	400	450	500	600	700	800	900	1000	1100	1200
1	V	0.0161	333.79	392.5	422.5	452.1	482.1	511.7	541.8	571.3	630.9	690.6	750.2	809.8	869.4	929.1	988.7
(101.76)	H	69.72	1105.2	1149.2	1171.9	1194.4	1217.3	1240.2	1263.5	1286.7	1333.9	1382.1	1431.0	1480.8	1531.4	1583.0	1635.4
	S	0.1326	1.9769	2.0491	2.0822	2.1128	2.1420	2.1694	2.1957	2.2206	2.2673	2.3107	2.3512	2.3892	2.4251	2.4592	2.4918
5	V	0.0164	73.600	78.17	84.24	90.21	96.26	102.19	108.23	114.16	126.11	138.05	149.99	161.91	173.83	185.80	197.72
(162.25)	H	130.13	1130.8	1148.3	1171.1	1193.6	1216.6	1239.8	1263.0	1286.1	1333.5	1381.8	1430.8	1480.6	1531.3	1582.9	1635.3
	S	0.2347	1.8437	1.8710	1.9043	1.9349	1.9642	1.9920	2.0182	2.0429	2.0898	2.1333	2.1738	2.2118	2.2478	2.2820	2.3146
10	V	0.0166	38.462	38.88	41.96	44.98	48.02	51.01	54.04	57.02	63.01	68.99	74.96	80.92	86.89	92.88	98.85
(193.21)	H	161.17	1143.3	1146.7	1170.2	1192.8	1216.0	1239.3	1262.5	1285.8	1333.3	1381.6	1430.6	1480.5	1531.2	1582.8	1635.2
	S	0.2834	1.7876	1.7928	1.8271	1.8579	1.8875	1.9154	1.9416	1.9665	2.0135	2.0570	2.0975	2.1356	2.1716	2.2058	2.2384
14.696	V	0.0167	26.828	....	28.44	30.52	32.61	34.65	36.73	38.75	42.83	46.91	50.97	55.03	59.09	63.19	67.25
(212.00)	H	180.07	1150.4	....	1169.2	1192.0	1215.4	1238.9	1262.1	1285.4	1333.0	1381.4	1430.5	1480.4	1531.1	1582.7	1635.1
	S	0.3120	1.7566	....	1.7838	1.8148	1.8446	1.8727	1.8989	1.9238	1.9709	2.0145	2.0551	2.0932	2.1292	2.1634	2.1960
15	V	0.0167	26.320	....	27.86	29.90	31.94	33.95	35.98	37.97	41.98	45.97	49.95	53.93	57.91	61.91	65.89
(213.03)	H	181.11	1150.7	....	1169.2	1192.0	1215.4	1238.9	1262.1	1285.4	1333.0	1381.4	1430.5	1480.4	1531.1	1582.7	1635.1
	S	0.3135	1.7548	....	1.7816	1.8126	1.8424	1.8705	1.8967	1.9216	1.9687	2.0123	2.0529	2.0910	2.1270	2.1612	2.1938
20	V	0.0168	20.110	....	20.81	22.36	23.91	25.43	26.95	28.45	31.46	34.46	37.44	40.43	43.42	46.43	49.41
(227.96)	H	196.16	1156.1	....	1168.0	1191.1	1214.8	1238.4	1261.6	1285.0	1332.7	1381.2	1430.3	1480.2	1531.0	1582.6	1635.1
	S	0.3356	1.7315	....	1.7485	1.7799	1.8101	1.8384	1.8646	1.8896	1.9368	1.9805	2.0211	2.0592	2.0952	2.1294	2.1620
25	V	0.0169	16.321	....	16.58	17.84	19.08	20.30	21.53	22.73	25.15	27.55	29.94	32.33	34.73	37.14	39.52
(240.07)	H	208.41	1160.4	....	1166.3	1190.2	1214.1	1237.9	1261.1	1284.6	1332.4	1381.0	1430.1	1480.0	1530.9	1582.5	1635.0
	S	0.3532	1.7137	....	1.7221	1.7570	1.7875	1.8160	1.8422	1.8673	1.9146	1.9584	1.9990	2.0371	2.0732	2.1074	2.1400
30	V	0.0170	13.763	....	....	14.82	15.87	16.89	17.91	18.92	20.94	22.94	24.94	26.93	28.93	30.94	32.93
(250.34)	H	218.83	1164.0	....	....	1189.2	1213.4	1237.4	1260.6	1284.2	1332.1	1380.8	1429.9	1479.9	1530.8	1582.4	1634.9
	S	0.3680	1.6992	....	....	1.7335	1.7643	1.7930	1.8192	1.8444	1.8918	1.9357	1.9763	2.0145	2.0506	2.0848	2.1174

35 (259.28)	V	0.0171	11.907	....	....	12.66	13.57	14.45	15.33	16.20	17.94	19.66	21.36	23.08	24.79	26.52	28.22
	H	227.92	1167.0	....	....	1188.2	1212.7	1236.9	1260.1	1283.8	1331.9	1380.6	1429.8	1479.8	1530.7	1582.3	1634.8
	S	0.3807	1.6869	....	....	1.7156	1.7468	1.7758	1.8020	1.8274	1.8750	1.9189	1.9596	1.9978	2.0339	2.0681	2.1007
40 (267.24)	V	0.0172	10.506	....	....	11.04	11.84	12.62	13.40	14.16	15.68	17.19	18.69	20.18	21.68	23.20	24.69
	H	236.02	1169.7	....	....	1187.1	1211.9	1236.4	1259.6	1283.4	1331.6	1380.4	1429.6	1479.6	1530.6	1582.2	1634.8
	S	0.3919	1.6763	....	....	1.6997	1.7313	1.7606	1.7868	1.8123	1.8600	1.9040	1.9447	1.9829	2.0191	2.0533	2.0860
45 (274.45)	V	0.0172	9.408	....	....	9.785	10.50	11.20	11.89	12.57	13.93	15.27	16.60	17.94	19.27	20.62	21.95
	H	243.38	1172.0	....	....	1185.9	1211.1	1235.8	1259.1	1283.0	1331.3	1380.1	1429.4	1479.4	1530.5	1582.1	1634.7
	S	0.4019	1.6668	....	....	1.6854	1.7175	1.7471	1.7734	1.7990	1.8468	1.8908	1.9315	1.9697	2.0059	2.0401	2.0728
50 (281.01)	V	0.0173	8.522	....	....	8.777	9.430	10.06	10.69	11.30	12.53	13.74	14.93	16.14	17.34	18.55	19.75
	H	250.09	1174.0	....	....	1184.6	1210.3	1235.2	1258.6	1282.6	1331.0	1379.9	1429.3	1479.3	1530.4	1582.0	1634.6
	S	0.4110	1.6583	....	....	1.6724	1.7051	1.7349	1.7613	1.7870	1.8349	1.8790	1.9198	1.9580	1.9942	2.0284	2.0611
55 (287.07)	V	0.0173	7.792	....	....	7.950	8.553	9.130	9.703	10.26	11.38	12.48	13.57	14.67	15.76	16.86	17.95
	H	256.30	1175.8	....	....	1183.2	1209.4	1234.6	1258.2	1282.2	1330.7	1379.7	1429.1	1479.2	1530.3	1581.9	1634.5
	S	0.4193	1.6506	....	....	1.6604	1.6938	1.7240	1.7507	1.7764	1.8244	1.8685	1.9093	1.9475	1.9837	2.0179	2.0512
60 (292.71)	V	0.0174	7.179	....	....	7.260	7.821	8.353	8.882	9.398	10.42	11.44	12.44	13.44	14.44	15.45	16.45
	H	262.10	1177.5	....	....	1181.8	1208.5	1234.0	1257.7	1281.8	1330.4	1379.5	1428.9	1479.0	1530.2	1581.8	1634.4
	S	0.4271	1.6437	....	....	1.6494	1.6834	1.7139	1.7407	1.7665	1.8146	1.8588	1.8996	1.9378	1.9741	2.0083	2.0410
65 (297.97)	V	0.0174	6.654	....	....	6.674	7.202	7.696	8.187	8.665	9.614	10.55	11.48	12.40	13.33	14.26	15.19
	H	267.51	1179.1	....	....	1180.4	1207.6	1233.4	1257.2	1281.4	1330.1	1379.3	1428.8	1478.9	1530.1	1581.7	1634.4
	S	0.4342	1.6374	....	....	1.6391	1.6738	1.7047	1.7316	1.7575	1.8057	1.8500	1.8909	1.9291	1.9654	1.9996	2.0323
70 (302.92)	V	0.0175	6.210	....	....	....	6.671	7.132	7.592	8.036	8.920	9.791	10.65	11.51	12.37	13.24	14.10
	H	272.61	1180.5	....	....	....	1206.7	1232.8	1256.7	1281.0	1329.9	1379.0	1428.6	1478.7	1530.0	1581.6	1634.3
	S	0.4409	1.6314	....	....	....	1.6647	1.6960	1.7230	1.7490	1.7974	1.8416	1.8826	1.9208	1.9572	1.9914	2.0241
75 (307.60)	V	0.0175	5.820	....	....	....	6.210	6.644	7.076	7.492	8.319	9.133	9.938	10.74	11.54	12.36	13.16
	H	277.44	1181.9	....	....	....	1205.8	1232.2	1256.2	1280.6	1329.6	1378.8	1428.4	1478.6	1529.8	1581.5	1634.2
	S	0.4472	1.6260	....	....	....	1.6563	1.6879	1.7150	1.7411	1.7896	1.8339	1.8749	1.9132	1.9495	1.9837	2.0164

† Reproduced by permission from "Steam Tables: Properties of Saturated and Superheated Steam," copyright 1940, Combustion Engineering, Inc.

Table D-2 Superheated Steam (Continued)

Abs press., psia (sat. temp.)	V H S	Sat. water	Sat. steam	Temperature, °F													
				340	360	380	400	420	450	500	600	700	800	900	1000	1100	1200
80 (312.03)	V	0.0176	5.476	5.720	5.889	6.055	6.217	6.384	6.623	7.015	7.793	8.558	9.313	10.07	10.82	11.58	12.33
	H	282.02	1183.1	1200.0	1211.0	1221.2	1231.5	1240.3	1255.7	1280.2	1329.3	1378.5	1428.2	1478.4	1529.7	1581.4	1634.1
	S	0.4532	1.6209	1.6424	1.6560	1.6683	1.6804	1.6905	1.7077	1.7339	1.7825	1.8268	1.8679	1.9062	1.9426	1.9768	2.0095
85 (316.25)	V	0.0176	5.169	5.368	5.528	5.685	5.839	5.995	6.226	6.594	7.329	8.050	8.762	9.472	10.18	10.90	11.61
	H	286.40	1184.3	1198.5	1210.0	1220.5	1230.7	1239.7	1255.1	1279.7	1329.0	1378.3	1428.0	1478.2	1529.6	1581.3	1634.0
	S	0.4587	1.6159	1.6339	1.6481	1.6608	1.6728	1.6831	1.7003	1.7266	1.7754	1.8198	1.8609	1.8992	1.9357	1.9699	2.0026
90 (320.27)	V	0.0177	4.898	5.055	5.208	5.357	5.504	5.653	5.869	6.220	6.916	7.599	8.272	8.943	9.626	10.29	10.96
	H	290.57	1185.4	1197.3	1209.0	1219.8	1230.0	1239.1	1254.5	1279.3	1328.7	1378.1	1427.9	1478.1	1529.5	1581.2	1634.0
	S	0.4641	1.6113	1.6264	1.6408	1.6538	1.6658	1.6763	1.6935	1.7200	1.7689	1.8134	1.8546	1.8929	1.9294	1.9636	1.9964
95 (324.13)	V	0.0177	4.653	4.773	4.921	5.063	5.205	5.346	5.552	5.886	6.547	7.195	7.834	8.481	9.117	9.751	10.38
	H	294.58	1186.4	1196.0	1208.0	1219.0	1229.3	1238.6	1254.0	1278.9	1328.4	1377.8	1427.7	1478.0	1529.4	1581.1	1633.9
	S	0.4692	1.6070	1.6191	1.6339	1.6472	1.6593	1.6700	1.6872	1.7138	1.7628	1.8073	1.8485	1.8869	1.9234	1.9576	1.9904
100 (327.83)	V	0.0177	4.433	4.520	4.663	4.801	4.936	5.070	5.266	5.589	6.217	6.836	7.448	8.055	8.659	9.262	9.862
	H	298.43	1187.3	1194.9	1207.0	1218.3	1228.4	1238.6	1253.7	1278.6	1327.9	1377.5	1427.5	1478.0	1529.2	1581.0	1633.7
	S	0.4741	1.6028	1.6124	1.6273	1.6409	1.6528	1.6645	1.6814	1.7080	1.7568	1.8015	1.8428	1.8814	1.9177	1.9520	1.9847
105 (331.38)	V	0.0178	4.232	4.292	4.429	4.562	4.691	4.820	5.007	5.316	5.916	6.507	7.090	7.670	8.245	8.819	9.391
	H	302.13	1188.2	1193.5	1205.9	1217.2	1227.6	1237.5	1252.9	1278.0	1327.6	1377.4	1427.3	1477.7	1529.2	1580.9	1633.7
	S	0.4787	1.5988	1.6055	1.6208	1.6344	1.6466	1.6580	1.6752	1.7020	1.7511	1.7960	1.8372	1.8757	1.9122	1.9464	1.9791
110 (334.79)	V	0.0178	4.050	4.084	4.217	4.345	4.469	4.592	4.773	5.069	5.643	6.208	6.765	7.319	7.869	8.417	8.963
	H	305.69	1189.0	1192.2	1204.9	1216.4	1226.9	1236.9	1252.4	1277.5	1327.4	1377.1	1427.1	1477.5	1529.1	1580.8	1633.6
	S	0.4832	1.5950	1.5990	1.6147	1.6286	1.6410	1.6525	1.6698	1.6966	1.7460	1.7908	1.8321	1.8706	1.9072	1.9414	1.9742
115 (338.08)	V	0.0179	3.882	....	4.022	4.146	4.266	4.384	4.558	4.843	5.393	5.935	6.469	6.999	7.525	8.049	8.572
	H	309.13	1189.8	....	1203.8	1215.6	1226.2	1236.3	1251.9	1277.1	1327.1	1376.9	1427.0	1477.4	1528.9	1580.7	1633.6
	S	0.4875	1.5915	....	1.6088	1.6230	1.6355	1.6471	1.6645	1.6915	1.7410	1.7859	1.8273	1.8658	1.9023	1.9366	1.9695

120	V	0.0179	3.728	....	3.845	3.963	4.079	4.194	4.361	4.635	5.165	5.685	6.197	6.705	7.210	7.713	8.215
(341.26)	H	312.46	1190.6	....	1202.7	1214.7	1225.4	1235.7	1251.4	1276.7	1326.8	1376.7	1426.8	1477.2	1528.8	1580.6	1633.5
	S	0.4916	1.5879	....	1.6028	1.6173	1.6299	1.6417	1.6592	1.6863	1.7359	1.7809	1.8223	1.8608	1.8974	1.9317	1.9646
125	V	0.0179	3.586	....	3.680	3.796	3.908	4.019	4.181	4.445	4.954	5.454	5.947	6.435	6.920	7.403	7.885
(344.34)	H	315.69	1191.3	....	1201.6	1213.7	1224.5	1235.0	1250.8	1276.3	1326.5	1376.4	1426.6	1477.1	1528.7	1580.5	1633.4
	S	0.4956	1.5846	....	1.5973	1.6119	1.6246	1.6367	1.6544	1.6817	1.7314	1.7764	1.8179	1.8565	1.8931	1.9274	1.9603
130	V	0.0180	3.455	....	3.528	3.641	3.750	3.857	4.013	4.268	4.760	5.242	5.716	6.186	6.653	7.117	7.581
(347.31)	H	318.81	1192.0	....	1200.4	1212.7	1223.6	1234.3	1250.3	1275.8	1326.1	1376.1	1426.4	1476.9	1528.6	1580.4	1633.3
	S	0.4995	1.5815	....	1.5918	1.6066	1.6194	1.6317	1.6496	1.6769	1.7267	1.7718	1.8134	1.8520	1.8887	1.9230	1.9559
135	V	0.0180	3.333	....	3.388	3.497	3.603	3.707	3.859	4.105	4.580	5.045	5.502	5.955	6.405	6.853	7.303
(350.21)	H	321.86	1192.7	....	1199.2	1211.7	1222.7	1233.6	1249.7	1275.4	1325.8	1375.9	1426.2	1476.8	1528.5	1580.3	1633.2
	S	0.5032	1.5784	....	1.5864	1.6015	1.6144	1.6269	1.6449	1.6724	1.7223	1.7674	1.8090	1.8476	1.8843	1.9186	1.9515
140	V	0.0180	3.220	....	3.258	3.364	3.467	3.567	3.715	3.954	4.413	4.862	5.303	5.741	6.175	6.607	7.037
(353.03)	H	324.83	1193.3	....	1198.0	1210.6	1221.8	1232.9	1249.1	1275.0	1325.5	1375.7	1426.0	1476.6	1528.4	1580.2	1633.2
	S	0.5069	1.5755	....	1.5813	1.5965	1.6097	1.6225	1.6406	1.6683	1.7183	1.7635	1.8051	1.8437	1.8804	1.9147	1.9476
145	V	0.0181	3.114	....	3.136	3.240	3.340	3.438	3.581	3.812	4.257	4.692	5.119	5.541	5.961	6.378	6.794
(355.76)	H	327.71	1193.9	....	1196.7	1209.5	1220.9	1232.2	1248.5	1274.5	1325.1	1375.4	1425.8	1476.5	1528.3	1580.1	1633.1
	S	0.5104	1.5726	....	1.5760	1.5914	1.6048	1.6178	1.6360	1.6638	1.7139	1.7592	1.8009	1.8396	1.8763	1.9106	1.9435
150	V	0.0181	3.016	....	....	3.124	3.221	3.317	3.456	3.681	4.112	4.533	4.946	5.355	5.761	6.164	6.567
(358.43)	H	330.53	1194.4	....	....	1208.4	1220.0	1231.4	1248.0	1274.1	1324.9	1375.1	1425.6	1476.3	1528.1	1580.0	1633.0
	S	0.5138	1.5698	....	....	1.5865	1.6002	1.6133	1.6319	1.6598	1.7101	1.7553	1.7970	1.8357	1.8724	1.9068	1.9397
155	V	0.0181	2.921	....	....	3.015	3.110	3.203	3.340	3.558	3.976	4.384	4.785	5.181	5.574	5.964	6.354
(361.02)	H	333.27	1195.0	....	....	1207.2	1219.1	1230.7	1247.5	1273.6	1324.5	1374.9	1425.4	1476.2	1528.0	1579.9	1632.9
	S	0.5172	1.5671	....	....	1.5818	1.5958	1.6091	1.6279	1.6558	1.7062	1.7516	1.7933	1.8321	1.8688	1.9032	1.9361
160	V	0.0182	2.834	....	....	2.913	3.006	3.097	3.230	3.443	3.849	4.245	4.633	5.018	5.398	5.777	6.155
(363.55)	H	335.95	1195.5	....	....	1206.0	1218.3	1230.0	1246.9	1273.2	1324.1	1374.7	1425.2	1476.0	1527.9	1579.8	1632.8
	S	0.5204	1.5646	....	....	1.5772	1.5917	1.6052	1.6241	1.6522	1.7026	1.7482	1.7899	1.8287	1.8655	1.8999	1.9328

Table D-2 Superheated Steam (Continued)

Abs press., psia (sat. temp.)	V H S	Sat. water	Sat. steam	Temperature, °F													
				400	**420**	**440**	**460**	**480**	**500**	**550**	**600**	**700**	**800**	**900**	**1000**	**1100**	**1200**
165	V	0.0182	2.752	2.909	2.997	3.084	3.170	3.251	3.334	3.533	3.729	4.114	4.491	4.864	5.234	5.601	5.967
(366.01)	H	338.55	1195.9	1217.4	1229.3	1241.1	1251.8	1262.4	1272.8	1298.5	1323.8	1374.5	1425.0	1475.9	1527.8	1579.7	1632.7
	S	0.5236	1.5619	1.5874	1.6011	1.6144	1.6262	1.6376	1.6486	1.6747	1.6991	1.7448	1.7865	1.8254	1.8622	1.8966	1.9295
170	V	0.0182	2.674	2.816	2.903	2.988	3.071	3.151	3.232	3.426	3.617	3.991	4.357	4.720	5.079	5.436	5.791
(368.42)	H	341.11	1196.3	1216.5	1228.4	1240.5	1251.3	1261.8	1272.3	1298.2	1323.5	1374.2	1424.9	1475.7	1527.6	1579.6	1632.7
	S	0.5266	1.5593	1.5832	1.5969	1.6105	1.6224	1.6337	1.6448	1.6711	1.6955	1.7412	1.7831	1.8219	1.8587	1.8931	1.9261
175	V	0.0182	2.601	2.730	2.814	2.897	2.979	3.057	3.136	3.325	3.510	3.875	4.231	4.584	4.932	5.279	5.625
(370.77)	H	343.61	1196.7	1215.6	1227.6	1239.9	1250.8	1261.3	1271.9	1297.8	1323.2	1374.0	1424.7	1475.6	1527.5	1579.5	1632.6
	S	0.5296	1.5569	1.5793	1.5931	1.6069	1.6189	1.6302	1.6414	1.6677	1.6922	1.7380	1.7799	1.8185	1.8553	1.8897	1.9227
180	V	0.0183	2.532	2.648	2.731	2.812	2.892	2.968	3.045	3.229	3.410	3.765	4.112	4.455	4.794	5.132	5.468
(373.08)	H	346.07	1197.2	1214.6	1226.8	1239.2	1250.2	1260.8	1271.5	1297.4	1322.8	1373.7	1424.5	1475.5	1527.4	1579.4	1632.5
	S	0.5325	1.5545	1.5751	1.5891	1.6030	1.6151	1.6265	1.6378	1.6641	1.6886	1.7345	1.7765	1.8154	1.8522	1.8866	1.9196
185	V	0.0183	2.466	2.570	2.651	2.731	2.809	2.884	2.958	3.139	3.315	3.661	3.999	4.333	4.664	4.992	5.319
(375.34)	H	348.47	1197.6	1213.7	1226.0	1238.4	1249.6	1260.3	1271.0	1297.4	1322.4	1373.4	1424.3	1475.3	1527.3	1579.3	1632.4
	S	0.5354	1.5522	1.5712	1.5853	1.5992	1.6115	1.6230	1.6343	1.6611	1.6853	1.7312	1.7733	1.8122	1.8491	1.8835	1.9165
190	V	0.0183	2.404	2.496	2.576	2.654	2.731	2.804	2.877	3.053	3.225	3.563	3.893	4.218	4.540	4.860	5.179
(377.55)	H	350.83	1198.0	1212.7	1225.1	1237.7	1249.0	1259.8	1270.5	1296.6	1322.1	1373.1	1424.1	1475.2	1527.1	1579.2	1632.3
	S	0.5382	1.5501	1.5674	1.5817	1.5959	1.6083	1.6199	1.6312	1.6577	1.6823	1.7282	1.7703	1.8093	1.8461	1.8806	1.9136
195	V	0.0184	2.344	2.426	2.505	2.581	2.656	2.728	2.799	2.972	3.140	3.470	3.791	4.109	4.423	4.735	5.046
(379.70)	H	353.13	1198.4	1211.7	1224.2	1237.0	1248.3	1259.3	1270.0	1296.2	1321.8	1372.9	1423.9	1475.0	1527.0	1579.1	1632.2
	S	0.5409	1.5479	1.5636	1.5780	1.5924	1.6048	1.6166	1.6279	1.6545	1.6792	1.7252	1.7673	1.8063	1.8432	1.8777	1.9107
200	V	0.0184	2.288	2.360	2.437	2.512	2.585	2.656	2.726	2.895	3.059	3.381	3.697	4.005	4.311	4.616	4.919
(381.82)	H	355.40	1198.7	1210.8	1223.7	1236.3	1247.9	1258.7	1269.4	1295.6	1321.4	1372.5	1423.9	1474.9	1526.6	1579.0	1632.1
	S	0.5436	1.5457	1.5599	1.5748	1.5889	1.6017	1.6133	1.6245	1.6511	1.6761	1.7221	1.7646	1.8035	1.8402	1.8749	1.9079

205	V	0.0184	2.235	2.297	2.372	2.446	2.518	2.587	2.656	2.821	2.982	3.297	3.604	3.906	4.205	4.502	4.798
(383.89)	H	357.61	1199.0	1209.7	1222.5	1235.4	1247.1	1258.2	1269.0	1295.4	1321.0	1372.4	1423.5	1474.7	1526.8	1578.9	1632.1
	S	0.5462	1.5436	1.5562	1.5709	1.5854	1.5983	1.6102	1.6216	1.6484	1.6731	1.7194	1.7616	1.8007	1.8377	1.8722	1.9052
210	V	0.0184	2.183	2.237	2.311	2.384	2.454	2.522	2.589	2.751	2.909	3.216	3.516	3.812	4.104	4.395	4.683
(385.93)	H	359.80	1199.4	1208.8	1221.8	1234.7	1246.5	1257.7	1268.5	1295.0	1320.7	1372.1	1423.3	1474.6	1526.6	1578.8	1632.0
	S	0.5488	1.5417	1.5527	1.5676	1.5821	1.5951	1.6071	1.6185	1.6454	1.6702	1.7165	1.7588	1.7980	1.8349	1.8695	1.9025
215	V	0.0185	2.134	2.179	2.252	2.324	2.393	2.460	2.526	2.685	2.839	3.140	3.433	3.722	4.008	4.292	4.574
(387.93)	H	361.95	1199.6	1207.8	1221.0	1234.0	1245.9	1257.2	1268.0	1294.6	1320.4	1371.9	1423.1	1474.4	1526.5	1578.7	1631.9
	S	0.5513	1.5395	1.5491	1.5643	1.5789	1.5920	1.6042	1.6156	1.6426	1.6675	1.7139	1.7562	1.7954	1.8324	1.8670	1.9000
220	V	0.0185	2.086	2.124	2.196	2.267	2.335	2.400	2.465	2.621	2.772	3.067	3.354	3.637	3.916	4.193	4.469
(389.89)	H	364.05	1199.9	1206.8	1220.1	1233.2	1245.2	1256.7	1267.5	1294.1	1320.0	1371.6	1422.9	1474.2	1526.4	1578.6	1631.8
	S	0.5538	1.5376	1.5457	1.5610	1.5757	1.5889	1.6013	1.6127	1.6397	1.6647	1.7112	1.7536	1.7928	1.8298	1.8644	1.8974
225	V	0.0185	2.042	2.072	2.142	2.212	2.279	2.344	2.407	2.560	2.708	2.997	3.278	3.555	3.828	4.100	4.369
(391.81)	H	366.11	1200.2	1205.8	1219.2	1232.3	1244.5	1256.2	1267.1	1293.7	1319.6	1371.4	1422.7	1474.1	1526.3	1578.5	1631.7
	S	0.5562	1.5358	1.5423	1.5577	1.5724	1.5858	1.5984	1.6099	1.6369	1.6619	1.7086	1.7510	1.7902	1.8272	1.8618	1.8948
230	V	0.0186	1.9989	2.021	2.091	2.160	2.226	2.289	2.352	2.502	2.647	2.930	3.205	3.477	3.744	4.010	4.274
(393.70)	H	368.16	1200.4	1204.9	1218.3	1231.6	1243.8	1255.6	1266.7	1293.3	1319.3	1371.1	1422.5	1474.0	1526.2	1578.4	1631.6
	S	0.5585	1.5337	1.5390	1.5544	1.5693	1.5827	1.5954	1.6071	1.6341	1.6592	1.7059	1.7484	1.7877	1.8247	1.8593	1.8923
235	V	0.0186	1.9573	1.973	2.042	2.110	2.175	2.237	2.298	2.446	2.589	2.866	3.136	3.402	3.664	3.924	4.182
(395.56)	H	370.17	1200.7	1203.9	1217.5	1230.8	1243.2	1255.0	1266.2	1292.9	1319.0	1370.9	1422.3	1473.8	1526.0	1578.3	1631.6
	S	0.5609	1.5320	1.5357	1.5513	1.5662	1.5798	1.5925	1.6043	1.6314	1.6566	1.7034	1.7459	1.7852	1.8222	1.8568	1.8899
240	V	0.0186	1.9176	....	1.995	2.062	2.126	2.187	2.247	2.392	2.532	2.805	3.069	3.330	3.586	3.841	4.095
(397.40)	H	372.16	1200.9	....	1216.6	1230.0	1242.5	1254.4	1265.7	1292.5	1318.6	1370.5	1422.1	1473.6	1525.9	1578.2	1631.5
	S	0.5632	1.5301	....	1.5482	1.5633	1.5770	1.5898	1.6017	1.6289	1.6541	1.7009	1.7435	1.7828	1.8199	1.8545	1.8876
245	V	0.0186	1.8797	....	1.950	2.015	2.078	2.139	2.198	2.341	2.479	2.746	3.006	3.261	3.513	3.762	4.011
(399.20)	H	374.11	1201.1	....	1215.6	1229.1	1241.8	1253.8	1265.2	1292.0	1318.3	1370.3	1421.9	1473.5	1525.8	1578.1	1631.4
	S	0.5654	1.5283	....	1.5450	1.5602	1.5742	1.5871	1.5991	1.6263	1.6517	1.6985	1.7411	1.7805	1.8176	1.8522	1.8853

Table D-2 Superheated Steam (Continued)

Abs press., psia (sat. temp.)	V H S	Sat. water	Sat. steam	Temperature, °F 420	440	460	480	500	520	550	600	700	800	900	1000	1100	1200
250	V	0.0187	1.8431	1.9065	1.9711	2.0334	2.0932	2.1515	2.2085	2.2920	2.4272	2.6897	2.9444	3.1949	3.4416	3.6867	3.9299
(400.97)	H	376.04	1201.4	1214.6	1228.3	1241.0	1253.2	1264.7	1274.5	1291.6	1317.9	1370.0	1421.7	1473.3	1525.6	1578.0	1631.3
	S	0.5677	1.5267	1.5419	1.5573	1.5713	1.5844	1.5965	1.6066	1.6238	1.6492	1.6961	1.7388	1.7782	1.8153	1.8500	1.8831
255	V	0.0187	1.8079	1.8686	1.9286	1.9899	2.0489	2.1065	2.1626	2.2447	2.3776	2.6354	2.8855	3.1313	3.3733	3.6138	3.8524
(402.71)	H	377.91	1201.6	1213.7	1227.5	1240.3	1252.6	1264.2	1274.2	1291.2	1317.5	1369.8	1421.5	1473.2	1525.5	1577.9	1631.2
	S	0.5698	1.5249	1.5388	1.5543	1.5684	1.5816	1.5938	1.6041	1.6212	1.6466	1.6937	1.7364	1.7759	1.8130	1.8477	1.8808
260	V	0.0187	1.7742	1.8246	1.8876	1.9482	2.0063	2.0631	2.1185	2.1991	2.3299	2.5833	2.8289	3.0701	3.3077	3.5437	3.7778
(404.43)	H	379.78	1201.8	1212.8	1226.6	1239.5	1252.0	1263.6	1273.8	1290.8	1317.1	1369.5	1421.3	1473.0	1525.4	1577.8	1631.1
	S	0.5720	1.5233	1.5359	1.5514	1.5656	1.5790	1.5912	1.6017	1.6188	1.6442	1.6914	1.7342	1.7737	1.8109	1.8456	1.8787
265	V	0.0187	1.7416	1.7858	1.8481	1.9080	1.9654	2.0213	2.0759	2.1554	2.2840	2.5331	2.7744	3.0114	3.2446	3.4761	3.7061
(406.12)	H	381.62	1202.0	1211.9	1225.7	1238.7	1251.2	1263.0	1273.4	1290.4	1316.8	1369.3	1421.1	1472.9	1525.3	1577.7	1631.1
	S	0.5741	1.5217	1.5330	1.5485	1.5628	1.5762	1.5886	1.5993	1.6164	1.6419	1.6892	1.7320	1.7715	1.8087	1.8434	1.8765
270	V	0.0188	1.7101	1.7486	1.8101	1.8692	1.9259	1.9810	2.0350	2.1131	2.2399	2.4847	2.7219	2.9548	3.1838	3.4112	3.6370
(407.79)	H	383.43	1202.2	1211.0	1224.9	1238.0	1250.6	1262.5	1273.0	1290.0	1316.4	1369.0	1420.9	1472.7	1525.1	1577.6	1631.0
	S	0.5761	1.5200	1.5301	1.5457	1.5601	1.5736	1.5861	1.5969	1.6140	1.6395	1.6869	1.7298	1.7693	1.8065	1.8413	1.8744
275	V	0.0188	1.6798	1.7127	1.7735	1.8318	1.8879	1.9422	1.9956	2.0725	2.1973	2.4382	2.6714	2.9002	3.1253	3.3486	3.5704
(409.44)	H	385.22	1202.3	1210.0	1224.1	1237.3	1250.0	1262.0	1272.6	1289.5	1316.1	1368.7	1420.7	1472.6	1525.0	1577.5	1630.9
	S	0.5782	1.5183	1.5271	1.5429	1.5574	1.5711	1.5837	1.5946	1.6116	1.6373	1.6847	1.7277	1.7673	1.8045	1.8393	1.8724
280	V	0.0188	1.6504	1.6780	1.7381	1.7957	1.8512	1.9048	1.9575	2.0334	2.1562	2.3932	2.6226	2.8475	3.0688	3.2883	3.5062
(411.06)	H	386.99	1202.5	1209.0	1223.2	1236.5	1249.4	1261.5	1272.2	1289.1	1315.7	1368.5	1420.5	1472.4	1524.9	1577.4	1630.8
	S	0.5802	1.5167	1.5241	1.5401	1.5547	1.5686	1.5813	1.5923	1.6093	1.6350	1.6826	1.7256	1.7652	1.8024	1.8372	1.8703
285	V	0.0188	1.6232	1.6446	1.7040	1.7610	1.8157	1.8687	1.9207	1.9955	2.1165	2.3499	2.5756	2.7968	3.0143	3.2300	3.4443
(412.66)	H	388.74	1202.7	1208.0	1222.3	1235.6	1248.7	1260.9	1271.8	1288.6	1315.4	1368.2	1420.3	1472.2	1524.7	1577.3	1630.7
	S	0.5822	1.5153	1.5214	1 5375	1.5521	1.5662	1.5790	1.5902	1.6071	1.6330	1.6806	1.7237	1.7633	1.8005	1.8353	1.8684

290	*V*	0.0189	1.5947	1.6122	1.6710	1.7273	1.7815	1.8338	1.8853	1.9590	2.0783	2.3080	2.5302	2.7478	2.9616	3.1738	3.3844
(414.24)	*H*	390.47	1202.9	1207.0	1221.4	1234.8	1248.0	1260.4	1271.4	1288.2	1315.0	1367.9	1420.1	1472.1	1524.6	1577.2	1630.6
	S	0.5841	1.5137	1.5184	1.5346	1.5493	1.5635	1.5766	1.5879	1.6048	1.6307	1.6784	1.7215	1.7612	1.7984	1.8332	1.8663
295	*V*	0.0189	1.5684	1.5809	1.6391	1.6948	1.7484	1.8001	1.8510	1.9236	2.0413	2.2677	2.4863	2.7004	2.9108	3.1195	3.3267
(415.80)	*H*	392.17	1203.0	1206.1	1220.5	1234.0	1247.4	1259.8	1271.0	1287.8	1314.7	1367.6	1419.9	1472.0	1524.5	1577.1	1630.5
	S	0.5861	1.5122	1.5157	1.5319	1.5467	1.5611	1.5742	1.5857	1.6026	1.6286	1.6763	1.7195	1.7593	1.7965	1.8313	1.8644
300	*V*	0.0189	1.5426	1.5506	1.6082	1.6634	1.7164	1.7677	1.8172	1.8896	2.0056	2.2286	2.4447	2.6547	2.8634	3.0670	3.2707
(417.33)	*H*	393.85	1203.2	1205.2	1219.5	1233.4	1246.6	1259.2	1270.5	1287.4	1314.4	1367.4	1419.7	1471.8	1524.4	1577.0	1630.4
	S	0.5879	1.5107	1.5130	1.5291	1.5443	1.5585	1.5718	1.5834	1.6004	1.6265	1.6742	1.7175	1.7572	1.7945	1.8294	1.8625
310	*V*	0.0189	1.4938	....	1.5495	1.6036	1.6555	1.7054	1.7546	1.8246	1.9375	2.1541	2.3631	2.5675	2.7682	2.9671	3.1645
(420.35)	*H*	397.16	1203.5	....	1217.8	1231.5	1245.3	1258.0	1269.6	1286.4	1313.5	1366.9	1419.3	1471.5	1524.1	1576.8	1630.3
	S	0.5917	1.5079	....	1.5240	1.5391	1.5539	1.5673	1.5793	1.5962	1.6224	1.6705	1.7138	1.7536	1.7909	1.8258	1.8590
320	*V*	0.0190	1.4479	....	1.4943	1.5473	1.5982	1.6472	1.6954	1.7637	1.8737	2.0844	2.2874	2.4857	2.6804	2.8735	3.0648
(423.29)	*H*	400.40	1203.8	....	1216.0	1229.9	1244.0	1256.8	1268.6	1285.6	1312.8	1366.3	1418.9	1471.2	1523.8	1576.6	1630.1
	S	0.5953	1.5052	....	1.5189	1.5342	1.5494	1.5629	1.5751	1.5922	1.6185	1.6667	1.7102	1.7501	1.7874	1.8224	1.8556
330	*V*	0.0190	1.4048	....	1.4424	1.4944	1.5445	1.5925	1.6397	1.7064	1.8138	2.0189	2.2163	2.4090	2.5981	2.7855	2.9712
(426.16)	*H*	403.56	1204.0	....	1214.1	1228.2	1242.5	1255.5	1267.6	1284.7	1312.1	1365.8	1418.4	1470.8	1523.6	1576.4	1630.0
	S	0.5988	1.5023	....	1.5136	1.5291	1.5445	1.5582	1.5707	1.5879	1.6144	1.6628	1.7063	1.7463	1.7837	1.8187	1.8520
340	*V*	0.0191	1.3640	....	1.3935	1.4446	1.4936	1.5409	1.5872	1.6525	1.7573	1.9572	2.1493	2.3368	2.5206	2.7027	2.8831
(428.96)	*H*	406.65	1204.2	....	1212.2	1226.5	1241.0	1254.2	1266.6	1283.8	1311.4	1365.2	1418.0	1470.5	1523.3	1576.2	1629.8
	S	0.6023	1.4997	....	1.5086	1.5243	1.5399	1.5538	1.5666	1.5839	1.6106	1.6591	1.7027	1.7428	1.7802	1.8152	1.8485
350	*V*	0.0191	1.3255	....	1.3472	1.3976	1.4460	1.4923	1.5377	1.6016	1.7041	1.8991	2.0863	2.2687	2.4475	2.6246	2.8000
(431.71)	*H*	409.70	1204.4	....	1210.3	1224.8	1239.5	1252.9	1265.5	1282.9	1310.6	1364.7	1417.6	1470.2	1523.0	1576.0	1629.6
	S	0.6057	1.4972	....	1.5038	1.5197	1.5355	1.5496	1.5626	1.5801	1.6069	1.6556	1.6993	1.7395	1.7769	1.8120	1.8453
360	*V*	0.0192	1.2889	....	1.3035	1.3532	1.4008	1.4463	1.4909	1.5536	1.6538	1.8441	2.0266	2.2044	2.3784	2.5506	2.7213
(434.39)	*H*	412.67	1204.5	....	1208.5	1223.1	1238.0	1251.5	1264.5	1282.0	1309.9	1364.1	1417.2	1469.9	1522.8	1575.8	1629.4
	S	0.6090	1.4946	....	1.4991	1.5151	1.5311	1.5453	1.5587	1.5763	1.6033	1.6521	1.6960	1.7362	1.7737	1.8088	1.8421

Table D-2 Superheated Steam (Continued)

Abs press., psia (sat. temp.)	V H S	Sat. water	Sat. steam	Temperature, °F													
				460	**480**	**500**	**520**	**540**	**560**	**580**	**600**	**700**	**800**	**900**	**1000**	**1100**	**1200**
370	V	0.0192	1.2545	1.3111	1.3579	1.4028	1.4466	1.4881	1.5286	1.5675	1.6063	1.7921	1.9703	2.1435	2.3131	2.4809	2.6471
(437.01)	H	415.58	1204.6	1221.4	1236.5	1250.2	1263.4	1275.2	1286.7	1298.3	1309.1	1363.6	1410.8	1469.6	1522.5	1575.6	1629.2
	S	0.6122	1.4921	1.5106	1.5268	1.5412	1.5548	1.5667	1.5781	1.5894	1.5997	1.6488	1.6928	1.7331	1.7706	1.8058	1.8391
380	V	0.0193	1.2217	1.2711	1.3173	1.3614	1.4045	1.4452	1.4850	1.5232	1.5612	1.7428	1.9168	2.0859	2.2512	2.4148	2.5768
(439.59)	H	418.45	1204.7	1219.8	1235.0	1248.8	1262.3	1274.2	1286.0	1297.5	1308.4	1363.0	1416.4	1469.2	1522.2	1575.4	1629.1
	S	0.6154	1.4897	1.5063	1.5226	1.5371	1.5510	1.5630	1.5747	1.5859	1.5963	1.6455	1.6896	1.7299	1.7675	1.8027	1.8361
390	V	0.0193	1.1904	1.2332	1.2788	1.3222	1.3647	1.4046	1.4436	1.4812	1.5184	1.6961	1.8661	2.0311	2.1925	2.3521	2.5101
(442.11)	H	421.27	1204.8	1218.0	1233.4	1247.4	1261.2	1273.2	1285.1	1296.7	1307.7	1362.5	1416.0	1468.9	1522.0	1575.2	1628.9
	S	0.6184	1.4872	1.5017	1.5183	1.5330	1.5472	1.5593	1.5711	1.5824	1.5929	1.6423	1.6865	1.7269	1.7646	1.7998	1.8332
400	V	0.0193	1.1609	1.1972	1.2422	1.2849	1.3269	1.3660	1.4042	1.4413	1.4777	1.6522	1.8179	1.9796	2.1367	2.2926	2.4475
(444.58)	H	424.02	1204.9	1216.5	1231.6	1245.9	1259.9	1272.4	1284.3	1295.8	1307.0	1362.1	1415.5	1468.6	1521.5	1574.8	1628.8
	S	0.6215	1.4850	1.4977	1.5140	1.5290	1.5434	1.5561	1.5678	1.5790	1.5897	1.6393	1.6835	1.7240	1.7615	1.7968	1.8304
410	V	0.0194	1.1327	1.1628	1.2071	1.2494	1.2906	1.3291	1.3669	1.4033	1.4390	1.6095	1.7722	1.9297	2.0837	2.2359	2.3864
(447.00)	H	426.74	1205.0	1214.6	1230.2	1244.5	1258.8	1271.2	1283.5	1295.1	1306.2	1361.4	1415.1	1468.3	1521.4	1574.8	1628.6
	S	0.6244	1.4828	1.4933	1.5101	1.5252	1.5399	1.5524	1.5646	1.5759	1.5865	1.6362	1.6806	1.7212	1.7589	1.7943	1.8277
420	V	0.0194	1.1058	1.1300	1.1738	1.2156	1.2561	1.2942	1.3312	1.3671	1.4021	1.5693	1.7285	1.8826	2.0332	2.1819	2.3290
(449.38)	H	429.42	1205.0	1213.0	1228.6	1243.1	1257.5	1270.2	1282.6	1294.3	1305.4	1360.8	1414.6	1468.0	1521.2	1574.6	1628.4
	S	0.6273	1.4805	1.4892	1.5060	1.5213	1.5361	1.5489	1.5612	1.5726	1.5832	1.6331	1.6776	1.7184	1.7561	1.7915	1.8249
430	V	0.0195	1.0800	1.0986	1.1419	1.1834	1.2233	1.2607	1.2972	1.3326	1.3670	1.5309	1.6869	1.8377	1.9850	2.1305	2.2742
(451.72)	H	432.05	1205.0	1211.2	1227.0	1241.7	1256.3	1269.1	1281.8	1293.5	1304.6	1360.3	1414.2	1467.6	1520.9	1574.4	1628.2
	S	0.6302	1.4782	1.4850	1.5020	1.5175	1.5326	1.5455	1.5581	1.5695	1.5801	1.6303	1.6748	1.7156	1.7534	1.7888	1.8222
440	V	0.0195	1.0554	1.0688	1.1116	1.1524	1.1918	1.2288	1.2648	1.2996	1.3334	1.4943	1.6472	1.7949	1.9390	2.0814	2.2220
(454.01)	H	434.63	1205.0	1209.6	1225.3	1240.2	1255.0	1268.0	1280.9	1292.6	1303.9	1359.7	1413.8	1467.3	1520.6	1574.1	1628.0
	S	0.6330	1.4762	1.4812	1.4981	1.5138	1.5291	1.5422	1.5550	1.5664	1.5772	1.6275	1.6722	1.7130	1.7508	1.7862	1.8197

450	V	0.0195	1.0318	1.0401	1.0824	1.1230	1.1617	1.1982	1.2337	1.2681	1.3013	1.4593	1.6092	1.7539	1.8951	2.0345	2.1720
(456.27)	H	437.18	1205.0	1207.9	1223.7	1238.7	1253.8	1266.9	1280.0	1291.8	1303.1	1359.1	1413.4	1467.0	1520.3	1573.9	1627.8
	S	0.6357	1.4739	1.4771	1.4941	1.5099	1.5255	1.5387	1.5517	1.5632	1.5740	1.6245	1.6694	1.7103	1.7481	1.7836	1.8171
460	V	0.0196	1.0092	....	1.0545	1.0946	1.1329	1.1690	1.2039	1.2379	1.2706	1.4258	1.5729	1.7147	1.8530	1.9896	2.1243
(458.48)	H	439.69	1205.0	....	1222.0	1237.2	1252.5	1265.8	1279.0	1291.0	1302.3	1358.6	1413.0	1466.6	1520.0	1573.7	1627.7
	S	0.6384	1.4719	....	1.4902	1.5062	1.5220	1.5354	1.5485	1.5602	1.5710	1.6217	1.6667	1.7076	1.7455	1.7811	1.8146
470	V	0.0196	0.9875	....	1.0278	1.0676	1.1053	1.1410	1.1755	1.2091	1.2412	1.3937	1.5381	1.6772	1.8127	1.9466	2.0785
(460.66)	H	442.17	1205.0	....	1220.2	1235.7	1251.2	1264.7	1278.0	1290.0	1301.5	1358.0	1412.5	1466.3	1519.8	1573.5	1627.5
	S	0.6411	1.4699	....	1.4862	1.5025	1.5185	1.5321	1.5453	1.5570	1.5680	1.6189	1.6639	1.7050	1.7429	1.7785	1.8120
480	V	0.0197	0.9668	....	1.0021	1.0416	1.0789	1.1141	1.1482	1.1813	1.2131	1.3630	1.5049	1.6413	1.7742	1.9054	2.0347
(462.80)	H	444.60	1205.0	....	1218.6	1234.2	1249.9	1263.5	1277.0	1289.1	1300.8	1357.5	1412.1	1466.0	1519.5	1573.3	1627.3
	S	0.6436	1.4679	....	1.4825	1.4989	1.5151	1.5288	1.5422	1.5539	1.5650	1.616.	1.6612	1.7023	1.7402	1.7758	1.8093
490	V	0.0197	.9466	....	0.9774	1.0166	1.0535	1.0884	1.1220	1.1548	1.1860	1.3335	1.4729	1.6067	1.7371	1.8659	1.9927
(464.91)	H	447.00	1204.9	....	1217.0	1232.7	1248.4	1262.3	1276.0	1288.3	1300.0	1356.9	1411.7	1465.6	1519.2	1573.1	1627.1
	S	0.6462	1.4659	....	1.4789	1.4954	1.5116	1.5256	1.5392	1.5511	1.5622	1.6135	1.6588	1.6999	1.7379	1.7736	1.8071
500	V	0.0197	0.9274	....	0.9538	0.9926	1.0290	1.0636	1.0969	1.1292	1.1600	1.3051	1.4417	1.5735	1.7016	1.8280	1.9532
(467.00)	H	449.40	1204.9	....	1215.3	1231.4	1246.6	1261.1	1275.0	1287.3	1299.3	1356.3	1411.2	1465.1	1518.8	1572.9	1627.3
	S	0.6488	1.4641	....	1.4752	1.4922	1.5079	1.5225	1.5363	1.5482	1.5596	1.6110	1.6564	1.6975	1.7356	1.7714	1.8052
510	V	0.0198	0.9090	....	0.9310	0.9695	1.0056	1.0397	1.0727	1.1046	1.1350	1.2780	1.4127	1.5418	1.6675	1.7915	1.9135
(469.05)	H	451.75	1204.8	....	1213.5	1229.6	1245.6	1259.9	1274.0	1286.6	1298.4	1355.7	1410.9	1465.0	1518.7	1572.6	1626.8
	S	0.6513	1.4621	....	1.4714	1.4883	1.5048	1.5192	1.5332	1.5454	1.5566	1.6082	1.6538	1.6951	1.7332	1.7689	1.8026
520	V	0.0198	0.8912	....	0.9091	0.9472	0.9829	1.0169	1.0494	1.0810	1.1110	1.2519	1.3844	1.5113	1.6347	1.7565	1.8763
(471.07)	H	454.07	1204.7	....	1211.8	1228.1	1244.2	1258.6	1272.9	1285.6	1297.6	1355.1	1410.4	1464.6	1518.4	1572.4	1626.6
	S	0.6537	1.4601	....	1.4677	1.4849	1.5015	1.5160	1.5302	1.5425	1.5539	1.6057	1.6514	1.6928	1.7310	1.7668	1.8005
530	V	0.0199	0.8741	....	0.8879	0.9258	0.9612	0.9948	1.0269	1.0582	1.0878	1.2267	1.3571	1.4818	1.6031	1.7228	1.8402
(473.05)	H	456.35	1204.6	....	1210.0	1226.5	1242.8	1257.3	1271.8	1284.8	1296.8	1354.6	1410.0	1464.3	1518.1	1572.2	1626.4
	S	0.6562	1.4584	....	1.4642	1.4816	1.4984	1.5130	1.5274	1.5400	1.5514	1.6035	1.6493	1.6908	1.7290	1.7648	1.7985

Table D-2 Superheated Steam (Continued)

Abs press., psia (sat. temp.)		Sat. water	Sat. steam	Temperature, °F													
				500	520	540	560	580	600	650	700	750	800	900	1000	1100	1200
540	V	0.0199	0.8576	0.9051	0.9401	0.9736	1.0054	1.0363	1.0655	1.1356	1.2025	1.2671	1.3309	1.4535	1.5727	1.6903	1.8056
(475.02)	H	458.62	1204.5	1225.0	1241.4	1256.1	1270.7	1283.8	1296.0	1325.6	1354.0	1382.1	1409.6	1463.9	1517.8	1572.0	1626.2
	S	0.6585	1.4565	1.4781	1.4950	1.5098	1.5243	1.5370	1.5486	1.5759	1.6009	1.6246	1.6469	1.6884	1.7266	1.7625	1.7962
550	V	0.0199	0.8416	0.8851	0.9198	0.9530	0.9846	1.0151	1.0441	1.1132	1.1791	1.2428	1.3055	1.4262	1.5434	1.6590	1.7724
(476.94)	H	460.83	1204.4	1223.4	1240.0	1254.8	1269.6	1282.9	1295.2	1324.9	1353.5	1381.6	1409.2	1463.6	1517.5	1571.7	1626.0
	S	0.6609	1.4548	1.4748	1.4919	1.5068	1.5215	1.5344	1.5461	1.5735	1.5987	1.6224	1.6447	1.6862	1.7244	1.7603	1.7940
560	V	0.0200	0.8263	0.8658	0.9003	0.9332	0.9644	0.9947	1.0233	1.0917	1.1566	1.2193	1.2810	1.3998	1.5151	1.6289	1.7403
(478.85)	H	463.04	1204.3	1221.8	1238.5	1253.5	1268.5	1282.0	1294.4	1324.2	1352.9	1381.1	1408.7	1463.2	1517.2	1571.5	1625.8
	S	0.6632	1.4530	1.4714	1.4886	1.5038	1.5187	1.5318	1.5436	1.5711	1.5964	1.6202	1.6425	1.6841	1.7224	1.7584	1.7921
570	V	0.0200	0.8114	0.8472	0.8814	0.9141	0.9450	0.9749	1.0033	1.0708	1.1348	1.1966	1.2575	1.3744	1.4879	1.5998	1.7093
(480.73)	H	465.22	1204.1	1220.2	1236.9	1252.2	1267.3	1281.0	1293.5	1323.5	1352.3	1380.6	1408.3	1462.9	1517.0	1571.3	1625.6
	S	0.6655	1.4512	1.4681	1.4853	1.5008	1.5156	1.5291	1.5410	1.5686	1.5940	1.6179	1.6403	1.6820	1.7204	1.7564	1.7901
580	V	0.0201	0.7968	0.8291	0.8631	0.8956	0.9263	0.9558	0.9839	1.0506	1.1137	1.1747	1.2347	1.3498	1.4616	1.5714	1.6794
(482.58)	H	467.37	1204.0	1218.6	1235.5	1250.9	1266.1	1280.0	1292.6	1322.8	1351.6	1380.0	1407.8	1462.5	1516.7	1571.0	1625.4
	S	0.6677	1.4494	1.4648	1.4822	1.4978	1.5128	1.5264	1.5384	1.5662	1.5916	1.6156	1.6381	1.6799	1.7183	1.7543	1.7881
590	V	0.0201	0.7831	0.8116	0.8455	0.8778	0.9082	0.9373	0.9653	1.0310	1.0934	1.1535	1.2128	1.3262	1.4360	1.5442	1.6505
(484.41)	H	469.50	1203.8	1217.0	1234.0	1249.6	1265.0	1278.9	1291.8	1322.1	1351.1	1379.5	1407.4	1462.2	1516.4	1570.8	1625.3
	S	0.6699	1.4477	1.4616	1.4791	1.4949	1.5101	1.5236	1.5359	1.5638	1.5894	1.6134	1.6360	1.6778	1.7162	1.7522	1.7861
600	V	0.0201	0.7695	0.7945	0.8284	0.8605	0.8907	0.9194	0.9471	1.0123	1.0738	1.1332	1.1915	1.3032	1.4115	1.5179	1.6224
(486.21)	H	471.59	1203.6	1215.6	1232.5	1248.3	1263.7	1278.1	1290.9	1321.4	1350.5	1379.0	1407.0	1461.8	1516.0	1570.5	1625.0
	S	0.6721	1.4460	1.4586	1.4760	1.4920	1.5072	1.5212	1.5334	1.5615	1.5871	1.6112	1.6339	1.6757	1.7141	1.7502	1.7841
610	V	0.0202	0.7565	0.7781	0.8120	0.8436	0.8736	0.9022	0.9296	0.9942	1.0548	1.1135	1.1708	1.2809	1.3878	1.4928	1.5964
(487.99)	H	473.67	1203.5	1213.8	1230.9	1246.9	1262.5	1276.8	1290.0	1320.6	1350.0	1378.5	1406.5	1461.5	1515.8	1570.3	1624.9
	S	0.6743	1.4444	1.4552	1.4728	1.4890	1.5044	1.5183	1.5309	1.5591	1.5850	1.6091	1.6318	1.6738	1.7123	1.7484	1.7823

620	V	0.0202	0.7438	0.7622	0.7960	0.8275	0.8572	0.8856	0.9127	0.9765	1.0364	1.0943	1.1505	1.2596	1.3648	1.4677	1.5707
(489.75)	H	475.72	1203.3	1212.2	1229.5	1245.5	1261.3	1275.8	1289.1	1319.9	1349.3	1377.9	1406.1	1461.2	1515.5	1570.1	1624.7
	S	0.6764	1.4427	1.4520	1.4698	1.4860	1.5016	1.5157	1.5284	1.5568	1.5827	1.6068	1.6296	1.6717	1.7102	1.7464	1.7803
630	V	0.0202	0.7316	0.7466	0.7802	0.8117	0.8413	0.8694	0.8963	0.9595	1.0187	1.0757	1.1312	1.2387	1.3423	1.4445	1.5449
(491.49)	H	477.75	1203.1	1210.6	1227.8	1244.1	1260.1	1274.7	1288.3	1319.2	1348.7	1377.4	1405.7	1460.8	1515.2	1569.9	1624.5
	S	0.6785	1.4410	1.4488	1.4665	1.4830	1.4988	1.5130	1.5260	1.5545	1.5805	1.6047	1.6276	1.6697	1.7083	1.7445	1.7784
640	V	0.0203	0.7197	0.7317	0.7651	0.7963	0.8258	0.8537	0.8804	0.9429	1.0015	1.0578	1.1124	1.2187	1.3210	1.4213	1.5193
(493.21)	H	479.79	1202.9	1209.0	1226.3	1242.7	1258.9	1273.6	1287.4	1318.5	1348.2	1376.8	1405.2	1460.5	1515.0	1569.7	1624.3
	S	0.6806	1.4394	1.4458	1.4636	1.4802	1.4962	1.5105	1.5236	1.5523	1.5785	1.6026	1.6256	1.6678	1.7065	1.7427	1.7766
650	V	0.0203	0.7082	0.7171	0.7504	0.7816	0.8107	0.8384	0.8648	0.9269	0.9846	1.0404	1.0944	1.1988	1.2999	1.3987	1.4958
(494.90)	H	481.73	1202.7	1207.3	1224.8	1241.3	1257.6	1272.5	1286.5	1317.8	1347.6	1376.3	1404.7	1460.1	1514.7	1569.4	1624.1
	S	0.6826	1.4379	1.4427	1.4607	1.4774	1.4935	1.5080	1.5213	1.5501	1.5764	1.6006	1.6236	1.6659	1.7046	1.7408	1.7748
660	V	0.0204	0.6969	0.7031	0.7361	0.7672	0.7962	0.8237	0.8499	0.9113	0.9686	1.0234	1.0769	1.1803	1.2797	1.3774	1.4727
(496.58)	H	483.77	1202.5	1205.7	1223.2	1240.0	1256.4	1271.4	1285.5	1317.1	1347.0	1375.8	1404.3	1459.7	1514.4	1569.2	1624.0
	S	0.6847	1.4363	1.4396	1.4576	1.4746	1.4908	1.5054	1.5188	1.5479	1.5742	1.5985	1.6216	1.6639	1.7027	1.7390	1.7730
670	V	0.0204	0.6861	0.6892	0.7224	0.7531	0.7820	0.8093	0.8354	0.8963	0.9527	1.0072	1.0599	1.1617	1.2600	1.3560	1.4503
(498.23)	H	485.61	1202.3	1204.0	1221.7	1238.7	1255.1	1270.2	1284.5	1316.3	1346.3	1375.3	1403.9	1459.4	1514.1	1569.0	1623.8
	S	0.6867	1.4349	1.4367	1.4549	1.4721	1.4883	1.5030	1.5166	1.5459	1.5723	1.5968	1.6200	1.6624	1.7012	1.7376	1.7716
680	V	0.0204	0.6757	....	0.7089	0.7397	0.7683	0.7954	0.8212	0.8814	0.9375	0.9912	1.0432	1.1440	1.2408	1.3357	1.4283
(499.87)	H	487.64	1202.1	....	1220.2	1237.3	1253.9	1269.1	1283.6	1315.6	1345.8	1374.7	1403.4	1459.0	1513.8	1568.7	1623.6
	S	0.6886	1.4332	....	1.4519	1.4692	1.4856	1.5004	1.5142	1.5437	1.5703	1.5947	1.6179	1.6603	1.6992	1.7356	1.7697
690	V	0.0205	0.6652	....	0.6956	0.7263	0.7549	0.7818	0.8075	0.8675	0.9225	0.9758	1.0272	1.1267	1.2223	1.3162	1.4075
(501.49)	H	489.56	1201.8	....	1218.5	1235.8	1252.5	1268.0	1282.7	1314.9	1345.1	1374.2	1402.8	1458.7	1513.6	1568.5	1623.4
	S	0.6906	1.4316	....	1.4488	1.4663	1.4828	1.4978	1.5118	1.5415	1.5681	1.5927	1.6159	1.6586	1.6975	1.7339	1.7680
700	V	0.0205	0.6552	....	0.6830	0.7133	0.7419	0.7687	0.7941	0.8534	0.9084	0.9608	1.0117	1.1096	1.2043	1.2965	1.3870
(503.09)	H	491.49	1201.6	....	1217.1	1234.7	1251.3	1266.8	1281.9	1314.3	1344.6	1373.7	1402.5	1458.2	1513.4	1568.2	1623.3
	S	0.6925	1.4301	....	1.4461	1.4638	1.4803	1.4953	1.5097	1.5396	1.5663	1.5908	1.6141	1.6567	1.6958	1.7321	1.7663

Table D-2 Superheated Steam (Continued)

Abs press., psia (sat. temp.)	V H S	Sat. water	Sat. steam	Temperature, °F 520	540	560	580	600	620	650	700	750	800	900	1000	1100	1200
725	V	0.0206	0.6314	0.6524	0.6827	0.7109	0.7373	0.7624	0.7864	0.8203	0.8740	0.9250	0.9745	1.0697	1.1612	1.2511	1.3383
(507.01)	H	496.2	1200.9	1212.8	1230.9	1248.0	1264.0	1279.1	1293.8	1312.3	1343.0	1372.3	1401.3	1457.4	1512.5	1567.7	1622.7
	S	0.6973	1.4263	1.4385	1.4568	1.4737	1.4892	1.5036	1.5173	1.5342	1.5612	1.5859	1.6094	1.6522	1.6913	1.7279	1.7621
750	V	0.0207	0.6091	0.6237	0.6538	0.6818	0.7080	0.7326	0.7561	0.7896	0.8419	0.8918	0.9399	1.0326	1.1212	1.2078	1.2928
(510.83)	H	500.8	1200.2	1208.8	1227.4	1244.9	1261.0	1276.6	1291.4	1310.5	1341.5	1371.0	1400.2	1456.5	1511.8	1567.1	1622.3
	S	0.7019	1.4225	1.4313	1.4501	1.4674	1.4830	1.4979	1.5117	1.5291	1.5564	1.5813	1.6049	1.6479	1.6871	1.7237	1.7580
775	V	0.0208	0.5882	0.5969	0.6268	0.6545	0.6803	0.7047	0.7278	0.7607	0.8119	0.8606	0.9073	0.9977	1.0838	1.1676	1.2505
(514.57)	H	505.3	1199.5	1204.7	1223.7	1241.5	1258.2	1271.9	1289.0	1308.6	1340.0	1369.7	1399.0	1455.6	1511.1	1566.6	1621.8
	S	0.7064	1.4189	1.4242	1.4434	1.4610	1.4772	1.4902	1.5062	1.5241	1.5518	1.5769	1.6006	1.6438	1.6832	1.7200	1.7543
800	V	0.0209	0.5685	0.5714	0.6013	0.6288	0.6545	0.6785	0.7013	0.7336	0.7838	0.8313	0.8770	0.9648	1.0486	1.1302	1.2105
(518.20)	H	509.7	1198.8	1200.3	1220.0	1238.2	1255.3	1271.4	1286.5	1306.8	1338.4	1368.5	1397.8	1454.9	1510.5	1566.0	1621.4
	S	0.7108	1.4155	1.4170	1.4369	1.4549	1.4715	1.4868	1.5009	1.5195	1.5473	1.5727	1.5964	1.6400	1.6794	1.7162	1.7506
825	V	0.0210	0.5500	....	0.5774	0.6046	0.6300	0.6539	0.6763	0.7081	0.7573	0.8038	0.8483	0.9338	1.0155	1.0950	1.1727
(521.75)	H	514.0	1198.0	....	1216.1	1234.9	1252.1	1268.5	1284.0	1304.8	1336.8	1367.0	1396.6	1453.8	1509.7	1565.4	1620.8
	S	0.7152	1.4121	....	1.4304	1.4490	1.4657	1.4813	1.4958	1.5148	1.5430	1.5685	1.5925	1.6362	1.6758	1.7127	1.7471
850	V	0.0210	0.5326	....	0.5545	0.5817	0.6070	0.6306	0.6528	0.6841	0.7323	0.7779	0.8213	0.9048	0.9845	1.0619	1.1375
(525.23)	H	518.3	1197.2	....	1212.4	1231.5	1249.1	1265.7	1281.4	1302.9	1335.2	1365.7	1395.4	1452.8	1509.0	1564.8	1620.4
	S	0.7194	1.4087	....	1.4240	1.4429	1.4600	1.4758	1.4905	1.5101	1.5386	1.5643	1.5883	1.6321	1.6720	1.7090	1.7435
875	V	0.0211	0.5162	....	0.5327	0.5601	0.5851	0.6085	0.6305	0.6615	0.7087	0.7535	0.7960	0.8773	0.9554	1.0306	1.1045
(528.62)	H	522.4	1196.4	....	1208.4	1228.0	1246.0	1263.0	1279.0	1300.9	1333.5	1364.4	1394.2	1451.9	1508.3	1564.3	1619.9
	S	0.7236	1.4056	....	1.4177	1.4371	1.4546	1.4708	1.4858	1.5058	1.5345	1.5606	1.5847	1.6287	1.6687	1.7058	1.7403
900	V	0.0212	0.5006	....	0.5123	0.5394	0.5644	0.5876	0.6094	0.6399	0.6866	0.7304	0.7720	0.8516	0.9277	1.0010	1.0727
(531.94)	H	526.6	1195.6	....	1204.0	1224.2	1242.6	1260.0	1276.5	1298.6	1331.8	1363.0	1392.9	1451.1	1507.8	1563.7	1619.3
	S	0.7276	1.4022	....	1.4106	1.4306	1.4484	1.4649	1.4803	1.5004	1.5296	1.5559	1.5801	1.6245	1.6647	1.7017	1.7362

925	V	0.0213	0.4858	....	0.4927	0.5199	0.5448	0.5678	0.5894	0.6196	0.6655	0.7085	0.7494	0.8272	0.9014	0.9731	1.0432
(535.20)	H	530.6	1194.7	....	1200.0	1220.8	1239.6	1257.0	1273.9	1296.7	1330.2	1361.7	1391.7	1450.0	1506.9	1563.1	1618.9
	S	0.7316	1.3991	....	1.4044	1.4250	1.4433	1.4599	1.4757	1.4965	1.5260	1.5526	1.5769	1.6214	1.6618	1.6990	1.7337
950	V	0.0214	0.4717	....	0.4741	0.5014	0.5262	0.5491	0.5705	0.6003	0.6456	0.6877	0.7277	0.8039	0.8766	0.9465	1.0148
(538.38)	H	534.6	1193.8	....	1195.8	1217.0	1236.4	1254.1	1271.3	1294.5	1328.5	1360.3	1390.5	1449.1	1506.1	1562.6	1618.5
	S	0.7355	1.3960	....	1.3980	1.4190	1.4378	1.4547	1.4708	1.4920	1.5220	1.5488	1.5733	1.6180	1.6584	1.6958	1.7305
975	V	0.0215	0.4583	....	....	0.4835	0.5083	0.5311	0.5524	0.5820	0.6266	0.6680	0.7073	0.7820	0.8533	0.9214	0.9880
(541.50)	H	538.5	1192.9	....	....	1213.0	1233.0	1251.2	1268.7	1292.1	1326.7	1358.8	1389.2	1448.2	1505.4	1562.0	1618.0
	S	0.7393	1.3929	....	....	1.4128	1.4322	1.4495	1.4659	1.4873	1.5178	1.5449	1.5695	1.6145	1.6551	1.6926	1.7274
1000	V	0.0216	0.4456	....	....	0.4665	0.4914	0.5141	0.5351	0.5639	0.6085	0.6492	0.6879	0.7611	0.8306	0.8974	0.9626
(544.56)	H	542.4	1191.9	....	....	1208.8	1229.4	1248.2	1265.8	1289.6	1324.9	1357.2	1388.0	1447.3	1504.7	1561.3	1617.5
	S	0.7431	1.3899	....	....	1.4066	1.4266	1.4445	1.4610	1.4827	1.5138	1.5411	1.5660	1.6113	1.6520	1.6895	1.7244
1025	V	0.0217	0.4334	....	....	0.4498	0.4751	0.4978	0.5188	0.5479	0.5913	0.6314	0.6695	0.7413	0.8095	0.8746	0.9384
(547.57)	H	546.1	1191.0	....	....	1204.5	1225.9	1245.2	1263.0	1287.6	1323.1	1355.7	1386.8	1446.4	1504.1	1560.6	1617.0
	S	0.7468	1.3871	....	....	1.4004	1.4212	1.4396	1.4562	1.4787	1.5100	1.5375	1.5627	1.6082	1.6491	1.6866	1.7216
1050	V	0.0218	0.4219	....	....	0.4345	0.4596	0.4822	0.5031	0.5320	0.5749	0.6143	0.6519	0.7223	0.7892	0.8531	0.9154
(550.52)	H	550.0	1190.0	....	....	1200.5	1222.4	1241.9	1260.2	1285.1	1321.3	1354.2	1385.6	1445.4	1503.4	1560.0	1616.5
	S	0.7504	1.3839	....	....	1.3942	1.4155	1.4341	1.4512	1.4739	1.5058	1.5336	1.5590	1.6047	1.6458	1.6833	1.7184
1075	V	0.0219	0.4108	....	....	0.4195	0.4446	0.4672	0.4879	0.5169	0.5592	0.5980	0.6349	0.7042	0.7696	0.8322	0.8933
(553.42)	H	553.7	1188.9	....	....	1196.3	1218.7	1238.8	1257.2	1282.8	1319.4	1352.6	1384.1	1444.3	1502.5	1559.3	1615.9
	S	0.7540	1.3810	....	....	1.3883	1.4100	1.4292	1.4464	1.4698	1.5020	1.5300	1.5555	1.6015	1.6428	1.6804	1.7156
1100	V	0.0219	0.4002	....	....	0.4054	0.4304	0.4530	0.4736	0.5027	0.5445	0.5828	0.6190	0.6871	0.7511	0.8125	0.8724
(556.26)	H	557.4	1187.8	....	....	1192.3	1214.9	1235.6	1254.5	1280.6	1317.8	1351.3	1383.0	1443.6	1501.7	1558.7	1615.4
	S	0.7575	1.3780	....	....	1.3824	1.4044	1.4241	1.4418	1.4656	1.4984	1.5267	1.5523	1.5986	1.6398	1.6776	1.7128
1125	V	0.0220	0.3902	....	....	....	0.4167	0.4392	0.4598	0.4883	0.5301	0.5678	0.6036	0.6706	0.7333	0.7938	0.8524
(559.07)	H	561.0	1186.7	....	....	....	1211.3	1232.2	1251.5	1277.7	1315.9	1349.6	1381.7	1442.6	1500.8	1558.2	1614.9
	S	0.7610	1.3752	....	....	....	1.3991	1.4190	1.4370	1.4610	1.4946	1.5231	1.5491	1.5956	1.6369	1.6749	1.7101

Table D-2 Superheated Steam (Continued)

Abs press., psia (sat. temp.)	V H S	Sat. water	Sat. steam	Temperature, °F 580	600	620	640	660	680	700	720	750	800	900	1000	1100	1200
1150	V	0.0221	0.3804	0.4035	0.4259	0.4468	0.4659	0.4839	0.5005	0.5165	0.5317	0.5537	0.5889	0.6549	0.7166	0.7760	0.8333
(561.81)	H	564.6	1185.6	1207.4	1228.7	1248.6	1266.9	1284.4	1299.5	1314.1	1327.9	1348.1	1380.4	1441.7	1500.2	1557.8	1614.5
	S	0.7644	1.3723	1.3934	1.4137	1.4323	1.4491	1.4649	1.4783	1.4910	1.5028	1.5197	1.5458	1.5926	1.6341	1.6723	1.7075
1175	V	0.0222	0.3710	0.3911	0.4133	0.4339	0.4530	0.4708	0.4874	0.5032	0.5183	0.5400	0.5747	0.6396	0.7005	0.7586	0.8149
(564.54)	H	568.2	1184.4	1203.6	1225.4	1245.4	1264.6	1281.5	1297.3	1312.1	1326.1	1346.4	1379.0	1440.6	1499.4	1557.0	1614.0
	S	0.7678	1.3694	1.3880	1.4088	1.4275	1.4451	1.4603	1.4743	1.4872	1.4991	1.5161	1.5425	1.5896	1.6313	1.6694	1.7048
1200	V	0.0223	0.3620	0.3793	0.4013	0.4219	0.4408	0.4585	0.4750	0.4907	0.5056	0.5271	0.5613	0.6251	0.6853	0.7423	0.7975
(567.19)	H	571.7	1183.2	1200.2	1222.1	1242.6	1261.5	1279.2	1295.3	1310.3	1324.4	1345.0	1377.7	1439.5	1499.0	1556.6	1613.6
	S	0.7712	1.3667	1.3831	1.4040	1.4232	1.4405	1.4565	1.4707	1.4838	1.4958	1.5131	1.5395	1.5867	1.6289	1.6671	1.7025
1225	V	0.0224	0.3534	0.3669	0.3895	0.4102	0.4290	0.4466	0.4631	0.4786	0.4934	0.5146	0.5483	0.6113	0.6702	0.7264	0.7806
(569.82)	H	575.1	1182.0	1195.6	1218.6	1239.7	1258.8	1276.7	1293.3	1308.3	1322.7	1343.3	1376.4	1438.6	1498.0	1556.0	1613.0
	S	0.7745	1.3640	1.3771	1.3990	1.4188	1.4363	1.4524	1.4671	1.4802	1.4925	1.5097	1.5365	1.5840	1.6262	1.6646	1.7000
1250	V	0.0225	0.3453	0.3549	0.3782	0.3991	0.4177	0.4354	0.4517	0.4672	0.4817	0.5027	0.5360	0.5980	0.6558	0.7113	0.7644
(572.39)	H	578.6	1180.8	1190.9	1215.0	1236.7	1256.0	1274.5	1291.2	1306.6	1321.0	1341.9	1375.2	1437.7	1497.1	1555.4	1612.6
	S	0.7777	1.3612	1.3710	1.3939	1.4142	1.4319	1.4486	1.4634	1.4768	1.4891	1.5066	1.5335	1.5813	1.6235	1.6620	1.6975
1275	V	0.0226	0.3371	0.3437	0.3672	0.3881	0.4068	0.4244	0.4406	0.4560	0.4705	0.4912	0.5241	0.5852	0.6420	0.6966	0.7488
(574.93)	H	582.0	1179.5	1186.5	1211.2	1233.5	1253.3	1272.0	1289.0	1304.6	1319.4	1340.3	1373.9	1436.7	1496.3	1554.7	1612.0
	S	0.7809	1.3584	1.3651	1.3887	1.4095	1.4277	1.4445	1.4596	1.4732	1.4858	1.5033	1.5305	1.5785	1.6207	1.6594	1.6950
1300	V	0.0227	0.3294	0.3329	0.3567	0.3776	0.3965	0.4140	0.4301	0.4453	0.4598	0.4803	0.5127	0.5730	0.6290	0.6826	0.7340
(577.43)	H	585.4	1178.3	1182.0	1207.7	1230.3	1250.6	1269.6	1287.0	1302.8	1317.8	1338.9	1372.6	1435.8	1495.7	1554.2	1611.6
	S	0.7840	1.3557	1.3593	1.3837	1.4049	1.4235	1.4406	1.4560	1.4698	1.4826	1.5002	1.5275	1.5758	1.6183	1.6570	1.6926
1325	V	0.0228	0.3220	....	0.3463	0.3673	0.3863	0.4037	0.4200	0.4350	0.4493	0.4696	0.5016	0.5611	0.6162	0.6689	0.7195
(579.89)	H	588.7	1177.0	....	1203.8	1227.0	1247.8	1267.0	1285.0	1300.8	1315.9	1337.3	1371.2	1434.8	1494.9	1553.4	1611.0
	S	0.7871	1.3530	....	1.3785	1.4002	1.4193	1.4366	1.4525	1.4663	1.4792	1.4971	1.5246	1.5731	1.6158	1.6545	1.6903

1350	*V*	0.0229	0.3147	....	0.3363	0.3576	0.3766	0.3940	0.4105	0.4252	0.4393	0.4594	0.4911	0.5497	0.6042	0.6559	0.7057
(582.32)	*H*	592.1	1175.8	....	1200.0	1223.8	1245.0	1264.6	1283.2	1299.1	1314.2	1335.8	1370.0	1433.8	1494.3	1552.8	1610.6
	S	0.7902	1.3504	....	1.3734	1.3957	1.4151	1.4328	1.4492	1.4631	1.4760	1.4941	1.5218	1.5705	1.6134	1.6521	1.6881
1375	*V*	0.0230	0.3078	....	0.3266	0.3480	0.3670	0.3847	0.4007	0.4154	0.4295	0.4494	0.4808	0.5387	0.5922	0.6432	0.6922
(584.71)	*H*	595.3	1174.5	....	1195.8	1220.2	1242.0	1262.1	1280.7	1296.8	1312.3	1334.1	1368.6	1432.8	1493.2	1552.0	1609.9
	S	0.7932	1.3477	....	1.3680	1.3908	1.4108	1.4289	1.4453	1.4593	1.4726	1.4908	1.5188	1.5678	1.6107	1.6496	1.6856
1400	*V*	0.0231	0.3011	....	0.3172	0.3388	0.3581	0.3760	0.3914	0.4063	0.4203	0.4401	0.4711	0.5283	0.5811	0.6313	0.6795
(587.07)	*H*	598.6	1173.2	....	1191.8	1216.9	1239.2	1260.1	1278.2	1294.9	1310.6	1332.8	1367.4	1431.9	1492.7	1551.7	1609.6
	S	0.7963	1.3452	....	1.3629	1.3863	1.4068	1.4256	1.4416	1.4562	1.4696	1.4882	1.5162	1.5654	1.6086	1.6476	1.6836
1425	*V*	0.0232	0.2947	....	0.3081	0.3297	0.3491	0.3668	0.3825	0.3972	0.4112	0.4308	0.4616	0.5180	0.5701	0.6195	0.6671
(589.40)	*H*	601.8	1171.8	....	1187.7	1213.2	1236.2	1257.2	1275.8	1292.9	1308.9	1331.1	1366.1	1431.0	1491.8	1551.0	1609.1
	S	0.7992	1.3425	....	1.3576	1.3814	1.4025	1.4215	1.4379	1.4528	1.4665	1.4850	1.5134	1.5629	1.6061	1.6453	1.6814
1450	*V*	0.0233	0.2885	....	0.2991	0.3211	0.3405	0.3580	0.3739	0.3885	0.4025	0.4220	0.4524	0.5082	0.5597	0.6083	0.6552
(591.70)	*H*	605.0	1170.5	....	1183.2	1209.7	1233.1	1254.3	1273.5	1290.6	1307.0	1329.7	1364.8	1430.1	1491.3	1550.4	1608.6
	S	0.8022	1.3401	....	1.3521	1.3769	1.3984	1.4175	1.4345	1.4493	1.4634	1.4824	1.5108	1.5606	1.6041	1.6432	1.6794
1475	*V*	0.0234	0.2824	....	0.2903	0.3126	0.3318	0.3495	0.3654	0.3801	0.3939	0.4129	0.4435	0.4986	0.5493	0.5973	0.6439
(593.97)	*H*	608.2	1169.1	....	1178.7	1206.2	1229.7	1251.5	1271.0	1288.8	1305.0	1327.4	1363.5	1429.1	1490.3	1549.8	1608.4
	S	0.8052	1.3375	....	1.3466	1.3723	1.3939	1.4135	1.4308	1.4463	1.4601	1.4789	1.5081	1.5582	1.6016	1.6410	1.6774
1500	*V*	0.0235	0.2765	....	0.2817	0.3044	0.3236	0.3413	0.3573	0.3721	0.3856	0.4042	0.4349	0.4894	0.5396	0.5869	0.6332
(596.20)	*H*	611.4	1167.7	....	1174.2	1202.5	1226.4	1248.5	1268.6	1286.8	1303.0	1325.4	1362.1	1428.1	1489.8	1549.3	1608.4
	S	0.8081	1.3350	....	1.3411	1.3676	1.3895	1.4094	1.4272	1.4431	1.4569	1.4757	1.5054	1.5558	1.5995	1.6390	1.6757
1525	*V*	0.0236	0.2708	....	....	0.2962	0.3158	0.3335	0.3495	0.3642	0.3777	0.3965	0.4265	0.4804	0.5299	0.5766	0.6218
(598.41)	*H*	614.5	1166.2	....	....	1198.6	1223.5	1245.8	1266.2	1284.7	1301.2	1324.1	1360.7	1427.0	1488.9	1548.6	1607.4
	S	0.8109	1.3323	....	....	1.3626	1.3855	1.4055	1.4236	1.4397	1.4538	1.4730	1.5026	1.5532	1.5971	1.6367	1.6732
1550	*V*	0.0237	0.2653	....	....	0.2883	0.3084	0.3261	0.3420	0.3567	0.3702	0.3891	0.4185	0.4719	0.5208	0.5669	0.6113
(600.59)	*H*	617.7	1164.8	....	....	1194.5	1220.5	1243.3	1264.0	1282.7	1299.6	1323.0	1359.4	1426.2	1488.3	1548.2	1607.0
	S	0.8138	1.3298	....	....	1.3576	1.3814	1.4020	1.4203	1.4365	1.4510	1.4706	1.5001	1.5511	1.5951	1.6348	1.6713

Table D-2 Superheated Steam (Continued)

Abs press., psia (sat. temp.)	V H S	Sat. water	Sat. steam	Temperature, °F													
				620	640	660	680	700	720	740	760	780	800	900	1000	1100	1200
1575	V	0.0238	0.2599	0.2804	0.3008	0.3186	0.3345	0.3492	0.3627	0.3755	0.3877	0.3993	0.4105	0.4633	0.5117	0.5573	0.6010
(602.74)	H	620.8	1163.4	1190.3	1216.9	1240.3	1261.3	1280.5	1297.6	1313.5	1328.8	1343.5	1357.8	1424.9	1487.3	1547.4	1606.2
	S	0.8166	1.3273	1.3524	1.3768	1.3979	1.4165	1.4332	1.4478	1.4612	1.4738	1.4858	1.4972	1.5485	1.5927	1.6325	1.6691
1600	V	0.0239	0.2548	0.2730	0.2935	0.3114	0.3274	0.3421	0.3555	0.3682	0.3802	0.3919	0.4031	0.4554	0.5032	0.5482	0.5914
(604.87)	H	623.9	1161.9	1186.3	1213.7	1237.6	1258.9	1278.4	1295.7	1311.8	1327.3	1342.2	1356.7	1424.1	1486.8	1547.0	1605.8
	S	0.8195	1.3249	1.3477	1.3728	1.3943	1.4132	1.4302	1.4449	1.4585	1.4713	1.4834	1.4950	1.5465	1.5909	1.6308	1.6674
1625	V	0.0240	0.2497	0.2656	0.2864	0.3044	0.3203	0.3348	0.3484	0.3610	0.3729	0.3845	0.3957	0.4474	0.4948	0.5391	0.5816
(606.97)	H	627.0	1160.4	1182.1	1210.2	1234.5	1256.2	1275.9	1293.5	1310.0	1325.5	1340.6	1355.2	1423.0	1486.0	1546.3	1605.2
	S	0.8222	1.3223	1.3425	1.3683	1.3902	1.4094	1.4266	1.4416	1.4555	1.4683	1.4806	1.4923	1.5440	1.5887	1.6287	1.6652
1650	V	0.0241	0.2448	0.2583	0.2794	0.2976	0.3136	0.3280	0.3417	0.3542	0.3661	0.3776	0.3887	0.4399	0.4867	0.5305	0.5724
(609.05)	H	630.0	1158.8	1177.6	1206.8	1231.7	1253.7	1273.6	1291.8	1308.4	1324.1	1339.3	1354.0	1422.1	1485.3	1545.7	1604.7
	S	0.8250	1.3198	1.3373	1.3641	1.3865	1.4060	1.4233	1.4389	1.4528	1.4658	1.4782	1.4899	1.5420	1.5867	1.6268	1.6634
1675	V	0.0242	0.2401	0.2511	0.2726	0.2909	0.3069	0.3214	0.3350	0.3474	0.3592	0.3706	0.3817	0.4324	0.4787	0.5220	0.5634
(611.10)	H	633.1	1157.2	1173.1	1203.2	1228.7	1251.0	1271.5	1289.7	1306.4	1322.2	1337.5	1352.5	1420.9	1484.4	1545.0	1604.1
	S	0.8278	1.3173	1.3321	1.3597	1.3827	1.4024	1.4203	1.4358	1.4499	1.4629	1 4754	1.4874	1.5396	1.5846	1.6248	1.6615
1700	V	0.0243	0.2354	0.2441	0.2659	0.2844	0.3006	0.3152	0.3286	0.3411	0.3528	0.3641	0.3750	0.4254	0.4711	0.5139	0.5549
(613.12)	H	636.1	1155.7	1168.1	1199.4	1225.7	1248.4	1269.4	1287.8	1304.7	1320.7	1336.2	1351.2	1420.1	1483.7	1544.4	1603.9
	S	0.8304	1.3147	1.3262	1.3549	1.3786	1.3987	1.4170	1.4327	1.4469	1.4602	1.4728	1.4848	1.5374	2.5825	1.6227	1.6597
1725	V	0.0244	0.2309	0.2384	0.2593	0.2780	0.2943	0.3088	0.3222	0.3346	0.3463	0.3575	0.3684	0.4183	0.4636	0.5058	0.5464
(615.13)	H	639.1	1154.1	1165.2	1195.6	1222.4	1245.8	1266.8	1285.5	1302.7	1318 9	1334.5	1349.5	1419.0	1482.9	1543.7	1603.4
	S	0.8332	1.3123	1.3226	1.3505	1.3747	1.3954	1.4136	1.4296	1.4441	1.4575	1.4702	1.4822	1.5352	1.5806	1.6208	1 6579
1750	V	0.0245	0.2265	0.2329	0.2529	0.2718	0.2882	0.3028	0.3162	0.3285	0.3402	0.3514	0.3622	0.4116	0.4564	0.4982	0.5383
(617.11)	H	642.1	1152.5	1162.0	1191.7	1219.2	1243.1	1264.5	1283.6	1300.9	1317.4	1333.1	1348.3	1418.1	1481.2	1543.3	1603.0
	S	0.8359	1.3099	1.3187	1.3460	1.3708	1.3919	1.4105	1.4269	1.4414	1.4550	1.4678	1.4800	1.5333	1.5780	1.6192	1.6562

1775	V	0.0246	0.2222	....	0.2466	0.2657	0.2822	0.2968	0.3102	0.3225	0.3340	0.3452	0.3559	0.4049	0.4493	0.4906	0.5302
(619.07)	H	645.0	1150.9	....	1187.6	1216.0	1240.4	1262.1	1281.5	1299.0	1315.5	1331.5	1346.8	1416.9	1481.3	1542.5	1602.3
	S	0.8386	1.3076	....	1.3413	1.3669	1.3885	1.4074	1.4239	1.4387	1.4523	1.4653	1.4775	1.5311	1.5767	1.6173	1.6544
1800	V	0.0247	0.2180	....	0.2405	0.2598	0.2764	0.2912	0.3045	0.3168	0.3283	0.3393	0.3499	0.3986	0.4425	0.4834	0.5224
(621.00)	H	648.0	1149.3	....	1183.7	1212.7	1237.6	1259.8	1279.6	1297.4	1313.9	1329.9	1345.3	1416.0	1480.6	1542.0	1601.8
	S	0.8412	1.3051	....	1.3367	1.3628	1.3848	1.4041	1.4211	1.4360	1.4497	1.4627	1.4750	1.5290	1.5748	1.6155	1.6526
1825	V	0.0248	0.2139	....	0.2345	0.2540	0.2708	0.2855	0.2991	0.3112	0.3225	0.3335	0.3441	0.3924	0.4357	0.4763	0.5147
(622.92)	H	650.9	1147.7	....	1179.5	1209.4	1234.9	1257.3	1277.8	1295.6	1312.2	1328.3	1343.9	1415.0	1479.6	1541.3	1601.1
	S	0.8439	1.3028	....	1.3319	1.3589	1.3815	1.4009	1.4185	1.4334	1.4471	1.4602	1.4727	1.5270	1.5728	1.6137	1.6508
1850	V	0.0249	0.2099	....	0.2285	0.2482	0.2651	0.2799	0.2936	0.3056	0.3170	0.3279	0.3384	0.3863	0.4293	0.4695	0.5075
(624.82)	H	653.9	1145.9	....	1175.1	1205.8	1231.8	1254.8	1275.6	1293.5	1310.5	1326.7	1342.4	1413.9	1479.0	1540.8	1600.8
	S	0.8465	1.3002	....	1.3269	1.3546	1.3776	1.3976	1.4154	1.4305	1.4445	1.4577	1.4703	1.5248	1.5710	1.6120	1.6492
1875	V	0.0251	0.2060	....	0.2225	0.2427	0.2597	0.2746	0.2882	0.3003	0.3115	0.3224	0.3328	0.3804	0.4229	0.4626	0.5003
(626.69)	H	656.9	1144.2	....	1170.5	1202.3	1229.0	1252.3	1273.5	1291.8	1308.7	1325.2	1340.8	1412.9	1478.0	1540.2	1600.3
	S	0.8491	1.2977	....	1.3218	1.3504	1.3741	1.3943	1.4124	1.4278	1.4418	1.4552	1.4677	1.5227	1.5689	1.6101	1.6475
1900	V	0.0252	0.2022	....	0.2165	0.2371	0.2543	0.2694	0.2828	0.2950	0.3063	0.3171	0.3274	0.3747	0.4170	0.4562	0.4934
(628.55)	H	659.9	1142.4	....	1165.6	1198.8	1225.9	1249.8	1271.0	1289.7	1307.0	1323.5	1339.4	1411.9	1477.5	1539.7	1599.8
	S	0.8517	1.2951	....	1.3163	1.3462	1.3702	1.3910	1.4091	1.4249	1.4392	1.4526	1.4653	1.5207	1.5672	1.6084	1.6457
1925	V	0.0253	0.1985	....	0.2107	0.2317	0.2491	0.2642	0.2776	0.2898	0.3010	0.3118	0.3221	0.3690	0.4109	0.4497	0.4864
(630.38)	H	662.8	1140.6	....	1160.9	1195.0	1223.1	1247.3	1268.7	1287.7	1305.1	1321.8	1337.8	1410.7	1476.6	1538.9	1599.0
	S	0.8543	1.2926	....	1.3111	1.3419	1.3667	1.3878	1.4061	1.4221	1.4364	1.4500	1.4628	1.5185	1.5652	1.6065	1.6438
1950	V	0.0254	0.1949	....	0.2049	0.2264	0.2440	0.2591	0.2726	0.2848	0.2960	0.3067	0.3170	0.3636	0.4052	0.4436	0.4800
(632.20)	H	665.8	1138.8	....	1155.8	1191.3	1220.0	1244.8	1266.5	1285.8	1303.4	1320.2	1336.3	1409.7	1476.0	1538.4	1598.8
	S	0.8569	1.2901	....	1.3056	1.3376	1.3630	1.3846	1.4031	1.4194	1.4339	1.4476	1.4604	1.5165	1.5635	1.6048	1.6424
1975	V	0.0256	0.1913	....	0.1992	0.2212	0.2391	0.2544	0.2678	0.2798	0.2910	0.3016	0.3117	0.3581	0.3994	0.4376	0.4737
(634.00)	H	668.7	1137.0	....	1150.5	1187.5	1217.0	1242.4	1264.4	1283.6	1301.5	1318.3	1334.4	1408.4	1475.1	1537.8	1598.4
	S	0.8595	1.2877	....	1.3000	1.3334	1.3595	1.3816	1.4004	1.4165	1.4313	1.4450	1.4579	1.5144	1.5617	1.6032	1.6409

Table D-2 Superheated Steam (Continued)

Abs press., psia (sat. temp.)	V H S	Sat. water	Sat. steam	Temperature, °F													
				660	**680**	**700**	**720**	**740**	**760**	**780**	**800**	**820**	**850**	**900**	**1000**	**1100**	**1200**
2000	V	0.0257	0.1879	0.2162	0.2344	0.2498	0.2633	0.2752	0.2862	0.2966	0.3067	0.3165	0.3305	0.3528	0.3940	0.4319	0.4678
(635.78)	H	671.7	1135.2	1183.7	1214.3	1240.0	1262.4	1281.8	1299.6	1316.4	1332.7	1348.6	1371.1	1407.2	1473.5	1537.4	1598.6
	S	0.8620	1.2851	1.3289	1.3560	1.3783	1.3975	1.4138	1.4285	1.4422	1.4552	1.4677	1.4851	1.5122	1.5592	1.6015	1.6395
2025	V	0.0258	0.1845	0.2112	0.2296	0.2452	0.2587	0.2707	0.2816	0.2919	0.3020	0.3116	0.3257	0.3478	0.3887	0.4262	0.4617
(637.54)	H	674.7	1133.3	1179.9	1211.1	1237.5	1260.3	1280.1	1297.9	1314.8	1331.3	1347.1	1370.1	1406.2	1473.8	1537.0	1598.1
	S	0.8646	1.2826	1.3246	1.3523	1.3752	1.3947	1.4114	1.4261	1.4398	1.4530	1.4654	1.4832	1.5103	1.5582	1.6001	1.6380
2050	V	0.0259	0.1812	0.2062	0.2248	0.2405	0.2541	0.2661	0.2769	0.2872	0.2972	0.3067	0.3208	0.3428	0.3833	0.4205	0.4556
(639.29)	H	677.6	1131.4	1175.7	1207.7	1234.6	1257.9	1278.0	1295.9	1313.0	1329.5	1345.3	1368.6	1405.1	1472.8	1536.2	1597.4
	S	0.8671	1.2800	1.3199	1.3483	1.3717	1.3916	1.4085	1.4233	1.4372	1.4504	1.4628	1.4808	1.5082	1.5562	1.5982	1.6362
2075	V	0.0261	0.1780	0.2013	0.2200	0.2358	0.2494	0.2615	0.2724	0.2828	0.2927	0.3022	0.3161	0.3374	0.3782	0.4151	0.4499
(641.02)	H	680.5	1129.5	1171.5	1204.3	1231.6	1255.2	1275.7	1294.0	1311.5	1328.1	1344.0	1367.3	1403.1	1472.1	1535.7	1597.0
	S	0.8697	1.2776	1.3154	1.3445	1.3682	1.3884	1.4056	1.4208	1.4350	1.4483	1.4608	1.4788	1.5056	1.5546	1.5967	1.6348
2100	V	0.0262	0.1748	0.1964	0.2152	0.2310	0.2447	0.2568	0.2679	0.2783	0.2882	0.2977	0.3114	0.3319	0.3730	0.4096	0.4441
(642.73)	H	683.4	1127.6	1167.3	1200.4	1228.2	1252.2	1273.0	1292.0	1309.6	1326.4	1342.6	1365.8	1400.8	1471.0	1534.8	1596.4
	S	0.8722	1.2751	1.3108	1.3401	1.3643	1.3848	1.4023	1.4180	1.4323	1.4458	1.4585	1.4764	1.5027	1.5525	1.5947	1.6330
2125	V	0.0263	0.1716	0.1918	0.2108	0.2268	0.2406	0.2527	0.2637	0.2739	0.2838	0.2933	0.3070	0.3280	0.3682	0.4045	0.4386
(644.43)	H	686.3	1125.6	1163.0	1197.3	1225.7	1250.2	1271.3	1290.2	1307.8	1324.8	1341.2	1364.7	1400.8	1470.5	1534.5	1596.0
	S	0.8747	1.2726	1.3060	1.3364	1.3611	1.3820	1.3994	1.4151	1.4294	1.4430	1.4559	1.4740	1.5011	1.5505	1.5929	1.6311
2150	V	0.0265	0.1685	0.1872	0.2064	0.2225	0.2364	0.2485	0.2594	0.2695	0.2793	0.2889	0.3026	0.3240	0.3633	0.3993	0.4331
(646.11)	H	689.2	1123.5	1158.6	1193.7	1222.7	1247.7	1269.1	1288.1	1305.7	1322.7	1339.4	1363.2	1400.5	1469.4	1533.8	1595.3
	S	0.8773	1.2700	1.3015	1.3326	1.3578	1.3792	1.3972	1.4129	1.4272	1.4408	1.4540	1.4723	1.5003	1.5492	1.5918	1.6300
2175	V	0.0266	0.1655	0.1828	0.2018	0.2182	0.2321	0.2442	0.2552	0.2654	0.2752	0.2847	0.2983	0.3196	0.3587	0.3943	0.4279
(647.77)	H	692.0	1121.5	1154.3	1189.7	1219.8	1244.9	1266.5	1286.1	1304.0	1321.3	1338.0	1361.9	1399.4	1468.9	1533.1	1595.0
	S	0.8798	1.2676	1.2971	1.3284	1.3546	1.3760	1.3942	1.4104	1.4250	1.4388	1.4519	1.4704	1.4985	1.5478	1.5903	1.6288

2200	V	0.0267	0.1626	0.1773	0.1972	0.2138	0.2277	0.2399	0.2509	0.2612	0.2710	0.2804	0.2939	0.3151	0.3540	0.3893	0.4226
(649.42)	H	695.0	1119.4	1148.0	1185.8	1216.2	1241.8	1263.8	1283.6	1302.9	1319.5	1336.2	1360.3	1398.0	1467.9	1532.3	1594.3
	S	0.8823	1.2649	1.2906	1.3239	1.3504	1.3723	1.3907	1.4072	1.4229	1.4362	1.4494	1.4679	1.4962	1.5458	1.5884	1.6269
2225	V	0.0269	0.1597	0.1727	0.1931	0.2097	0.2237	0.2360	0.2471	0.2574	0.2671	0.2765	0.2900	0.3109	0.3495	0.3847	0.4175
(651.06)	H	697.9	1117.4	1142.7	1182.1	1213.3	1239.1	1261.7	1282.0	1300.5	1318.0	1335.2	1359.2	1397.0	1467.0	1531.9	1593.7
	S	0.8848	1.2625	1.2852	1.3201	1.3472	1.3693	1.3883	1.4051	1.4201	1.4341	1.4476	1.4661	1.4944	1.5441	1.5871	1.6255
2250	V	0.0270	0.1569	0.1680	0.1889	0.2055	0.2197	0.2320	0.2432	0.2535	0.2632	0.2726	0.2860	0.3066	0.3449	0.3800	0.4125
(652.67)	H	700.8	1115.3	1137.2	1178.0	1209.7	1236.5	1259.2	1279.8	1298.6	1316.3	1333.5	1357.8	1395.5	1465.8	1531.2	1593.1
	S	0.8873	1.2599	1.2795	1.3156	1.3432	1.3661	1.3852	1.4022	1.4175	1.4317	1.4452	1.4640	1.4922	1.5421	1.5854	1.6239
2275	V	0.0272	0.1542	0.1630	0.1848	0.2018	0.2159	0.2281	0.2393	0.2496	0.2594	0.2688	0.2821	0.3026	0.3407	0.3755	0.4077
(654.27)	H	703.8	1113.2	1130.8	1174.3	1207.1	1233.9	1256.7	1277.6	1296.8	1314.8	1332.1	1356.5	1394.4	1465.1	1530.7	1592.8
	S	0.8898	1.2573	1.2731	1.3116	1.3401	1.3630	1.3822	1.3995	1.4151	1.4295	1.4431	1.4619	1.4903	1.5405	1.5839	1.6225
2300	V	0.0274	0.1514	0.1580	0.1807	0.1980	0.2120	0.2241	0.2353	0.2457	0.2556	0.2649	0.2781	0.2986	0.3365	0.3709	0.4029
(655.87)	H	706.7	1111.0	1124.2	1169.9	1204.0	1230.9	1253.9	1275.0	1294.5	1312.9	1330.3	1354.9	1393.2	1464.3	1529.9	1592.2
	S	0.8923	1.2547	1.2665	1.3070	1.3366	1.3596	1.3790	1.3964	1.4123	1.4270	1.4407	1.4597	1.4884	1.5388	1.5823	1.6210
2325	V	0.0275	0.1488	0.1530	0.1766	0.1941	0.2083	0.2207	0.2319	0.2423	0.2520	0.2613	0.2744	0.2948	0.3324	0.3665	0.3983
(657.74)	H	709.7	1108.8	1117.1	1165.7	1200.4	1228.2	1252.0	1273.3	1293.0	1311.5	1329.0	1353.6	1392.1	1463.5	1529.2	1591.7
	S	0.8948	1.2521	1.2595	1.3025	1.3327	1.3565	1.3765	1.3941	1.4101	1.4249	1.4387	1.4577	1.4866	1.5372	1.5807	1.6196
2350	V	0.0277	0.1462	0.1479	0.1725	0.1901	0.2046	0.2172	0.2285	0.2389	0.2484	0.2576	0.2706	0.2910	0.3282	0.3621	0.3936
(659.00)	H	712.6	1106.5	1109.7	1161.2	1196.6	1225.2	1249.8	1271.7	1291.6	1309.8	1327.3	1352.1	1391.0	1462.3	1528.3	1590.9
	S	0.8974	1.2495	1.2524	1.2980	1.3287	1.3532	1.3739	1.3920	1.4082	1.4227	1.4365	1.4557	1.4848	1.5354	1.5791	1.6180
2375	V	0.0278	0.1436	....	0.1686	0.1863	0.2010	0.2137	0.2250	0.2353	0.2450	0.2541	0.2671	0.2873	0.3244	0.3580	0.3892
(660.55)	H	715.6	1104.0	....	1156.9	1193.0	1222.4	1247.5	1269.5	1289.6	1308.3	1325.8	1350.8	1389.8	1461.7	1527.8	1590.4
	S	0.9000	1.2467	....	1.2935	1.3249	1.3501	1.3712	1.3893	1.4057	1.4207	1.4344	1.4537	1.4830	1.5340	1.5777	1.6166
2400	V	0.0280	0.1410	....	0.1646	0.1824	0.1974	0.2101	0.2214	0.2317	0.2415	0.2506	0.2636	0.2836	0.3205	0.3538	0.3848
(662.09)	H	718.5	1101.4	....	1152.2	1189.1	1219.4	1244.7	1267.1	1287.3	1306.5	1324.2	1349.6	1388.5	1460.8	1526.9	1589.8
	S	0.9025	1.2438	....	1.2887	1.3208	1.3467	1.3680	1.3865	1.4030	1.4183	1.4323	1.4519	1.4810	1.5323	1.5761	1.6152

Table D-2 Superheated Steam (Continued)

Abs press., psia (sat. temp.)	V H S	Sat. water	Sat. steam	Temperature, °F													
				680	**700**	**720**	**740**	**760**	**780**	**800**	**820**	**850**	**900**	**950**	**1000**	**1100**	**1200**
2450	V	0.0283	0.1360	0.1567	0.1750	0.1902	0.2032	0.2147	0.2250	0.2347	0.2440	0.2569	0.2766	0.2954	0.3130	0.3458	0.3764
(665.12)	H	724.6	1096.3	1142.3	1181.3	1212.9	1239.5	1262.8	1283.5	1302.9	1321.4	1347.1	1386.3	1423.8	1459.0	1525.6	1588.9
	S	0.9076	1.2381	1.2787	1.3127	1.3397	1.3620	1.3813	1.3981	1.4136	1.4282	1.4481	1.4774	1.5045	1.5290	1.5732	1.6125
2500	V	0.0287	0.1313	0.1488	0.1680	0.1834	0.1967	0.2083	0.2188	0.2285	0.2375	0.2503	0.2700	0.2884	0.3058	0.3381	0.3683
(668.10)	H	730.7	1091.0	1131.9	1173.9	1206.8	1234.5	1258.6	1280.1	1299.9	1318.3	1344.2	1384.3	1421.8	1457.3	1524.2	1587.9
	S	0.9127	1.2322	1.2683	1.3048	1.3329	1.3562	1.3761	1.3936	1.4095	1.4240	1.4440	1.4740	1.5011	1.5258	1.5701	1.6097
2550	V	0.0291	0.1264	0.1408	0.1606	0.1766	0.1902	0.2020	0.2125	0.2223	0.2315	0.2440	0.2634	0.2817	0.2989	0.3308	0.3604
(671.03)	H	736.7	1085.6	1120.0	1164.7	1199.8	1228.9	1254.0	1276.0	1296.5	1315.6	1341.5	1381.8	1419.8	1455.6	1522.9	1586.7
	S	0.9179	1.2265	1.2568	1.2957	1.3257	1.3502	1.3709	1.3888	1.4052	1.4202	1.4402	1.4704	1.4979	1.5228	1.5674	1.6070
2600	V	0.0295	0.1219	0.1323	0.1541	0.1706	0.1842	0.1961	0.2066	0.2164	0.2256	0.2380	0.2573	0.2754	0.2924	0.3237	0.3530
(673.91)	H	743.1	1080.1	1106.0	1157.0	1194.0	1223.9	1249.7	1272.2	1293.1	1312.6	1338.9	1379.7	1418.1	1454.2	1521.6	1585.8
	S	0.9232	1.2205	1.2433	1.2877	1.3193	1.3444	1.3657	1.3840	1.4008	1.4161	1.4364	1.4670	1.4947	1.5199	1.5645	1.6044
2650	V	0.0300	0.1173	0.1235	0.1469	0.1640	0.1782	0.1903	0.2009	0.2108	0.2200	0.2322	0.2514	0.2691	0.2859	0.3169	0.3458
(676.75)	H	749.5	1074.5	1090.1	1146.8	1186.4	1218.3	1245.1	1268.3	1289.9	1309.9	1336.2	1377.5	1415.9	1452.2	1520.2	1584.7
	S	0.9287	1.2147	1.2284	1.2777	1.3116	1.3384	1.3606	1.3794	1.3967	1.4125	1.4328	1.4637	1.4914	1.5167	1.5618	1.6018
2700	V	0.0305	0.1123	....	0.1402	0.1581	0.1725	0.1846	0.1954	0.2053	0.2147	0.2269	0.2458	0.2632	0.2798	0.3105	0.3389
(679.54)	H	756.1	1068.3	....	1137.5	1179.7	1213.0	1240.3	1264.5	1286.5	1307.2	1334.0	1375.7	1414.0	1450.6	1519.2	1583.8
	S	0.9342	1.2082	....	1.2684	1.3045	1.3325	1.3550	1.3747	1.3923	1.4086	1.4293	1.4606	1.4882	1.5137	1.5592	1.5993
2750	V	0.0310	0.1077	....	0.1335	0.1520	0.1670	0.1794	0.1902	0.2000	0.2094	0.2215	0.2402	0.2576	0.2740	0.3043	0.3322
(682.28)	H	763.0	1061.8	....	1127.1	1172.2	1207.4	1236.0	1260.8	1283.0	1304.2	1331.4	1373.3	1412.2	1449.2	1518.0	1582.8
	S	0.9399	1.2016	....	1.2583	1.2969	1.3265	1.3501	1.3703	1.3881	1.4048	1.4258	1.4572	1.4853	1.5110	1.5566	1.5969
2800	V	0.0316	0.1032	....	0.1267	0.1461	0.1615	0.1741	0.1851	0.1950	0.2045	0.2166	0.2351	0.2521	0.2685	0.2983	0.3258
(684.98)	H	770.0	1054.6	....	1115.9	1164.5	1201.5	1231.3	1257.0	1279.8	1301.7	1329.3	1371.5	1410.4	1447.9	1517.0	1581.9
	S	0.9458	1.1944	....	1.2476	1.2892	1.3203	1.3449	1.3658	1.3840	1.4013	1.4226	1.4542	1.4823	1.5084	1.5542	1.5945

2850	V	0.0322	0.0986	....	0.1198	0.1404	0.1563	0.1690	0.1801	0.1903	0.1995	0.2117	0.2298	0.2469	0.2629	0.2925	0.3197
(687.65)	H	777.5	1046.6	....	1103.5	1156.6	1195.8	1226.5	1252.9	1276.9	1298.4	1326.8	1368.9	1408.7	1446.0	1515.8	1581.1
	S	0.9521	1.1866	....	1.2359	1.2813	1.3143	1.3397	1.3611	1.3803	1.3973	1.4192	1.4507	1.4795	1.5055	1.5517	1.5923
2900	V	0.0329	0.0941	....	0.1126	0.1348	0.1511	0.1641	0.1754	0.1855	0.1949	0.2069	0.2250	0.2418	0.2578	0.2870	0.3138
(690.26)	H	785.2	1038.1	....	1089.2	1148.3	1189.5	1221.6	1249.1	1273.3	1295.7	1324.2	1367.1	1406.9	1444.8	1514.8	1580.2
	S	0.9586	1.1785	....	1.2228	1.2733	1.3080	1.3345	1.3569	1.3762	1.3939	1.4159	1.4480	1.4767	1.5032	1.5495	1.5902
2950	V	0.0337	0.0895	....	0.1052	0.1292	0.1460	0.1593	0.1708	0.1809	0.1902	0.2023	0.2202	0.2369	0.2527	0.2815	0.3081
(692.83)	H	793.6	1028.9	....	1073.2	1139.5	1183.1	1216.6	1245.1	1269.8	1292.5	1321.7	1364.9	1405.1	1443.2	1513.5	1579.4
	S	0.9655	1.1697	....	1.2080	1.2647	1.3014	1.3290	1.3522	1.3720	1.3898	1.4124	1.4448	1.4738	1.5004	1.5469	1.5879
3000	V	0.0346	0.0849	....	0.0972	0.1236	0.1410	0.1546	0.1661	0.1763	0.1856	0.1978	0.2155	0.2322	0.2478	0.2763	0.3027
(695.37)	H	802.6	1019.3	....	1054.0	1130.3	1176.4	1211.4	1240.6	1266.0	1289.0	1319.0	1362.5	1403.4	1441.7	1512.4	1578.9
	S	0.9731	1.1607	....	1.1907	1.2559	1.2947	1.3236	1.3474	1.3677	1.3858	1.4090	1.4416	1.4711	1.4978	1.5447	1.5860
3050	V	0.0357	0.0804	....	0.0868	0.1183	0.1361	0.1499	0.1617	0.1719	0.1812	0.1935	0.2111	0.2275	0.2431	0.2712	0.2973
(697.84)	H	812.9	1007.7	....	1025.0	1121.2	1169.6	1205.9	1236.3	1262.4	1285.7	1316.4	1360.4	1401.3	1440.2	1511.1	1577.9
	S	0.9818	1.1501	....	1.1650	1.2472	1.2879	1.3179	1.3437	1.3635	1.3819	1.4056	1.4386	1.4681	1.4952	1.5422	1.5837
3100	V	0.0372	0.0752	....	....	0.1128	0.1312	0.1456	0.1576	0.1680	0.1771	0.1891	0.2068	0.2231	0.2385	0.2663	0.2921
(700.29)	H	824.6	994.0	....	....	1110.8	1162.4	1200.9	1232.5	1259.4	1282.7	1313.4	1358.2	1399.5	1438.7	1509.9	1577.0
	S	0.9916	1.1376	....	....	1.2374	1.2808	1.3126	1.3383	1.3599	1.3782	1.4019	1.4355	1.4653	1.4926	1.5398	1.5815
3150	V	0.0392	0.0691	....	....	0.1075	0.1266	0.1410	0.1532	0.1636	0.1729	0.1850	0.2024	0.2187	0.2341	0.2615	0.2871
(702.69)	H	841.3	976.3	....	....	1100.3	1155.3	1194.7	1227.6	1254.9	1279.2	1310.5	1355.5	1397.5	1437.2	1508.6	1576.1
	S	1.0056	1.1217	....	....	1.2276	1.2738	1.3064	1.3331	1.3550	1.3741	1.3983	1.4320	1.4623	1.4900	1.5373	1.5792
3200	V	0.0443	0.0596	....	....	0.1024	0.1217	0.1368	0.1493	0.1596	0.1687	0.1810	0.1985	0.2145	0.2296	0.2570	0.2822
(705.04)	H	871.3	946.6	....	....	1089.8	1146.9	1189.3	1223.4	1251.1	1275.2	1307.7	1353.8	1395.7	1435.2	1507.6	1575.3
	S	1.0311	1.0958	....	....	1.2180	1.2660	1.3011	1.3288	1.3510	1.3699	1.3950	1.4296	1.4598	1.4874	1.5353	1.5774
3206.2‡	V	0.0541	0.0541	....	....	0.1018	0.1211	0.1363	0.1488	0.1591	0.1682	0.1805	0.1980	0.2140	0.2290	0.2564	0.2816
(705.34)	H	910.3	910.3	....	....	1088.5	1145.8	1188.5	1222.9	1250.6	1274.8	1307.3	1353.4	1395.3	1434.8	1507.3	1575.1
	S	1.0645	1.0645	....	....	1.2170	1.2652	1.3005	1.3284	1.3506	1.3696	1.3947	1.4293	1.4596	1.4872	1.5351	1.5773

‡ Critical pressure.

ANSWERS TO SELECTED PROBLEMS

1-1 373.15(K)
1-2 $A = 3.9199 \times 10^{-3}(°C)^{-1}$
$B = -5.9022 \times 10^{-7}(°C)^{-2}$
1-6 233.15(K)
1-7 419.57(K)

2-1 501(atm); 48.78(°C)
2-2 417.7(atm)
2-4 222.1(atm)
2-8 $3.213(m)/(s)^2$
2-9 $147.2(lb_m)$; $25(lb_f)$
2-11 19,700 and $14,700(lb_f)/(in)^2$

3-1 -1.508×10^4 and $-6.104 \times 10^5(ft)(lb_f)$
3-6 $W_{12} = -15{,}120$; $W_{23} = 8{,}640$; $W_{13} = 0$; and $W_{1231} = -6{,}480(ft)(lb_f)$
3-7 $35,240(ft)(lb_f)$
3-8 (*b*) $250(ft)(lb_f)$
3-10 522.7(J)
3-13 (*a*) 181(J)
(*b*) 0.09832 and 29.5(J)
(*c*) −181.1 and −210.5(J)

4-6 (*a*) 60(Btu)
(*b*) −70(Btu)
(*c*) 50 and 10(Btu)

5-6 $180(lb_m)$
5-7 (*b*) $6,183(cm)^3$; −1,254(J)
5-8 (*a*) 571.1(K)
(*b*) 150(m)
5-9 $0.005041(lb_m)$
5-11 (*a*) 8,951(J)
(*b*) 594.6(K)
(*c*) 3,227(K)
(*d*) $91.18 \times 10^3(J)$
5-13 (*c*) −9.8(K)/(km)
5-14 344.7(K); 0.65(bar)
5-19 $Q = -1{,}541$; $W = 2{,}191$; and $\Delta U = 650$(Btu)
5-23 (*a*) $Q = \Delta U = 1{,}650$(Btu) and $W = 0$
(*b*) $Q = 0$ and $W = \Delta U = -1{,}650$(Btu)
(*c*) $Q = -W = -1{,}283$(Btu) and $\Delta U = 0$
5-24 (*a*) $Q = -W = -9{,}749$(J)/(mol) and $\Delta U = 0$
(*b*) $Q = -W = -3{,}923$(J)/(mol) and $\Delta U = 0$
(*c*) $Q = -W = -5{,}488$(J)/(mol) and $\Delta U = 0$
5-26 $Q = 1.526 \times 10^4$(Btu)
5-27 57.65(J)/(mol)(K)
5-28 165.3(kJ)

6-5 (*a*) $Q = 362.2$(J)
(*b*) $W = -362.2$(J)
(*c*) $Q = -W = -12.589$(J) and $\eta = 96.52$ percent

7-9 70.0(Btu)
7-10 (*a*) $W = -58.33$(Btu)
(*b*) $W = -58.33$(Btu)

8-4 (*a*) Zero
(*b*) 8.329(J)/(K)
(*c*) 5.787(J)/(K)
(*d*) 5.787(J)/(K)

8-5 (*a*) 1,305.9; −1,121.7; and 184.2(J)/(K)
(*b*) 97.4(J)/(K)
8-7 P = 1.5(atm); T = 500(R); $\Delta S_{total} = 9.245 \times 10^{-4}$(Btu)/(R)
8-12 $\Delta S_{gas} = -4.575$(cal)/(K); $\Delta S_{reservoir}$ = 7.319(cal)/(K); ΔS_{total} = 2.744 (cal)/(K)
8-13 ΔS = 0.6877(Btu)/(R)
8-14 44.18(J)/(mol)(K)
8-15 (*a*) 6.277(J)/(K)
(*b*) 1.385(J)/(K)
(*c*) 3.635(J)/(K)
8-16 (*a*) Zero
(*b*) 1.83×10^{-5}(J)/(K)
8-17 278.1(J)/(K)
8-18 −16.02(J)/(K)
8-20 (*a*) 0.0364; −0.1280; and 0.0916 (Btu)/(R)
(*b*) The same
8-25 400(K)
8-26 (*a*) 519.67(R)
(*b*) 685.71(R)
(*c*) 2.639(atm)
(*d*) 830.2(Btu)
(*e*) 1.386(Btu)/(R) for both
8-29 (*a*) −70(Btu)
(*b*) −110(Btu)
(*c*) 80(Btu)

9-3 Q = −9,629(cal)
9-4 ΔS = 1.445(Btu)/(lb_m)(R); ΔU = 897.4(Btu)/(lb_m)
9-5 30.99(atm)
9-6 (*b*) 0.06338
(*c*) 0.71(J)/(K)
9-7 $W = Q = \Delta U = \Delta H = 0$; ΔS = 19.14(J)/(K); $\Delta A = \Delta G$ = −5,743(J)
9-10 (*a*) −88.03(J)
(*b*) 51.04(J)
(*c*) −36.99(J)
(*d*) 0.697(K)
9-12 −0.45; 0.60; and 3.65(K)
9-18 (*a*) 254.4(atm)
(*b*) ΔU = −18.57(J); ΔH = 360.7(J); ΔS = −0.06868(J)/(K); Q = −18.76(J); W = 0.19(J)
9-22 (*a*) Q = −72.7(Btu); W = 0
(*b*) Q = −98.0(Btu); W = 23.3(Btu)
9-23 ΔH = 5,008 vs. 5,224(Btu); ΔU = 3,857 vs. 4,032(Btu); ΔS = −0.0396 vs. +0.0271(Btu)/(R)
9-24 (*a*) 384.3 and 400(°F)
(*b*) 34.01(ft)^3/(lb_m)
(*c*) 0.1835(Btu)/(lb_m)(R)
9-25 334.9(bar)
9-26 0.200(ft)^3/(lb_m); 29.4(Btu)/(lb_m); 0.0254(Btu)/(lb_m)(R)
9-27 −300(Btu)
9-28 (*a*) W = −110.9(Btu); Q = 114.6(Btu)
(*b*) t_2 = 289.2(°F); W = −73.3 (Btu)
9-29 (*a*) 39.99 percent
(*b*) 895.1(Btu)/(lb_m)
(*c*) 1.1393(Btu)/(lb_m)(R)
9-30 281.01(°F); 1.4747(Btu)/(lb_m)(R); 970.7(Btu)/(lb_m)
9-31 17.0 percent
9-32 56.3×10^3(Btu)
9-33 Q = −1,050(Btu); P = 186.2 (psia)
9-39 Z = 0.987; $\Delta H'$ = 235(J)/(mol); $\Delta S'$ = 0.304(J)/(mol)(K)
9-40 (*a*) ΔH = −5,008(Btu)/(lb mol); ΔS = 0.040(Btu)/(lb mol)(R)
(*b*) ΔH = −4,990(Btu)/(lb mol); ΔS = 0.067(Btu)/(lb mol)(R)
9-46 (*a*) Z = 0.8848; V = 3,435(cm)^3/(mol)
(*b*) Z = 0.902; V = 3,503(cm)^3/(mol)
(*c*) Z = 0.9481; V = 3,681(cm)^3/(mol)
9-47 W = −134.8(Btu); Q = −841.3 (Btu)
9-49 −1,428(cm)^3/(mol)
9-50 (*a*) 0.976
(*b*) 224(J)/(mol)
(*c*) 0.474(J)/(mol)(K)
9-51 12.5(lb_m)

10-1 5.966×10^5(Btu)
10-2 566(°F)
10-3 (*e*) 0.544 and 0.75(lb mol)
10-5 0.53(Btu)
10-8 461(Btu)
10-10 158(°F)
10-12 Q = −8,900(Btu)/(min)

10-13 2.14(Btu)/(lb$_m$)
10-14 29,850(lb$_m$)/(h); 86.6 percent
10-15 (*a*) 4.40(lb$_m$)
10-18 63.9(ft)
10-19 43.3(min)
10-20 799(hp); 1,329(Btu)/(lb$_m$); 1.7640 (Btu)/(lb$_m$)(R); 602.6(°F); 6.564 (ft)^3/(lb$_m$)
10-23 $Q = -101.3$(kJ)
10-24 181(°F); 2.46(atm)
10-30 0.066(lb$_m$)
10-31 67.7(hp); 42,500(Btu)/(min)
10-32 60.8(lb$_m$)/(h)
10-34 (*a*) 1,327(ft)/(s)
(*b*) 0.795(in)^2
10-35 100(°F)

11-9 551(Btu)
11-16 (*a*) 0.315
(*b*) 0.613
(*c*) 61,200(J)/(mol)
11-17 94,600(J)/(mol)
11-19 2
11-20 −49,241(cal)/(mol)
11-22 (*a*) $y_1 = 0.318$
11-23 (*b*) $K = 2$
11-25 766(K)
11-28 $Q = -6,830$(Btu)/(h)
11-29 −46,490 and −44,800(cal)/(mol)
11-32 80.3×10^{-9}; 5.0×10^{-9}; and 75.3×10^{-9}(in)(lb$_f$)

12-1 $\sigma = 3 \times 10^{-9}$(cal); $\sigma/U = 3.3 \times 10^{-12}$

INDEX

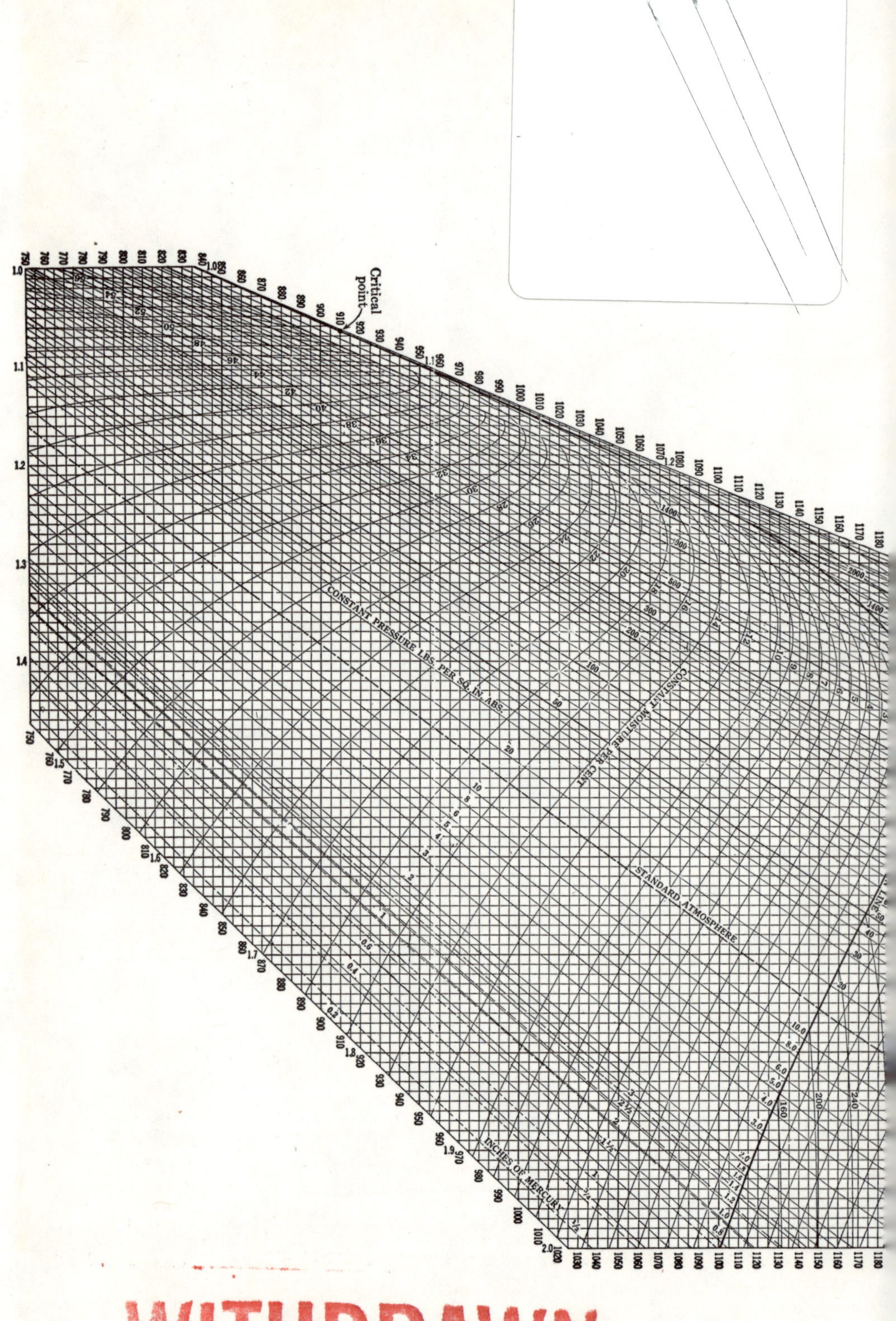
Critical point
CONSTANT PRESSURE LBS. PER SQ. IN. ABS.
CONSTANT MOISTURE PER CENT
STANDARD ATMOSPHERE
INCHES OF MERCURY